Quantum Statistical Mechanics in Classical Phase Space

Quantum Statistical Mechanics in Classical Phase Space

Phil Attard

IOP Publishing, Bristol, UK

ISBN 978-0-7503-4055-7 (ebook)
ISBN 978-0-7503-4053-3 (print)
ISBN 978-0-7503-4056-4 (myPrint)
ISBN 978-0-7503-4054-0 (mobi)

DOI 10.1088/978-0-7503-4055-7

Version: 20211101

IOP ebooks

British Library Cataloguing-in-Publication Data: A catalogue record for this book is available from the British Library.

Published by IOP Publishing, wholly owned by The Institute of Physics, London

IOP Publishing, Temple Circus, Temple Way, Bristol, BS1 6HG, UK

US Office: IOP Publishing, Inc., 190 North Independence Mall West, Suite 601, Philadelphia, PA 19106, USA

Contents

Author biography

Phil Attard

Phil Attard researches broadly in statistical mechanics, quantum mechanics, thermodynamics, and colloid science. He has held academic positions in Australia, Europe, and North America, and he was a Professorial Research Fellow of the Australian Research Council. He has authored some 120 papers, 10 review articles, and 4 books, with over 7000 citations.

As an internationally recognized researcher he has made seminal contributions to the theory of electrolytes and the electric double layer, to measurement techniques for atomic force microscopy and colloid particle interactions, and to computer simulation and integral equation algorithms for condensed matter. Attard is perhaps best known for his discovery of nanobubbles and for establishing their nature.

Recent research has focused on non-equilibrium systems. He has discovered a new entropy—the second entropy—as the basis for non-equilibrium thermodynamics, hydrodynamics, and chemical kinetics, and he has derived the probability distribution for non-equilibrium statistical mechanics. The theory provides a coherent approach to non-equilibrium systems and to irreversible processes, and it has lead to the development of stochastic molecular dynamics and non-equilibrium Monte Carlo computer simulation algorithms.

Attard has formulated quantum statistical mechanics in classical phase space, for its conceptual insight into quantum mechanics, for its account of the transition to our classical world, and for its potential for efficient computational approaches to many-body condensed matter systems.

His previous books are:

Attard P 2002 *Thermodynamics and Statistical Mechanics: Equilibrium by Entropy Maximisation* (London: Academic)

Attard P 2012 *Non-Equilibrium Thermodynamics and Statistical Mechanics: Foundations and Applications* (Oxford: Oxford University Press)

Attard P 2015 *Quantum Statistical Mechanics: Equilibrium and Non-Equilibrium Theory from First Principles* (Bristol: IOP Publishing)

Attard P 2018 *Entropy Beyond the Second Law. Thermodynamics and Statistical Mechanics for Equilibrium, Non-Equilibrium, Classical, and Quantum Systems* (Bristol: IOP Publishing)

Chapter 1

Introduction

There are compelling arguments in favor of formulating quantum statistical mechanics in classical phase space. The approach has many potential applications in condensed matter, and it has already led to a new explanation for the λ-transition and superfluidity in liquid helium. However, for the idea to reach its full potential, significant challenges to its practical computational implementation must be met.

1.1 Why phase space?

At the conceptual level, one motivation for seeking a phase space formulation of quantum systems is to gain new insight into the structure of the physical universe. Specifically, mathematical derivations that give a quantitative characterization of quantum systems, but in the setting of classical phase space, would surely help close the puzzling gap between the quantum nature of the sub-microscopic world and the classical nature of the macroscopic world. One can no more deny the validity of quantum mechanics as the law of the Universe, than one can close one eyes to the classical behavior of the world around us. Indeed the great triumph of Newton's mechanics from the seventeenth century was that they encapsulated the observed laws of nature. Thus the shock in the early twentieth century when it was discovered that at the fundamental level the Universe is actually governed by quantum mechanics. This tension between the fundamental quantum behavior at the sub-microscopic level and the classical behavior of macroscopic everyday phenomena remains present today.

The simplistic explanation for the origin of classical behavior—that quantum mechanics applies to sub-microscopic particles and classical mechanics applies to macroscopic objects—begs the question. It offers neither physical insight into how our world comes to be, nor any new quantitative methods for treating it. What would actually be useful is a mechanistic and mathematical treatment for the transition from quantum to classical behavior. In answering the question of how classical behavior arises, one might hope to delineate quantitatively the respective

doi:10.1088/978-0-7503-4055-7ch1

limits of quantum and classical behavior, and perhaps even to develop methods for characterizing that gray boundary region where both coexist to greater or lesser degree.

Of course quantum mechanics can be used to describe—in principle but not in practice—the behavior of many-particle systems in any regime. However to treat systems that are unambiguously classical using the direct methods of quantum mechanics (i.e., operators, wave functions, Schrödinger's equation, eigenstates, particle statistics, symmetrization, etc) rather than the established methods of classical statistical mechanics (phase space, probability distributions, particle dynamics, etc) would be both unnecessary and so numerically inefficient as to be beyond the powers of present-day or even foreseeable computers.

Currently then there is a curious dichotomy wherein reliable exact methods exist for the treatment of either few-body fully quantum systems, or else many-body fully classical systems. But, despite many approximate approaches explored over the years (see below), there is arguably a dearth of practical reliable tools to deal with many-body systems between these two extremes.

Here lies the practical motivation for seeking methods to describe quantitatively those systems that are primarily classical, but which have small but measurable quantum effects. Almost all terrestrial condensed matter systems are classed in this category. As a quantitative example, the quantum contribution to the thermodynamic properties of water at standard temperature and pressure is on the order of one part in ten thousand. Other properties of water (e.g., spectral lines, hydrogen bonds) have larger quantum contributions. The degree to which it is necessary to account for quantum effects in characterizing materials, technologies, or industrial processes of course varies with the system being studied and the property that is sought. Although as argued above it is not feasible to use direct quantum methods, equally one should pause before treating terrestrial condensed matter systems solely with the methods of classical statistical mechanics. In any particular case such classical calculations are of doubtful reliability and accuracy unless there is some means of calculating or bounding the quantum contribution.

A practical quantitative description of real-world, many-particle systems must lie between approaches that invoke the axioms of quantum mechanics directly, and uncorrected methods of classical statistical mechanics that neglect quantum contributions entirely. The former is computationally prohibitive, and the latter is possibly an over-simplification that leaves out what could be important effects. Calling attention to the predominantly classical nature of the observed world opens the door to new approaches or more efficient procedures that are better designed for describing it. After all, the fact that the world of common experience is predominantly classical, with a soupçon of quantum, surely suggests that a theory mixed from similar ingredients would be useful.

To this end one might envisage a form of quantum perturbation theory, with classical statistical mechanics as the leading order term. Depending upon the structure of the theory that emerges, one could imagine a systematic expansion in which higher order terms account for successive quantum contributions. In those circumstance it is tempting to move the goal posts, with more ambitious aims to treat many-particle phenomena in which quantum behavior is not just a perturbation but is a significant or even dominant contribution. The point is that although one might begin with, and be motivated by, quantum perturbations to classical statistical mechanics, the final applications of the approaches that are developed could well extend deep into the quantum regime (cf the treatment of the λ-transition and superfluidity in liquid helium in chapter 5).

Ultimately the only limit on the application of the theory should be its accuracy and reliability. Perhaps in the fullness of time algorithms based in classical phase space can be developed to the point where they are more efficient computationally for a fully quantum many-body system than direct quantum methods based on finding the full system eigenstates and symmetrized eigenfunctions.

1.2 Why not direct quantum methods?

In the preceding subsection the main argument advanced for formulating quantum statistical mechanics in classical phase space was that the most useful applications of science lie in the terrestrial sphere, and in this regime classical behavior is predominant. Even though in a formal mathematical sense such classical behavior is ultimately due to and expressible by the axioms and equations of quantum mechanics, it is arguably better to develop approximate or exact theories from a starting point that is in close proximity to the eventual application. Such a base is provided by classical phase space.

To anticipate the detailed results set out in the text, the starting point is a formally exact transformation of quantum statistical mechanics in which the specific quantum rules and axioms are embodied in particular functions of phase space. The evaluation of these phase functions may require approximate techniques, and different approximations may suit different applications. The key questions are of course how general are the approximations, how efficient are the computational algorithms, and how reliable are the final results. Specific examples and applications of such algorithms are given in the text.

For brevity, I shall refer to alternatives to this classical phase space formulation as direct quantum mechanics, such as methods based on the direct evaluation of the energy eigenstates and symmetrized eigenfunctions of the system, or other wave function methods, with statistical averages being a weighted sum of expectation values. Before listing a number of such direct methods, I paint in broad strokes the generic impediments in practice to the application of direct quantum methods to many-particle condensed matter systems.

The computational challenges for the quantum mechanics of many-particle systems primarily arise from the difficulty in finding the energy eigenfunctions and eigenvalues of the system, and from the difficulty in enforcing the boson and fermion occupancy rules. In addition to the technical barriers that inhibit the accurate numerical description of many-particle systems, it is the rapid increase in computational cost with system size that can be prohibitive (Bloch *et al* 2008, Hernando and Vaníček 2013). Common partial differential equation algorithms, for example, grow exponentially with system size (Morton and Mayers 2005).

One can see this with simple scaling arguments. Consider a quantum system consisting of N particles in d dimensions. If one stores the eigenfunctions on a grid of M points per axis, then the amount of memory required grows as M^{Nd}. Such exponential growth severely limits the size of a system that can be treated directly. Further, suppose that the system either exactly or approximately can be expressed in terms of single-particle energy states. If these have fixed spacing Δ_E, then for a temperature T the number of accessible excited single-particle states can be expected to be on the order of $k_B T/\Delta_E \gtrsim 1$. Realistically, then, one would require on the order of $(k_B T/\Delta_E)^N$ eigenvalues and orthogonal eigenfunctions in order to characterize the whole system. Again this grows exponentially with the size of the system. Finally, it is essential that the eigenfunctions be fully symmetrized for bosons, or anti-symmetrized for fermions, which requires $N! \sim N^N$ operations. These estimates show how challenging a direct quantum calculation is for a large many-particle system.

Of course these scaling arguments provide rather crude estimates. Various and sophisticated attempts to ameliorate the difficulties with direct quantum methods have been made. The breadth of such attempts can be illustrated by a small selection: imaginary-time non-uniform mesh methods (Hernando and Vaníček 2013), pseudo-potential and mean-field methods (Dalfovo *et al* 1995, Savenko *et al* 2013, Rogel-Salazar 2013), density functional theory (Parr 1994, McMahon *et al* 2012), quantum Monte Carlo methods (Pollet 2012, McMahon *et al* 2012, Mallory and Mandelshtam 2015, Ancilotto *et al* 2017), lattice Gaussian approaches (Dimler *et al* 2009, Shimshovitz and Tannor 2012, Machnes *et al* 2016), collocation methods (Kosloff 1988), discrete variable representation methods (Harris *et al* 1965), and variational Gaussian wave-packet methods (Frantsuzov *et al* 2003, Frantsuzov and Mandelshtam 2004, Georgescu and Mandelshtam 2010, Georgescu and Mandelshtam 2011, Georgescu *et al* 2011). These methods usually invoke one or more approximations (e.g., variational functionals, discrete grids, truncated eigenvalue spectrum or even restriction to the ground state, neglect symmetrization, etc), which can limit their individual reliability and range of application.

Despite in many cases the proven merits of these algorithms and their demonstrated improvement over, for example, partial differential equation methods, the size of quantum many-body systems that can currently be treated computationally remains perhaps two orders of magnitude smaller than classical many-body systems. Faced with the many serious challenges inherent in direct treatments of quantum systems, workers are faced with two choices: either they can focus on a small system and develop techniques to solve it exactly up to a certain level, or else they can look to use approximate approaches on larger systems.

As an example of the exact but small system approach, Hernando and Vaníček (2013) found the first 50 energy eigenvalues and eigenfunctions for five Lennard–Jones particles using a non-uniform discretization of space and time. The algorithm is said to be considerably more efficient than the exponential scaling of standard finite difference methods. Nevertheless limiting the system to five particles and obtaining no more than 50 energy levels gives an indication of the severity of the computational challenges with the method. (Compare this with the results in table 9.1 below, where 10^4 energy states were found to be necessary for a four particle system with similar interaction strength at standard temperature.)

As an example of the approximate but large system approach, Georgescu and Mandelshtam (2011) used unsymmetrized Gaussian wave packets in a variational approximation to study the ground state of 6500 Lennard–Jones atoms. They estimated that the scaling with system size was reduced from $\mathcal{O}(N^3)$ to $\mathcal{O}(N^2)$. Although the size of the system is impressive, it comes at the price of a variational approximation, the neglect of wave function symmetrization, and the neglect of excited states.

This brief survey of direct quantum methods for many-body systems does not do justice to the efforts that have been expended by many workers in pursuit of practical algorithms for realistic systems. It does not detail the very real improvements that have been made over the years. (For comprehensive reviews of the state of the art in computational methods see the two papers just mentioned.) But in many ways the point is not whether one can make incremental improvements in the computational efficiency of direct approaches, but rather whether the problem is inherently so difficult that a fundamentally different viewpoint is required.

1.3 Advantages and challenges of phase space

Why are quantum many-body systems so difficult to treat computationally? The question becomes all the more pertinent when one draws the contrast with classical many-body systems. Whereas the full direct quantum treatment of many-body systems scales exponentially with particle number, accurate and reliable computer simulations for classical many-body systems scale *sub-linearly* with particle number (this is for fixed statistical relative error with cut-off potentials).

Of course, this favorable scaling of classical systems is no guarantee that transforming quantum statistical mechanics to classical phase space will enjoy the same computational advantages. After all, the system remains fundamentally quantum, and if the transformation is formally exact (which it is), then one would want to know how it is that the quantum phase functions that are introduced in the process avoid the exponential scaling with system size (which they do). But even before explaining these things, one can appreciate the motivation not only for looking for alternatives to direct quantum methods, but also for looking in the direction of classical phase space.

In order to understand at a general conceptual level why classical phase space might be advantageous, and perhaps to narrow down the search for efficient algorithms, we need to figure out more precisely just what it is about quantum systems that make them so computationally challenging. We also need to understand how it is that fundamental quantum axioms do not automatically preclude a classical phase space formulation, and how a classical phase space formulation can accommodate these quantum axioms.

To these ends one can pinpoint five key generic differences between quantum and classical systems:

- quantization
- superposition
- non-commutation
- symmetrization
- non-localization.

1.3.1 Quantization

Quantization, or the discrete nature of system states, first lead to the discovery of quantum mechanics. Two phenomena that come to mind are optical spectra and the photo-electric effect. In addition one can mention the ultraviolet catastrophe—the divergence of the energy density of black-body radiation—which was resolved when Planck (1900) postulated the quantization of electromagnetic radiation. The issue quantization raises with the classical phase space formulation of quantum statistical mechanics is that the classical particle momenta are continuous. (The positions are continuous in both quantum and classical mechanics.) In quantum mechanics for a system with edge length L in which the wave function does not go to zero at the boundaries, the momentum eigenvalues must be discrete and separated by $\Delta_{\rm p} = 2\pi\hbar/L$ per particle per dimension in order for the momentum operator to be Hermitian (Messiah 1961, chapter V, section 11). Since we are interested in macroscopic systems and $L \sim N^{1/3}$, one sees that this spacing is infinitesimal.

One property of the continuum momentum formulation requires attention, namely that the phase space integral accords vanishing weight to the point of zero momentum in two- or three-dimensional space. For non-interacting bosons this point is the energy ground state, which is of specific interest in, for example, Bose–Einstein condensation. To describe such phenomena with non-interacting bosons, the point of zero momentum has to be treated explicitly and additionally to the integral over phase space (section 5.2).

In the case of direct quantum methods, quantization is a major challenge because the discrete states must be individually characterized, and the number of such states grows exponentially with system size. Even for as few as four particles, the data in table 9.1 in the text shows that more than 10^4 energy states may be needed for a reliable description at standard temperature. Furthermore, higher energy eigenfunctions need to be made orthogonal to all the lower order ones, a process that grows quadratically with the number of states.

1.3.2 Superposition

The second quantum feature listed above is the superposition of states. This stems from the fact that any linear combination of wave functions that are solutions to Schrödinger's equation is itself a solution. Since any wave function is a linear combination of eigenfunctions, the expectation value of an operator for a system in a particular wave state is the correspondingly weighted sum over the elements of the matrix representation of the operator,

$$\langle\psi|\hat{A}|\psi\rangle = \sum_{n,m} \psi_n^* \psi_m A_{nm}.$$

The off-diagonal elements in this sum correspond to the superposition of the respective states. The physical interpretation is in essence that the system exists in the different eigenstates simultaneously, and these interfere with each other. This is contrary to our experience of the classical world in which systems occupy one, and only one, state at a time.

In addressing the impact of superposition on the classical phase space formulation of quantum systems we first slightly digress to discuss the general physical reasons that superposition effects are absent in the classical world. Take a cat, for example, which in common observation is in a single state at any one time. If the cat is placed in an opaque box, it stretches credulity to assert that it then exists in a superposition of states. This has never been observed, possibly because the box is opaque, but more likely because it never happens. There can only be one outcome when Occam's razor is applied to Schrödinger's cat (Attard 2018).

It is essential to accept, as a matter of observation and as a matter of principle, that superposition effects do not occur in the classical world. In general superposition effects occur on scales of the order of Planck's constant, $\hbar$. For example, in the two-slit experiment, the spacing between the slits must be less than about the de Broglie wavelength of the particle in order for the interference pattern to be visible. One can use this fact to explain the absence of observed superposition effects in the classical world.

The mathematical argument relies upon the construction of a set of orthogonal non-overlapping wave packets of width $\mathcal{O}(\hbar)$, indexed by a grid of points in classical phase space with spacing greater than Planck's constant (Attard 2018). This set makes a low pass filter for changes in position and momentum and it provides a complete basis for measurements with no better than classical resolution. With this basis only the diagonal terms contribute to the expectation value, which is to say that there is no interference between the superposed states. This explains the suppression of the superposition of states in the classical world.

Of course, as a matter of logic and of mathematics, the fact that superposition effects cannot be measured if the resolution is sufficiently coarse does not of itself mean that indirect effects of the superposition of states are absent. Nor in fact is any argument provided that one is justified is restricting phase space to such a coarse grid. It is possible, even likely, that the quantum phase functions that occur in the transformation to classical phase space display structure on scales smaller than Planck's constant, and serious errors would be introduced by the introduction of such a coarse-grained filter. At the end of the day the aim is to obtain a formally exact transformation to classical phase space and in such an exact transformation all quantum properties must be present until and unless additional approximations are later introduced. In other words, although the above arguments correctly explain the absence of quantum superposition in the classical world, they do not themselves justify neglecting the superposition of states in the formally exact transformation to classical phase space.

In fact the suppression of superposition effects already occurs at the quantum level. Indeed it is the reason why the present book focuses on quantum statistical mechanics rather than on quantum mechanics itself. In chapter 12 it is shown that the exchange of a conserved variable between the subsystem of interest (a so-called open quantum system) and the environment or reservoir collapses the entangled wave function into a mixture of pure states. Such collapse has been explored in the quantum theory of measurement (Zurek 1982, Joos and Zeh 1985, Zeh 2001, Zurek 2003, 2004, Schlosshauer 2005). Indeed, open quantum systems have also been studied with versions of the Schrödinger equation that include Langevin-type dissipative and fluctuation terms (Breuer and Petruccione 2002, Weiss 2012), including those that result from the projection from the reservoir onto the subsystem of the density matrix (Spohn 1980, Gardiner 1991) or operator (Gardiner 1988, Iles-Smith *et al* 2013).

In chapter 12 it is shown that wave function collapse is the reason that the von Neumann partition function of quantum statistical mechanics is of the form

$$Z = \mathrm{TR}'\left\{e^{-\beta\hat{\mathcal{H}}}\right\} = \sum_{\mathbf{n}}{}' e^{-\beta\mathcal{H}_{\mathbf{nn}}}.$$

Due to the collapse of the wave function there are no superposed states and the only contributions to the partition function and to statistical averages comes from the diagonal elements of the various operator matrices. Since this is the starting point for the transformation to classical phase space, superposition effects need not be considered.

1.3.3 Non-commutation

The third distinction between quantum and classical systems listed above is non-commutativity. In the context of the transformation to classical phase space the relevant commutator relation is that for the position and momentum operators,

$$[\hat{q}, \hat{p}] = i\hbar.$$

From this the Heisenberg uncertainty relation for the variance of their expectation values follows, $\Delta q \; \Delta p \geqslant \hbar$. The commutator relation is often interpreted to mean that one cannot measure the position and momentum of a particle simultaneously, and the uncertainty relation is often said to mean that a particle does not possess an exact position and momentum.

However one should be cautious in applying everyday language to quantum phenomena. Generally the former has a well-accepted meaning and intuitive interpretation inevitably based on classical experience. There is no reason to suppose that such words apply at the quantum level in the same way as at the macroscopic level. When it comes to quantum mechanics one should be guided by the axioms and specific equations, and one should eschew the temptation to interpret these in everyday terms. Specifically, there is no ambiguity or dispute with the mathematics of an expectation value; but the meaning of measurement at the quantum level is highly contestable. Whereas it is straightforward to take the expectation value of the position or momentum operators, or to apply the commutator relation to them, or to deduce the uncertainty principle for them, it makes no sense to couple any of these to the word 'measurement' if the latter cannot be expressed in agreed mathematical terms. Conversely, if measurement is taken to be isomorphic to a specific mathematical operation in quantum mechanics, then the name of that mathematical operation should be used in preference to the word 'measurement' because there is no ambiguity or classical baggage with the former, unlike the latter.

This discussion of the proper use of language in the quantum context is not merely an academic exercise. It has direct consequences for the present program of transforming quantum statistical mechanics to classical phase space. Since the latter is classical, we can say without hesitation that a phase space point gives the 'position' and the 'momentum' of every particle in the system. These words have their everyday meaning, and amongst other things it implies that we can attribute to each particle values of position and momentum simultaneously. Moreover, since phase space is a continuum, we can do so with infinite precision. (It also implies that that we can label each particle, the consequences of which will be discussed in the following subsection on symmetrization.) The point is, however, that there is no reason to suppose that this violates the axioms of quantum mechanics because the words we use at the macroscopic level have no meaning at the quantum level — unless we specify exactly the axiom, equation, or symbol to which they refer.

For the case of the classical position one can identify at least four different quantum mathematical symbols that it could be the analog of: the position representation coordinate, the position eigenvalue, the position expectation value, and the position operator. These are quite distinct from each other, and the classical word 'position' should not be used to refer to more than one of them. To cut a long story short, in the text it will be shown that the position coordinate of classical phase space corresponds to the quantum position eigenvalue.

Exactly the same issue arises for momentum and exactly the same conclusion applies: the momentum coordinate of classical phase space corresponds to the quantum momentum eigenvalue. (As was discussed above, the actual discretization of momentum eigenvalues versus the continuous variation of the momentum coordinate in phase space can be assumed to be unimportant in the limit of a macroscopic system—with one exception.) Once one realizes that phase space points correspond to position and momentum eigenvalues the difficulties raised above disappear. Since eigenvalues are scalars, they commute. There is no uncertainty relation for eigenvalues. And eigenvalues can be specified simultaneously (which is not the same thing as saying that their respective operators can be applied simultaneously—see chapter 7). Whether or not the usual interpretation of the uncertainty relation (the position and momentum of a particle cannot be measured simultaneously, or a particle does not possess an exact position and momentum) has any meaning is irrelevant to the present phase space formulation because the uncertainty relation and the commutator equation do not apply to position and momentum eigenvalues.

Having dealt with the most common objection to using classical phase space for quantum systems, it is important to add that the non-commutativity of the position and momentum operators does in fact create significant computational challenges for the transformation of quantum statistical mechanics to classical phase space. It would be fair to say that dealing with this issue has taken more time and effort than all the other issues combined.

In a nut-shell the problem arises from the Maxwell–Boltzmann operator, $e^{-\beta \hat{\mathcal{H}}}$. Since the Hamiltonian or energy operator is the sum of the kinetic energy operator, which is the sum of the squares of the momentum operators, and the potential energy, which depends upon the positions, the power series expansion of the exponential mixes the two in all possible ways. Non-commutativity prevents one writing this as the product of two exponentials, which one could do in the classical case.

The easy part of the problem is to define a phase function by a certain asymmetric expectation value of the Maxwell–Boltzmann operator that occurs in the exact transformation to classical phase space (chapter 7). I call this the commutation function; its defining equation is essentially the same as that given by Wigner (1932) and by Kirkwood (1933). The hard part of the problem lies in finding practical, efficient, and accurate ways to compute this function. To this end four different approaches have been tried: high temperature expansions (chapter 8), including the fluctuation series (section 8.5), nested commutator expansions (chapter 9), local state expansions (chapter 10), and many-body expansions (chapter 11). These have been tested for specific systems and they are illustrated with numerical results in the text.

The high temperature expansion, which is little different to an expansion in powers of Planck's constant, extends the results of Wigner (1932) and Kirkwood (1933). It is systematic, and each term comprises purely classical functions: the momenta, and the potential energy and its gradients. In so far as the gradients of the interaction potential are more short-ranged than the potential itself, then this expansion for the commutation function is local. Perhaps the most promising expansion derived in chapter 8 is the one formulated for the commutation function exponent, because this is an extensive function. However there are several limitations of the high temperature expansion method: the derivation is inelegant, the individual terms rapidly become complicated to derive and numerically challenging to evaluate, and the series appears slowly converging for temperatures typical of terrestrial condensed matter. One cannot help but compare the high temperature expansion for the commutation function to the virial expansion for the pressure in classical statistical mechanics: it is formally exact, it is of some historical interest, but it is questionable whether it will ever prove a viable numerical approach for realistic systems.

The fluctuation series, section 8.5, writes the commutation function in terms of the fluctuations of the phase space expectation value of the energy operator. The terms are readily obtained from a recursion relation and are directly related to the temperature expansion coefficients. A modified exponentiated series for the commutation function exponent has extensive terms that are essentially the fluctuations of the fluctuations; alas, to date the computational implementation of these has proved wanting. Nevertheless the fluctuation series provides a physical interpretation for the terms in the temperature expansion and a rational basis for believing in the convergence of the series for the commutation function.

The nested commutator expansion for the commutation function, chapter 9, factorizes the Maxwell–Boltzmann operator into exponentials of the potential energy operator, the kinetic energy operator, and a residual operator. The latter expresses the correction to the classical Maxwell–Boltzmann factor as a series of nested commutators weighted by powers of inverse temperature, with explicit expressions being given for second, third, and fourth order terms. One method given for deriving the commutation function exponent operator is via a temperature derivative, and there is consequently some overlap between the high temperature expansion and the nested commutator expansion. One advantage of the latter is that the individual terms appear in a highly compact form, and a great deal of internal cancelation occurs in them. Like the high temperature expansion, the expectation value of the commutation function exponent operator involves gradients of the potential energy, and as such is highly localized.

The local state and the many-body approaches to the commutation function are summarized in section 1.3.5 below.

In summary, the consequences of the non-commutativity of the position and momentum operators poses less a conceptual challenge—how classical particles possess position and momentum simultaneously and with infinite precision—than a numerical one: how to compute the commutation phase function that arises efficiently, accurately, and reliably.

1.3.4 Symmetrization

It is a fundamental postulate of quantum mechanics that the wave function must be fully symmetric (bosons) or fully anti-symmetric (fermions) with respect to permutations of identical particles (Messiah 1961, chapter xiv). This leads to particle state occupancy statistics, in which more than one boson, but not more than one fermion, can occupy the same single-particle state (Merzbacher 1970, chapter 20). Arguably the formulation in terms of symmetrized wave functions is more fundamental than is the formulation in terms of particle statistics because not all operators have eigenstates that are single-particle states, but all eigenfunctions must be symmetrized. Wave function symmetrization implies particle occupancy statistics in the single-particle state case (chapter 2). In the case of multi-particle states, such as the energy states for interacting particles, wave function symmetrization gives rise to a so-called symmetrization factor, which is useful in evaluating any sum over states, and which may be regarded as the generalization of particle occupancy statistics to multi-particle states (chapter 3).

A second issue with the particle statistics formulation is that it is difficult to see how the occupancy rules for single-particle states can be formulated or enforced in the continuum limit that the spacing between states goes to zero. In the present phase space formulation of quantum statistical mechanics, both the position and the momentum eigenvalues describe single-particle states in the continuum limit, and so particle occupancy statistics are potentially an issue here. Fortunately however, the symmetrization factor alluded to above carries over in the continuum limit to become a symmetrization function for phase space. This is the way that the fundamental rule for wave function symmetrization is enforced in classical phase space.

Beyond particle statistics, a point needs to be made on the computational cost of symmetrization, which will be expanded upon in the context of non-localization, section 1.3.5 below. In discussing the challenges faced by direct quantum methods in the subsection above, it was pointed out that the symmetrization of the wave function required the permutation of all N particles in the system, $N!$ operations in total. This of course is prohibitive for a large system. The question therefore arises, both for direct methods and for the proposed phase space formulation, whether it is possible to reduce this number, and perhaps restrict the symmetrization to a subset or subsets of the particles in the system. If valid, the computational savings would be enormous.

Messiah (1961, chapter xiv, section 8) addresses a similar question. He points out that *if* the total system wave function factorizes into the product of wave functions for groups of particles confined to respective domains, and *if* these wave functions do not overlap, then it is possible to permute the particles within each group independently. These are pretty big ifs. It is hard to see how they would be satisfied within the usual sorts of systems treated by quantum statistical mechanics. Symmetrization appears to pose insurmountable difficulties for direct quantum methods.

As mentioned, the formulation proposed here introduces the so-called symmetrization function of classical phase space. The highly favorable feature of the derivation is that it is formulated as a systematic permutation loop expansion. The terms involve Fourier exponents of the momenta and separations of specific groups of particles, with closely separated particles dominating. It is shown that this gives the known analytic results for the quantum ideal gas and for non-interacting quantum harmonic oscillators, and leads to a novel physical interpretation and a more efficient computational formulation of the latter, chapter 4. It also leads to a new interpretation and physical mechanism for the λ-transition in liquid helium, and consequently for Bose–Einstein condensation and superfluidity in general, chapter 5. The terms in the loop expansion are effectively ordered in powers of fugacity, which means that the error due to truncating the infinite series decreases with decreasing density. The computational resources required for obtaining each term appear to scale linearly with N. Separating commutation and symmetrization effects into two phase functions enables an infinite order re-summation of the latter that accelerates its convergence, chapter 3.

The final point to be made on symmetrization was alluded to above: the formally exact transformation to classical phase space gives a label to each particle. The label is arbitrary, but it serves to unambiguously associate a specific position and momentum with each particle. In quantum mechanics the symmetrization of the wave function is said to make identical particles indistinguishable (Messiah 1961, chapter xiv). In truth we need not inquire into the meaning of the word 'indistinguishable' or to the physical picture that it is meant to convey. We only have to make sure that we follow the mathematical consequences of wave function symmetrization. Since all symmetrization effects are subsumed exactly into the symmetrization phase function, we can assert that it is correct and proper for particles to be distinguished in classical phase space by labels. Whatever 'indistinguishable' might actually mean in quantum theory, the full mathematical consequences of wave function symmetrization are conveyed by the symmetrization function using labeled particles in phase space.

1.3.5 Non-localization

Quantum mechanics is a non-local theory. The most eye-opening if not most common manifestation of this is the Einstein–Podolsky–Rosen (EPR) paradox (Einstein *et al* 1935, Mermin 1985). The simplest realization of this paradox concerns a spin zero particle that decays into two spin half particles, which fly apart until their spins are subsequently measured by detectors aligned at possibly different angles. Upon measurement the entangled wave function collapses. The quantum prediction is such that prior to collapse the properties of the wave function cannot be considered local to each particle but rather global to the system as a whole, even when the detectors are separated by macroscopic distances. The collapse is apparently instantaneous, and certainly super-luminal. The quantum prediction of non-locality has been unambiguously and quantitatively confirmed by laboratory measurement (Aspect *et al* 1981).

In the context of the proposed program to perform quantum statistical mechanics in classical phase space, the non-locality of quantum mechanics is directly relevant in explaining the ultimate reason why direct quantum methods are not computationally feasible, and in underscoring the significant challenge faced by classical phase space methods to do better.

The most common, almost mundane, manifestation of quantum non-locality is quantization itself. The discrete solutions to Schrödinger's equation, the eigenstates, ultimately arise from the boundary conditions enforced on the wave function. Such boundary conditions must be applied even when the boundaries are separated by macroscopic distances. In general it is not possible to subdivide a system into subsystems and to solve Schrödinger's equation independently on each. Of course one could match the wave function and its gradient across the boundaries of the subsystems, but ultimately one has to impose conditions at the boundary of the system itself, which vitiates the point of creating the subsystems in the first place. The discrete eigenstates of the system are those states that satisfy the boundary conditions of the full system, and there is no general method that gives these as the product of eigenstates of quasi-independent subsystems. The necessity of solving for the system as a whole means that all N particles contribute, and so it is not possible to avoid the exponential scaling with system size that was discussed above.

The non-local nature of quantum mechanics in necessitating eigenfunctions for the system as a whole also explains why wave function symmetrization requires $N!$ operations. For a system of interacting particles one cannot assume that the wave function breaks up into products of non-overlapping wave functions of distinct groups of particles, and so one cannot confine the permutations to within such groups.

From these arguments one can see the problems faced by direct quantum methods in developing feasible computational algorithms. But what of methods based in classical phase space? If the transformation of quantum statistical mechanics is formally exact, how does one preclude the non-local phenomena just elucidated from being equally prohibitive?

The detailed answer to this question must await the mathematical analysis of the following chapters, specifically chapters 2 and 3 for wave function symmetrization. But at this introductory stage one can give an overview that the reader might take on trust until the details are filled in.

First we can return to the concept of non-locality and ask why the EPR paradox is so surprising. The answer is of course that in the classical world everything is local, and all of our intuition and experience is based on classical macroscopic phenomena. One can perhaps therefore appreciate why the transformation to classical phase space has a chance of overcoming the global scaling problems that impede direct quantum methods: the starting point is described by a local theory. As an example, reliable computational methods are known for sub-dividing classical phase space, even for particles with long-range interaction potentials, which is the reason that computer simulations of classical systems are able to scale sub-linearly with system size.

This does not entirely answer the question because two quantum phase functions remain after the transformation to classical phase space: the commutation function and the symmetrization function. These supposedly embody completely and exactly quantum statistical mechanics, including its non-local nature.

Turning first to the symmetrization function, it was mentioned above that this was formulated in terms of a loop expansion. The loops are compact and localized, and small loops dominate (but see chapter 5). If all terms in the series were kept, it would require $N!$ operations to evaluate. But keeping only the first few or several terms, and exponentially resumming those that are kept, is much more tractable computationally. Of course this is an approximation, but the point of such a systematic expansion is that error estimates of the neglected terms can be made.

For the commutation function four computational algorithms are derived in the text. These divide naturally into two classes: in the first class are the high temperature expansion (chapter 8), including the fluctuation series (section 8.5), and the nested commutator expansion (chapter 9); in the second class are the local state expansion (chapter 10), and the many-body expansion (chapter 11). In the brief outlines of the high temperature series, the fluctuation series, and the nested commutator approaches given above it was pointed out that they yielded classical phase space functions that invoked the gradients of the potential energy and that as such they were highly localized.

The many-body expansion, chapter 11, is a series for the commutation function that is formally exact for an N-particle system if N terms are retained. The hope is that truncating the series after a few or several terms will yield accurate results. In order to calculate the n-body term, one has to obtain the energy eigenvalues and eigenfunctions of an n-particle system. The energy spectrum depends upon the specified boundary conditions, which are non-local. This presents a conceptual issue for the approach that remains to be satisfactorily resolved. The promise of this many-body approach is that it casts the commutation function as an effective potential and all of the known results for classical statistical mechanics can be carried over directly. Numerical data obtained to date indicate that terms beyond the pair level of the expansion are required for reliable results.

Finally, the local state expansion for the commutation function, chapter 10, as the name implies, is specifically designed to deal with the inherently non-local nature of quantum mechanics. Although the direct effect of position and momentum non-commutativity is highly localized, occurring in a volume in phase space of Planck's constant per particle per dimension, because the expression for the commutation function involves the exponential of the energy operator, the non-local nature of the energy eigenstates and eigenfunctions have a bearing on it if it is reformulated as a sum over such states. (The high temperature expansion, fluctuation series, and the nested commutator expansion remain local because they are not formulated as a sum over energy eigenstates.)

The local state expansion decomposes the total system phase space configuration into independent clusters of one or more quantum particles in an effective potential field due to their fixed configuration neighbors. This relies upon the distinction that was alluded to above between the position representation coordinate and the position eigenvalue. The effective local field differs from the actual local field at each particle, and is designed to maximize the analytic cancelation of terms in the expansion. One can formulate increasingly sophisticated approximations, namely singlet, pair, …, n-particle clusters. For a given cluster size, temperature-dependent corrections to the effective local energy operator may be developed that systematically improve the corresponding effective Maxwell–Boltzmann operator. The n-particle eigenvalues and eigenfunctions of the effective local energy operator are used to calculate the n-particle commutation functions, which are combined to form the full N-particle commutation function. Numerical results in chapter 10 using just the first term of the singlet effective field expansion already appear quite accurate.

Without going into mathematical detail, the effective local field is a purely classical construct, and as such it is local in nature. This is how one gets around the conventional quantum prohibition on localization, and which allows the sub-linear scaling with system size of classical statistical mechanics to be exploited.

This effective local field offers exciting possibilities for developing other novel approaches to quantum statistical mechanics. At the conceptual level it explains how the non-local nature of quantum mechanics is transformed to the local behavior of classical statistical mechanics. In practical terms, the localized commutation function that results gives confidence that the formally exact transformation of quantum statistical mechanics to classical phase space provides a viable basis for developing practical and reliable computational approaches to many-particle quantum systems.

1.4 Old applications, new perspectives

One of the strongest arguments given above for formulating quantum statistical mechanics in classical phase space is that the terrestrial world is predominantly classical, and that successive quantum corrections are just that. Beyond this it was argued that the formulation might actually or eventually prove useful even for systems that are fully quantum in nature. Both of these arguments raise the question: apart from its undoubted formal and conceptual interest, how credible is it that this new approach to quantum statistical mechanics might actually have useful application or lead to new discoveries?

There are three things to look for in concrete applications. First, whether it reproduces well-known and broadly accepted quantum results, thereby raising confidence in the phase space approach. Second, whether it provides new insight or novel interpretation in what are already well-tilled fields. And third, whether the application to a concrete problem reveals specific techniques otherwise overlooked in the formal and generic development of the method.

I shall not pretend that this book is replete with problems or worked examples; only three chapters are focused on applications (chapters 4, 5, and 6). It is early days in the development of the classical phase space approach, and it takes time to build up a collection of methods and applications. Nevertheless to date a limited repertoire has been gathered, and already enough results, both old and new, and some quite significant, have been established to provide examples for each of the three points listed above.

One of the first applications is to the quantum ideal gas, section 4.1, where it is shown that the phase space formulation yields the same analytic expressions as the fugacity expansion given in the usual textbooks (Point 1). It is also shown that the terms in the fugacity expansion lie in a one-to-one relationship with the terms in the permutation loop expansion (Point 2).

A second example is that of independent harmonic oscillators, section 4.2. As the very first quantum application—applied by Planck (1900) to the problem of black-body radiation—many would be surprised that anything new remains to be said on the subject. But here not only is it shown that the phase space formulation reproduces the textbook result for the single oscillator (Point 1), but also a new infinite series is given for a collection of oscillators. The new series can be mathematically transformed into the conventional infinite series (Point 1), whilst the terms have a novel physical interpretation as permutation loops that arise from wave function symmetrization (Point 2).

A third example, also found in most quantum and solid state textbooks, is that of the heat capacity due to lattice vibrations, section 6.1. These were originally treated as quantum harmonic oscillators, first by Einstein (1907), and later superseded by Debye (1912). Here benchmark results are given for a one-dimensional harmonic crystal (nearest neighbor interactions), for which an analytic solution to the quantum many-particle problem is obtained, section 6.2. Contrary to popular wisdom, here it is found that at absolute zero the heat capacity goes to zero exponentially, in agreement with Einstein (1907), but in contradiction to the power law decay predicted by Debye (1912) (Point 2). (Of the two, Debye's theory can be fitted more accurately to the exact analytic results in the intermediate temperature regime.)

A fourth example, numerical in nature, is for interacting Lennard–Jones atoms in one-dimension, section 5.3. In this case the Monte Carlo simulations in classical phase space are shown to agree over a limited regime with benchmark results obtained by more conventional quantum methods (Hernando and Vaníček 2013) (Point 1). They also offer insight into the roles that position–momentum non-commutativity and wave function symmetrization play in a realistic system of interacting particles (Point 2).

Not all applications treated in the book fulfill one or other of the three aims listed above. In some cases I have used examples from other texts with minor changes and limited insight. In other cases I present numerical results that stand on their own as there is no conventional benchmarks to compare them with. Also there are a few examples where the application is generic but concrete (such as the primary quantum departure from classical statistical mechanics, section 8.2.2), which correct the currently accepted expression (Point 2), and which stand or fall on the validity of their mathematical derivation (Point 1).

1.4.1 λ-Transition and superfluidity

By far the most significant application studied here is the λ-transition in liquid helium and its relation to Bose–Einstein condensation and superfluidity, chapter 5. This turns out to provide examples of all three points listed above. The system is analyzed using five methods: the conventional ideal gas model originally put forward by London (1938) (section 5.2), an exact summation over states for the ideal gas (section 5.3), Monte Carlo simulations in classical phase space using Lennard–Jones interactions between the helium atoms (section 5.4), a factorization method on the far side of the λ-transition (section 5.5), and an integral equation method using a Markov superposition approximation (section 6.3). The first and second methods provide an illustration of Point 3: the exact summation confirms the ansatz that underscores London's (1938) treatment of the ideal gas model—for bosons the ground state has to be added separately and explicitly to the integral over the continuum of excited states—thereby revealing an important issue that is otherwise overlooked by the continuum phase space formulation of quantum statistical mechanics. The Monte Carlo simulations give a detailed molecular picture of the λ-transition from the high temperature side (Point 2), and, in agreement with the exact enumeration (Point 1), gives a picture of Bose–Einstein condensation in terms of the growth and decay of permutation loops that offers a new physical picture of the λ-transition and Bose–Einstein condensation (Point 2). The simulations are limited in a number of aspects (neglect of the ground state, dilute permutation loop approximation, 'only' the Lennard–Jones interaction potential), but nevertheless qualitatively and by one quantitative measure the prediction for the λ-transition in liquid helium-4 is closer to the experimentally measured transition than that of the ideal gas model.

The study of the λ-transition offers a new physical understanding of Bose–Einstein condensation and superfluidity, section 5.6. It is found, for example, that it is not the occupation of the energy ground state *per se* that is primarily responsible for the λ-transition. Rather, the λ-transition marks the explosive growth in number, size, and extent of position permutation loops. These nucleate the condensation of bosons into low-lying momentum states, which provides the concrete mechanism for superfluidity that has arguably been missing to date.

The two currently accepted theories for superfluidity, the two-fluid ideal gas model (Tisza 1938), and the two-fluid phonon-roton model (Landau 1941), both assert that the superfluid comprises bosons in the energy ground state that are transparent to the normal fluid (Balibar 2014). But neither offer a mechanism for this, and both suffer the long-standing criticism of Landau (1941) himself—that bosons in excited states must interact and collide with each other, thereby creating friction and viscosity, which would destroy the superfluid state. The present calculations for momentum states overcome this objection, since they emphasize that collisions that destroy momentum permutation loops and momentum state occupancy are statistically suppressed because they decrease loop entropy. The present picture explains superfluidity as due in part to the non-local nature of the momentum loops, since in micro-pores this leads directly to inviscid plug flow. In addition, it is pointed out that there is a transverse momentum gap in pores or thin films that suppresses momentum transfer with the walls; the absence of momentum transfer implies superfluid flow (section 5.6.2). The theory's prediction of the critical flow velocity, above which superfluidity is destroyed, is in agreement with measured values, and it provides a detailed molecular explanation of superfluidity.

Of course one can debate the merits of these insights, the reliability of the interacting boson approximations, or the molecular mechanisms proposed for Bose–Einstein condensation and superfluidity. But if nothing else this example of the λ-transition in liquid helium demonstrates the applicability of classical phase space methods to systems that are predominantly quantum. And they show that this novel formulation of quantum mechanics for condensed matter systems can lead to new discoveries.

References

Ancilotto F, Barranco M, Coppens F, Eloranta J, Halberstadt N, Hernando A, Mateo D and Pi M 2017 Density functional theory of doped superfluid liquid helium and nanodroplets *Int. Rev. Phys. Chem.* **36** 621

Aspect A, Grangier P and Roger G 1981 Experimental tests of realistic local theories via Bell's theorem *Phys. Rev. Lett.* **47** 460

Attard P 2018 Herding Shrödinger's supposéd cat. How the classical world emerges from the quantum universe arXiv:1809.08757

Balibar S 2014 Superfluidity: how quantum mechanics became visible *History of Artificial Cold, Scientific, Technological and Cultural Issues* ed K Gavroglu (Dordrecht: Springer) pp 93–117

Bloch I, Dalibard J and Zwerger W 2008 Many-body physics with ultracold gases *Rev. Mod. Phys.* **80** 885

Breuer H P and Petruccione F 2002 *The Theory of Open Quantum Systems* (Oxford: Oxford University Press)

Dalfovo F, Lastri A, Pricaupenko L, Stringari S and Treiner J 1995 Structural and dynamical properties of superfluid helium: a density-functional approach *Phys. Rev.* B **52** 1193
Dimler F, Fechner S, Rodenberg A, Brixner T and Tannor D J 2009 Accurate and efficient implementation of the von Neumann representation for laser pulses with discrete and finite spectra *New J. Phys.* **11** 105052
Debye P 1912 Zur theorie der spezifischen wärme (The theory of specific heat) *Ann. Phys.* **344** 789
Einstein A 1907 Plancksche theorie der strahlung und die theorie der spezifischen wärme (Planck's theory of radiation and the theory of specific heat) *Ann. Phys.* **22** 180
Einstein A, Podolsky B and Rosen N 1935 Can quantum-mechanical descriptionof physical reality be considered complete? *Phys. Rev.* **47** 777
Frantsuzov P A, Neumaier A and Mandelshtam V A 2003 Gaussian resolutions for equilibrium density matrices *Chem. Phys. Lett.* **381** 117
Frantsuzov P A and Mandelshtam V A 2004 Quantum statistical mechanics with Gaussians: equilibrium properties of van der Waals clusters *J. Chem. Phys.* **121** 9247
Gardiner C W 1988 Quantum noise and quantum Langevin equations *IBM J. Res. Dev.* **32** 127
Gardiner C W 1991 *Quantum Noise* (Berlin: Springer)
Georgescu I and Mandelshtam V A 2010 Molecular dynamics with quantum fluctuations *Phys. Rev.* B **82** 094305
Georgescu I and Mandelshtam V A 2011 A fast variational Gaussian wave-packet method: size-induced structural transitions in large neon clusters *J. Chem. Phys.* **135** 154106
Georgescu I, Deckman J, Fredrickson L J and Mandelshtam V A 2011 Thermal Gaussian molecular dynamics for quantum dynamics simulations of many-body systems: application to liquid para-hydrogen *J. Chem. Phys.* **134** 174109
Harris D O, Engerholm G G and Gwinn W D 1965 Calculation of matrix elements for one-dimensional quantum-mechanical problems and the application to anharmonic oscillators *J. Chem. Phys.* **43** 1515
Hernando A and Vaníček J 2013 Imaginary-time nonuniform mesh method for solving the multidimensional Schrödinger equation: fermionization and melting of quantum Lennard–Jones crystals *Phys. Rev.* A **88** 062107
Iles-Smith J, Lambert N and Nazir A 2013 Environmental dynamics, non-Gaussianity, and the emergence of noncanonical equilibrium states in open quantum systems *Phys. Rev.* A **90** 032114
Joos E and Zeh H D 1985 The emergence of classical properties through interaction with the environment *Z. Phys.* **59** 223
Kirkwood J G 1933 Quantum statistics of almost classical particles *Phys. Rev.* **44** 31
Kosloff R 1988 Time-dependent quantum-mechanical methods for molecular dynamics *J. Phys. Chem.* **92** 2087
Landau L D 1941 Two-fluid model of liquid helium II *J. Phys. USSR* **5** 71
Landau L D 1941 Theory of the superfluidity of helium II *Phys. Rev.* **60** 356
London F 1938 The λ-phenomenon of liquid helium and the Bose–Einstein degeneracy *Nature* **141** 643
Mallory J D and Mandelshtam V A 2015 Binding energies from diffusion Monte Carlo for the MB-pol H2O and D2O dimer: a comparison to experimental values *J. Chem. Phys.* **143** 144303
Machnes S, Assémat E and Tannor D 2016 Quantum dynamics in phase space using the biorthogonal von Neumann bases: algorithmic considerations arXiv:1603.03963
McMahon J M, Morales M A, Pierleoni C and Ceperley D M 2012 The properties of hydrogen and helium under extreme conditions *Rev. Mod. Phys.* **84** 1607

Mermin N D 1985 Is the Moon there when nobody looks? Reality and the quantum theory *Physics Today* **38** 38
Merzbacher E 1970 *Quantum Mechanics* 2nd edn (New York: Wiley)
Messiah A 1961 *Quantum Mechanics* vols 1 and 2 (Amsterdam: North-Holland)
Morton K and Mayers D 2005 *Numerical Solution of Partial Differential equations, An Introduction* 2nd edn (Cambridge: Cambridge University Press)
Parr R G and Yang W 1994 *Density-Functional Theory of Atoms and Molecules* 2nd edn (Oxford: Oxford University Press)
Pollet L 2012 Recent developments in quantum Monte-Carlo simulations with applications for cold gases *Rep. Prog. Phys.* **75** 094501
Planck M 1900 Über eine verbesserung der wienschen spektralgleichung (On an improvement of Wien's equation for the spectrum) *Verhandl. Deut. Physik. Ges.* **2** 202, 237
Planck M 1914 *The Theory of Heat Radiation* translated by Masius MP (Philadelphia, PA: Blakiston's Sons and Co.)
Rogel-Salazar J 2013 The Gross–Pitaevskii equation and Bose–Einstein condensates *Eur. J. Phys.* **34** 247
Savenko I G, Liew T C H and Shelykh I A 2013 Stochastic Gross–Pitaevskii equation for the dynamical thermalization of Bose–Einstein condensates *Phys. Rev. Lett.* **110** 127402
Schlosshauer M 2005 Decoherence, the measurement problem, and interpretations of quantum mechanics *Rev. Mod. Phys.* **76** 1267
Shimshovitz A and Tannor D J 2012 Phase-space approach to solving the time-independent Schrödinger equation *Phys. Rev. Lett.* **109** 070402
Spohn H 1980 Kinetic equations from Hamiltonian dynamics: Markovian limits *Rev. Mod. Phys.* **52** 569
Tisza L 1938 Transport phenomena in helium II *Nature* **141** 913
Weiss U 2012 Quantum Dissipative Systems *Quantum Dissipative Systems* (New Jersey: World Scientific)
Wigner E 1932 On the quantum correction for thermodynamic equilibrium *Phys. Rev.* **40** 749
Zeh H D 2001 *The Physical Basis of the Direction of Time* 4th edn (Berlin: Springer)
Zurek W H 1982 Environment-induced superselection rules *Phys. Rev.* D **26** 1862
Zurek W H 2003 Decoherence, einselection, and the quantum origins of the classical *Rev. Mod. Phys.* **75** 715
Zurek W H 2004 Probabilities from entanglement, Born's rule from envariance *Phys. Rev.* A **71** 052105

IOP Publishing

Quantum Statistical Mechanics in Classical Phase Space

Phil Attard

Chapter 2

Wave packet formulation

Points in classical phase space are shown to correspond to the position and momentum eigenvalues of the particles in the system. In general for single-particle states, particle statistics and state occupancy rules can be enforced by a symmetrization factor that allows a sum over unique, allowed states to be written as an unrestricted, weighted sum over all states. The symmetrization factor gives an internal entropy for the quantum microstates.

These ideas are illustrated with wave packets, which are approximately simultaneous eigenfunctions of the position, momentum, and energy operators. They are more exact in the classical low density limit in which phase space can be discretized coarsely. The leading order correction due to wave packet symmetrization is shown to give a temperature-dependent effective potential, leading to a Gaussian pair probability in phase space.

2.1 Introduction

In the real world, physical systems are complicated, and to describe them it is ultimately necessary to resort to simplifying approximations. The best practice for developing approximate approaches is to formulate the problem in a formally exact fashion as far as possible, and to delay the introduction of the approximate part of the approach until near the end. The reasons for this are primarily that approximations can often be specific to the problem at hand, whereas exact approaches are more universal and have a wider range of applicability. Also, approximations can be *ad hoc*, their domain of validity can be unclear, and they can lack systematic corrections or improvements. In contrast proceeding from a formally exact basis often opens the door to a systematic expansion with terms becoming successively smaller in specified limits.

So much for the theoretical ideal. In actual practice approximations are usually introduced at an early stage in the analysis, often motivated by physical considerations. Although this approach has the limitations adumbrated in the preceding

doi:10.1088/978-0-7503-4055-7ch2

paragraph, it can have compensating virtues, such as making for a simpler analysis, finding consonance with physical intuition, and providing a pedagogic introduction to the problem.

This chapter takes the second, approximate, approach to the problem of formulating quantum statistical mechanics in classical phase space, and defers to chapters 3 and 7 the more formal exact analysis.

The specific aim here is to get quickly to the classical limit from an elementary starting point in quantum mechanics. Beginning with Gaussian wave packets as an approximation to the energy eigenfunctions of a multi-particle system, the von Neumann trace over the density matrix for the statistical average and partition function is transformed to an integral over classical phase space. The point of the analysis is to identify the physical conditions in which the classical limit holds.

Three concepts—commutation, symmetrization, localization—raised in this chapter will recur throughout the book. First, in quantum mechanics the position and momentum operators do not commute, but in classical phase space both values are specified simultaneously. The wave packets used in this chapter enable the transformation to classical phase space within the approximation that they are simultaneous eigenfunctions of the two operators.

Second, the symmetrization of the quantum mechanical wave function usually necessitates specifying the system by quantum state occupancy rather than by distinguishable particle labels, but in the classical limit labeled particles are at a point in the continuum that is phase space; there are no discrete states capable of being unambiguously occupied or unoccupied. In this chapter the symmetrization of wave packets is used to show how particle statistics of bosons and fermions can be reconciled with the phase space continuum and distinguishable particles, and the primary correction in a systematic expansion of symmetrization effects is calculated.

And third, the quantization of states in general arises from the imposition of boundary conditions; the eigenfunctions of a macroscopic system are global and in general they cannot be obtained by sub-division into local regions. Similarly the symmetrization just alluded to refers to the wave function of all particles in the system, and to the occupancy of whole-system states. In contrast classical mechanics and classical statistical mechanics are local theories in which such global requirements do not apply. In the present chapter the problem of localizing a quantum system is avoided by simply invoking wave packets unconstrained by specific boundary conditions without addressing explicitly the validity of doing so. In later chapters the problem of localization will turn out to be one of the main challenges in the formally exact transformation of quantum statistical mechanics to classical phase space.

The present chapter suffers from two limitations: first the analysis is approximate, and the results are motivated more by physical considerations than by mathematical rigor. And second, in arriving at the final destination of the classical limit, nothing is said about the fly over states, namely those intermediate regimes which are primarily classical, but where quantum effects may be measurable. Or those regimes where quantum phenomena are possibly significant, or even dominant. These lacunae are filled by the more formal procedures in the rest of the book.

2.2 Wave packets as eigenfunctions in the classical limit

2.2.1 Definition

The primary goal here is to show how to get from the quantum picture to classical statistical mechanics, which is of course formulated in the classical phase space of the particles' positions and momenta. To this end we use wave packets because they are localized simultaneously in position and momentum. Some care must be taken with using these as a basis, as there are issues with their orthogonality and completeness. Also the meaning of simultaneous localization must be reconciled with the Heisenberg uncertainty principle.

Consider a system consisting of N particles in three dimensions, with position representation coordinates $\mathbf{r} = \{\mathbf{r}_1, \mathbf{r}_2, \dots, \mathbf{r}_N\}$, $\mathbf{r}_j = \{r_{jx}, r_{jy}, r_{jz}\}$, $j = 1, 2, \dots, N$. Define analogous vectors for the position label $\mathbf{q}$, and the momentum label $\mathbf{p}$. It will sometimes be convenient to write the combined labels as $\mathbf{\Gamma} \equiv \{\mathbf{q}, \mathbf{p}\}$. The representation coordinates $\mathbf{r}$ belong to the real continuum. The labels are discretized, for example $q_{j\alpha} = \ell_{\mathrm{q},\, j\alpha}\Delta_{\mathrm{q}}$ and $p_{j\alpha} = \ell_{\mathrm{p},\, j\alpha}\Delta_{\mathrm{p}}$, with the ℓ_{q} and ℓ_{p} being integers. Below $\mathbf{\Gamma}$ and ℓ are used interchangeably to label the system state. The meaning of position and momentum labels emerges from the following analysis.

At this stage we do not need to be more precise about the grid spacing or the system size. Messiah (1961, chapter V, section 11) insists that in order for the momentum operator to be Hermitian periodic boundary conditions must be imposed on all wave functions, and discrete momentum eigenvalues must have spacing $\Delta_{\mathrm{p}} = 2\pi\hbar/L$ where L is the system edge length. This is true if the wave function does not go to zero at the boundaries. In any case, here we do not explicitly use this (but we consistently do so in later chapters), but instead observe that for a macroscopic system, $L \to \infty$, the present grid spacing could be made an integer multiple of this.

Spin could be included in the set of commuting dynamical variables, with $\mathbf{x}_j = \{\mathbf{r}_j, \sigma_j\}$, where $\sigma_j \in \{-S, -S+1, \dots, S\}$ is the z-component of the spin of particle j. See Messiah (1961, section 14.1), or Merzbacher (1970, section 20.5), or section 6.4 below. However, we do not include spin in the present analysis.

The minimum uncertainty wave packet for the system is (Messiah 1961, Merzbacher 1970)

$$\begin{aligned} \zeta_{\mathbf{q},\,\mathbf{p}}(\mathbf{r}) &\equiv \frac{1}{C}\exp\left\{\frac{-(\mathbf{r}-\mathbf{q})\cdot(\mathbf{r}-\mathbf{q})}{4\xi^2} - \frac{1}{\mathrm{i}\hbar}\mathbf{p}\cdot(\mathbf{r}-\mathbf{q})\right\} \\ &= \prod_{j=1}^{N} \zeta^{(1)}_{\mathbf{q}_j,\,\mathbf{p}_j}(\mathbf{r}_j) \\ &= \prod_{j=1}^{N}\prod_{\alpha=x,y,z} \frac{1}{(2\pi\xi^2)^{1/4}} e^{-(r_{j\alpha}-q_{j\alpha})^2/4\xi^2} e^{-p_{j\alpha}(r_{j\alpha}-q_{j\alpha})/\mathrm{i}\hbar}, \end{aligned} \tag{2.1}$$

where ξ is the packet width, and $C^2 = (2\pi\xi^2)^{3N/2}$. This is normalized,

$$\langle \zeta_{\mathbf{q},\,\mathbf{p}} | \zeta_{\mathbf{q},\,\mathbf{p}} \rangle = \int \mathrm{d}\mathbf{r}\ \zeta_{\mathbf{q},\,\mathbf{p}}(\mathbf{r})^*\ \zeta_{\mathbf{q},\,\mathbf{p}}(\mathbf{r}) = 1. \tag{2.2}$$

Note how the wave packet for the full multi-particle system is the product of independent single-particle wave packets.

In general these wave packets do not form a strictly orthonormal set. Instead one has

$$\left\langle \zeta_{\mathbf{q}',\,\mathbf{p}'} \middle| \zeta_{\mathbf{q},\,\mathbf{p}} \right\rangle = \exp\left\{ -\frac{(\mathbf{q}' - \mathbf{q})^2}{8\xi^2} - \frac{\xi^2(\mathbf{p}' - \mathbf{p})^2}{2\hbar^2} - \frac{\mathbf{q}' \cdot \mathbf{p} - \mathbf{q} \cdot \mathbf{p}'}{\mathrm{i}\hbar} \right\}. \tag{2.3}$$

This is a Gaussian in position separation and momentum separation, with an oscillatory modulation. In the continuum limit, it may be regarded as the product of soft Dirac-delta functions, $\delta_{2\xi}(\mathbf{q}' - \mathbf{q})\, \delta_{\hbar/\xi}(\mathbf{p}' - \mathbf{p})$, neglecting the modulation. Alternatively, with the discretization of the labels, it can be seen as the product of approximate Kronecker-deltas, $\delta_{\ell_q',\, \ell_q}\, \delta_{\ell_p',\, \ell_p}$.

The interpretation as a Kronecker-delta can be made more exact by taking the spacing between phase space grid points to be much larger than the variance of the Gaussians,

$$\Delta_{\mathrm{q}} \gg 2\xi, \quad \text{and} \quad \Delta_{\mathrm{p}} \gg \hbar/\xi. \tag{2.4}$$

In this case one sees that the imaginary part of the exponent is zero when $\ell_q' = \ell_q$, and $\ell_p' = \ell_p$. Conversely, when the imaginary part is non-zero, the Gaussians themselves are zero. The orthogonality of non-overlapping wave packets confined to coarse grid points is sketched in figure 2.1. This condition for the orthonormality of the wave

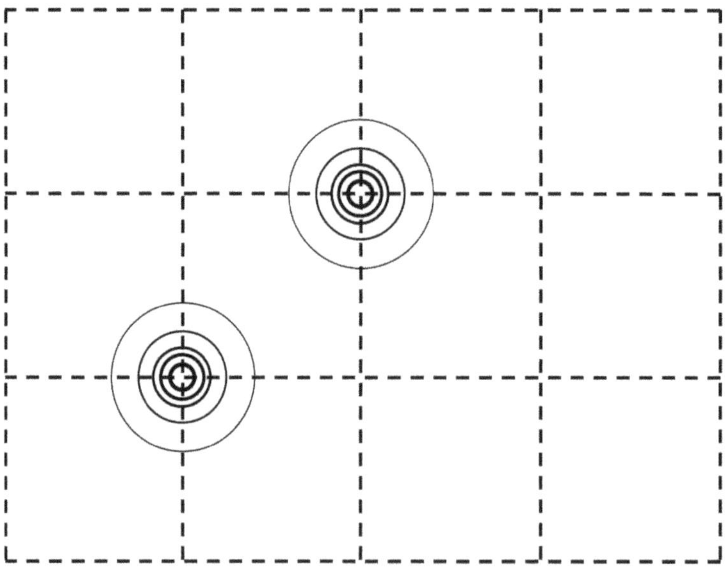

Figure 2.1. Orthogonality of wave packets on a coarse grid.

packets with discrete labels, $\Delta_q \Delta_p \gg 2\hbar$, is ultimately a reflection of the Heisenberg uncertainty principle.

Our goal is to apply the analysis to statistical averages in classical phase space. In practice this means transforming from a sum over grid points to an integral over the phase space continuum. This raises the question of the sensitivity to the choice of wave packet width, since the limits $\xi \to 0$ and $\xi \to \infty$ appear to pose difficulties. For fixed width ξ, the orthonormality condition equation (2.4) would appear to preclude the continuum limit $\Delta_q \to 0$, $\Delta_p \to 0$. We shall not address these questions here, but we do point out that in practical terms such a transformation from the sum to an integral is valid without actually taking the zero width limit if the phase space function whose average is sought is slowly varying over the volume of the grid cells,

$$\begin{aligned} f(\mathbf{q} + \mathbf{\Delta}_q, \mathbf{p} + \mathbf{\Delta}_p) &= f(\mathbf{q}, \mathbf{p}) + \mathbf{\Delta}_q \cdot \nabla_q f(\mathbf{q}, \mathbf{p}) + \mathbf{\Delta}_p \cdot \nabla_p f(\mathbf{q}, \mathbf{p}) + \cdots \\ &= f(\mathbf{q}, \mathbf{p}) + \mathcal{O}(\Delta_q, \Delta_p). \end{aligned} \tag{2.5}$$

For the transformation to the continuum to work, the second equality needs to be a reasonable approximation in those regions of phase space that contribute significantly to the statistical average. We shall refer to the restriction on the grid widths, equation (2.4), as the classical limit, and we shall attempt to elucidate the physical conditions for which this limit is likely valid.

In addition to orthogonality, an important property of basis functions is that they form a complete set, which is to say that any wave function in the Hilbert space can be expanded in a series of them. We note that the first factor of the wave packet, the Gaussian in position peaked at $\mathbf{q}$, can be used to fit the amplitude of a wave function $\psi(\mathbf{r})$ at $\mathbf{r} \approx \mathbf{q}$, and the second factor, the plane wave with wave number $\mathbf{p}$, can be used to fit its phase. If one can tolerate the discretization error in the phase, the wave packet can be used to fit any wave function on the position grid points. However in the classical limit the wave packets go to zero between grid points in position, and so they cannot fit an arbitrary wave function between such points. They are therefore complete only in the subspace of wave functions that are sampled exclusively at grid points. This is sufficient for slowly varying phase functions, and this initial formulation of quantum statistical mechanics in classical phase space is only valid in the classical limit regime where such functions can be taken to be the only relevant ones.

2.2.2 Eigenfunctions

The probability amplitude in the position representation is

$$\zeta_{\mathbf{q},\mathbf{p}}(\mathbf{r})^* \zeta_{\mathbf{q},\mathbf{p}}(\mathbf{r}) = \frac{1}{C^2} e^{-(\mathbf{r}-\mathbf{q})\cdot(\mathbf{r}-\mathbf{q})/2\xi^2}. \tag{2.6}$$

This is a Gaussian of width ξ (per particle, per direction), peaked at the position label $\mathbf{q}$. Similarly in the momentum representation, the probability amplitude has width $\hbar/2\xi$ and is peaked at the momentum label $\mathbf{p}$. To within an error of these magnitudes, the minimum uncertainty wave function is approximately a simultaneous eigenfunction of the position and momentum operators,

$$\hat{\mathbf{q}}\, \zeta_{\mathrm{q,\,p}}(\mathbf{r}) = \mathbf{r}\, \zeta_{\mathrm{q,\,p}}(\mathbf{r}) \approx \mathbf{q}\, \zeta_{\mathrm{q,\,p}}(\mathbf{r}), \tag{2.7}$$

and

$$\hat{\mathbf{p}}\, \zeta_{\mathrm{q,\,p}}(\mathbf{r}) = \left[\frac{i\hbar}{2\xi^2}(\mathbf{r}-\mathbf{q}) + \mathbf{p}\right]\zeta_{\mathrm{q,\,p}}(\mathbf{r}) \approx \mathbf{p}\, \zeta_{\mathrm{q,\,p}}(\mathbf{r}). \tag{2.8}$$

(Recall that the momentum operator is $\hat{\mathbf{p}} = -i\hbar\nabla_{\mathbf{r}}$.) Since the probability amplitude is sharply peaked about $\{\mathbf{q},\, \mathbf{p}\}$, the approximation in these is to neglect contributions $\mathcal{O}(\xi)$ and $\mathcal{O}(\hbar/2\xi)$, respectively.

Because the wave packet is an approximate simultaneous position and momentum eigenfunction, for any sufficiently slowly varying phase function one can write $f(\hat{\mathbf{q}},\, \hat{\mathbf{p}})\zeta_{\mathrm{q,\,p}}(\mathbf{r}) \approx f(\mathbf{q},\, \mathbf{p})\zeta_{\mathrm{q,\,p}}(\mathbf{r})$. In particular, the wave packet is an approximate energy eigenfunction,

$$\hat{H}(\mathbf{r})\, \zeta_{\mathrm{q,\,p}}(\mathbf{r}) \equiv H(\hat{\mathbf{q}},\, \hat{\mathbf{p}})\, \zeta_{\mathrm{q,\,p}}(\mathbf{r}) \approx H(\mathbf{q},\, \mathbf{p})\, \zeta_{\mathrm{q,\,p}}(\mathbf{r}). \tag{2.9}$$

The eigenvalue is the classical Hamiltonian energy function of phase space, which is usually split into kinetic and potential energies,

$$H(\mathbf{q},\, \mathbf{p}) = \mathcal{K}(\mathbf{p}) + U(\mathbf{q}), \quad \mathcal{K}(\mathbf{p}) = \frac{p^2}{2m} = \frac{1}{2m}\sum_{j,\alpha} p_{j\alpha}^2. \tag{2.10}$$

Here identical particles of mass m are assumed.

The position and momentum labels, $\mathbf{q}$ and $\mathbf{p}$, occur naturally in the Hamiltonian function of classical mechanics, and they therefore play precisely the same role as the classical positions and momenta. Of course the discretization of phase space on a grid of width $\Delta_{\mathrm{q}}\Delta_{\mathrm{p}} \gg 2\hbar$, equation (2.4), is inherently a reflection of the uncertainty principle. However, when we eventually transform the various sums over the position and momentum labels to an integral over continuum phase space, the grid width, the wave packet width, and their relation to the uncertainty principle will disappear.

It should be noted in the above that the position labels are distinct from the position representation coordinates. Similarly the momentum labels are distinct from the momentum representation coordinates. This distinction between position eigenvalues and position representation coordinates will play a key role in the formally exact local state expansion that is derived in chapter 10.

2.3 Wave packet symmetrization and overlap

A fundamental axiom of quantum mechanics is that the wave function must be either fully symmetric (bosons) or fully anti-symmetric (fermions) with respect to interchange of identical particles (Messiah 1961, Merzbacher 1970). For the present wave packets, the symmetrized form is

$$\begin{aligned} \zeta_\Gamma^\pm(\mathbf{r}) &\equiv \frac{1}{\sqrt{N!\chi_\Gamma^\pm}} \sum_{\hat{\mathrm{P}}} (\pm 1)^p \, \zeta_\Gamma(\hat{\mathrm{P}}\mathbf{r}) \\ &\equiv \frac{1}{\sqrt{N!\chi_\Gamma^\pm}} \sum_{\hat{\mathrm{P}}} (\pm 1)^p \, \zeta_{\hat{\mathrm{P}}\Gamma}(\mathbf{r}). \end{aligned} \tag{2.11}$$

The meaning of particle permutation can be seen from the fact that a vector is an ordered set. In the present case the first element of the vector of coordinates, or of the vector of position–momentum labels, is associated with the first particle, the second element with the second particle, etc. Hence the particle permutator $\hat{\mathrm{P}}$ can be applied to one vector relative to the other.

In the above expression for the symmetrized wave packet, the sum is over all $N!$ permutations of the N particles. Recall that we sometimes write the combined position and momentum labels as $\mathbf{\Gamma} \equiv \{\mathbf{q}, \mathbf{p}\}$. Here and below $p(\hat{\mathrm{P}})$ is the number of pair transpositions that comprise the permutation operator $\hat{\mathrm{P}}$ (or simply their parity). The plus sign is for bosons and the minus sign is for fermions.

By inspection one sees that the symmetrized wave functions show the mandated behavior for bosons and fermions under particle interchange,

$$\zeta_\Gamma^\pm(\hat{\mathrm{P}}\mathbf{r}) = (\pm 1)^p \zeta_\Gamma^\pm(\mathbf{r}). \tag{2.12}$$

That is, for bosons it is unchanged, but for fermions it changes sign if the permutation has odd parity (i.e., an odd number of pair transpositions).

Note how the permutation operator can be equivalently applied to the representation coordinates, or to the position $\mathbf{q}$ and momentum $\mathbf{p}$ labels. (Technically $\zeta_{\hat{\mathrm{P}}\Gamma}(\mathbf{r}) = \zeta_\Gamma(\hat{\mathrm{P}}^{-1}\mathbf{r})$, but this is unimportant when summed over all permutations.) Because these labels are discrete and are the product of single-particle states, we can speak sensibly of a quantum state being occupied or unoccupied, and of the number of occupying particles. For the general case of exact energy eigenfunctions, the quantum states are not single-particle states, and it makes no sense to discuss the number occupancy of a given state (but see section 3.3 below).

The normalization of the symmetrized wave function is ensured by taking

$$\begin{aligned} \chi_\Gamma^\pm &= \frac{1}{N!} \sum_{\hat{\mathrm{P}},\hat{\mathrm{P}}'} (\pm 1)^{p+p'} \langle \zeta_\Gamma(\hat{\mathrm{P}}\mathbf{r}) | \zeta_\Gamma(\hat{\mathrm{P}}'\mathbf{r}) \rangle \\ &= \sum_{\hat{\mathrm{P}}} (\pm 1)^p \langle \zeta_\Gamma(\hat{\mathrm{P}}\mathbf{r}) | \zeta_\Gamma(\mathbf{r}) \rangle \\ &= \sum_{\hat{\mathrm{P}}} (\pm 1)^p \int \mathrm{d}\mathbf{r} \; \zeta_\Gamma(\hat{\mathrm{P}}\mathbf{r})^* \zeta_\Gamma(\mathbf{r}). \end{aligned} \tag{2.13}$$

The quantity $\chi_\Gamma^\pm$ is here and below called the symmetrization factor. It was originally called the overlap factor because in the general case of non-orthonormal wave packets it reflects the overlap between them (Attard 2016). In fact it is more general than the original application, as it also works for orthonormal eigenfunctions and for either single-particle or multi-particles states. It also works in the continuum limit. In these more general cases the symmetrization factor tells how much symmetrization counts the same microstate multiple times, as is discussed in chapter 3. In the simplest case, such as the present orthonormal wave packets that are composed of discrete single-particle states, the symmetrization factor enforces particle statistics and occupancy rules, as is shown in section 2.3.2 below.

2.3.1 Pair transposition

Before exploring the general properties of the symmetrization factor, it may be worth illustrating the nature of particle permutations with a simple example.

Because the wave function is the product of individual particle wave packets, it is simplest to focus on a single pair transposition, since all permutations can be decomposed into consecutive pair transpositions. Let $\hat{P}_{jk}$ transpose particles j and k, so that

$$\begin{aligned}\zeta_{\Gamma}(\hat{\mathbf{P}}_{jk}\mathbf{r}) &= \zeta_{\Gamma_1 \ldots \Gamma_j \ldots \Gamma_k \ldots}(\mathbf{r}_1 \ldots \mathbf{r}_k \ldots \mathbf{r}_j \ldots) \\ &= \zeta_{\Gamma_1}^{(1)}(\mathbf{r}_1) \ldots \zeta_{\Gamma_j}^{(1)}(\mathbf{r}_k) \ldots \zeta_{\Gamma_k}^{(1)}(\mathbf{r}_j) \ldots\end{aligned} \tag{2.14}$$

The so-called dimer symmetrization or overlap factor consists of the inner product of the original and the transposed wave function,

$$\begin{aligned}\chi_{jk}^{\pm,\,(2)} &\equiv \pm \left\langle \zeta_{\Gamma}(\hat{\mathbf{P}}_{jk}\mathbf{r}) \middle| \, \zeta_{\Gamma}(\mathbf{r}) \right\rangle \\ &= \pm \left\langle \zeta_{\Gamma_j}^{(1)}(\mathbf{r}_k)\zeta_{\Gamma_k}^{(1)}(\mathbf{r}_j) \middle| \, \zeta_{\Gamma_j}^{(1)}(\mathbf{r}_j)\zeta_{\Gamma_k}^{(1)}(\mathbf{r}_k) \right\rangle \\ &= \pm \left| \left\langle \zeta_{\Gamma_j}^{(1)} | \zeta_{\Gamma_k}^{(1)} \right\rangle \right|^2 \\ &= \frac{\pm 1}{(2\pi\xi^2)^3} \left| \int \mathrm{d}\mathbf{r}_j e^{-(\mathbf{r}_j - \mathbf{q}_k)^2/4\xi^2} e^{\mathbf{p}_k \cdot (\mathbf{r}_j - \mathbf{q}_k)/i\hbar} \right. \\ &\quad \left. \times\, e^{-(\mathbf{r}_j - \mathbf{q}_j)^2/4\xi^2} e^{-\mathbf{p}_j \cdot (\mathbf{r}_j - \mathbf{q}_j)/i\hbar} \right|^2 \\ &= \pm \exp\left\{ \frac{-1}{4\xi^2}(\mathbf{q}_k - \mathbf{q}_j)^2 - \frac{\xi^2}{\hbar^2}(\mathbf{p}_k - \mathbf{p}_j)^2 \right\}.\end{aligned} \tag{2.15}$$

The second equality follows because the unpermuted single-particle wave packets are normalized and so their respective inner product each gives a factor of unity. The third equality follows because what remains in the product of two integrals over the dummy variables $\mathbf{r}_j$ and $\mathbf{r}_k$, respectively, and these integrals are the complex conjugate of each other. The final equality shows that for the general wave packet the dimer overlap factor is evidently an un-normalized Gaussian in position and momentum that ties the two particles together.

The positive sign, for bosons, and the negative sign, for fermions, will ultimately add or subtract from the phase space weight for the pair of particles in this configuration. Even at the dimer level, and before detailed analysis, one can already see how the symmetrization factor might convert particle occupancy statistics into an effective potential, either attractive or repulsive, between particles in classical phase space.

In general the single-particle wave packets do not form an orthonormal set, and so the right hand side lies between 0 and ±1. In the classical limit, equation (2.4), the wave packets are orthonormal, and in this case

$$\left\langle \zeta_{\Gamma}(\hat{\mathrm{P}}_{jk}\mathbf{r}) \middle| \zeta_{\Gamma}(\mathbf{r}) \right\rangle = \delta_{\ell_j, \ell_k}, \quad \Delta_{\mathrm{q}} \gg 2\xi, \quad \text{and} \quad \Delta_{\mathrm{p}} \gg \hbar/\xi. \tag{2.16}$$

Note that we use the vector of integers $\boldsymbol{\ell}$ interchangeably with the vector of real numbers $\boldsymbol{\Gamma} = \{\mathbf{q}, \mathbf{p}\}$ to label the quantum state. The inner product is one if the two particles are in the same position–momentum one-particle state, and it is zero otherwise.

The symmetrization factor is most clearly explained in terms of one-particle states, which is the case with the present wave packets. However, the formalism itself holds in general for multi-particle states, which will be treated in chapter 3.

2.3.2 Symmetrization and occupancy of single-particle states

Particle statistics, or state occupancy, for bosons and for fermions arise from the rules for wave function symmetrization. The purpose of this section is to show that the symmetrization factor defined above from the normalization of the symmetrized wave function, equation (2.13), is exactly the correct weight factor that is required to replace an arbitrary sum over occupied states by a sum over all states whether the particles be bosons or fermions. This is one example of the utility of the symmetrization factor. Two other advantages are that it generalizes the occupancy rules to multi-particle states, and to overlapping states (i.e., those with non-orthogonal wave functions).

The usual texts on quantum mechanics treat the subject of wave function symmetrization by invoking quantum states that comprise single, identical particle, states (Messiah 1961, Merzbacher 1970, Pathria 1972). Indeed the familiar concept that an arbitrary number of bosons, but at most one fermion, can occupy the same state is predicated upon, and only makes sense, if the state referred to is a single-particle state. Since the present basis consists of wave packets, whose states are single-particle, we need to show that the symmetrization factor is related to these usual rules of particle occupancy. In section 3.3, the analysis is generalized to systems in which the concept of occupancy is inapplicable because the relevant states are not single-particle.

The position–momentum states are discrete so that $\boldsymbol{\Gamma}_j$ is equivalent to ℓ_j; it is the single-particle state occupied by particle j. The state of the system is $\boldsymbol{\Gamma} = \{\boldsymbol{\Gamma}_1, \boldsymbol{\Gamma}_2, \dots, \boldsymbol{\Gamma}_N\}$, which is equivalently $\boldsymbol{\ell} = \{\ell_1, \ell_2, \dots, \ell_N\}$. We may order the possible single-particle states $a = 1, 2, \dots, A$, and say equivalently that particle j is in the state ℓ_j, or else that it is in the ath state, with $a = a(\ell_j)$. Let $m_a(\boldsymbol{\ell}) = \sum_j \delta_{a(\ell_j), a}$ be the number of particles in the single-particle state a when the system is in state $\boldsymbol{\ell}$. We may regard m_a as a component of the A-dimensional vector $\mathbf{m}(\boldsymbol{\ell})$, which tells the occupancy numbers of the possible single-particle states when the system is in the labeled particle state ℓ. Clearly there is a many to one mapping from the system state ℓ to the occupancy state **m**, since the latter doesn't distinguish which particle or particles are in the given single-particle state.

A function of the state of the system may be written $f(\ell)$, or as $f(\mathbf{m})$, the latter being short-hand for the more precise $f_s(\mathbf{m}(\ell)) = f(\ell)$. Since the particles are identical, and since the labels are arbitrary, either description should suffice, provided that the rules are properly accounted for.

Define the maximum possible number of particles in a single-particle state to be $N_+ = \infty$ (for bosons) and $N_- = 1$ (for fermions). With these the sum of a function in the occupancy picture may be written

$$
\begin{aligned}
F^{\pm} &\equiv \sum_{\mathbf{m}}{}^{\pm} f(\mathbf{m}) \\
&= \sum_{m_1=0}^{N_\pm} \sum_{m_2=0}^{N_\pm} \cdots \sum_{m_A=0}^{N_\pm} f(\mathbf{m}).
\end{aligned}
\tag{2.17}
$$

Each possible state appears once, and once only, in this sum. (This sum over distinct, allowed, states is just the primed trace TR$'$.)

One would like to write this instead as a sum over the states of the labeled particles, ℓ,

$$
\begin{aligned}
F^{\pm} &\equiv \frac{1}{N!} \sum_{\ell} \chi_\ell^{\pm} f(\ell) \\
&= \frac{1}{N!} \sum_{\ell_1} \sum_{\ell_2} \cdots \sum_{\ell_N} \chi_\ell^{\pm} f(\ell).
\end{aligned}
\tag{2.18}
$$

The sum over the states of each particle here is unrestricted. This means that it includes forbidden states occupied by more than one fermion, and it includes the same state (i.e., ones that differ only by a permutation of the labels of bosons in the same single-particle state) multiple times. We need to show that the symmetrization factor $\chi_\ell^{\pm}$ corrects for this and makes this agree with the preceding expression. Figure 2.2 illustrates the idea, with the weight referred to being $\chi_\ell^{\pm}/N!$.

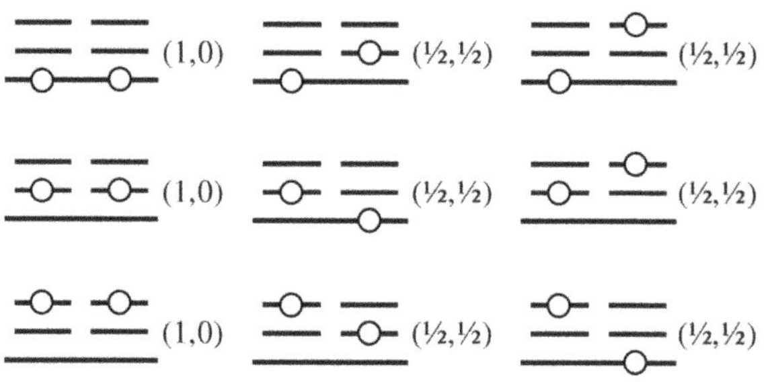

Figure 2.2. The nine possible configurations of a system with two identical particles and three single-particle states. To the right of each configuration is the weight $(\chi_\ell^+/N!, \chi_\ell^-/N!)$ for bosons (first number) and for fermions (second number).

From equation (2.13), the symmetrization factor is

$$\chi_\ell^\pm \equiv \sum_{\hat{\mathrm{P}}}(\pm 1)^p\langle\hat{\mathrm{P}}\ell|\ell\rangle = \sum_{\hat{\mathrm{P}}}(\pm 1)^p\prod_{j=1}^{N}\delta_{\ell_j,\ell_j'}, \tag{2.19}$$

where ℓ_j' is the state that particle j is in after the permutation $\hat{\mathrm{P}}$. (Recall that there is a one-to-one map from ℓ to $\mathbf{\Gamma} = \{\mathbf{q}, \mathbf{p}\}$.)

For the present single-particle states with orthogonal wave packets, the inner product $\langle\hat{\mathrm{P}}\ell|\ell\rangle$ is zero for any permutation that swaps particles in different states. Conversely it is unity for any transposition of particles in the same state, and for any permutation composed solely of such same state transpositions.

2.3.2.1 Bosons

For bosons, the sum over all particle states, $\sum_\ell$, counts the same occupancy state multiple times. There are $N!/\prod_a m_a!$ ways of re-arranging the N labeled particles without changing the occupancy state $\mathbf{m}(\ell)$. To correct for this double counting, it is necessary to weight each state by the inverse of this, namely

$$\begin{aligned} F^+ &= \sum_{\mathbf{m}} f(\mathbf{m}) \\ &= \sum_\ell \frac{\prod_a m_a(\ell)!}{N!} f(\ell) \\ &= \frac{1}{N!}\sum_\ell \chi_\ell^+ f(\ell). \end{aligned} \tag{2.20}$$

Here $m_a(\ell)$ tells how many particles are each in the same single-particle state a when the system is in the state ℓ. One can easily confirm the final equality directly from the definition of the symmetrization factor, equation (2.19): for bosons it is just the number of permutations amongst particles in the same state, $\chi_\ell^+ = \prod_a m_a(\ell)!$.

2.3.2.2 Fermions

To show that the formula works for fermions, it must be proven that $\chi_\ell^- = 0$ if any $m_a \geqslant 2$. The proof is as follows.

If $m_a(\ell) = 0$ or 1 for all possible single-particle states $a = 1, 2, \ldots, A$, then any permutation of the actual particle states must make the inner product zero, $\langle\hat{\mathrm{P}}\ell|\ell\rangle = 0$, $\hat{\mathrm{P}} \neq \hat{\mathrm{I}}$. Hence only the identity permutation survives in the sum over permutations and $\chi_\ell^- = 1$ in this case.

Contrariwise consider a particular single-particle state a occupied by more than one particle, $m_a(\ell) \geqslant 2$, when the system is in state ℓ. The permutations of particles within the state a have $\langle\hat{\mathrm{P}}^{(a)}\ell|\ell\rangle = 1$, where the superscript indicates that the permutation is restricted to one of the $m_a!$ permutations amongst those particles in the state a. The symmetrization factor will vanish in this case if the number of even permutations equals the number of odd permutations, since these have opposite sign in the sum over the permutations.

To see that these two are indeed equal, suppose that $\{\hat{\mathrm{P}}_\alpha\}_{\alpha=1}^{M!}$ is any set of permutation operators giving all the $M!$ permutations of M objects. Let $\hat{\mathrm{P}}'_\alpha = \hat{\mathrm{P}}_{jk}\hat{\mathrm{P}}_\alpha$, where $\hat{\mathrm{P}}_{jk}$ transposes objects j and k. Clearly $\hat{\mathrm{P}}'_\alpha$ has opposite parity to $\hat{\mathrm{P}}_\alpha$. Since $\{\hat{\mathrm{P}}'_\alpha\}_{\alpha=1}^{M!}$ is also a set giving all the $M!$ permutations of M objects, it follows that the number of even permutations of M objects must equal the number of odd permutations.

From this one can conclude that for fermions, χ_ℓ^- must be zero for any ℓ with two or more particles in the same state. This is because there are an equal number of odd and even permutations, each yielding unity for the inner product, and each with opposite sign in the sum over permutations. That is

$$\begin{aligned}
\chi_\ell^- &= \sum_{\hat{\mathrm{P}}} (-1)^p \langle \hat{\mathrm{P}}\ell | \ell \rangle \\
&= \prod_a{}^{m_a(\ell)<2} 1 \times \prod_a{}^{m_a(\ell)\geqslant 2} \sum_{\hat{\mathrm{P}}(m_a)} (-1)^{p(m_a)} \\
&= \prod_a{}^{m_a(\ell)<2} 1 \times \prod_a{}^{m_a(\ell)\geqslant 2} \left\{ N^{\text{even}}(m_a) - N^{\text{odd}}(m_a) \right\} \\
&= \begin{cases} 1, & \text{all } m_a < 2, \\ 0, & \text{any } m_a \geqslant 2. \end{cases}
\end{aligned} \tag{2.21}$$

Here $\hat{\mathrm{P}}(m_a)$ is a permutation of the $m_a(\ell)$ particles in the single-particle state a, and $N^{\text{even/odd}}(m_a)$ is the number of such permutations with even or odd parity.

These two results for bosons and fermions confirm that the expression for the symmetrization factor, equation (2.19), enables the unweighted sum of a function of occupancy numbers to be written as in equation (2.18). The latter is a weighted sum of the function of particle states over all particle states, including forbidden states and the same state counted multiple times, with the correct weight being the symmetrization factor. In general terms it is much more convenient to evaluate an unrestricted sum than a restricted one. Also, the definition of the symmetrization factor as the normalizing constant for the symmetrized wave function, equation (2.19), is general, and it holds also for wave functions that do not form an orthonormal set (such as the present wave packets not in the classical limit), for wave functions that represent multi-particle states (such as the exact energy eigenfunctions for a system of interacting particles), and for a continuum of states (such as classical phase space). In these three cases it would be difficult if not impossible to write a statistical average, for example, as a sum over occupancy numbers. It is plausible that the present definition also works in these cases. Plausibility is transformed into certainty with the proof for multi-particle states in section 3.3.

2.4 Statistical averages in phase space

2.4.1 Partition function

We consider a canonical equilibrium system, where the subsystem has number of particles N and volume V, and the reservoir has temperature T. Instead of the latter we usually exhibit the inverse temperature, $\beta \equiv 1/k_B T$, where k_B is Boltzmann's constant. The partition function is (Messiah 1961, Merzbacher 1970)

$$Z^{\pm}(N, V, T) = \mathrm{TR}' \, e^{-\beta \hat{H}}. \tag{2.22}$$

This is derived in chapter 12, and it is used in chapter 7 as the starting point for a formally exact transformation of quantum statistical mechanics to classical phase space, valid in all regions of the phase diagram. Here that transformation is performed using wave packets, and it is strictly valid only in the classical limit.

The prime on the trace in the above formula is quite important as it signifies that only allowed quantum states should be included, and that these should be distinct and each counted once only. Messiah (1961, chapter XIV, sections 6 and 7) makes a similar point that the trace should be performed on a restricted subspace containing only allowed distinct states. The symmetrized wave function normalization factor given by him in the case of one-particle, orthogonal states is equivalent to the symmetrization factor given here, at least for the same case.

Unfortunately not all workers avert to the need to restrict the trace. Pathria (1972, equation (9.6.2)), gives the partition function as the Boltzmann weighted sum over energy states, with the implication being that these are all states (the issue of distinct states is not raised), but no symmetrization correction is exhibited.

As mentioned in the previous section it is usually most convenient to invoke an unrestricted sum, together with a weight factor that is zero for forbidden states and in inverse proportion to the number of times an allowed distinct state is counted. It was shown that the symmetrization factor had these properties.

In the present classical limit, we can take the wave packets to be the energy eigenfunctions of the subsystem. The partition function can be derived directly using the symmetrized wave packets,

$$\begin{aligned} Z^{\pm}(N, V, T) &= \frac{1}{N!} \sum_{\ell} \chi_{\ell}^{\pm} \langle \zeta_{\ell}^{\pm} | \, e^{-\beta \hat{\mathcal{H}}} \, | \zeta_{\ell}^{\pm} \rangle \\ &= \frac{1}{N!} \sum_{\ell} \chi_{\ell}^{\pm} \frac{1}{N!} \sum_{\hat{\mathrm{P}}', \hat{\mathrm{P}}''} \frac{(\pm 1)^{p'+p''}}{\sqrt{\chi_{\hat{\mathrm{P}}'\ell}^{\pm} \chi_{\hat{\mathrm{P}}''\ell}^{\pm}}} \langle \zeta_{\hat{\mathrm{P}}'\ell} | \, e^{-\beta \hat{\mathcal{H}}} \, | \zeta_{\hat{\mathrm{P}}''\ell} \rangle \\ &= \frac{1}{N!} \sum_{\ell} \chi_{\ell}^{\pm} \sum_{\hat{\mathrm{P}}} \frac{(\pm 1)^{p}}{\chi_{\ell}^{\pm}} e^{-\beta \mathcal{H}(\mathbf{\Gamma})} \langle \zeta_{\hat{\mathrm{P}}\ell} | \zeta_{\ell} \rangle \\ &= \frac{1}{N!} \sum_{\ell} \chi^{\pm}(\mathbf{\Gamma}) e^{-\beta \mathcal{H}(\mathbf{\Gamma})}. \end{aligned} \tag{2.23}$$

Note that we use ℓ interchangeably with $\mathbf{\Gamma} = \{\mathbf{q}, \mathbf{p}\}$ to label the quantum state. For the sum over states we use the former integer label, whereas when we transform to the continuum we shall use the latter real argument. The first equality here reflects the argument given in the preceding subsection, namely that an unrestricted sum over all states can be used provided that they are weighted with the symmetrization factor. The third equality arises because the symmetrization factor is unchanged by a permutation of its label, $\chi^{\pm}_{\hat{\mathrm{P}}\ell} = \chi^{\pm}_{\ell} = \chi^{\pm}(\mathbf{\Gamma})$. For this equality also is invoked the fact that in general for identical particles the Hamiltonian operator is unchanged by a permutation of the particles,

$$\hat{\mathcal{H}}(\mathbf{r}) = \hat{\mathcal{H}}(\hat{\mathrm{P}}\mathbf{r}). \tag{2.24}$$

The reason that this must be the case is that it ensures that a symmetrized or anti-symmetrized wave function conserves its symmetry during its time evolution, which is to say that bosons remain bosons, and fermions remain fermions. It also means that an eigenfunction of the energy operator remains an eigenfunction when it is symmetrized, and it has unchanged eigenvalue.

The symmetrization factor as it appears here provides a weight for phase space, and as such it represents the internal entropy for phase space points,

$$S^{\mathrm{s},\,\pm}(\mathbf{\Gamma}) = k_{\mathrm{B}} \ln \chi^{\pm}(\mathbf{\Gamma}). \tag{2.25}$$

The nature of its derivation as arising from the sum over permutations confirms that it counts the internal configurations of the quantum microstates.

2.4.2 Averages

Consider now the statistical average of an operator that is an ordinary function of the position and momentum operators, $\hat{O} = O(\hat{\mathbf{q}}, \hat{\mathbf{p}})$. As discussed in section 2.2.2, in the classical limit the wave packets are approximate simultaneous eigenfunctions of the position and momentum operators, and therefore also of this operator $\hat{O}\,\zeta_{\mathbf{q},\,\mathbf{p}}(\mathbf{r}) \approx O(\mathbf{q}, \mathbf{p})\zeta_{\mathbf{q},\,\mathbf{p}}(\mathbf{r})$. Therefore the analysis for the partition function carries through unchanged for the statistical average

$$\begin{aligned}
\langle \hat{O} \rangle_{\mathrm{stat}} &= \frac{1}{Z^{\pm}} \mathrm{TR}' \left\{ \hat{O} e^{-\beta \hat{\mathcal{H}}} \right\} \\
&= \frac{1}{Z^{\pm} N!} \sum_{\ell} \chi^{\pm}_{\ell} \langle \zeta^{\pm}_{\ell} \mid \hat{O} e^{-\beta \hat{\mathcal{H}}} \mid \zeta^{\pm}_{\ell} \rangle \\
&= \frac{1}{Z^{\pm} N!} \sum_{\ell} \chi^{\pm}(\mathbf{\Gamma}) O(\mathbf{\Gamma}) e^{-\beta \mathcal{H}(\mathbf{\Gamma})}.
\end{aligned} \tag{2.26}$$

2.4.3 Phase space

The transformation of these results to classical phase space is straightforward. With $q_{j\alpha} = \ell_{\mathrm{q},\,j\alpha}\Delta_{\mathrm{q}}$ and $p_{j\alpha} = \ell_{\mathrm{p},\,j\alpha}\Delta_{\mathrm{p}}$, the volume element is

$$\mathrm{d}\mathbf{\Gamma} = (\Delta_{\mathrm{q}}\Delta_{\mathrm{p}})^{3N} \mathrm{d}\ell \to h^{3N} \mathrm{d}\ell. \tag{2.27}$$

In taking the classical limit, equation (2.4), we have set $\Delta_{\mathrm{q}} = \sqrt{2\pi}\,\xi$ and $\Delta_{\mathrm{p}} = \sqrt{2\pi}\,\hbar/\xi$. We could have used arbitrary multiples of these, since constant factors have no physical effect on the partition function or averages. We could also set Planck's constant equal to unity with no physical effect on classical averages or classical thermodynamics (Attard 2002). In any case it is significant that the volume element is independent of the width of the wave packets that were used to derive the results.

In view of this result one can transform directly the sum over quantum states to an integral over classical phase space. The partition function becomes

$$Z^{\pm}(N, V, T) = \frac{1}{h^{3N}N!} \int \mathrm{d}\mathbf{\Gamma}\; e^{-\beta\mathcal{H}(\mathbf{\Gamma})}\chi^{\pm}(\mathbf{\Gamma}), \tag{2.28}$$

and the statistical average becomes

$$\langle O\rangle^{\pm}_{\text{stat}} = \frac{1}{h^{3N}N!Z^{\pm}} \int \mathrm{d}\mathbf{\Gamma}\; e^{-\beta\mathcal{H}(\mathbf{\Gamma})}\chi^{\pm}(\mathbf{\Gamma})O(\mathbf{\Gamma}). \tag{2.29}$$

It is worth reiterating that this result relied upon the classical limit equation (2.4). (Strictly speaking, we have only demonstrated that the symmetrization factor enforces the particle occupancy rules in the case that the wave packets form an orthonormal set.) Transforming the sum over position–momentum states to the phase space integral is valid when the integrand is slowly varying between grid points. The Maxwell–Boltzmann factor is slowly varying when the exponent is small, which is the case at high temperatures, $\beta \to 0$. Also, in general the particle interaction potential contains a sharp, rapidly varying repulsion at small separations that defines the particle size and keeps them from interpenetrating each other. At low densities (equivalently, low pressures) the particles are on average well-separated and only access the slowly varying tail of the potential. In this case again the Maxwell–Boltzmann factor varies but slowly, which makes the phase space integral also valid in low density regime.

It remains to elucidate the rôle of the symmetrization factor $\chi^{\pm}(\mathbf{\Gamma})$ in the classical regime.

2.4.4 Symmetrization factor in the classical limit

The symmetrization factor will be formally analyzed in chapter 3. Here we discuss it less rigorously with a view to understanding its behavior in the classical limit.

The symmetrization factor is a sum over all permutations of the inner product of the permuted wave function and the original wave function. Since here the wave function is the product of single-particle wave packets, for any permutation the unpermuted particles contribute unity to the product of inner products. The inner product of the remaining permuted particles is zero unless the wave packets of consecutive particles in each permutation loop that comprise it overlap (cf. the discussion of the dimer transposition in section 2.3.1). This means that the particles involved must be close together in order for the contribution to the symmetrization factor from a particular permutation to be non-zero.

As was discussed above, the transformation to the phase space integral is valid at low densities. In this regime particles are only rarely close together. The probability of two particles being close together is small, and the probability of three or four particles being close is even smaller. Therefore in the low density regime one can neglect permutations that involve many particles. In the sum over all permutations for the symmetrization factor one need only retain the identity permutation, and all permutations of two particles,

$$\begin{aligned}\chi_\Gamma^\pm &= \sum_{\hat{\mathrm{P}}}(\pm 1)^p \langle \zeta_\Gamma(\hat{\mathrm{P}}\mathbf{r})|\, \zeta_\Gamma(\mathbf{r})\rangle \\ &\approx \langle \zeta_\Gamma(\hat{\mathrm{I}}\mathbf{r})|\, \zeta_\Gamma(\mathbf{r})\rangle \pm \sum_{j<k}^{N} \left\langle \zeta_\Gamma(\hat{\mathrm{P}}_{jk}\mathbf{r}) \middle|\, \zeta_\Gamma(\mathbf{r})\right\rangle + \cdots \\ &= 1 \pm \sum_{j<k}^{N} \left| \left\langle \zeta_{\Gamma_j}^{(1)} | \zeta_{\Gamma_k}^{(1)} \right\rangle \right|^2 \\ &= 1 \pm \sum_{j<k}^{N} e^{-(\mathbf{q}_k-\mathbf{q}_j)^2/4\xi^2} e^{-\xi^2(\mathbf{p}_k-\mathbf{p}_j)^2/\hbar^2}.\end{aligned} \tag{2.30}$$

In the second equality permutations involving 3, 4 particles, etc, have been neglected. Only the identity and all permutations involving 2 particles have been retained. Equation (2.15) has been used for the final two equalities. The final term of the final equality is evidently a sum of Gaussians in position separation and in momentum separation with variance $\sqrt{2}\,\xi$ and $\hbar/\sqrt{2}\,\xi$, respectively. This Gaussian form for the dimer overlap factor holds for the general wave packet, not just the classical limit (although the latter has been invoked in the transformations to classical phase space).

As mentioned, at low densities configurations with two particles close enough for their wave packets to overlap are exceedingly rare. Thus the leading order term of unity dominates the dimer contribution in this regime, leaving $\chi^\pm(\mathbf{\Gamma}) = 1$. In this limit there is no distinction between bosons and fermions. With this result, the partition function and statistical average given in the preceding subsection become

$$Z(N, V, T) = \frac{1}{h^{3N}N!} \int \mathrm{d}\mathbf{\Gamma}\; e^{-\beta\mathcal{H}(\mathbf{\Gamma})}, \tag{2.31}$$

and

$$\langle O \rangle_{\text{stat}} = \frac{1}{h^{3N}N!Z} \int \mathrm{d}\mathbf{\Gamma}\; e^{-\beta\mathcal{H}(\mathbf{\Gamma})} O(\mathbf{\Gamma}). \tag{2.32}$$

These are the classical results.

It remains to establish parameters for each given system that give quantitative meaning to ‘high’ in high temperature, and to ‘low’ in low density. And it remains to formulate explicitly quantum perturbations in regimes approaching the classical limit, and quantum effects far from the classical limit. These tasks are approached from a formally exact analysis in the succeeding chapters.

References

Attard P 2002 *Thermodynamics and Statistical Mechanics: Equilibrium by Entropy Maximisation* (London: Academic)
Attard P 2016 Expansion for quantum statistical mechanics based on wave function symmetrization arXiv:1603.07757
Merzbacher E 1970 *Quantum Mechanics* 2nd edn (New York: Wiley)
Messiah A 1961 *Quantum Mechanics* vols 1 and 2 (Amsterdam: North-Holland)
Pathria R K 1972 *Statistical Mechanics* (Oxford: Pergamon)

IOP Publishing

Quantum Statistical Mechanics in Classical Phase Space

Phil Attard

Chapter 3

Symmetrization factor and permutation loop expansion

The formally exact transformation of quantum statistical mechanics to classical phase space is derived. The symmetrization factor generalizes state occupancy rules (i.e., particle statistics) to multi-particle states, and allows the trace to be written as an unrestricted sum. This theorem enables the partition function to be written in terms of multi-particle energy states. The symmetrization factor as a sum over permutations gives the internal entropy for the subsystem microstate.

Examples of localized permuted states motivate writing the sum over particle permutators as a series of products of disjoint permutation loops naturally ordered in terms of increasing size and number. These permutation loops play important roles in succeeding chapters. For single-particle states, the permutation loop series for the grand partition function can be analytically resummed giving the loop expansion for the grand potential.

3.1 Introduction

In the preceding chapter wave packets were invoked as approximate energy eigenfunctions, and the quantum statistical average and partition function were transformed to classical phase space. Although approximate, the analysis did show how the classical limit emerges, and it did raise general issues such as the simultaneous specification of position and momentum, the symmetrization of the wave function and its implications for the counting of distinct, allowed states, and the localization of eigenfunctions, eigenstates, and symmetrization effects. All of these recur in the formally exact transformation of quantum statistical mechanics to classical phase space, which is the task set for this chapter.

For a canonical equilibrium quantum system, the partition function may be written symbolically as

$$Z(T) = \mathrm{TR}'\, e^{-\beta\hat{\mathcal{H}}}. \tag{3.1}$$

Feynman (1998) describes this as the summit of statistical mechanics, explaining that the latter consists either of the descent, where the law is applied to various cases, or else the ascent, where the law is derived and the fundamental thermodynamic and statistical concepts are elucidated.

In this book the ascent is deferred to chapter 12, where the law is derived by analyzing the collapse of the wave function in an open quantum system. The descent—applications—occupies chapters 4, 5, and 6. Most of the book, including this chapter, is spent at the summit itself, manipulating and reformulating the expression for the partition function with the aim of transforming it to classical phase space, and with developing practical algorithms and expansions that can be applied not just in the near-classical regime where quantum effects act as a mere perturbation, but also to the regime where quantum phenomena are dominant.

3.2 Partition function

Conceptually, the partition function is the total number of quantum states. Technically this means the total states of the total system. In fact it is only the subsystem states that are summed over, with the Boltzmann weight providing the number of states of the reservoir or environment that depend on each subsystem state (Attard 2002). I use the word ‘subsystem’ in order to explicitly denote that part of the total system that is of direct interest; others call this an ‘open system’ or simply ‘system’. The remainder of the total system is the reservoir or environment (Attard 2002). In any case, it is obvious that for the partition function one should count only allowed states, and each such allowed state should be distinct and counted once only. These two restrictions are signified by the prime on the trace.

The formulation of the partition function as a trace is deliberately generic. It can be explicitly expressed as a sum over quantum states of any complete operator of the subsystem. We shall return to this general option later. Here we begin with the concrete case of energy states.

Let $\phi_{\mathbf{n}}(\mathbf{r})$ be an energy eigenfunction in the position representation,

$$\hat{\mathcal{H}}(\mathbf{r})\,\phi_{\mathbf{n}}(\mathbf{r}) = E_{\mathbf{n}}\phi_{\mathbf{n}}(\mathbf{r}). \tag{3.2}$$

We shall sometimes write the energy eigenfunction as $|\phi_{\mathbf{n}}\rangle$ or even as $|\mathbf{n}\rangle$. For simplicity we do not deal with spin variables here; a brief discussion of these in the present context is given in section 6.4. We suppose that the energy eigenfunctions form a complete orthonormal set

$$\langle\phi_{\mathbf{n}'}|\phi_{\mathbf{n}''}\rangle = \delta_{\mathbf{n}',\mathbf{n}''}, \quad \text{and} \quad \sum_{\mathbf{n}} \phi_{\mathbf{n}}(\mathbf{r}')\phi_{\mathbf{n}}(\mathbf{r}'')^{*} = \delta(\mathbf{r}', \mathbf{r}''). \tag{3.3}$$

In general the potential energy in the Hamiltonian operator is the sum of one-body, two-body, etc, potentials, and as such it is a many-body operator. This means that the energy states $\mathbf{n}$ are multi-particle states that cannot be decomposed into single-particle states. It is possible that the energy state label $\mathbf{n}$ is a $3N$-dimensional vector. But even so it is not possible to identify the individual components with a single particle. This is a significant development on the analysis of wave packets in chapter 2, which consist of the products of single-particle wave functions and quantum states.

It is a fundamental axiom of quantum mechanics that the actual wave function of the system (in our language subsystem) must be either fully symmetric (bosons) or else fully antisymmetric (fermions) with respect to particle interchange (Messiah 1961, Merzbacher 1970). Using a plus sign for bosons and a minus sign for fermions, the symmetrized energy eigenfunctions are as in equation (2.11),

$$\phi_{\mathbf{n}}^{\pm}(\mathbf{r}) \equiv \frac{1}{\sqrt{N!\chi_{\mathbf{n}}^{\pm}}} \sum_{\hat{\mathrm{P}}} (\pm 1)^{p}\, \phi_{\mathbf{n}}(\hat{\mathrm{P}}\mathbf{r}). \tag{3.4}$$

Recall that $\hat{\mathrm{P}}$ is one of the $N!$ particle permutations and $p(\hat{\mathrm{P}})$ is its parity. It is not meaningful to apply the particle permutator to the multi-particle energy state label $\mathbf{n}$.

As in equation (2.13), normalization is ensured by taking

$$\chi_{\mathbf{n}}^{\pm} = \frac{1}{N!} \sum_{\hat{\mathrm{P}},\hat{\mathrm{P}}'} (\pm 1)^{p+p'} \langle \phi_{\mathbf{n}}(\hat{\mathrm{P}}\mathbf{r}) | \phi_{\mathbf{n}}(\hat{\mathrm{P}}'\mathbf{r}) \rangle = \sum_{\hat{\mathrm{P}}} (\pm 1)^{p} \langle \phi_{\mathbf{n}}(\hat{\mathrm{P}}\mathbf{r}) | \phi_{\mathbf{n}}(\mathbf{r}) \rangle. \tag{3.5}$$

Since the symmetrization factor depends upon the wave function, it would be clearer to show this explicitly, as in $\chi^{\pm}[\phi_{\mathbf{n}}]$, but for simplicity we do not do this unless required for disambiguation.

In chapter 2, section 2.3.2, it was shown that the symmetrization factor enabled the sum over distinct allowed quantum states to be replaced by an unrestricted sum over all states. This was particularly useful in formulating the partition function as a sum over given quantum states. However, the proof of the validity of the formulation was specific to single-particle states, and it needs to be modified or carried out anew for the multi-particle case. Such a proof is given in subsection 3.3.2. Anticipating that proof, here we simply assume that the symmetrization factor correctly accounts for particle statistics and that in effect it generalizes the concept of state occupation to multi-particle states. In this case the partition function can be written as an unrestricted sum over multi-particle states, just as for single-particle states,

$$\begin{aligned} Z^{\pm}(N, V, T) &= \mathrm{TR}'\, e^{-\beta\hat{\mathcal{H}}} \\ &= \frac{1}{N!} \sum_{\mathbf{n}} \chi_{\mathbf{n}}^{\pm} \left\langle \phi_{\mathbf{n}}^{\pm} \,|\, e^{-\beta\hat{\mathcal{H}}} \,|\, \phi_{\mathbf{n}}^{\pm} \right\rangle \\ &= \frac{1}{N!} \sum_{\mathbf{n}} \sum_{\hat{\mathrm{P}}} (\pm 1)^{p} \left\langle \phi_{\mathbf{n}}(\mathbf{r}) \right| e^{-\beta\hat{\mathcal{H}}} \left| \phi_{\mathbf{n}}(\hat{\mathrm{P}}\mathbf{r}) \right\rangle. \end{aligned} \tag{3.6}$$

The form of the second equality leads to the interpretation that the symmetrization factor can be considered to be an internal entropy for the subsystem microstate, $S_{\mathbf{n}}^{\mathrm{s},\pm} = k_{\mathrm{B}} \ln \chi_{\mathbf{n}}^{\pm}$. The final equality corrects an analogous expression given by Kirkwood (1933, equation (10)). Here V is the volume of the subsystem which contains the N particles, and the inverse temperature is $\beta \equiv 1/k_{\mathrm{B}}T$. The Hamiltonian operator is necessarily symmetric, $\hat{\mathcal{H}}(\hat{\mathrm{P}}\mathbf{r}) = \hat{\mathcal{H}}(\mathbf{r})$ (Messiah 1961, Merzbacher 1970). This formula applies to any complete set of states. In the particular case of energy states, $\hat{\mathcal{H}}|\mathbf{n}\rangle = E_{\mathbf{n}}|\mathbf{n}\rangle$, one has $\langle\phi_{\mathbf{n}}^{\pm}|e^{-\beta\hat{\mathcal{H}}}|\phi_{\mathbf{n}}^{\pm}\rangle = e^{-\beta E_{\mathbf{n}}}$.

The partition function as written here is for the canonical equilibrium system, which is a subsystem that can exchange energy with a reservoir or environment of temperature T (Attard 2002). In the treatment of symmetrization of energy eigenfunctions comprising single-particle states in section 3.4.3 below, we shall analyze a grand canonical system, in which the subsystem can exchange both energy and particles with a reservoir of temperature T and chemical potential μ (Attard 2002). The grand canonical partition function essentially sums the canonical partition function over number

$$\Xi^{\pm}(\mu, V, T) = \sum_{N=0}^{\infty} \frac{z^N}{N!} \sum_{\mathbf{n}} \chi_{\mathbf{n}}^{\pm} \left\langle \phi_{\mathbf{n}}^{\pm} \mid e^{-\beta\hat{\mathcal{H}}} \mid \phi_{\mathbf{n}}^{\pm} \right\rangle, \tag{3.7}$$

where the fugacity is $z = e^{\beta\mu}$, μ being the chemical potential.

3.3 Symmetrization and occupancy for multi-particle states

In section 2.3.2, it was shown that the symmetrization factor for single-particle states correctly gave the occupancy rules for bosons and fermions. This meant that a sum over allowed occupied states could be written as an unrestricted sum over all states, with the symmetrization factor providing the correct weight. In the present case of multi-particle states, it is necessary to show likewise that the unrestricted sum over all states with symmetrization factor weight gives the correct result. We begin by setting out the formal properties of basis functions that are the product of single-particle functions, which corresponds to the case that the quantum states of the subsystem can be identified with the occupancy of single-particle states.

3.3.1 Single-particle states

Let $\zeta_\ell(\mathbf{r}) = \prod_{j=1}^{N} \zeta_{\ell_j}^{(1)}(\mathbf{r}_j)$ be a complete set of orthonormal basis functions that consist of the product of single-particle functions and states. One has the completeness and orthonormality conditions of the full functions on the full Hilbert space,

$$\langle\zeta_{\ell'}|\zeta_{\ell''}\rangle = \delta_{\ell',\ell''} \quad \text{and} \quad \sum_{\ell} \zeta_\ell(\mathbf{r}')\zeta_\ell(\mathbf{r}'')^* = \delta(\mathbf{r}' - \mathbf{r}''). \tag{3.8}$$

The symmetrized form is

$$\zeta_\ell^{\pm}(\mathbf{r}) \equiv \frac{1}{\sqrt{N!\chi_\ell^{\pm}}} \sum_{\hat{\mathrm{P}}} (\pm 1)^p\, \zeta_\ell(\hat{\mathrm{P}}\mathbf{r}), \tag{3.9}$$

with

$$\chi_\ell^{\pm} = \frac{1}{N!} \sum_{\hat{\mathrm{P}},\hat{\mathrm{P}}'} (\pm 1)^{p+p'} \langle\zeta_\ell(\hat{\mathrm{P}}\mathbf{r})|\zeta_\ell(\hat{\mathrm{P}}'\mathbf{r})\rangle = \sum_{\hat{\mathrm{P}}} (\pm 1)^p \langle\zeta_\ell(\hat{\mathrm{P}}\mathbf{r})|\zeta_\ell(\mathbf{r})\rangle. \tag{3.10}$$

In case of ambiguity we shall write $\chi^{\pm}[\zeta_\ell]$. The symmetrization factor allows a sum over allowed, distinct states to be written as an unrestricted sum over all states,

$$F^{\pm} = \mathrm{TR}'\hat{f} = \frac{1}{N!}\sum_{\ell} \chi_\ell^{\pm} \langle \zeta_\ell^{\pm} | \hat{f} | \zeta_\ell^{\pm} \rangle. \tag{3.11}$$

This was proven to be consistent with the well-known particle occupancy rules in section 2.3.2.

To recapitulate that proof, let $m_a(\ell)$ be the number of particles in the single-particle state $a = 1, 2, \ldots$ when the subsystem is in the state ℓ. For bosons, the number of permutations of particles that leave the subsystem state unchanged is $N!/\prod_a m_a(\ell)!$. The unrestricted sum over all states counts each unique state this many times. To correct for this, each term has to be weighted by the inverse of this, namely $\chi_\ell^{+}/N! = \prod_a m_a(\ell)!/N!$, or $\chi_\ell^{+} = \prod_a m_a(\ell)!$. This can be seen to be true directly from the definition.

For fermions, only zero or one particle may be in any one single-particle state. Hence χ_ℓ^{-} must vanish if $m_a(\ell) \geqslant 2$ for any a. In section 2.3.2, this was indeed shown to be the case. There remains $N!$ as the number of times the sum over all states counts each permutation of the particles. Since the occupied states have at most one particle in them, there is no need to correct for any internal permutations within a single-particle state.

The $\zeta_\ell^{\pm}(\mathbf{r})$ belong to the fully symmetric or fully antisymmetric sub-space of the Hilbert space. We now want to show in what sense they respectively form a complete orthonormal basis for these sub-spaces.

From the orthonormality of the full functions, $\delta_{\ell',\ell''} = \langle \zeta_{\ell'} | \zeta_{\ell''} \rangle$, one has

$$\begin{aligned} \sum_{\hat{\mathrm{P}}',\hat{\mathrm{P}}''} (\pm 1)^{p'+p''} \delta_{\hat{\mathrm{P}}'\ell',\hat{\mathrm{P}}''\ell''} &= \sum_{\hat{\mathrm{P}}',\hat{\mathrm{P}}''} (\pm 1)^{p'+p''} \langle \zeta_{\hat{\mathrm{P}}'\ell'} | \zeta_{\hat{\mathrm{P}}''\ell''} \rangle \\ &= \sum_{\hat{\mathrm{P}}',\hat{\mathrm{P}}''} (\pm 1)^{p'+p''} \langle \zeta_{\ell'}(\hat{\mathrm{P}}'\mathbf{r}) | \zeta_{\ell''}(\hat{\mathrm{P}}''\mathbf{r}') \rangle \\ &= N! \sqrt{\chi_{\ell'}^{\pm} \chi_{\ell''}^{\pm}} \langle \zeta_{\ell'}^{\pm} | \zeta_{\ell''}^{\pm} \rangle, \end{aligned} \tag{3.12}$$

or

$$\langle \zeta_{\ell'}^{\pm} | \zeta_{\ell''}^{\pm} \rangle = \frac{1}{\sqrt{\chi_{\ell'}^{\pm} \chi_{\ell''}^{\pm}}} \sum_{\hat{\mathrm{P}}} (\pm 1)^{p} \delta_{\hat{\mathrm{P}}\ell',\ell''} \equiv \frac{1}{\chi_{\ell'}^{\pm}} \delta_{\ell',\ell''}^{\pm}. \tag{3.13}$$

The symmetrized Kronecker-δ picks out subsystem states that are related by a particle permutation.

Conversely from the completeness condition, $\delta(\mathbf{r}' - \mathbf{r}'') = \sum_\ell \zeta_\ell(\mathbf{r}')\zeta_\ell(\mathbf{r}'')^*$, one has

$$\begin{aligned} \sum_{\hat{\mathrm{P}}',\hat{\mathrm{P}}''} (\pm 1)^{p'+p''} \delta(\hat{\mathrm{P}}'\mathbf{r}' - \hat{\mathrm{P}}''\mathbf{r}'') &= \sum_{\hat{\mathrm{P}}',\hat{\mathrm{P}}''} (\pm 1)^{p'+p''} \sum_{\ell} \zeta_\ell(\hat{\mathrm{P}}'\mathbf{r}')\zeta_\ell(\hat{\mathrm{P}}''\mathbf{r}'')^* \\ &= N! \sum_{\ell} \chi_\ell^{\pm} \zeta_\ell^{\pm}(\mathbf{r}')\zeta_\ell^{\pm}(\mathbf{r}'')^*, \end{aligned} \tag{3.14}$$

or

$$\sum_{\ell} \chi_{\ell}^{\pm} \zeta_{\ell}^{\pm}(\mathbf{r}')\zeta_{\ell}^{\pm}(\mathbf{r}'')^{*} = \sum_{\hat{\mathrm{P}}}(\pm 1)^{p}\delta(\hat{\mathrm{P}}\mathbf{r}' - \mathbf{r}'') \equiv \delta^{\pm}(\mathbf{r}' - \mathbf{r}''). \tag{3.15}$$

This symmetrized Dirac-δ function picks out positions that are the permutation of each other.

We now need to obtain the form for the expansion of any symmetrized wave function as a series of symmetrized basis functions. The normalized projection onto a symmetrized sub-space of any wave function is

$$\begin{aligned}
\psi^{\pm}(\mathbf{r}) &= \frac{1}{\sqrt{N!\chi^{\pm}[\psi]}} \sum_{\hat{\mathrm{P}}}(\pm 1)^{p}\psi(\hat{\mathrm{P}}\mathbf{r}) \\
&= \frac{1}{\sqrt{N!\chi^{\pm}[\psi]}} \sum_{\hat{\mathrm{P}}}(\pm 1)^{p}\sum_{\ell}\langle\zeta_{\ell}|\psi\rangle\, \zeta_{\ell}(\hat{\mathrm{P}}\mathbf{r}) \\
&= \frac{1}{\sqrt{\chi^{\pm}[\psi]}} \sum_{\ell} \sqrt{\chi^{\pm}[\zeta_{\ell}]}\, \langle\zeta_{\ell}|\psi\rangle\, \zeta_{\ell}^{\pm}(\mathbf{r}).
\end{aligned} \tag{3.16}$$

This uses the completeness of the set of full basis functions. This obviously has the correct symmetries. Now if we take ψ itself to be symmetrized, $\psi(\mathbf{r}) \Rightarrow \psi^{\pm}(\mathbf{r})$, then $\chi^{\pm}[\psi^{\pm}] = N!$ and this becomes

$$\begin{aligned}
\psi^{\pm}(\mathbf{r}) &= \frac{1}{\sqrt{\chi^{\pm}[\psi^{\pm}]}} \sum_{\ell} \sqrt{\chi^{\pm}[\zeta_{\ell}]}\, \langle\zeta_{\ell}|\psi^{\pm}\rangle\, \zeta_{\ell}^{\pm}(\mathbf{r}) \\
&= \frac{1}{\sqrt{N!}} \sum_{\ell} \sqrt{\chi^{\pm}[\zeta_{\ell}]}\, \sqrt{\frac{\chi^{\pm}[\zeta_{\ell}]}{N!}}\, \langle\zeta_{\ell}^{\pm}|\psi^{\pm}\rangle\, \zeta_{\ell}^{\pm}(\mathbf{r}) \\
&= \frac{1}{N!} \sum_{\ell} \chi^{\pm}[\zeta_{\ell}]\, \langle\zeta_{\ell}^{+}|\psi^{\pm}\rangle\, \zeta_{\ell}^{+}(\mathbf{r}).
\end{aligned} \tag{3.17}$$

Using the fact that $\langle\zeta_{\ell}(\hat{\mathrm{P}}\mathbf{r})|\psi^{\pm}(\mathbf{r})\rangle = \langle\zeta_{\ell}(\mathbf{r})|\psi^{\pm}(\hat{\mathrm{P}}^{-1}\mathbf{r})\rangle = (\pm 1)^{p}\langle\zeta_{\ell}(\mathbf{r})|\psi^{\pm}(\mathbf{r})\rangle$, the second equality follows since

$$\langle\zeta_{\ell}|\psi^{\pm}\rangle = \frac{1}{N!} \sum_{\hat{\mathrm{P}}}(\pm 1)^{p}\langle\zeta_{\ell}(\hat{\mathrm{P}}\mathbf{r})|\psi^{\pm}(\mathbf{r})\rangle = \sqrt{\frac{\chi^{\pm}[\zeta_{\ell}]}{N!}}\, \langle\zeta_{\ell}^{\pm}|\psi^{\pm}\rangle. \tag{3.18}$$

Equation (3.17) is the desired expansion of any symmetrized wave function as a series of symmetrized basis functions. Notice how the symmetrization factor allows this to be written as an unrestricted sum over all states.

It is a useful exercise to confirm that this expression for the expansion of a symmetrized function is consistent with the above results for completeness and orthonormality. Using the above result for completeness the right-hand side is

$$\begin{aligned}\frac{1}{N!}\sum_{\ell}\chi^{\pm}[\zeta_\ell]\,\langle\zeta_\ell^{\pm}|\psi^{\pm}\rangle\,\zeta_\ell^{\pm}(\mathbf{r}) &= \frac{1}{N!}\int \mathrm{d}\mathbf{r}'\;\psi^{\pm}(\mathbf{r}')\delta^{\pm}(\mathbf{r}'-\mathbf{r})\\ &= \sum_{\hat{\mathrm{P}}}(\pm 1)^{p}\frac{1}{N!}\int \mathrm{d}\mathbf{r}'\;\psi^{\pm}(\mathbf{r}')\delta(\hat{\mathrm{P}}\mathbf{r}'-\mathbf{r})\\ &= \sum_{\hat{\mathrm{P}}}(\pm 1)^{p}\frac{1}{N!}\psi^{\pm}(\hat{\mathrm{P}}\mathbf{r})\\ &= \psi^{\pm}(\mathbf{r}).\end{aligned} \tag{3.19}$$

This is the necessary result. For future reference this shows that the symmetrized completeness expression is an identity that can be inserted into the inner product of symmetrized wave functions. For example

$$\begin{aligned}\langle\psi_1^{\pm}|\psi_2^{\pm}\rangle &= \sum_{\ell}\langle\psi_2^{\pm}|\zeta_\ell\rangle\,\langle\zeta_\ell|\psi_2^{\pm}\rangle\\ &= \frac{1}{N!}\sum_{\ell}\chi^{\pm}[\zeta_\ell]\,\langle\psi_2^{\pm}|\zeta_\ell^{\pm}\rangle\,\langle\zeta_\ell^{\pm}|\psi_2^{\pm}\rangle.\end{aligned} \tag{3.20}$$

The first equality uses the completeness of the full eigenfunctions on the full Hilbert space, and the second uses the completeness just established of the symmetrized eigenfunctions on the symmetrized sub-space.

Using the above result for orthogonality one has

$$\begin{aligned}\langle\zeta_{\ell'}^{\pm}|\psi^{\pm}\rangle &= \frac{1}{N!}\sum_{\ell}\chi^{\pm}[\zeta_\ell]\,\langle\zeta_\ell^{\pm}|\psi^{\pm}\rangle\,\langle\zeta_{\ell'}^{\pm}|\zeta_\ell^{\pm}\rangle\\ &= \frac{1}{N!}\sum_{\ell}\chi^{\pm}[\zeta_\ell]\,\langle\zeta_\ell^{\pm}|\psi^{\pm}\rangle\,\frac{1}{\sqrt{\chi^{\pm}[\zeta_{\ell'}]\chi^{\pm}[\zeta_\ell]}}\delta^{\pm}_{\ell',\ell}\\ &= \frac{1}{N!}\sum_{\ell}\chi^{\pm}[\zeta_\ell]\,\langle\zeta_\ell^{\pm}|\psi^{\pm}\rangle\,\frac{1}{\sqrt{\chi^{\pm}[\zeta_{\ell'}]\chi^{\pm}[\zeta_\ell]}}\sum_{\hat{\mathrm{P}}}(\pm 1)^{p}\delta_{\hat{\mathrm{P}}\ell',\ell}\\ &= \frac{1}{N!}\sum_{\hat{\mathrm{P}}}(\pm 1)^{p}\chi^{\pm}[\zeta_{\hat{\mathrm{P}}\ell'}]\,\langle\zeta_{\hat{\mathrm{P}}\ell'}^{\pm}|\psi^{\pm}\rangle\,\frac{1}{\sqrt{\chi^{\pm}[\zeta_{\ell'}]\chi^{\pm}[\zeta_{\hat{\mathrm{P}}\ell'}]}}\\ &= \langle\zeta_{\ell'}^{\pm}|\psi^{\pm}\rangle,\end{aligned} \tag{3.21}$$

since $\chi^{\pm}[\zeta_{\hat{\mathrm{P}}\ell'}] = \chi^{\pm}[\zeta_{\ell'}]$. Note that $\chi_\ell^{\pm}$ is generally complex. Again this is the necessary result.

3.3.2 Multi-particle states

The particle permutator $\hat{\mathrm{P}}$ gives a permutation of N objects. Primarily it is applied to the position representation coordinates, $\mathbf{r}$, which is an N-dimensional vector, with each entry being a 3-dimensional vector. Single-particle states, such as the position-momentum states that were used for wave packets in chapter 2, or momentum states in general, have a vector label with elements in one-to-one correspondence with the particles. As such the particle permutator can be applied to the single-particle state label, as was done above on occasion.

In the case of multi-particle states, such as the energy states for a subsystem of interacting particles, the states do not correspond to individual particles, and so the particle permutator cannot be applied to the vector label that denotes the subsystem state. With this limitation in mind, we have to show explicitly that the correct formula for an unrestricted sum over multi-particle states is formally weighted with the multi-particle symmetrization factor, just as is the case of single-particle states. Preliminary to this, we establish the general properties of the multi-particle basis functions and their symmetrized projections.

Let $\phi_{\mathbf{n}}(\mathbf{r})$ be one of a complete set of orthonormal eigenfunctions that forms a basis for the Hilbert space. Here $\mathbf{n}$ labels a multi-particle state. One has

$$\langle \phi_{\mathbf{n}'} | \phi_{\mathbf{n}''} \rangle = \delta_{\mathbf{n}',\mathbf{n}''} \quad \text{and} \quad \sum_{\mathbf{n}} \phi_{\mathbf{n}}(\mathbf{r}')\phi_{\mathbf{n}}(\mathbf{r}'')^{*} = \delta(\mathbf{r}' - \mathbf{r}''). \tag{3.22}$$

The symmetrized form is

$$\phi_{\mathbf{n}}^{\pm}(\mathbf{r}) \equiv \frac{1}{\sqrt{N!\chi_{\mathbf{n}}^{\pm}}} \sum_{\hat{\mathrm{P}}} (\pm 1)^{p}\, \phi_{\mathbf{n}}(\hat{\mathrm{P}}\mathbf{r}), \tag{3.23}$$

with

$$\chi_{\mathbf{n}}^{\pm} = \frac{1}{N!} \sum_{\hat{\mathrm{P}},\hat{\mathrm{P}}'} (\pm 1)^{p+p'} \langle \phi_{\mathbf{n}}(\hat{\mathrm{P}}\mathbf{r}) | \phi_{\mathbf{n}}(\hat{\mathrm{P}}'\mathbf{r}) \rangle = \sum_{\hat{\mathrm{P}}} (\pm 1)^{p} \langle \phi_{\mathbf{n}}(\hat{\mathrm{P}}\mathbf{r}) | \phi_{\mathbf{n}}(\mathbf{r}) \rangle. \tag{3.24}$$

When there is a risk of ambiguity we shall write the symmetrization factor as $\chi^{\pm}[\phi_{\mathbf{n}}]$.

The derivation of the completeness condition given above for single-particle symmetrized eigenfunctions carries over directly to the multi-particle case, and one ends up with

$$\sum_{\mathbf{n}} \chi_{\mathbf{n}}^{\pm}\, \phi_{\mathbf{n}}^{\pm}(\mathbf{r}')\phi_{\mathbf{n}}^{\pm}(\mathbf{r}'')^{*} = \sum_{\hat{\mathrm{P}}} (\pm 1)^{p} \delta(\hat{\mathrm{P}}\mathbf{r}' - \mathbf{r}'') \equiv \delta^{\pm}(\mathbf{r}' - \mathbf{r}''). \tag{3.25}$$

The derivation of this only requires the permutator to be applied to the particle representation coordinates.

We are required to prove that even in the case of multi-particle states one can use an unrestricted sum over all states with the symmetrization factor as the weight,

$$F^{\pm} \equiv \mathrm{TR}'\, \hat{f} = \frac{1}{N!} \sum_{\mathbf{n}} \chi_{\mathbf{n}}^{\pm}\, \langle \phi_{\mathbf{n}}^{\pm} | \hat{f}\, | \phi_{\mathbf{n}}^{\pm} \rangle. \tag{3.26}$$

The multi-particle symmetrized eigenfunctions may be expressed as a series of single-particle symmetrized basis functions, equation (3.17),

$$\phi_{\mathbf{n}}^{\pm}(\mathbf{r}) = \frac{1}{N!} \sum_{\ell} \chi^{\pm}[\zeta_{\ell}]\, \langle \zeta_{\ell}^{\pm} | \phi_{\mathbf{n}}^{\pm} \rangle\, \zeta_{\ell}^{\pm}(\mathbf{r}). \tag{3.27}$$

Using this, the right-hand side of the result to be proven may be written

$$
\begin{aligned}
&\frac{1}{N!}\sum_{\mathbf{n}}\chi^{\pm}[\phi_{\mathbf{n}}]\,\langle\phi_{\mathbf{n}}^{\pm}|\hat{f}\,|\phi_{\mathbf{n}}^{\pm}\rangle \\
&= \frac{1}{N!}\sum_{\mathbf{n}}\chi^{\pm}[\phi_{\mathbf{n}}]\frac{1}{(N!)^2}\sum_{\ell',\ell''}\chi^{\pm}[\zeta_{\ell'}]\chi^{\pm}[\zeta_{\ell''}]\,\langle\phi_{\mathbf{n}}^{\pm}|\zeta_{\ell''}^{\pm}\rangle\,\langle\zeta_{\ell''}^{\pm}|\hat{f}\,|\zeta_{\ell'}^{\pm}\rangle\,\langle\zeta_{\ell'}^{\pm}|\phi_{\mathbf{n}}^{\pm}\rangle \\
&= \frac{1}{N!}\sum_{\hat{\mathrm{P}}}(\pm 1)^{p}\frac{1}{(N!)^2}\sum_{\ell',\ell''}\chi^{\pm}[\zeta_{\ell'}]\chi^{\pm}[\zeta_{\ell''}]\,\langle\zeta_{\ell''}^{\pm}|\hat{f}\,|\zeta_{\ell'}^{\pm}\rangle\,\langle\zeta_{\ell'}^{\pm}(\hat{\mathrm{P}}\mathbf{r})|\zeta_{\ell''}^{\pm}(\mathbf{r})\rangle \\
&= \frac{1}{(N!)^2}\sum_{\ell',\ell''}\chi^{\pm}[\zeta_{\ell'}]\chi^{\pm}[\zeta_{\ell''}]\,\langle\zeta_{\ell''}^{\pm}|\hat{f}\,|\zeta_{\ell'}^{\pm}\rangle\,\langle\zeta_{\ell'}^{\pm}(\mathbf{r})|\zeta_{\ell''}^{\pm}(\mathbf{r})\rangle \\
&= \frac{1}{(N!)^2}\sum_{\ell',\ell''}\frac{\chi^{\pm}[\zeta_{\ell'}]\chi^{\pm}[\zeta_{\ell''}]}{\sqrt{\chi^{\pm}[\zeta_{\ell'}]\chi^{\pm}[\zeta_{\ell''}]}}\sum_{\hat{\mathrm{P}}}(\pm 1)^{p}\delta_{\hat{\mathrm{P}}\ell',\ell''}\langle\zeta_{\ell''}^{\pm}|\hat{f}\,|\zeta_{\ell'}^{\pm}\rangle \\
&= \frac{1}{N!}\sum_{\ell'}\chi^{\pm}[\zeta_{\ell'}]\,\langle\zeta_{\ell'}^{\pm}|\hat{f}\,|\zeta_{\ell'}^{\pm}\rangle \\
&= F^{\pm},
\end{aligned}
\tag{3.28}
$$

as required. Recall that the validity of the expression as an unrestricted sum over single-particle states was proven in section 2.3.2.

In can be mentioned that there is another way to interpret the unrestricted, weighted sum over all states that avoids explicitly invoking particle statistics or state occupancy rules. One has

$$
\begin{aligned}
\mathrm{TR}'\,\hat{f} &= \frac{1}{N!}\sum_{\mathbf{n}}\chi_{\mathbf{n}}^{\pm}\,\langle\phi_{\mathbf{n}}^{\pm}|\hat{f}\,|\phi_{\mathbf{n}}^{\pm}\rangle \\
&= \frac{1}{(N!)^2}\sum_{\mathbf{n}}\sum_{\hat{\mathrm{P}}',\hat{\mathrm{P}}''}(\pm 1)^{p'+p''}\langle\phi_{\mathbf{n}}(\hat{\mathrm{P}}'\mathbf{r})|\hat{f}\,|\phi_{\mathbf{n}}(\hat{\mathrm{P}}''\mathbf{r})\rangle.
\end{aligned}
\tag{3.29}
$$

The second equality, which follows the first upon insertion of the definition of the symmetrization factor, explicitly symmetrizes the basis functions, which necessarily embodies, and is more fundamental than, the particle statistics and state occupancy rules. Summing over permutations over-counts the number of distinct states by a factor of $(N!)^2$; hence this factor in the denominator. (The neglect of this factor is one contribution to the error in a widely used result for the primary quantum correction to classical statistical mechanics (Kirkwood 1933). This is discussed in section 8.2.2.) The second equality has a direct physical interpretation that does not require reference to particle state occupancy, and so it applies directly to multi-particle states as well as to single-particle states. Of course from the mathematical point of view it is entirely equal to the first equality in which the symmetrization factor appears explicitly.

3.4 Symmetrization expansion of the partition function

Shortly a formally exact series expansion will be made for the symmetrization factor. An ordering of the terms and a resummation of the series is made based on certain arguments about the relative size of various contributions. These rely upon the notion that the symmetrized permutations of particles are localized in space. Before proceeding with the formal analysis this localization idea is discussed and demonstrated in two simple models, section 3.4.1.

Following this is an exact analysis of the symmetrization factor, section 3.4.2, where it is written as a series expansion, the terms of which are products of closed permutation loops. In section 3.4.3, the analysis is applied to energy eigenstates that are single-particle states. In this case it is shown that in the thermodynamic limit an infinite resummation can be performed that enables the grand potential to be written as a series of loop grand potentials, each term of which is extensive with the subsystem volume. This result is used in chapter 4 to obtain the known analytic expressions for the free energy of a quantum ideal gas, section 4.1, and for non-interacting quantum harmonic oscillators, section 4.2.

The analysis of the symmetrization factor for single-particle states will prove directly relevant to the transformation to classical phase space, chapter 7, where momentum states will be used, although of course these are not energy eigenstates. However before then, in section 3.4.4 a short digression is taken to explore the localization of the symmetrization loops in the case of multi-particle energy eigenstates.

3.4.1 Localization of permuted states

3.4.1.1 Wave packets

In section 2.3.1, using wave packets as the basis functions, it was shown that the dimer symmetrization factor, which results from the transposition of two particles and which might therefore be expected to be the leading symmetrization correction to classical results, was a Gaussian in position separation and in momentum separation. Basically this says that the effects of particle permutations are only important if the particles involved are close enough for their single-particle wave functions to overlap. Obviously wave packets are an approximation to the multi-particle energy eigenfunctions for a real subsystem of interacting particles, but nevertheless the clustering picture that results contains an element of reality.

3.4.1.2 Non-interacting particles. I.

It might be thought that this localization of particle permutations is a specific consequence of the localized nature of wave packets and that it would not hold for the exact eigenfunctions of the system. After all, the discrete nature of the exact quantum eigenstates of a system arises directly from the boundary conditions applied to the wave function; these are necessarily global, not local. However we can demonstrate that symmetrization localization also holds for non-localized basis functions by analyzing the case of ideal, non-interacting particles. For these the energy operator consists solely of the kinetic energy

$$\hat{\mathcal{H}}^{\mathrm{id}}(\mathbf{r}) = \hat{\mathcal{K}}(\mathbf{r}) = \frac{-\hbar^2}{2m}\nabla^2 = \frac{-\hbar^2}{2m}\sum_{j=1}^{N}\sum_{\alpha=x,y,z}\partial_{j\alpha}^2. \tag{3.30}$$

The energy eigenfunctions are just plane waves,

$$\phi_{\mathbf{p}}(\mathbf{r}) = \frac{e^{-\mathbf{p}\cdot\mathbf{r}/i\hbar}}{L^{3N/2}} = \prod_{j=1}^{N} \phi_{\mathbf{p}_j}^{(1)}(\mathbf{r}_j), \quad \phi_{\mathbf{p}_j}^{(1)}(\mathbf{r}_j) = \frac{e^{-\mathbf{p}_j\cdot\mathbf{r}_j/i\hbar}}{L^{3/2}}, \tag{3.31}$$

where L is the subsystem edge-length. This is the product of single-particle eigenfunctions and the subsystem state is a vector of single-particle momentum states, $\mathbf{p} = \{\mathbf{p}_1, \mathbf{p}_2, \dots, \mathbf{p}_N\}$. (In fact, the eigenfunctions are just wave packets with $\xi \to \infty$.) The eigenvalue equation is

$$\hat{\mathcal{H}}^{\mathrm{id}}(\mathbf{r})\phi_{\mathbf{p}}(\mathbf{r}) = \frac{p^2}{2m}\phi_{\mathbf{p}}(\mathbf{r}). \tag{3.32}$$

Obviously because they don't decay, the plane wave eigenfunctions are not localized in position space. (Admittedly, they are highly localized in momentum space.)

As in section 2.3.1 where the dimer symmetrization factor for wave packets was derived, let $\hat{P}_{jk}$ transpose particles j and k (more precisely, transpose the jth and kth elements of the relevant vector), so that

$$\begin{aligned}\phi_{\mathbf{p}}(\hat{\mathrm{P}}_{jk}\mathbf{r}) &= \phi_{\mathbf{p}_1\dots\mathbf{p}_j\dots\mathbf{p}_k\dots}(\mathbf{r}_1\dots\mathbf{r}_k\dots\mathbf{r}_j\dots)\\ &= \phi_{\mathbf{p}_1}^{(1)}(\mathbf{r}_1)\dots\phi_{\mathbf{p}_j}^{(1)}(\mathbf{r}_k)\dots\phi_{\mathbf{p}_k}^{(1)}(\mathbf{r}_j)\dots\end{aligned} \tag{3.33}$$

The dimer symmetrization factor is the inner product of the original and the transposed wave function,

$$\begin{aligned}\chi_{jk}^{\pm,(2)} &\equiv \pm\left\langle \phi_{\mathbf{p}}(\hat{\mathrm{P}}_{jk}\mathbf{r})\middle| \phi_{\mathbf{p}}(\mathbf{r})\right\rangle\\ &= \pm\left\langle \phi_{\mathbf{p}_j}^{(1)}(\mathbf{r}_k)\phi_{\mathbf{p}_k}^{(1)}(\mathbf{r}_j)\middle| \phi_{\mathbf{p}_j}^{(1)}(\mathbf{r}_j)\phi_{\mathbf{p}_k}^{(1)}(\mathbf{r}_k)\right\rangle\\ &= \pm\left|\langle\phi_{\mathbf{p}_j}^{(1)}|\phi_{\mathbf{p}_k}^{(1)}\rangle\right|^2\\ &= \frac{\pm 1}{L^6}\left|\int_{-L/2}^{L/2} \mathrm{d}\mathbf{r}_j\; e^{\mathbf{p}_k\cdot\mathbf{r}_j/i\hbar}e^{-\mathbf{p}_j\cdot\mathbf{r}_j/i\hbar}\right|^2\\ &= \frac{\pm 1}{L^6}\prod_{\alpha=x,y,z}\frac{\hbar^2}{[p_{k\alpha}-p_{j\alpha}]^2}\left| e^{[p_{k\alpha}-p_{j\alpha}]L/2i\hbar} - e^{-[p_{k\alpha}-p_{j\alpha}]L/2i\hbar}\right|^2\\ &= \pm\,\delta_{\mathbf{p}_k,\,\mathbf{p}_j}^2.\end{aligned} \tag{3.34}$$

This is a Kronecker-δ function if the momentum components are quantized, $p_{j\alpha} = 2\pi n_{j\alpha}\hbar/L$. Since ultimately $L \to \infty$, we can take it to go over to a Dirac-δ function. This result shows that for non-interacting particles the energy eigenfunction dimer symmetrization factor vanishes unless the two particles are close together in momentum space.

3.4.1.3 Non-interacting particles. II.

Perhaps even more instructive is to analyze the effect of symmetrization on the position density for this ideal subsystem. To this end the complete orthogonal set of position eigenfunctions are just Dirac-δ functions,

$$|\mathbf{q}\rangle = \delta(\mathbf{r} - \mathbf{q}). \tag{3.35}$$

One can take the symmetrized position basis functions to be

$$|\mathbf{q}^{\pm}\rangle = \frac{1}{\sqrt{N!}} \sum_{\hat{\mathrm{P}}} (\pm 1)^{p} \delta(\hat{\mathrm{P}}\mathbf{r} - \mathbf{q}). \tag{3.36}$$

This sets $\chi_{\mathbf{q}}^{\pm} = 1$, which gives the correct orthonormalization,

$$\begin{aligned}
\langle \mathbf{q}'^{\pm} | \mathbf{q}''^{\pm} \rangle &= \sum_{\hat{\mathrm{P}}', \hat{\mathrm{P}}''} (\pm 1)^{p'+p''} \frac{1}{N!} \int \mathrm{d}\mathbf{r}\ \delta(\hat{\mathrm{P}}'\mathbf{r} - \mathbf{q}') \delta(\hat{\mathrm{P}}''\mathbf{r} - \mathbf{q}'') \\
&= \frac{1}{N!} \sum_{\hat{\mathrm{P}}', \hat{\mathrm{P}}''} (\pm 1)^{p'+p''} \delta(\hat{\mathrm{P}}'\mathbf{q}' - \hat{\mathrm{P}}''\mathbf{q}'') \\
&= \sum_{\hat{\mathrm{P}}} (\pm 1)^{p} \delta(\hat{\mathrm{P}}\mathbf{q}' - \mathbf{q}'').
\end{aligned} \tag{3.37}$$

The right-hand side is zero unless $\mathbf{q}''$ is a permutation of $\mathbf{q}'$. In the latter case, if, for example, $\hat{\mathrm{P}}_0\mathbf{q}' = \mathbf{q}''$, then one term on the right-hand side is non-zero, namely $(\pm 1)^{p_0}\delta(\mathbf{q}'' - \mathbf{q}'')$. This is the correct normalization for the continuum.

There are two reasons for believing that there is at most one permutation that transforms one position label into the other. First, in the continuum there is vanishingly small mathematical probability for two particles being at the same position point. And second, on physical grounds particles must have a repulsive core that precludes the individual Dirac-δ functions from overlap.

We can therefore take it as axiomatic that for physical position eigenstates $j \neq k \Leftrightarrow \mathbf{q}_j \neq \mathbf{q}_k$. This means that the occupancy of each single-particle position state is zero or one, and therefore that there are no internal permutations within a single-particle position state. In other words, each subsystem position state $\mathbf{q}$ belongs to an equivalence class with exactly $N!$ members, which explains why $\chi_{\mathbf{q}}^{\pm} = 1$.

In chapter 12, the Maxwell–Boltzmann operator $e^{-\beta\hat{\mathcal{H}}}$ is shown to give the reservoir or environment weight to the otherwise equally weighted states of the subsystem. Accordingly we can identify the quantum probability density for the superposition of position to be (we neglect proportionality constants that don't depend upon position)

$$\begin{aligned}
W_N(\mathbf{q}'', \mathbf{q}') &\propto \langle \mathbf{q}''^{\pm} | e^{-\beta\hat{\mathcal{H}}} | \mathbf{q}'^{\pm} \rangle \\
&= \sum_{\mathbf{p}} \chi_{\mathbf{p}}^{\pm} \langle \mathbf{q}''^{\pm} | e^{-\beta\hat{\mathcal{H}}} | \phi_{\mathbf{p}}^{\pm} \rangle \langle \phi_{\mathbf{p}}^{\pm} | \mathbf{q}'^{\pm} \rangle \\
&= \sum_{\mathbf{p}} \chi_{\mathbf{p}}^{\pm} e^{-\beta p^2/2m} \langle \mathbf{q}''^{\pm} | \phi_{\mathbf{p}}^{\pm} \rangle \langle \phi_{\mathbf{p}}^{\pm} | \mathbf{q}'^{\pm} \rangle.
\end{aligned} \tag{3.38}$$

Now

$$\begin{aligned}\langle \mathbf{q}''^{\pm}|\phi_{\mathbf{p}}^{\pm}\rangle &= \frac{1}{N!\sqrt{\chi_{\mathbf{p}}^{\pm}}} \sum_{\hat{\mathrm{P}}',\hat{\mathrm{P}}''} (\pm 1)^{p'+p''} \langle \delta(\hat{\mathrm{P}}''\mathbf{r} - \mathbf{q}'')|\phi_{\mathbf{p}}(\hat{\mathrm{P}}'\mathbf{r})\rangle \\ &= \frac{1}{\sqrt{\chi_{\mathbf{p}}^{\pm}}} \sum_{\hat{\mathrm{P}}'} (\pm 1)^{p'} \langle \delta(\mathbf{r} - \mathbf{q}'')|\phi_{\mathbf{p}}(\hat{\mathrm{P}}'\mathbf{r})\rangle \\ &= \frac{1}{\sqrt{\chi_{\mathbf{p}}^{\pm}}} \sum_{\hat{\mathrm{P}}'} (\pm 1)^{p'} \phi_{\mathbf{p}}(\hat{\mathrm{P}}'\mathbf{q}'').\end{aligned} \tag{3.39}$$

With this and transforming to the momentum continuum, one sees that the position probability density is proportional to

$$\begin{aligned}W_N(\mathbf{q}'', \mathbf{q}') &\propto \sum_{\hat{\mathrm{P}}',\hat{\mathrm{P}}''} (\pm 1)^{p'+p''} \int \mathrm{d}\mathbf{p}\, e^{-\beta p^2/2m} \phi_{\mathbf{p}}(\hat{\mathrm{P}}''\mathbf{q}'') \phi_{\mathbf{p}}(\hat{\mathrm{P}}'\mathbf{q}')^* \\ &\propto \sum_{\hat{\mathrm{P}}',\hat{\mathrm{P}}''} (\pm 1)^{p'+p''} \int \mathrm{d}\mathbf{p}\, e^{-\beta p^2/2m} e^{-\mathbf{p}\cdot(\hat{\mathrm{P}}''\mathbf{q}'')/i\hbar} e^{\mathbf{p}\cdot(\hat{\mathrm{P}}'\mathbf{q}')/i\hbar} \\ &\propto \sum_{\hat{\mathrm{P}}} (\pm 1)^{p} \int \mathrm{d}\mathbf{p}\, e^{-(\beta p^2/2m) + \mathbf{p}\cdot(\mathbf{q}' - \hat{\mathrm{P}}\mathbf{q}'')/i\hbar} \\ &= \sum_{\hat{\mathrm{P}}} (\pm 1)^{p} \int \mathrm{d}\mathbf{p}\, e^{-\beta[\mathbf{p} + i(m/\beta\hbar)(\mathbf{q}' - \hat{\mathrm{P}}\mathbf{q}'')]^2/2m} e^{-m(\mathbf{q}' - \hat{\mathrm{P}}\mathbf{q}'')^2/2\beta\hbar^2} \\ &\propto \sum_{\hat{\mathrm{P}}} (\pm 1)^{p} e^{-\pi(\mathbf{q}' - \hat{\mathrm{P}}\mathbf{q}'')^2/\Lambda^2}.\end{aligned} \tag{3.40}$$

Here and throughout the thermal wavelength is $\Lambda \equiv \sqrt{2\pi\hbar^2/mk_{\mathrm{B}}T}$. The diagonal element of the particle position density is therefore

$$\begin{aligned}W_N(\mathbf{q}, \mathbf{q}) &= \sum_{\hat{\mathrm{P}}} (\pm 1)^{p} e^{-\pi(\mathbf{q} - \hat{\mathrm{P}}\mathbf{q})^2/\Lambda^2} \\ &= \sum_{\hat{\mathrm{P}}} (\pm 1)^{p} \prod_{j=1}^{N} e^{-\pi(\mathbf{q}_j - \mathbf{q}_{j'})^2/\Lambda^2},\end{aligned} \tag{3.41}$$

where j' is the label of particle j after the permutation $\hat{\mathrm{P}}$.

The nature of the Gaussians that appear here means that the particle position density is dominated by permutations amongst particles that are close together (since in order for $(\mathbf{q}_j - \mathbf{q}_{j'})^2$ to be small, originally particle j must have been close to particle j'). To leading order this is the identity permutation, $\hat{\mathrm{I}}$, with $j' = j$, followed by a single transposition, say $\hat{\mathrm{P}}_{jk}$, with $j' = k$ and $k' = j$, followed by the permutation of three particles, $\hat{\mathrm{P}}_{jk\ell} = \hat{\mathrm{P}}_{\ell j}\hat{\mathrm{P}}_{jk}$, with $j' = \ell$, $k' = j$, and $\ell' = k$, etc. Hence one can expand the density as

$$W_N(\mathbf{q}, \mathbf{q}) = 1 \pm \sum_{j<k} e^{-2\pi q_{jk}^2/\Lambda^2} + \sum_{j<k<\ell} e^{-\pi q_{jk}^2/\Lambda^2} e^{-\pi q_{k\ell}^2/\Lambda^2} e^{-\pi q_{\ell j}^2/\Lambda^2} + \cdots \tag{3.42}$$

where $\mathbf{q}_{jk} = \mathbf{q}_j - \mathbf{q}_k$ and $q_{jk}^2 = \mathbf{q}_{jk} \cdot \mathbf{q}_{jk}$. The Gaussian in q_{jk} vanishes when $q_{jk} \gg \Lambda$. In the low density $N/V \to 0$ and/or high temperature $\Lambda \to 0$ limits, one has $N\Lambda^3/V \ll 1$. Since the spacing between particles is on the order of the inverse of the cube root of the number density, one sees that in the high temperature, low density limit the corrections due to quantum symmetrization are negligible. This decay of symmetrization effects with distance illustrates how it is localized to clusters of neighboring particles.

In the low density, high temperature limit the single pair transpositions dominate the permutation sum. In this case one can write the position probability density in Maxwell–Boltzmann form with an effective pair potential

$$v(q_{jk}) = -k_B T \ln\left[1 \pm e^{-2\pi q_{jk}^2/\Lambda^2}\right]. \tag{3.43}$$

This potential is for ideal particles and it arises from wave function symmetrization. The upper sign is for bosons and the lower sign is for fermions. This is graphed in figure 3.1. One can see that wave function symmetrization induces an effective attraction between bosons and an effective repulsion between fermions. This is how particle occupancy statistics manifest themselves in the continuum. Although this effective potential was derived for ideal non-interacting particles, one will not go too far wrong in thinking of real interacting particles as having a similar effective potential.

The conclusion that symmetrization effects are localized was made on the basis of two approximate calculations: wave packets, which are themselves localized but approximate eigenfunctions for interacting particles, and plane waves, which are non-localized wave functions that are the exact eigenfunctions for non-interacting particles. The consonance of these two different calculations is rather persuasive that symmetrization effects are in general only important between neighboring particles, and that therefore they can be usefully ordered in terms of clusters of increasing size. (Localization here can be in either momentum space or else in position space. In

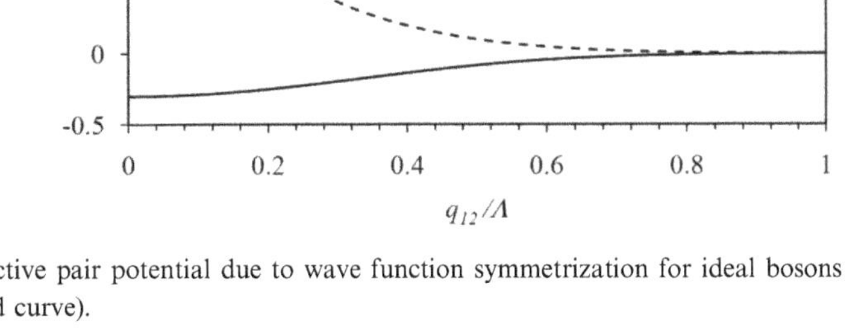

Figure 3.1. Effective pair potential due to wave function symmetrization for ideal bosons (solid curve) and fermions (dashed curve).

later chapters, specifically chapter 5, the momentum permutation loops are non-localized in the position continuum and fully localized for the quantized momentum states.)

3.4.2 Loop expansion of the permutation series

The symmetrization factor, which enables the partition function to be written as an unrestricted sum over all quantum states, requires the sum over all permutations of the particle positions, equation (3.5). We now obtain a systematic series expansion for this function.

Any permutation can be cast as the product of disconnected loops (figure 3.2). A loop is the cyclic permutation of a set of particles. This can be written as a sequence of connected pair transpositions, which is to say that the end label of one transposition is the start label of the next transposition. A monomer is a one-particle loop (the identity permutation). A dimer is a two-particle loop (a single transposition), for example, $1 \rightarrow 2 \rightarrow 1$, which is equivalent to the single transposition $\hat{P}_{21}$. The trimer, or three-particle loop, $1 \rightarrow 2 \rightarrow 3 \rightarrow 1$, is equivalent to the double transposition $\hat{P}_{32}\hat{P}_{21}$. This permutation can also be achieved as $\hat{P}_{21}\hat{P}_{13}$ and as $\hat{P}_{31}\hat{P}_{12}$. The tetramer $1 \rightarrow 2 \rightarrow 3 \rightarrow 4 \rightarrow 1$, is equivalent to the triple transposition $\hat{P}_{43}\hat{P}_{32}\hat{P}_{21}$, and others besides. In general an l-mer is an l-particle loop. An l-mer of particles $1, 2, \ldots, l$ in order can be written as the application of $l - 1$ successive transpositions, $\hat{P}^{l-1} \equiv \hat{P}_{l,l-1} \ldots \hat{P}_{32}\hat{P}_{21}$. Because one has to sum over all permutations, the nodes of the loops can be re-labeled as convenient.

Evidently the parity of a loop is the parity of the number of nodes minus one. That is, an l-mer has parity $l - 1$. We usually restrict the word 'loop' to dimers and above, and use the word 'monomer' for the case that the identity permutation is applied.

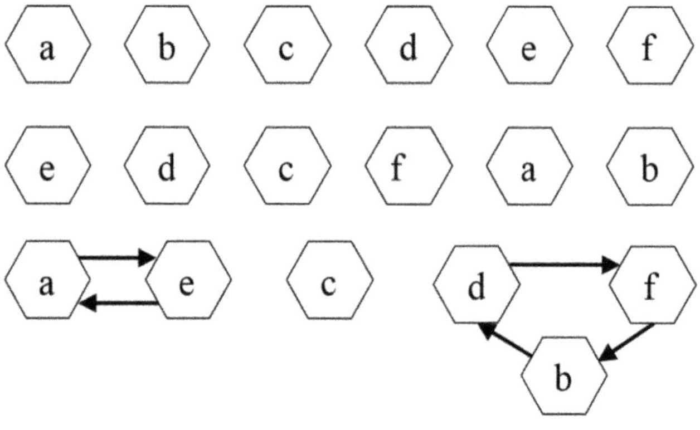

Figure 3.2. Six objects (top row) and a permutation of their order (second row). The third row shows the permutation factored into its loops: the dimer $a \rightarrow e \rightarrow a$, the monomer $c \rightarrow c$, and the trimer $d \rightarrow f \rightarrow b \rightarrow d$.

The sum over permutation operators can be written as the sum over all possible loop factors, with the smallest loops, and the smallest number of factors ordered first,

$$\sum_{\hat{\mathrm{P}}}(\pm 1)^p\, \hat{\mathrm{P}} = \hat{\mathrm{I}} \pm \sum_{i,j}{}' \hat{\mathrm{P}}_{ij} + \sum_{i,j,k}{}' \hat{\mathrm{P}}_{ij}\hat{\mathrm{P}}_{jk} + \sum_{i,j,k,l}{}' \hat{\mathrm{P}}_{ij}\hat{\mathrm{P}}_{kl} \pm \ldots \tag{3.44}$$

The prime on the sums restrict them to unique loops, with each index being different. The first term is just the identity. The second term is a dimer loop, the third term is a trimer loop, and the fourth term is the product of two different dimers, etc.

3.4.3 Exact expansion for single-particle states

The present subsection focuses upon single-particle states. In this case it will be shown that the individual terms in the loop expansion of the permutation sum just derived factorize into independent products of monomers and permutation loops. This enables a powerful resummation of the partition function into a simple series for the grand potential that is formally exact in the thermodynamic limit.

Apart from this infinite resummation, the treatment of the overlap factors that occur in the permutation loops is rather useful for single-particle states in general. In the case, for example, of momentum states, the eigenfunctions are just plane waves, and the symmetrization or loop overlap factors can be calculated explicitly for a given momentum state. These will play a key role in the formulation of quantum statistical mechanics in classical phase space in chapter 7.

3.4.3.1 Single-particle states

Define the relevant unsymmetrized subsystem basis functions for the present case of single-particle states to be

$$\phi_{\mathbf{n}}(\mathbf{r}) = \prod_{j=1}^{N} \phi^{(1)}_{\mathbf{n}_j}(\mathbf{r}_j). \tag{3.45}$$

These could be momentum states and eigenfunctions, but they don't have to be. The equations in terms of momentum states $\mathbf{p}$ used above are now invoked using generic single-particle states $\mathbf{n}$ in their place.

In equation (3.44), the sum over all particle permutators was systematically decomposed into sums over permutation loops and their products. The symmetrization or overlap factor, $\chi^{\pm}_{\mathbf{n}} = \sum_{\hat{\mathrm{P}}}(\pm 1)^p \langle \phi_{\mathbf{n}}(\hat{\mathrm{P}}\mathbf{r})|\phi_{\mathbf{n}}(\mathbf{r})\rangle$, is the sum of the expectation values of these loops. The monomer symmetrization factor is unity,

$$\chi^{\pm,(1)}_{\mathbf{n}} = \langle \phi_{\mathbf{n}}(\mathbf{r})|\phi_{\mathbf{n}}(\mathbf{r})\rangle = 1. \tag{3.46}$$

The dimer overlap factor for particles j and k in the state $\mathbf{n}$ is

$$\begin{aligned}\chi^{\pm,(2)}_{\mathbf{n};jk} &= \pm \langle \phi_{\mathbf{n}}(\hat{\mathrm{P}}_{jk}\mathbf{r})|\phi_{\mathbf{n}}(\mathbf{r})\rangle \\ &= \pm \langle \phi^{(1)}_{\mathbf{n}_j}(\mathbf{r}_k)|\phi^{(1)}_{\mathbf{n}_k}(\mathbf{r}_k)\rangle\, \langle \phi^{(1)}_{\mathbf{n}_k}(\mathbf{r}_j)|\phi^{(1)}_{\mathbf{n}_j}(\mathbf{r}_j)\rangle \\ &= \pm \langle \mathbf{n}_j|\mathbf{n}_k\rangle\, \langle \mathbf{n}_k|\mathbf{n}_j\rangle.\end{aligned} \tag{3.47}$$

This dimer overlap factor for the permutation of particles j and k only involves their respective single-particle wave functions.

Similarly the trimer overlap factor is

$$\begin{aligned}\chi^{\pm,(3)}_{\mathbf{n};jkl} &= \langle \phi_{\mathbf{n}}(\hat{\mathrm{P}}_{jk}\hat{\mathrm{P}}_{kl}\mathbf{r})|\phi_{\mathbf{n}}(\mathbf{r})\rangle \\ &= \langle \phi^{(3)}_{\mathbf{n}_j,\,\mathbf{n}_k,\,\mathbf{n}_l}(\mathbf{r}_l,\,\mathbf{r}_j,\,\mathbf{r}_k)|\phi^{(3)}_{\mathbf{n}_j,\,\mathbf{n}_k,\,\mathbf{n}_l}(\mathbf{r}_j,\,\mathbf{r}_k,\,\mathbf{r}_l)\rangle \\ &= \langle \phi^{(1)}_{\mathbf{n}_k}(\mathbf{r}_j)|\phi^{(1)}_{\mathbf{n}_j}(\mathbf{r}_j)\rangle\ \langle \phi^{(1)}_{\mathbf{n}_j}(\mathbf{r}_l)|\phi^{(1)}_{\mathbf{n}_l}(\mathbf{r}_l)\rangle\ \langle \phi^{(1)}_{\mathbf{n}_l}(\mathbf{r}_k)|\phi^{(1)}_{\mathbf{n}_k}(\mathbf{r}_k)\rangle \\ &= \langle \mathbf{n}_k|\mathbf{n}_j\rangle\ \langle \mathbf{n}_j|\mathbf{n}_l\rangle\ \langle \mathbf{n}_l|\mathbf{n}_k\rangle.\end{aligned} \tag{3.48}$$

Again since the expectation values of the unpermuted monomers each contribute a factor of unity, only the permuted particle's and their states contribute to this.

In general for single-particle states the l-loop symmetrization or overlap factor is

$$\begin{aligned}\chi^{\pm,(l)}_{\mathbf{n};j_1j_2\ldots j_l} &= (\pm 1)^{l-1}\langle \mathbf{n}_{j_1}|\mathbf{n}_{j_2}\rangle\ \langle \mathbf{n}_{j_2}|\mathbf{n}_{j_3}\rangle\ldots\langle \mathbf{n}_{j_l}|\mathbf{n}_{j_1}\rangle \\ &= (\pm 1)^{l-1}\delta_{\mathbf{n}_{j_1},\mathbf{n}_{j_2}}\ \delta_{\mathbf{n}_{j_2},\mathbf{n}_{j_3}}\ldots\delta_{\mathbf{n}_{j_l},\mathbf{n}_{j_1}}.\end{aligned} \tag{3.49}$$

The product of Kronecker-deltas means that this is only non-zero if all the particles in the permutation loop occupy the same single-particle state. This result is exact, correct, and necessary, even for fermions. It is only the sum over all permutations (i.e., over all permutation loops and products thereof) that must come to zero if more than one fermion is in the same single-particle state. In section 4.2 it will be shown by explicit calculation that this expression gives the known textbook result for independent harmonic oscillators for both bosons and fermions.

Since the particle states are independent, disjoint permutation loops factorize into the product of their corresponding expectation values. Hence the overlap factor for the product of dimer loops reduces to the product of dimer overlap factors,

$$\begin{aligned}\chi^{\pm,(2,\,2)}_{\mathbf{n};ij,\,kl} &= \langle \phi_{\mathbf{n}}(\hat{\mathrm{P}}_{ij}\hat{\mathrm{P}}_{kl}\mathbf{r})|\phi_{\mathbf{n}}(\mathbf{r})\rangle \\ &= \langle \phi_{\mathbf{n}}(\hat{\mathrm{P}}_{ij}\mathbf{r})|\phi_{\mathbf{n}}(\mathbf{r})\rangle\ \langle \phi_{\mathbf{n}}(\hat{\mathrm{P}}_{kl}\mathbf{r})|\phi_{\mathbf{n}}(\mathbf{r})\rangle \\ &= \chi^{\pm,(2)}_{\mathbf{n};ij}\ \chi^{\pm,(2)}_{\mathbf{n};kl}.\end{aligned} \tag{3.50}$$

By definition of distinct permutations, the i, j, k, and l must all be different, in which case this factorization is exact. By this same argument, a similar factorization holds for the overlap factors of all products of permutation loops.

Because the overlap factor $\chi_{\mathbf{n}}$ is the sum over all permutations, it can be rewritten as the sum over all possible monomers and loops. This gives the loop expansion for the overlap factor for use in the partition function as

$$\chi^{\pm}_{\mathbf{n}} = 1 + \sum_{ij}{}' \chi^{\pm,(2)}_{\mathbf{n};ij} + \sum_{ijk}{}' \chi^{\pm,(3)}_{\mathbf{n};ijk} + \sum_{ijkl}{}' \chi^{\pm,(2)}_{\mathbf{n};ij}\chi^{\pm,(2)}_{\mathbf{n};kl} + \cdots \tag{3.51}$$

Note that the parity factor for fermions and bosons, $(\pm 1)^{l-1}$, has been incorporated into the definition of the $\chi^{\pm,(l)}_{\mathbf{n}}$.

3.4.3.2 Energy eigenstates

It is convenient to work in the grand canonical system. The grand canonical partition function is

$$\Xi^{\pm}(\mu, V, T) = \sum_{N=0}^{\infty} z^N Z^{\pm}(N, V, T) = \sum_{N=0}^{\infty} \frac{z^N}{N!} \sum_{\mathbf{n}} \chi_{\mathbf{n}}^{\pm} e^{-\beta \mathcal{H}_{\mathbf{n}}}, \tag{3.52}$$

where the fugacity is $z = e^{\beta\mu}$, μ being the chemical potential, and the energy eigenvalue is $\mathcal{H}_{\mathbf{n}} = \sum_{j=1}^{N} \mathcal{H}_{\mathbf{n}_j}^{(1)}$. This is the first point in the current subsection 3.4.3 that we assume that the quantum states $\mathbf{n}$ are energy eigenstates. It is assumed that the set of single-particle energy eigenstates is complete; contrariwise, sum over the principle and the degenerate states.

One can insert into this the loop expansion of the overlap factor and evaluate the terms one at a time. The monomer term is

$$\begin{aligned} \Xi_1 &= \sum_{N=0}^{\infty} \frac{z^N}{N!} \sum_{\mathbf{n}} e^{-\beta \mathcal{H}_{\mathbf{n}}} = \sum_{N=0}^{\infty} \frac{z^N}{N!} \prod_{j=1}^{N} \left[\sum_{\mathbf{n}_j} e^{-\beta \mathcal{H}_{\mathbf{n}_j}^{(1)}} \right] \\ &\equiv \sum_{N=0}^{\infty} \frac{z^N Z_1^N}{N!} \\ &\equiv e^{-\beta \Omega_1}. \end{aligned} \tag{3.53}$$

This does not distinguish bosons from fermions, and, in the continuum limit, is just the classical expression (assuming that the single-particle eigenvalues $\mathcal{H}_{\mathbf{n}_j}^{(1)}$ have a classical counterpart). The monomer grand potential is given by $-\beta\Omega_1 = z \sum_{\mathbf{n}_j} e^{-\beta \mathcal{H}_{\mathbf{n}_j}^{(1)}}$, the particle label j being arbitrary.

The single dimer term is

$$\begin{aligned} \Xi_2^{\pm} &= \sum_{N=2}^{\infty} \frac{z^N}{N!} \sum_{\mathbf{n}} e^{-\beta \mathcal{H}_{\mathbf{n}}} \sum_{ij}{}' \chi_{\mathbf{n};ij}^{\pm,(2)} \\ &= \sum_{N=2}^{\infty} \frac{z^N N(N-1)}{2N!} \sum_{\mathbf{n}} e^{-\beta \mathcal{H}_{\mathbf{n}}} \chi_{\mathbf{n};12}^{\pm,(2)} \\ &= \sum_{N=2}^{\infty} \frac{z^N N(N-1)}{2N!} \prod_{j=3}^{N} \left[\sum_{\mathbf{n}_j} e^{-\beta \mathcal{H}_{\mathbf{n}_j}^{(1)}} \right] \sum_{\mathbf{n}_1, \mathbf{n}_2} e^{-\beta \mathcal{H}_{\mathbf{n}_1,\mathbf{n}_2}^{(2)}} \chi_{\mathbf{n};12}^{\pm,(2)} \\ &= \sum_{N=2}^{\infty} \frac{z^{N-2}}{(N-2)!} Z_1^{N-2} \frac{z^2}{2} \sum_{\mathbf{n}_1, \mathbf{n}_2} e^{-\beta \mathcal{H}_{\mathbf{n}_1,\mathbf{n}_2}^{(2)}} \chi_{\mathbf{n};12}^{\pm,(2)} \\ &= \Xi_1 \frac{z^2}{2} \sum_{\mathbf{n}_1, \mathbf{n}_2} e^{-\beta \mathcal{H}_{\mathbf{n}_1,\mathbf{n}_2}^{(2)}} \chi_{\mathbf{n};12}^{\pm,(2)} \\ &\equiv (-\beta \Omega_2^{\pm}) \Xi_1. \end{aligned} \tag{3.54}$$

The second equality follows because the sum over all states makes all dimer pairs equivalent. Obviously $\mathcal{H}^{(2)}_{\mathbf{n}_1,\,\mathbf{n}_2} = \mathcal{H}^{(1)}_{\mathbf{n}_1} + \mathcal{H}^{(1)}_{\mathbf{n}_2}$, and similarly below.

Similarly the single trimer term is

$$\begin{aligned}
\Xi_3^{\pm} &= \sum_{N=3}^{\infty} \frac{z^N}{N!} \sum_{\mathbf{n}} e^{-\beta\mathcal{H}_{\mathbf{n}}} \sum_{jkl}{}' \chi^{\pm,(3)}_{\mathbf{n};jkl} \\
&= \sum_{N=3}^{\infty} \frac{2!z^N}{3!(N-3)!} \sum_{\mathbf{n}} e^{-\beta\mathcal{H}_{\mathbf{n}}} \chi^{\pm,(3)}_{\mathbf{n};123} \\
&= \sum_{N=2}^{\infty} \frac{z^{N-3}}{(N-3)!} Z_1^{N-3} \frac{z^3}{3} \sum_{\mathbf{n}_1,\mathbf{n}_2,\mathbf{n}_3} e^{-\beta\mathcal{H}^{(3)}_{\mathbf{n}_1,\mathbf{n}_2,\mathbf{n}_3}} \chi^{\pm,(3)}_{\mathbf{n};123} \\
&= \Xi_1 \frac{z^3}{3} \sum_{\mathbf{n}_1,\mathbf{n}_2,\mathbf{n}_3} e^{-\beta\mathcal{H}^{(3)}_{\mathbf{n}_1,\mathbf{n}_2,\mathbf{n}_3}} \chi^{\pm,(3)}_{\mathbf{n};123} \\
&\equiv (-\beta\Omega_3^{\pm})\Xi_1 .
\end{aligned} \tag{3.55}$$

The combinatorial pre-factor comes from two contributions. In general there are $N!/l!(N-l)!$ ways of choosing l different particles from N particles. And there are $(l-1)!$ distinct ways of arranging these clockwise in a loop of l particles. Obviously $l = 3$ here.

The double dimer product term is

$$\begin{aligned}
\Xi_{22}^{\pm} &= \sum_{N=4}^{\infty} \frac{z^N}{N!} \sum_{\mathbf{n}} e^{-\beta\mathcal{H}_{\mathbf{n}}} \sum_{ijkl}{}' \chi^{\pm,(2,\,2)}_{\mathbf{n};ij,\,kl} \\
&= \sum_{N=4}^{\infty} \frac{z^N}{N!} \sum_{\mathbf{n}} e^{-\beta\mathcal{H}_{\mathbf{n}}} \sum_{ijkl}{}' \chi^{\pm,(2)}_{\mathbf{n};ij} \chi^{\pm,(2)}_{\mathbf{n};kl} \\
&= \sum_{N=4}^{\infty} \frac{z^N}{N!} \frac{N(N-1)(N-2)(N-3)}{2^3} \sum_{\mathbf{n}} e^{-\beta\mathcal{H}_{\mathbf{n}}} \chi^{\pm,(2)}_{\mathbf{n};12} \chi^{\pm,(2)}_{\mathbf{n};34} \\
&= \sum_{N=4}^{\infty} \frac{z^{N-4}}{(N-4)!} Z_1^{N-4} \frac{z^4}{2^3} \sum_{\mathbf{n}_1,\mathbf{n}_2} e^{-\beta\mathcal{H}^{(2)}_{\mathbf{n}_1,\mathbf{n}_2}} \chi^{\pm,(2)}_{\mathbf{n};12} \sum_{\mathbf{n}_3,\mathbf{n}_4} e^{-\beta\mathcal{H}^{(2)}_{\mathbf{n}_3,\mathbf{n}_4}} \chi^{\pm,(2)}_{\mathbf{n};34} \\
&= \Xi_1 \frac{1}{2} \left[\frac{z^2}{2} \sum_{\mathbf{n}_1,\mathbf{n}_2} e^{-\beta\mathcal{H}^{(2)}_{\mathbf{n}_1,\mathbf{n}_2}} \chi^{\pm,(2)}_{\mathbf{n};12} \right]^2 \\
&= \frac{1}{2} (-\beta\Omega_2^{\pm})^2 \Xi_1 .
\end{aligned} \tag{3.56}$$

In the third equality the combinatorial factor gives the number of distinct ways for allocating four particles to two dimers.

Continuing in this fashion it is clear that

$$\begin{aligned}\Xi^{\pm}(\mu, V, T) &= \Xi_1 + (-\beta\Omega_2^{\pm})\Xi_1 + (-\beta\Omega_3^{\pm})\Xi_1 + \frac{1}{2}(-\beta\Omega_2^{\pm})^2\Xi_1 + \cdots \\ &= \Xi_1 \sum_{\{m_l\}} \prod_{l=2}^{\infty} \frac{1}{m_l!}(-\beta\Omega_l^{\pm})^{m_l} \\ &= \Xi_1 \prod_{l=2}^{\infty} \sum_{m_l=0}^{\infty} \frac{1}{m_l!}(-\beta\Omega_l^{\pm})^{m_l} \\ &= \Xi_1 \prod_{l=2}^{\infty} e^{-\beta\Omega_l^{\pm}},\end{aligned} \tag{3.57}$$

where m_l is the number of loops of l particles, and the loop grand potential for $l \geqslant 2$ is given by

$$\begin{aligned}-\beta\Omega_l^{\pm} &= \frac{z^l}{l} \sum_{\mathbf{n}_1,\ldots,\mathbf{n}_l} e^{-\beta\mathcal{H}^{(l)}_{\mathbf{n}_1\ldots\mathbf{n}_l}} \chi^{\pm,(l)}_{\mathbf{n}_1\ldots\mathbf{n}_l} \\ &= \frac{(\pm 1)^{l-1} z^l}{l} \sum_{\mathbf{n}_1,\ldots,\mathbf{n}_l} \prod_{j=1}^{l} \left[e^{-\beta\mathcal{H}^{(1)}_{\mathbf{n}_j}} \delta_{\mathbf{n}_j,\mathbf{n}_{j+1}} \right], \quad \mathbf{n}_{l+1} \equiv \mathbf{n}_1 \\ &= \frac{(\pm 1)^{l-1} z^l}{l} \sum_{\mathbf{n}_1} e^{-l\beta\mathcal{H}^{(1)}_{\mathbf{n}_1}}.\end{aligned} \tag{3.58}$$

This formula also holds for the monomer grand potential, $l = 1$.

Notice how equation (3.49) for the loop symmetrization factor collapses the l sums over all loop particles' single-particle states to a single sum over the single-particle states, since the loop symmetrization factor is only non-zero when all the loop particles are in the same single-particle state. This makes the loop expansion very efficient. If one includes M single-particle states and L-loop potentials, then there are just $L \times M$ terms in the series for the grand potential. The contribution of successive loops decays exponentially with l, so L need not be large. Contrast this with the direct calculation of the partition function for $\bar{N}$ particles, which would require on the order of $\bar{N}^M$ terms.

The grand potential is given by the logarithm of the partition function, $\Omega^{\pm}(\mu, V, T) = -k_B T \ln \Xi^{\pm}(\mu, V, T)$, which is just the sum of the loop potentials,

$$\begin{aligned}-\beta\Omega^{\pm}(\mu, V, T) &= -\beta \sum_{l=1}^{\infty} \Omega_l^{\pm}. \\ &= \sum_{l=1}^{\infty} \frac{(\pm 1)^{l-1} z^l}{l} \sum_{\mathbf{n}_1} e^{-l\beta\mathcal{H}^{(1)}_{\mathbf{n}_1}} \\ &= \mp \sum_{\mathbf{n}_1} \ln\left[1 \mp z e^{-\beta\mathcal{H}^{(1)}_{\mathbf{n}_1}}\right].\end{aligned} \tag{3.59}$$

Even with permutation particle symmetrization, this reduces to a sum over single-particle states, as one might expect for independent, non-interacting particles.

In the case that the particles can move throughout the volume of the subsystem V, the loop potentials are extensive. That is, they scale with the volume of the subsystem (assuming no externally applied potential). Hence one can define the loop grand potential density, $\Omega_l^{\mp}(\mu, V, T)/V$. The reason that the loop potentials are extensive is that the center of mass of the l-mer is free to roam throughout the homogeneous volume V. But the particles of the loop must remain in the vicinity of the center of mass because of the connectivity of the loop. It is essential that the loop potentials are extensive because the grand potential of the quantum system has to be extensive.

In the general case where the particles interact via an interaction potential, the single-particle state basis functions are not eigenfunctions of the energy operator. However in the absence of inter-particle interactions, single-particle energy eigenfunctions exist, as was assumed from equation (3.52) onwards, and in this case one can factorize and resum the symmetrization factor as here. The energy eigenstates comprise single-particle states when the subsystem is composed of non-interacting particles. In chapter 4, some examples of these are treated, such as the ideal gas, section 4.1, and independent harmonic oscillators, section 4.2.

3.4.4 Approximate localization for multi-particle states

In the preceding subsection a loop expansion for the symmetrization factor was given that in the case of single-particle energy states allowed an infinite resummation of the partition function that gave the grand potential as a simple series of loop potentials. Here is briefly discussed an approximate treatment of multi-particle energy states that allows a similar factorization and resummation. Here we focus on a dimer permutation, which is sufficient to illustrate the idea.

For N particles in the multi-particle energy state $\mathbf{n}$, the transposition of particles j and k gives a dimer symmetrization or overlap factor

$$\begin{aligned} \chi^{\pm,(2)}_{\mathbf{n};jk} &= \pm \langle \zeta_{\mathbf{n}}(\hat{P}_{jk}\mathbf{r}) | \zeta_{\mathbf{n}}(\mathbf{r}) \rangle \\ &= \pm \int \mathrm{d}\mathbf{r}\; \zeta_{\mathbf{n}}(\ldots\mathbf{r}_k\ldots\mathbf{r}_j\ldots)^* \zeta_{\mathbf{n}}(\ldots\mathbf{r}_j\ldots\mathbf{r}_k\ldots). \end{aligned} \tag{3.60}$$

The question addressed in the present subsection is to what extent the states $\mathbf{n}$ that give a non-zero overlap factor correspond to a localization of the permuted particles' states. For example, ideally the inner product of the non-permuted particles would cancel these particles out leaving just a two-particle state,

$$\chi^{\pm,(2)}_{\mathbf{n};jk} \approx \pm \int \mathrm{d}\mathbf{r}_j\, \mathrm{d}\mathbf{r}_k\; \zeta_{\mathbf{n}_{jk}}(\mathbf{r}_k, \mathbf{r}_j)^* \zeta_{\mathbf{n}_{jk}}(\mathbf{r}_j, \mathbf{r}_k). \tag{3.61}$$

Obviously this would be exact if the eigenfunction were the product of one-particle eigenfunctions, as for the non-interacting subsystem analyzed above in sections 3.4.1.2 and 3.4.3.

Perhaps less drastically, one could identify a quasi-independent, localized cluster $2 \leqslant c \ll N$ of particles in a multi-particle state $\mathbf{n}_c \in \mathbf{n}$ that contain the permuted particles,

$$\chi_{\mathbf{n};jk}^{\pm,(2)} \approx \pm \int \mathrm{d}\mathbf{r}^c \zeta_{\mathbf{n}_c}(\hat{P}_{jk}\mathbf{r}^c)^* \zeta_{\mathbf{n}_c}(\mathbf{r}^c). \tag{3.62}$$

The reason for discussing the possibility of localized clusters is that the product of closed permutation loops occurs in the permutation expansion of the symmetrization factor. If each permutation loop can be considered to be contained within a localized cluster, then their states are independent and the overlap of the product factorizes into the product of the individual permutation loop overlap factors. As for the single-particle states above, this considerably simplifies the subsequent analysis.

A shown in equation (3.42) above for non-interacting particles, the quantum correction due to symmetrization decays rapidly with separation. The conclusion is that quantum permutation is only significant when the particles are in the same locality in position space.

A similar point about localization is made by Messiah (1961, chapter XIV, section 8). He shows that if the particles are represented by wave packets, then the wave function need not be symmetrized with respect to interchange of particles in non-overlapping wave packets. This is another way of viewing the present clusters.

From the point of view of formally exact analysis, it is unclear how one would actually set about obtaining the exact wave function of a cluster that in some way fitted with the wave function of particles outside of the cluster. To this end the local state expansion given in chapter 10 is perhaps the best way to implement the idea quantitatively.

A detailed analysis of symmetrization loops in the case of liquid helium in chapter 5 shows how localization varies with temperature. It also shows that it is actually the non-locality of momentum permutation loops that is responsible for Bose–Einstein condensation and for superfluidity.

References

Attard P 2002 *Thermodynamics and Statistical Mechanics: Equilibrium by Entropy Maximisation* (London: Academic)

Feynman R P 1998 *Statistical Mechanics* 2nd edn (Boulder, CO: Westview)

Kirkwood J G 1933 Quantum statistics of almost classical particles *Phys. Rev.* **44** 31

Merzbacher E 1970 *Quantum Mechanics* 2nd edn (New York: Wiley)

Messiah A 1961 *Quantum Mechanics* vol 1 and 2 (Amsterdam: North-Holland)

Chapter 4

Applications with single-particle states

This chapter presents some simple, concrete applications of quantum statistical mechanics. Although the final results can be found in standard textbooks, here the emphasis is placed on those aspects relevant to classical phase space, or to the loop formulation of wave function symmetrization and particle occupation statistics. The applications are the ideal gas (section 4.1), a system of independent harmonic oscillators (section 4.2), the occupancy of single-particle states (section 4.3), and ideal fermions (section 4.4).

4.1 Ideal gas

This section derives the fugacity expansion for the quantum ideal gas. This is a system in which the particles do not interact with each other, nor with any external potential. The purpose is to illustrate and to verify the permutation loop expansion, both for single-particle states, and for the classical phase space formulation.

4.1.1 Single-particle states

For ideal particles, the potential energy is zero, $U(\mathbf{r}) = 0$, and the energy operator is just the kinetic energy operator,

$$\hat{\mathcal{H}}^{\mathrm{id}}(\mathbf{r}) = \hat{\mathcal{K}}(\mathbf{r}) = \frac{-\hbar^2}{2m}\nabla^2 = \frac{-\hbar^2}{2m}\sum_{j=1}^{N}\nabla_j^2. \tag{4.1}$$

The energy eigenfunction equation is

$$\hat{\mathcal{H}}^{\mathrm{id}}(\mathbf{r})\zeta_{\mathbf{n}}(\mathbf{r}) = \mathcal{H}_{\mathbf{n}}\zeta_{\mathbf{n}}(\mathbf{r}). \tag{4.2}$$

For the ideal gas, the energy eigenfunctions are just momentum eigenfunctions, which are the product of plane waves,

$$\zeta_{\mathbf{n}}(\mathbf{r}) = \frac{1}{V^{N/2}} \prod_{j=1}^{N} e^{\mathrm{i}\mathbf{p}_j\cdot\mathbf{r}_j/\hbar}, \quad \mathbf{p}_j = \frac{2\pi\hbar}{L}\mathbf{n}_j, \quad n_{j\alpha} = 0, \pm1, \pm2, \ldots. \tag{4.3}$$

Here $\mathbf{p}_j$ is the momentum eigenvalue of of particle j, $\mathbf{n}$ is a 3N-dimensional integer vector, $\alpha \in \{x, y, z\}$, and $V = L^3$ is the volume of the subsystem. The energy eigenvalues are

$$\mathcal{H}_{\mathbf{n}} = \frac{1}{2m}\sum_{j=1}^{N} p_j^2. \tag{4.4}$$

The symmetrization or overlap factor for an l-loop, equation (3.49), is

$$\begin{aligned} \chi_{\mathbf{n}}^{\pm,(l)} &= (\pm1)^{l-1}\langle\zeta_{\mathbf{n}}(\mathbf{r}_2, \mathbf{r}_3, \ldots, \mathbf{r}_l, \mathbf{r}_1)|\zeta_{\mathbf{n}}(\mathbf{r}_1, \mathbf{r}_2, \ldots, \mathbf{r}_l)\rangle \\ &= (\pm1)^{l-1}\frac{1}{V^l}\int \mathrm{d}\mathbf{r} \prod_{j=1}^{l} e^{\mathrm{i}\mathbf{p}_j\cdot[\mathbf{r}_j-\mathbf{r}_{j-1}]/\hbar}, \end{aligned} \tag{4.5}$$

where the indices are counted mod l (i.e., $\mathbf{r}_0 \equiv \mathbf{r}_l$). This reduces to a Kronecker-δ, as in equation (3.49), but it is better to leave it as an integral as the continuum limit for the energy will now be taken.

The loop potential, equation (3.58), for $l \geqslant 2$ is given by

$$\begin{aligned} -\beta\Omega_l^{\pm,\mathrm{id}} &= \frac{z^l}{l}\sum_{\mathbf{n}^l} e^{-\beta\mathcal{H}_{\mathbf{n}^l}^{(l)}}\chi_{\mathbf{n}^l}^{\pm,(l)} \\ &= \frac{z^l}{l}\frac{V^l}{(2\pi\hbar)^{3l}}\int \mathrm{d}\mathbf{p}^l\, e^{-\beta\mathcal{H}_{\mathbf{p}}}\chi_{\mathbf{p}}^{\pm,(l)} \\ &= \frac{z^l}{l}\frac{(\pm1)^{l-1}}{(2\pi\hbar)^{3l}}\int \mathrm{d}\mathbf{p}^l \prod_{j=1}^{l} e^{-(\beta/2m)p_j^2}\int \mathrm{d}\mathbf{r}^l \prod_{j=1}^{l} e^{\mathrm{i}\mathbf{p}_j\cdot[\mathbf{r}_j-\mathbf{r}_{j-1}]/\hbar} \\ &= \frac{z^l}{l}\frac{(\pm1)^{l-1}}{(2\pi\hbar)^{3l}}\int \mathrm{d}\mathbf{p}^l \int \mathrm{d}\mathbf{r}^l \prod_{j=1}^{l} e^{-(\beta/2m)p_j^2+\mathrm{i}\mathbf{p}_j\cdot\mathbf{r}_j/\hbar-\mathrm{i}\mathbf{p}_j\cdot\mathbf{r}_{j-1}/\hbar} \\ &= \frac{z^l}{l}\frac{(\pm1)^{l-1}}{(2\pi\hbar)^{3l}}\int \mathrm{d}\mathbf{p}^l \prod_{j=1}^{l} e^{-(\beta/2m)[\mathbf{p}_j+\mathrm{i}(m/\beta)\mathbf{r}_j-\mathrm{i}(m/\beta)\mathbf{r}_{j-1}]^2} \\ &\quad\times \int \mathrm{d}\mathbf{r}^l \prod_{j=1}^{l} e^{-(m/2\beta\hbar^2)[\mathbf{r}_j-\mathbf{r}_{j-1}]^2} \\ &= \frac{z^l}{l}\frac{(\pm1)^{l-1}}{(2\pi\hbar)^{3l}}\left[\frac{2\pi m}{\beta}\right]^{3l/2}\int \mathrm{d}\mathbf{r}_1'\ldots\mathrm{d}\mathbf{r}_{l-1}'\,\mathrm{d}\mathbf{r}_l\, e^{-(m/2\beta\hbar^2)\underline{\underline{G}}^{(l)}:\mathbf{r}'\mathbf{r}'} \\ &= \frac{(\pm1)^{l-1}z^l}{l\Lambda^{3l}}\left(\frac{2\pi\beta\hbar^2}{m}\right)^{3(l-1)/2} |\,\underline{\underline{G}}^{(l)}\,|^{-3/2}\, V \\ &= \frac{(\pm1)^{l-1}z^l V}{l\Lambda^3}\,|\,\underline{\underline{G}}^{(l)}\,|^{-3/2}. \end{aligned} \tag{4.6}$$

Here $z = e^{\beta\mu}$ is the fugacity, $\mathbf{r}'_j \equiv \mathbf{r}_j - \mathbf{r}_l$, $j = 1, 2, \ldots, l-1$, $\Lambda = \sqrt{2\pi\hbar^2/mk_BT}$ is the thermal wavelength, and $\underline{\underline{G}}^{(l)}$ is an $(l-1) \times (l-1)$ tridiagonal matrix with 2 on the main diagonal and -1 just above and just below the main diagonal, with all other entries 0. It is straightforward to show that the determinant is

$$\begin{aligned} |\underline{\underline{G}}^{(l)}| &= 2\,|\underline{\underline{G}}^{(l-1)}| - |\underline{\underline{G}}^{(l-2)}| \\ &= l. \end{aligned} \tag{4.7}$$

With this the loop grand potential for $l \geqslant 2$ is given by

$$-\beta\Omega_l^{\pm,\mathrm{id}} = (\pm 1)^{l-1} z^l V \Lambda^{-3} l^{-5/2}. \tag{4.8}$$

This expression can also be shown to hold for the monomer grand potential, $l = 1$.

In thermodynamics, the grand potential is just the pressure times the volume, $\Omega = -pV$. Hence this result for the loop potential gives the fugacity expansion for the pressure of the quantum ideal gas as

$$\beta p^{\pm,\mathrm{id}} \Lambda^3 = \sum_{l=1}^{\infty} \frac{-\beta\Omega_l^{\pm,\mathrm{id}}\Lambda^3}{V} = \sum_{l=1}^{\infty} (\pm 1)^{l-1} z^l l^{-5/2}. \tag{4.9}$$

What is here derived as a loop expansion agrees with the known fugacity expansion for the pressure (Pathria 1972, equations (7.1.7) and (D5)).

The relationship between density and fugacity for the quantum ideal gas is readily obtained from the thermodynamic result (Pathria 1972, Attard 2002),

$$\bar{N}(z, V, T) = \frac{-\partial\Omega(z, V, T)}{\partial\mu}. \tag{4.10}$$

Applying this to the loop expansion gives the fugacity expansion of the quantum ideal gas density,

$$\begin{aligned} \rho^{\pm,\mathrm{id}}(z, T) &= \frac{1}{\beta}\frac{\partial\beta p^{\pm,\mathrm{id}}(z, T)}{\partial z}\frac{\partial z}{\partial\mu} \\ &= \Lambda^{-3}\sum_{l=1}^{\infty}(\pm 1)^{l-1} z^l l^{-3/2}. \end{aligned} \tag{4.11}$$

Using the fugacity as a parameter, these give the equation of state of the ideal gas, which is plotted in figure 4.1. The classical result is just the monomer contribution, $l = 1$, which is $\beta p^{\mathrm{id}} = \rho$. It can be seen that the equation of state for bosons lies below the classical curve, and that of fermions lies above it. As well, the density and pressure for bosons extends to higher values than for fermions. The fugacity expansions do not converge for $z > 1$.

It should be mentioned that in deriving these last two results via the integral over momentum states, the zero momentum contribution, which is the ground state, has been given zero weight by the three-dimensional volume element $\mathrm{d}\mathbf{k}_j = 4\pi k_j^2 \mathrm{d}k_j$.

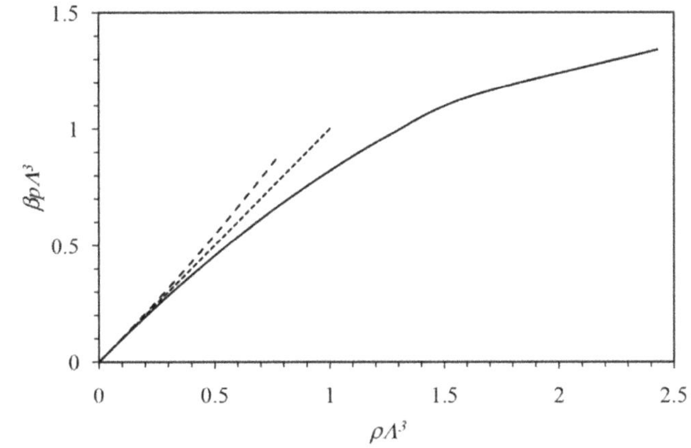

Figure 4.1. Equation of state for the ideal gas for bosons (full curve), for fermions (long-dashed curve), and for classical particles (short-dashed curve) for fugacity $z \in [0, 1]$.

In some circumstances, such as bosons at low temperatures, this suppressed contribution can be significant and should be added explicitly (see section 5.2).

4.1.2 Classical phase space

Now the derivation is carried out using the classical phase space formalism.

For an ideal gas the potential energy is zero, $U(\mathbf{r}) = 0$, and the classical Hamiltonian is just the kinetic energy, $\mathcal{H}^{\text{id}}(\mathbf{q}, \mathbf{p}) = \mathcal{K}_{\mathbf{p}} = p^2/2m$. The momentum eigenfunctions, $\phi_{\mathbf{p}}(\mathbf{r}) = V^{-1/2}e^{-\mathbf{p}\cdot\mathbf{r}/i\hbar}$, are also eigenfunctions of the energy operator, $\hat{\mathcal{H}}^{\text{id}}(\mathbf{r})\phi_{\mathbf{p}}(\mathbf{r}) = \mathcal{K}_{\mathbf{p}}\phi_{\mathbf{p}}(\mathbf{r})$.

With these, the defining equation for the commutation function, equation (7.7), becomes

$$\begin{aligned} e^{-\beta\mathcal{H}^{\text{id}}(\mathbf{r},\,\mathbf{p})}\omega^{\text{id}}(\mathbf{r},\,\mathbf{p})\phi_{\mathbf{p}}(\mathbf{r}) &= e^{-\beta\hat{\mathcal{H}}^{\text{id}}(\mathbf{r})}\phi_{\mathbf{p}}(\mathbf{r}) \\ &= e^{-\beta\mathcal{K}_{\mathbf{p}}}\phi_{\mathbf{p}}(\mathbf{r}). \end{aligned} \tag{4.12}$$

Evidently the commutation function for the ideal gas equals unity,

$$\omega^{\text{id}}(\mathbf{r},\,\mathbf{p}) = 1. \tag{4.13}$$

This is not unexpected as there is no potential energy in this system, and so there are no non-commutativity issues for the kinetic energy operator and the potential energy in any expansion of the Maxwell–Boltzmann operator.

The loop grand potential is given by equation (7.36). For the present ideal gas with commutation function unity and using equation (7.29) for the loop symmetrization function it is

$$
\begin{aligned}
& -\beta\Omega_{\omega_p}^{\pm,\mathrm{id},(l)} \\
&= \left\langle \frac{N!}{(N-l)!l} \eta_{q;1\ldots l}^{\pm(l)} \right\rangle_{\omega_p=1,1} \\
&= \frac{1}{\Xi_{1,1}} \sum_{N=l}^{\infty} \frac{z^N}{h^{3N}N!} \frac{N!}{(N-l)!l} \int \mathrm{d}\mathbf{p}\, \mathrm{d}\mathbf{q}\; e^{-\beta\mathcal{K}_{\mathbf{p}}} \eta_{q;1\ldots l}^{\pm(l)} \\
&= \frac{(\pm 1)^{l-1} z^l}{h^{3l} l} \int \mathrm{d}\mathbf{p}^l\, \mathrm{d}\mathbf{q}^l \prod_{j=1}^{l} [e^{-\beta p_j^2/2m}] e^{\mathbf{q}_{1,l}\cdot\mathbf{p}_l/i\hbar} \prod_{j=1}^{l-1} [e^{\mathbf{q}_{j+1,j}\cdot\mathbf{p}_j/i\hbar}].
\end{aligned}
\tag{4.14}
$$

The final equality here is identical to the third equality in the expression for the loop potential given in the preceding subsection, equation (4.6). Hence the present phase space formulation of the quantum ideal gas gives the same result for the fugacity expansion as the single-particle state formulation.

4.2 Independent harmonic oscillators

The following treatment of a system of independent harmonic oscillators is a specific example of the exact expansion for general single-particle states given in section 3.4.3.

4.2.1 General single-particle energy states

Let

$$
\phi_{\mathbf{n}}(\mathbf{r}) = \prod_{j=1}^{N} \phi_{\mathbf{n}_j}^{(1)}(\mathbf{r}_j) \tag{4.15}
$$

be one of a complete orthogonal set of unsymmetrized energy eigenfunctions, with the eigenvalue equation being

$$
\hat{\mathcal{H}}(\mathbf{r})\phi_{\mathbf{n}}(\mathbf{r}) = E_{\mathbf{n}}\phi_{\mathbf{n}}(\mathbf{r}), \quad E_{\mathbf{n}} = \sum_{j=1}^{N} \varepsilon_{\mathbf{n}_j}. \tag{4.16}
$$

These are evidently single-particle energy states. Since the single-particle states are orthonormal, $\langle \mathbf{n}'(\mathbf{r}_j)|\mathbf{n}''(\mathbf{r}_j)\rangle = \delta_{\mathbf{n}',\mathbf{n}''}$, equation (3.49) for the l-loop symmetrization factor applies,

$$
\begin{aligned}
\chi_{\mathbf{n}_1\ldots\mathbf{n}_l}^{\pm,(l)} &= (\pm 1)^{l-1} \langle \mathbf{n}_1|\mathbf{n}_2\rangle \langle \mathbf{n}_2|\mathbf{n}_3\rangle \ldots \langle \mathbf{n}_l|\mathbf{n}_1\rangle \\
&= (\pm 1)^{l-1} \delta_{\mathbf{n}_1,\mathbf{n}_2} \delta_{\mathbf{n}_2,\mathbf{n}_3} \ldots \delta_{\mathbf{n}_l,\mathbf{n}_1}.
\end{aligned}
\tag{4.17}
$$

Following the analysis of single-particle energy states in section 3.4.3, the loop grand potential, equation (3.58), is given by

$$
-\beta\Omega^{\pm,(l)} = \frac{(\pm 1)^{l-1} z^l}{l} \sum_{\mathbf{n}_1} e^{-\beta l \varepsilon_{\mathbf{n}_1}}. \tag{4.18}
$$

The $\mathbf{n}_1$ labels a single-particle state. It is worth emphasizing that this result holds only because all the particles of the loop are in the same single-particle state, even fermions. The grand potential itself is

$$\begin{aligned} -\beta\Omega^{\pm} &= \sum_{l=1}^{\infty} \frac{(\pm 1)^{l-1} z^l}{l} \sum_{\mathbf{n}_1} e^{-l\beta\varepsilon_{\mathbf{n}_1}} \\ &= \mp \sum_{\mathbf{n}_1} \ln\left[1 \mp z e^{-\beta\varepsilon_{\mathbf{n}_1}}\right]. \end{aligned} \tag{4.19}$$

4.2.2 Simple harmonic oscillators

For the case of the simple harmonic oscillator of frequency ω in d-dimensions, the well-known single-particle energy is (Messiah 1961, Merzbacher 1970)

$$\varepsilon_{\mathbf{n}_j} = \left[\frac{d}{2} + n_{jx} + n_{jy} + \cdots + n_{jd}\right]\hbar\omega, \tag{4.20}$$

with $n_{j\alpha} = 0, 1, 2, \ldots$. Using the above result the grand potential for a system on non-interacting simple harmonic oscillators is therefore

$$\begin{aligned} -\beta\Omega^{\pm} &= \sum_{l=1}^{\infty} \frac{(\pm 1)^{l-1} z^l}{l} e^{-dl\beta\hbar\omega/2} \prod_{\alpha=x}^{d} \sum_{n_\alpha=0}^{\infty} e^{-l\beta\hbar\omega n_\alpha} \\ &= \sum_{l=1}^{\infty} \frac{(\pm 1)^{l-1} z^l}{l} \left[\frac{e^{-l\beta\hbar\omega/2}}{1 - e^{-l\beta\hbar\omega}}\right]^d \\ &= \sum_{l=1}^{\infty} \frac{(\pm 1)^{l-1} z^l}{l} \left[e^{l\beta\hbar\omega/2} - e^{-l\beta\hbar\omega/2}\right]^{-d}. \end{aligned} \tag{4.21}$$

Note that this diverges for $z > e^{d\beta\hbar\omega/2}$.

The average energy of this system of independent harmonic oscillators is

$$\begin{aligned} \langle\hat{\mathcal{H}}\rangle_{z,T}^{\pm} &= \left(\frac{\partial(\beta\Omega^{\pm})}{\partial\beta}\right)_z \\ &= \frac{d\hbar\omega}{2} \sum_{l=1}^{\infty} (\pm 1)^{l-1} z^l \frac{e^{l\beta\hbar\omega/2} + e^{-l\beta\hbar\omega/2}}{\left[e^{l\beta\hbar\omega/2} - e^{-l\beta\hbar\omega/2}\right]^{d+1}}. \end{aligned} \tag{4.22}$$

This is plotted in figure 4.2, where the monomer $l = 1$ term may also be called the result for classical or distinguishable particles. It can be seen that at fixed temperature the average energy increases with increasing fugacity. In general the average energy for bosons is greater than that for fermions, and the monomer energy lies between the two. At low values of the fugacity all three energies converge. As $z \to e^{d\beta\hbar\omega/2}$, the boson energy diverges.

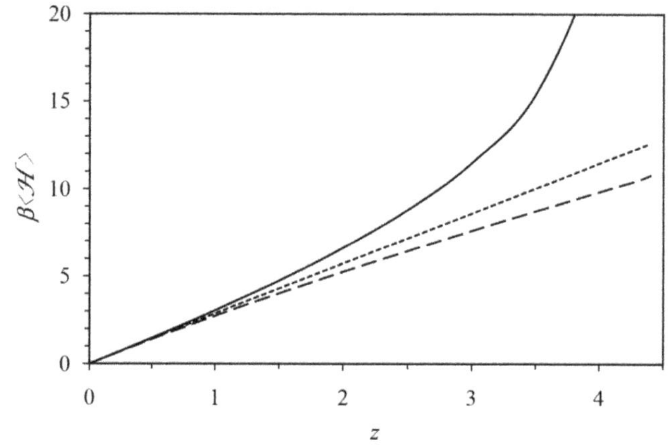

Figure 4.2. Average energy of a system of independent harmonic oscillators at $\beta\hbar\omega = 1$ and $d = 3$ for bosons (full curve), for fermions (long-dashed curve), and for monomers (short-dashed curve).

The average number of oscillators in the open system is given by the fugacity derivative of the grand potential. Hence the average number of oscillators in a permutation loop of size l is

$$\begin{aligned}\langle N\rangle_{z,T}^{\pm,(l)} &= -\beta z\left(\frac{\partial\Omega^{\pm,(l)}}{\partial z}\right)_T \\ &= (\pm1)^{l-1}z^l\sum_{\mathbf{n}_1} e^{-\beta l\varepsilon_{\mathbf{n}_1}} \\ &= (\pm1)^{l-1}z^l e^{-dl\beta\hbar\omega/2}\prod_{\alpha=x}^{d}\sum_{n_\alpha=0}^{\infty} e^{-l\beta\hbar\omega n_\alpha} \\ &= (\pm1)^{l-1}z^l[e^{l\beta\hbar\omega/2} - e^{-l\beta\hbar\omega/2}]^{-d}.\end{aligned} \tag{4.23}$$

The average number of oscillators in total is

$$\langle N\rangle_{z,T}^{\pm} = -\beta z\left(\frac{\partial\Omega^{\pm}}{\partial z}\right)_T = \sum_{l=1}^{\infty}(\pm1)^{l-1}z^l[e^{l\beta\hbar\omega/2} - e^{-l\beta\hbar\omega/2}]^{-d}. \tag{4.24}$$

It can be seen in figure 4.3 that the average number increases monotonically with increasing fugacity. Fractional numbers of course result from the averaging process. At low values of the fugacity particle statistics again have no influence, and all three cases—bosons, fermions, and monomers—coincide. In general there are more bosons for a given fugacity than fermions, and the number of monomers lies between the two. The average number of bosons diverges in the limit $z \to e^{d\beta\hbar\omega/2}$. The average energy per particle, $\langle\hat{\mathcal{H}}\rangle_{z,T}^{\pm}/\langle N\rangle_{z,T}^{\pm}$, at constant temperature with increasing fugacity decreases monotonically for bosons, increases monotonically for fermions, and is a constant (equal to the ratio of the first terms in the respective series) for monomers that lies between the two.

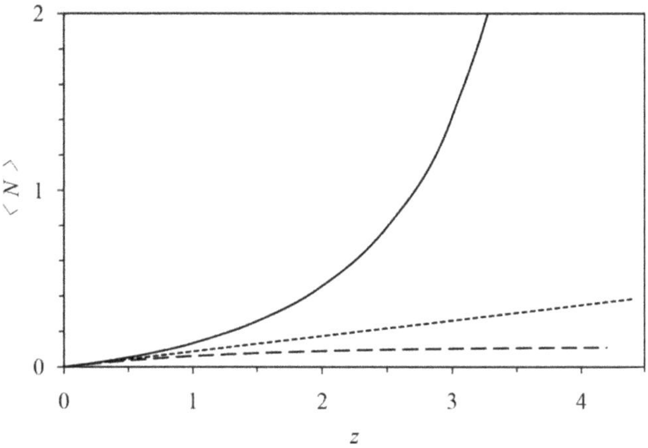

Figure 4.3. Average number of independent oscillators in the system at $\beta\hbar\omega = 1$ and $d = 3$ for bosons (full curve), for fermions (long-dashed curve), and for monomers (short-dashed curve).

4.2.3 Conventional derivation and interpretation

Although the present derivation of the general result for non-interacting particles via the loop expansion is different to the approach usually taken in textbooks, it may be shown to agree with those standard results. The loop series does provide a novel interpretation of how symmetrization effects contribute to the grand potential.

Following Pathria (1972, section 6.2), single-particle states labeled by ε can be occupied by $N_\varepsilon = 0, 1, \ldots, N_\pm$ particles, with $N_+ = \infty$ for bosons, and $N_- = 1$ for fermions. The grand partition function is the weighted sum over all possible occupancies of each state,

$$\begin{aligned}
\Xi^\pm(z) &= \sum_{N_0=0}^{N_\pm} \sum_{N_1=0}^{N_\pm} \ldots \sum_{N_\infty=0}^{N_\pm} z^{\sum'_\varepsilon N_\varepsilon} e^{-\beta \sum'_\varepsilon \varepsilon N_\varepsilon} \\
&= \prod_{\varepsilon=0}^{\infty}{}' \left\{ \sum_{N_\varepsilon=0}^{N_\pm} z^{N_\varepsilon} e^{-\beta\varepsilon N_\varepsilon} \right\} \\
&= \prod_{\varepsilon=0}^{\infty}{}' [1 \mp z e^{-\beta\varepsilon}]^{\mp 1} \\
&= \prod_{n_x=0}^{\infty} \prod_{n_y=0}^{\infty} \ldots \prod_{n_d=0}^{\infty} [1 \mp z e^{-\beta\varepsilon_{\mathbf{n}}}]^{\mp 1}.
\end{aligned} \tag{4.25}$$

The sums and products over energy is an abbreviated notation that visits each state once, as is explicit in the final equality. This is the expression given by Pathria (1972). The grand potential is given by the logarithm of this, a subsequent expansion of which yields

$$\begin{aligned} -\beta\Omega^{\pm} &= \mp \sum_{\mathbf{n}_1} \ln[1 \mp z e^{-\beta\varepsilon_{\mathbf{n}_1}}] \\ &= \pm \sum_{\mathbf{n}_1} \sum_{l=1}^{\infty} \frac{(\pm z)^l}{l} e^{-l\beta\varepsilon_{\mathbf{n}_1}} \\ &= \sum_{l=1}^{\infty} \frac{(\pm 1)^{l-1} z^l}{l} \sum_{\mathbf{n}_1} e^{-l\beta\varepsilon_{\mathbf{n}_1}}. \end{aligned} \tag{4.26}$$

This agrees with the above expression based on symmetrization loops, equation (4.19).

Again it must be emphasized that in the loop expansion all particles in the loop must be in the same single-particle state in order for the symmetrization factor to be non-zero. This holds even for fermions. From this perspective it is remarkable that the loop expansion approach gives the same result as the textbook particle occupancy approach, which explicitly excludes system states with more than one fermion in the same single-particle state.

For the case of a system of independent simple harmonic oscillators, the first equality in equation (4.26) is

$$-\beta\Omega^{\pm} = \mp \sum_{n_x=0}^{\infty} \dots \sum_{n_d=0}^{\infty} \ln[1 \mp z e^{-\beta\hbar\omega d/2} e^{-\beta\hbar\omega n_x} \dots e^{-\beta\hbar\omega n_d}]. \tag{4.27}$$

This multi-dimensional sum over possible energy states is not as neat as the one-dimensional sum over symmetrization loops, equation (4.21). That latter appears to be rapidly converging and computationally efficient.

Although the present approach and the textbook approach arrive at the same result for ideal states, the present approach proceeds from a rather different view-point, namely that the symmetrization of the wave function is the fundamental axiom, and that the occupancy of states (multiple for bosons, single for fermions) is a quantity derived from it. The present approach is extremely useful when the results are transformed to the continuum that is classical phases space. In the case of the continuum it is impossible to define unambiguously discrete states and the occupancy thereof, whereas the symmetrization of the wave function itself remains a valid concept.

4.3 Occupancy of single-particle states

4.3.1 Loop grand potential and derivatives

The independent harmonic oscillators analyzed in the preceding section belong to a general class of systems wherein the energy states are single-particle states. The occupancy statistics of such states has been analyzed by Pathria (1972, section 6.3), which treatment is largely followed here.

Equation (3.59) expresses the grand potential for single-particle state systems as a series of loop potentials. As mentioned above, following the analysis of single-particle energy states in section 3.4.3, the loop grand potential, equation (3.58), is given by

$$-\beta\Omega_l^{\pm} = \frac{(\pm 1)^{l-1} z^l}{l} \sum_n e^{-\beta l \varepsilon_n}. \tag{4.28}$$

Here n labels a single-particle state, and ε_n is its energy, which is the energy of a particle in that state. The fugacity is related to the chemical potential as $z = e^{\beta\mu}$. The grand potential itself is the sum of these loop potentials

$$\begin{aligned} -\beta\Omega^{\pm} &= \sum_{l=1}^{\infty} \frac{(\pm 1)^{l-1} z^l}{l} \sum_n e^{-l\beta\varepsilon_n} \\ &= \mp \sum_n \ln\left[1 \mp z e^{-\beta\varepsilon_n}\right]. \end{aligned} \tag{4.29}$$

As derived in equation (4.26), the final equality is the expression for the grand potential of single-particle state systems more usually to be found in textbooks (Pathria 1972, section 6.2).

The $l = 1$ term in the loop expansion corresponds to the classical case. The particles in this case can be called distinguishable, since it arises from the neglect of all permutations, and it manifests no difference between bosons and fermions. The classical grand potential is given by $-\beta\Omega_1^{\pm} = z \sum_n e^{-\beta\varepsilon_n}$.

From the usual equations of equilibrium thermodynamics (Pathria 1972, Attard 2002), the average number of particles in the open system is given by the fugacity derivative of the grand potential

$$\begin{aligned} \langle N \rangle_{z,V,T}^{\pm} &= -\beta z \left(\frac{\partial \Omega^{\pm}}{\partial z} \right)_{V,T} \\ &= \sum_{l=1}^{\infty} (\pm 1)^{l-1} z^l \sum_n e^{-l\beta\varepsilon_n} \\ &= \sum_n \frac{z e^{-\beta\varepsilon_n}}{1 \mp z e^{-\beta\varepsilon_n}}. \end{aligned} \tag{4.30}$$

(This is the average rather than the most likely quantity that usually results from a thermodynamic derivative because we are using the grand potential as a proxy for the logarithm of the grand partition function, and derivatives of the latter yield statistical averages (Attard 2002). The two are equal in the thermodynamic limit.) The second equality can be interpreted for bosons as the sum of the most likely number of particles belonging to permutation loops of each size. Similarly the average energy is

$$\begin{aligned} \langle E \rangle_{z,V,T}^{\pm} &= \left(\frac{\partial (\beta\Omega^{\pm})}{\partial \beta} \right)_{z,V} \\ &= \sum_{l=1}^{\infty} (\pm 1)^{l-1} z^l \sum_n \varepsilon_n e^{-l\beta\varepsilon_n} \\ &= \sum_n \frac{z \varepsilon_n e^{-\beta\varepsilon_n}}{1 \mp z e^{-\beta\varepsilon_n}}. \end{aligned} \tag{4.31}$$

The pressure is given by $\beta p V = -\Omega^{\pm}(z, V, T)$.

4.3.2 Average occupation number

The average occupation number of level j is given

$$\begin{aligned} \left\langle N_j \right\rangle^{\pm}_{z,\,V,\,T} &= \beta^{-1}\left(\frac{\partial(\beta\Omega^{\pm})}{\partial\varepsilon_j}\right)_{z,V,\varepsilon_{k\neq j}} \\ &= \sum_{l=1}^{\infty}(\pm1)^{l-1}z^l e^{-l\beta\varepsilon_j} \\ &= \frac{ze^{-\beta\varepsilon_j}}{1 \mp ze^{-\beta\varepsilon_j}}. \end{aligned} \tag{4.32}$$

Summing this, and ε_j times this, over the energy levels gives the above results for the average number and energy. In the classical case average occupation number is

$$\left\langle N_j \right\rangle^{\text{cl}}_{z,\,V,\,T} = \beta^{-1}\left(\frac{\partial(\beta\Omega_1^{\pm})}{\partial\varepsilon_j}\right)_{z,V,\varepsilon_{k\neq j}} = ze^{-\beta\varepsilon_j}. \tag{4.33}$$

Figure 4.4 shows the average occupancy number of a single-particle state with energy ε as a function of the parameter $\beta(\varepsilon - \mu)$. The latter is the negative of the exponent in $ze^{-\beta\varepsilon} = e^{-\beta(\varepsilon-\mu)}$. It can be seen that the occupancy decreases with increasing parameter value (decreasing chemical potential). For $\beta(\varepsilon - \mu) \gg 1$, there is on average less than one particle in the state, which is what one would expect at low densities. In this regime there is no difference between bosons, fermions, and classical particles, which is what one would expect, since wave function symmetrization and particle statistics are only important at high densities.

As the chemical potential approaches the energy of the state from below, $\mu \to \varepsilon^-$, figure 4.4 shows that the average number of bosons in that state (i.e., with that value

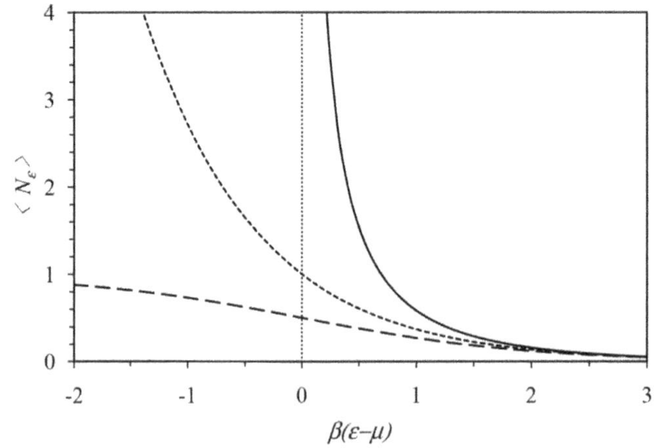

Figure 4.4. Average occupation number of energy level ε for ideal bosons (solid curve), fermions (long-dashed curve), and classical particles (short-dashed curve). The dotted line is a guide to the eye.

of single-particle energy) diverges. This is the remarkable phenomenon of Bose–Einstein condensation.

As the parameter becomes more negative, $\mu \gg \varepsilon$ (or else $\mu > \varepsilon$ and $\beta \to \infty$), it is increasingly likely that the state will be occupied by a fermion. The average occupancy number must be less than one because not more than one fermion can be in the single-particle state with that energy. This means that for the fermionic system as a whole, as the chemical potential is increased, (or the temperature is decreased for $\mu > \varepsilon$), the states will fill successively from the ground state up, each with a single fermion, until near the Fermi surface, where $\mu = \varepsilon$.

For classical particles, figure 4.4 shows that the average occupancy number continues to increase with increasingly negative values of the parameter (increasing chemical potential).

4.3.3 Occupancy fluctuations

It is of interest to analyze the fluctuations in the occupancy of the single-particle states. In general fluctuations are given by the second derivative of the grand potential (Pathria 1972, Attard 2002). In the present case we have (dropping the subscripts on the averages)

$$\begin{aligned} \left\langle N_j^2 \right\rangle^{\pm} - \left\langle N_j \right\rangle^{\pm 2} &= \beta^{-2} \left[\frac{\partial^2(-\beta\Omega_1^{\pm})}{\partial \varepsilon_j^2} \right]_{z,V,\varepsilon_{k\neq j}} \\ &= -\beta^{-1} \left[\frac{\partial \left\langle N_j \right\rangle^{\pm}}{\partial \varepsilon_j} \right]_{z,V} . \end{aligned} \tag{4.34}$$

Hence the mean square relative fluctuation is

$$\begin{aligned} \frac{\left\langle N_j^2 \right\rangle^{\pm} - \left\langle N_j \right\rangle^{\pm 2}}{\left\langle N_j \right\rangle^{\pm 2}} &= \frac{-1}{\beta \left\langle N_j \right\rangle^{+}} \left[\frac{\partial \ln \left\langle N_j \right\rangle^{\pm}}{\partial \varepsilon_j} \right]_{z,V} \\ &= \frac{1}{\left\langle N_j \right\rangle^{\pm}} \pm \frac{1}{\left\langle N_j \right\rangle^{\pm}} \frac{z e^{-\beta\varepsilon_j}}{1 \mp z e^{-\beta\varepsilon_j}} \\ &= \frac{1}{\left\langle N_j \right\rangle^{\pm}} \pm 1. \end{aligned} \tag{4.35}$$

The first term on the right-hand side is the classical result, (without the superscript $\pm$). At low occupancy, $\left\langle N_j \right\rangle^{\pm} \to 0$, this dominates, as is expected. It shows that the fluctuations in the occupancy of any single-particle state are relatively large in this regime, again as expected.

The quantum correction for bosons, $+1$, shows that the fluctuations are of the same order as the occupancy number itself in the high occupancy regime, $\mu \to \varepsilon_j^-$, $\left\langle N_j \right\rangle^{\pm} \to \infty$. In contrast, the quantum correction for fermions, -1, cancels with the classical contribution in the full occupancy regime, $\mu \gg \varepsilon_j$, $\left\langle N_j \right\rangle^{-} \to 1$, which is to say that the fluctuations vanish. In this regime there is a high probability that the single-particle state is always occupied by a fermion, and a correspondingly low probability that it is ever unoccupied.

4.3.4 Occupancy probability distribution

The statistics of the occupancy numbers for the different types of particles may be further elucidated by examining the probability distribution. For the case of classical particles, the probability that N particles have energy ε is

$$\wp(N|\varepsilon) = \frac{z^N e^{-\beta\varepsilon N}}{N!Z(\varepsilon)}, \quad Z(\varepsilon) \equiv \exp[ze^{-\beta\varepsilon}]. \tag{4.36}$$

This is proportional to the corresponding term in the expression for the monomer grand partition function, equation (3.53), with $\mathcal{H}^{(1)}_{\mathbf{n}_j} \Rightarrow \varepsilon$. Using the classical average occupancy, equation (4.33), $\langle N\rangle_\varepsilon = ze^{-\beta\varepsilon}$, this may be written

$$\wp(N|\varepsilon) = \frac{\langle N\rangle_\varepsilon^N}{N!} e^{-\langle N\rangle_\varepsilon}. \tag{4.37}$$

The classical occupancy probability is just a Poisson distribution, with the optimal occupancy number (i.e., that for which the distribution is a maximum) equal to the average occupancy number. The ratio of successive terms, $\wp(N|\varepsilon)/\wp(N-1|\varepsilon) = \langle N\rangle_\varepsilon/N$, decreases monotonically with N. For N larger than the average value, this ratio is less than unity.

For the case of fermions, the average occupancy number can be written in terms of the probability distribution as

$$\langle N\rangle_\varepsilon^- = \sum_{N=0}^{\infty} N\wp^-(N|\varepsilon) = \wp^-(1|\varepsilon), \tag{4.38}$$

since $\wp^-(N|\varepsilon) = 0$, $N > 1$. Hence one has

$$\wp^-(1|\varepsilon) = \langle N\rangle_\varepsilon^-, \quad \text{and} \quad \wp^-(0|\varepsilon) = 1 - \langle N\rangle_\varepsilon^-. \tag{4.39}$$

The fact that the occupancy probability vanishes for $N \geqslant 2$ may be interpreted as saying that a fermion in a single-particles state strongly repels other fermions from occupying the same state.

For the case of bosons, equation (2.20) gives the symmetrization factor for bosons as $\chi_\ell^+ = \prod_a m_a(\ell)!$, where $m_a(\ell)$ is the number of particles in the same single-particle state a when the system is in the state ℓ. With this the grand partition function equation (3.52) for bosons in the present notation is

$$\begin{aligned}\Xi^+(\mu, V, T) &= \sum_{N=0}^{\infty} \frac{z^N}{N!} \sum_{\boldsymbol{\varepsilon}} \chi_{\boldsymbol{\varepsilon}}^+ e^{-\beta\mathcal{H}_{\boldsymbol{\varepsilon}}} \\ &= \sum_{N=0}^{\infty} \frac{1}{N!} \sum_{\varepsilon_1=0}^{\infty} \sum_{\varepsilon_2=0}^{\infty} \cdots \sum_{\varepsilon_N=0}^{\infty} \prod_{\varepsilon} N_\varepsilon! z^{N_\varepsilon} e^{-\beta N_\varepsilon \varepsilon}.\end{aligned} \tag{4.40}$$

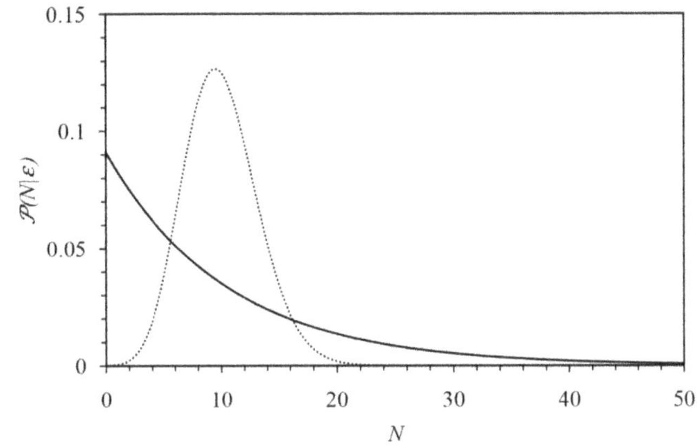

Figure 4.5. Single-particle state occupation number probability distribution for $\langle N\rangle_\varepsilon = 10$ for bosons (full curve) and for classical particles (dotted curve).

Here the state $\boldsymbol{\varepsilon} = \{\varepsilon_1, \varepsilon_2, \dots, \varepsilon_N\}$ gives the energy of each particle, and N_ε gives the number of particles with energy ε; obviously $\sum_\varepsilon N_\varepsilon = N$.

The quantity $\chi_{\boldsymbol{\varepsilon}}^{+}/N! = \prod_\varepsilon N_\varepsilon!/N!$ is the combinatorial correction that makes the sum over all states equivalent to the sum over unique states. The remaining part of the summand, $z^{N_\varepsilon}e^{-\beta N_\varepsilon \varepsilon}$ is the weight of the state ε when it is occupied by N_ε particles. Hence the normalized occupancy probability distribution for bosons is

$$\begin{aligned}\wp^{+}(N|\varepsilon) &= z^N e^{-\beta N\varepsilon}(1 - ze^{-\beta\varepsilon}) \\ &= \frac{(\langle N\rangle_\varepsilon^{+})^N}{(1 + \langle N\rangle_\varepsilon^{+})^{N+1}}.\end{aligned} \tag{4.41}$$

The ratio of successive terms, $\wp^{+}(N|\varepsilon)/\wp^{+}(N-1|\varepsilon) = \langle N\rangle_\varepsilon^{+}/(1 + \langle N\rangle_\varepsilon^{+})$, is less than unity and independent of N. This contrasts with the behavior of the occupancy probability for classical particles, wherein the ratio decreases with increasing N (see figure 4.5). The fact that the probability for classical particles decays increasingly rapidly, whereas that for bosons decays at a constant rate, might be interpreted to mean that, due to the symmetrization of the wave function, bosons have an excess attraction to already occupied states.

4.4 Ideal fermions

4.4.1 Loop grand potential, energy, number

The loop grand potential for ideal fermions, equation (4.8), is

$$-\beta\Omega^{-,(l)} = (-1)^{l-1} g z^l V \Lambda^{-3} l^{-5/2}, \quad l \geqslant 1. \tag{4.42}$$

This result was derived by an integral over the energy (i.e., momentum) continuum for the ideal gas. As such it neglects the ground state contribution, (cf section 5.2). This is significant for bosons at low temperatures, where multiple, eventually most, bosons occupy the ground state. This is the phenomenon of Bose–Einstein condensation, which is treated in detail in chapter 5 in the context of the λ-transition in liquid helium.

Since no more than one fermion can occupy the same single-particle state (two for spin-1/2 particles, if we mean energy states), for them the ground state can never become dominant. Hence neglecting the ground state contribution, as in the derivation of the above expression, is of no significance for fermions.

This expression includes g, which is a weight factor for the internal configurations that have not been counted such as spin; it is often set equal to 2 for spin-1/2 particles (Pathria 1972, section 8.1). The pressure due to loops of size l is $p^{-,(l)} = -\Omega^{-,(l)}/V$. The average number of fermions in such loops is $\bar{N}^{-,(l)} = z(\partial(-\beta\Omega^{-,(l)})/\partial z)_{T,V}$, which, as for bosons, is just l times the loop grand potential itself,

$$\bar{N}^{-,(l)} = (-1)^{l-1} g z^l V \Lambda^{-3} l^{-3/2}. \tag{4.43}$$

The loop energy for these ideal fermions is

$$\bar{E}^{-,(l)} = \left(\frac{\partial(\beta\Omega^{-,(l)})}{\partial\beta}\right)_{V,z} = \frac{3}{2\beta}(-1)^{l-1} g\Lambda^{-3} V z^l l^{-5/2}. \tag{4.44}$$

As for all versions of the ideal gas, this is proportional to the loop pressure, $\bar{E}^{-,(l)} = 2p_l^- V/3$.

Summing over all loops gives

$$\beta p^- = g\Lambda^{-3} f_{5/2}(z), \tag{4.45}$$

and

$$\bar{N}^- = g\Lambda^{-3} V f_{3/2}(z). \tag{4.46}$$

Here the Fermi–Dirac integral is (Pathria 1972, appendix E)

$$\begin{aligned} f_n(z) &= \frac{1}{\Gamma(n)} \int_0^\infty \mathrm{d}x \, \frac{x^{n-1} z e^{-x}}{1 + z e^{-x}} \\ &= \sum_{l=1}^{\infty} (-1)^{l-1} z^l l^{-n}. \end{aligned} \tag{4.47}$$

A feature of the permutation loop analysis of wave function symmetrization is that it provides a physical interpretation for the otherwise purely mathematical terms in the fugacity expansion of the Fermi–Dirac integral.

4.4.2 Heat capacity, free energy, and entropy

Now

$$\left(\frac{\partial \bar{N}^-}{\partial z}\right)_{T,V} = g\Lambda^{-3} V z^{-1} f_{1/2}(z), \tag{4.48}$$

since $f_n'(z) = z^{-1} f_{n-1}(z)$, and

$$\left(\frac{\partial \bar{N}^-}{\partial T}\right)_{z,V} = \frac{3}{2} T^{-1} g \Lambda^{-3} V f_{3/2}(z). \tag{4.49}$$

From the average loop energy found above, the average energy is

$$\bar{E}^-(z, V, T) = \sum_{l=1}^{\infty} \bar{E}^{-,(l)}(z, V, T) = \frac{3}{2\beta} g \Lambda^{-3} V f_{5/2}(z). \tag{4.50}$$

Inserting these into the general expression for the heat capacity at constant number and volume, section 5.2.3, equation (5.18), gives

$$\begin{aligned}
\frac{C_V^-}{k_B N^-}
&= \frac{1}{k_B N^-} \left(\frac{\partial \bar{E}^-}{\partial T}\right)_{N,V} \\
&= \frac{1}{k_B N^-} \left\{ \left(\frac{\partial \bar{E}^-}{\partial T}\right)_{z,V} - \left(\frac{\partial \bar{E}^-}{\partial z}\right)_{T,V} \left(\frac{\partial \bar{N}^-}{\partial T}\right)_{z,V} \left(\frac{\partial \bar{N}^-}{\partial z}\right)_{T,V}^{-1} \right\} \\
&= \frac{1}{k_B g \Lambda^{-3} V f_{3/2}(z)} \left\{ \frac{15 k_B}{4} g \Lambda^{-3} V f_{5/2}(z) \right. \\
&\quad \left. - \frac{3 k_B T}{2} g \Lambda^{-3} V z^{-1} f_{3/2}(z) \frac{(3/2) T^{-1} g \Lambda^{-3} V f_{3/2}(z)}{g \Lambda^{-3} V z^{-1} f_{1/2}(z)} \right\} \\
&= \frac{15 f_{5/2}(z)}{4 f_{3/2}(z)} - \frac{9 f_{3/2}(z)}{4 f_{1/2}(z)}.
\end{aligned} \tag{4.51}$$

The Helmholtz free energy is the thermodynamic potential of the canonical equilibrium system (specified subsystem number N, volume V, and reservoir temperature T) (Attard 2002). It can be obtained from the grand potential as (Pathria 1972)

$$F^-(N, V, T) = \Omega^-(\mu, V, T) + N\mu = N k_B T \left\{ \ln z - \frac{f_{5/2}(z)}{f_{3/2}(z)} \right\}. \tag{4.52}$$

Here the fugacity is $z = \exp \beta\mu = \bar{z}^-(N, V, T)$ (or else $N = \bar{N}^-(z, V, T)$). This uses the above results for the most likely pressure and number. The total unconstrained entropy for the canonical equilibrium system is $S_{\text{tot}}(N, V, T) = -F^-(N, V, T)/T$ (Attard 2002). Since the subsystem-dependent part of the reservoir entropy is $S_r(E_s|T) = -E_s/T$, the subsystem entropy is

$$S_s^-(N, V, T) = \frac{1}{T}\{\bar{E}^- - F^-(N, V, T)\} = N k_B T \left\{ \frac{5 f_{5/2}(z)}{2 f_{3/2}(z)} - \ln z \right\}. \tag{4.53}$$

The classical limit is either high temperature $\beta \to 0$ or low fugacity $z \to 0$. In the latter case the loop series show $f_n(z) \sim z$, $z \to 0$. This is the same as keeping just the monomer contribution. Hence in the low fugacity limit one has to leading order

$$\begin{aligned} \beta p^- &\sim g\Lambda^{-3}z, \\ \bar{N}^- &\sim g\Lambda^{-3}Vz, \\ \bar{E}^-(z, V, T) &\sim \frac{3}{2\beta}g\Lambda^{-3}Vz, \quad \text{and} \\ \frac{C_V^-}{k_B N^-} &\sim \frac{3}{2}. \end{aligned} \tag{4.54}$$

If one were to eliminate the fugacity in favor of the number density, $\rho^- = \bar{N}^-/V$, one would confirm that these are the same as for the classical ideal gas. In this limit the gas is said to be nondegenerate. Basically there are no consequences of wave function symmetrization evident because the probability of two fermions occupying the same state are so small: high temperature means many accessible states, and low density means relatively few fermions.

4.4.3 Fermi energy

In the opposite limit, $N^-\Lambda^3/gV \gg 1$, the gas of fermions is said to be degenerate. Equation (4.32) gives the average occupation number for single-particle states of energy ε; for fermions it reads $\bar{N}_\varepsilon^- = ze^{-\beta\varepsilon}/(1 + ze^{-\beta\varepsilon})$. Figure 4.6 plots this for different temperatures. The transition at the Fermi energy that was identified in figure 4.4 becomes increasingly sharp as the temperature is decreased. In fact it tends to a Heaviside step function $\bar{N}_\varepsilon^-/g \to \Theta(\mu - \varepsilon)$ at low temperatures, $\beta \to \infty$. As mentioned for the fermionic system as a whole, as the chemical potential is increased, low energy states fill first, each with a single fermion. As the temperature

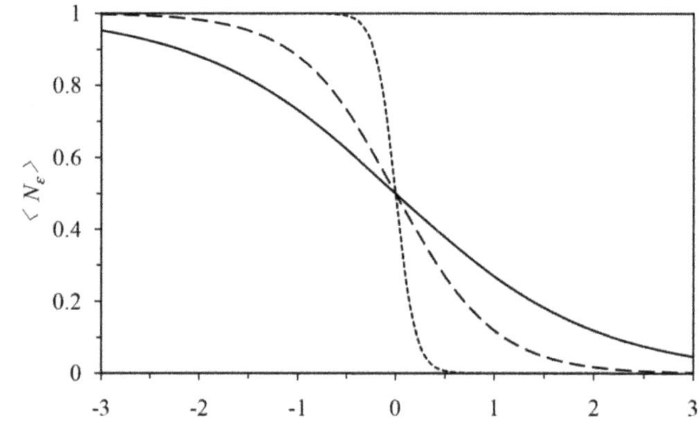

Figure 4.6. Average occupation number of single-particle energy level ε for ideal fermions for $\beta\mu_0 = 1$ (solid curve), 2 (long-dashed curve), and 10 (short-dashed curve). This sets $g = 1$.

is decreased for $\mu > \varepsilon$, low energy states also first fill. It is only near the Fermi surface, where $\varepsilon = \mu$, that the states are on average partially occupied.

The Fermi energy is the upper limit on occupied states at absolute zero. It is related to the Fermi momentum as $\varepsilon_{\rm F} = p_{\rm F}^2/2m$. This is of course non-relativistic. With $\Delta_p = 2\pi\hbar/L$ being the spacing between momentum states per particle per dimension, counting every state as occupied by g fermions gives (Pathria 1972)

$$N = \frac{4\pi g}{\Delta_p^3}\int_0^{p_{\rm F}} {\rm d}p\, p^2 = \frac{4\pi g V}{3h^3} p_{\rm F}^3. \tag{4.55}$$

Hence $p_{\rm F} = (3h^3\rho/4\pi g)^{1/3}$ and $\varepsilon_{\rm F} = (3h^3\rho/4\pi g)^{2/3}/2m$.

The ground state or zero point energy of the system is (Pathria 1972)

$$E_0 = \frac{4\pi g}{\Delta_p^3}\int_0^{p_{\rm F}} {\rm d}p\, p^2(p^2/2m) = \frac{2\pi g V}{5\, mh^3} p_{\rm F}^5. \tag{4.56}$$

From this and the preceding expression the ground state energy per particle is

$$\frac{E_0^-}{N} = \frac{3p_{\rm F}^2}{10\, m} = \frac{3}{5}\varepsilon_{\rm F}. \tag{4.57}$$

The ground state pressure is

$$p_0^- = \frac{2E_0^-}{3V} = \frac{2}{5}\rho\varepsilon_{\rm F} \propto \rho^{5/3}. \tag{4.58}$$

That the fermionic pressure is non-zero in the ground state is due to the forced filling of single-particle energy states up to the Fermi surface. It is positive because increasing the volume decreases the spacing between states, which reduces the subsystem energy and increases the reservoir entropy. The situation is very different than for bosons, which all condense into the ground energy state at absolute zero. Figure 4.6 shows that at absolute zero the state occupancy is a Heaviside step function, $N_\varepsilon^-/g = \Theta(\varepsilon_{\rm F} - \varepsilon)$. At finite temperatures, departures from this occur only when $\beta|\varepsilon_{\rm F} - \varepsilon| = \mathcal{O}(1)$, which is to say in a small number of states on either side of the Fermi energy. The fraction of fermions in on average partially occupied states is $\mathcal{O}(1/\beta\varepsilon_{\rm F})$. Only these are capable of changing state and therefore of contributing to a change in thermodynamic properties with temperature.

For the degenerate case $N^-\Lambda^3/gV \gg 1$, the Fermi–Dirac integrals can be expanded in powers of $1/\ln z$. One has (Pathria 1972)

$$\begin{aligned}
f_{5/2}(z) &= \frac{8}{15\sqrt{\pi}}(\ln z)^{5/2}\left\{1 + \frac{5\pi^2}{8}(\ln z)^{-2} + \cdots\right\}\\
f_{3/2}(z) &= \frac{4}{3\sqrt{\pi}}(\ln z)^{3/2}\left\{1 + \frac{\pi^2}{8}(\ln z)^{-2} + \cdots\right\}\\
f_{1/2}(z) &= \frac{2}{\sqrt{\pi}}(\ln z)^{1/2}\left\{1 - \frac{\pi^2}{24}(\ln z)^{-2} + \cdots\right\}.
\end{aligned} \tag{4.59}$$

These give for the average number (Pathria 1972)

$$\bar{N}^{-}=g\Lambda^{-3}V\frac{4}{3\sqrt{\pi}}(\ln z)^{3/2}\left\{1+\frac{\pi^2}{8}(\ln z)^{-2}+\cdots\right\}. \tag{4.60}$$

To leading order this gives

$$\beta^{-1}\ln z \equiv \mu = \beta^{-1}(3\rho\Lambda^3\sqrt{\pi}/4g)^{2/3} = \varepsilon_{\mathrm{F}}. \tag{4.61}$$

This agrees with the result derived above at absolute zero. To next order one has

$$\beta^{-1}\ln z \equiv \mu = \varepsilon_{\mathrm{F}}\left\{1-\frac{\pi^2}{12}(\beta\varepsilon_{\mathrm{F}})^{-2}+\cdots\right\}. \tag{4.62}$$

Similarly one can show that the average energy per fermion is (Pathria 1972)

$$\frac{\bar{E}^{-}}{\bar{N}^{-}}=\frac{3}{5}\varepsilon_{\mathrm{F}}\left\{1+\frac{5\pi^2}{12}(\beta\varepsilon_{\mathrm{F}})^{-2}+\cdots\right\}, \tag{4.63}$$

and that the pressure is

$$p^{-}=\frac{2\bar{E}^{-}}{3V}=\frac{2}{5}\rho\varepsilon_{\mathrm{F}}\left\{1+\frac{5\pi^2}{12}(\beta\varepsilon_{\mathrm{F}})^{-2}+\cdots\right\}. \tag{4.64}$$

The specific heat is just the temperature derivative of the average energy

$$\frac{C_V^{-}}{\bar{N}^{-}k_{\mathrm{B}}}=\frac{\pi^2}{2\beta\varepsilon_{\mathrm{F}}}+\cdots \tag{4.65}$$

This vanishes linearly with temperature at absolute zero.

The heat capacity for this system of ideal fermions is shown in figure 4.7. At high temperatures it tends to the classical value, $C_{\mathrm{V}}^{\mathrm{id,cl}} = 3Nk_{\mathrm{B}}/2$. It vanishes linearly with

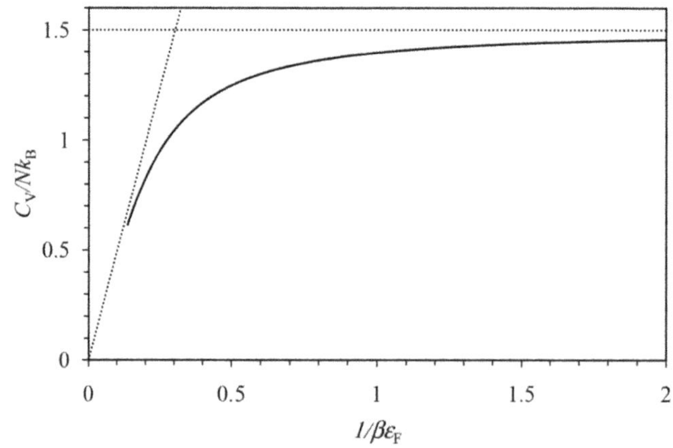

Figure 4.7. The specific heat capacity for ideal fermions as a function of temperature ($g = 2$). The dotted lines are the asymptotes.

temperature approaching absolute zero, $C_V^{-,\mathrm{id}} \sim \pi^2 N k_B / 2\beta\varepsilon_F$. It is worth mentioning that the results in the figure were obtained by the numerical evaluation of the Fermi–Dirac integrals. The utility of the loop expansion is limited to $z < 1$ ($\beta\varepsilon_F \lesssim 1$, $\rho\Lambda^3 \lesssim 1.5$). In contrast to bosons, the fugacity for fermions becomes large without limit approaching absolute zero, in which case the loop expansion is of no direct use.

References

Attard P 2002 *Thermodynamics and Statistical Mechanics: Equilibrium by Entropy Maximisation* (London: Academic)

Merzbacher E 1970 *Quantum Mechanics* 2nd edn (New York: Wiley)

Messiah A 1961 *Quantum Mechanics* vols 1 and 2 (Amsterdam: North-Holland)

Pathria R K 1972 *Statistical Mechanics* (Oxford: Pergamon)

Chapter 5

The λ-transition and superfluidity in liquid helium

The λ-transition in helium-4, which is an example of Bose–Einstein condensation, is modeled by an ideal gas, sections 5.2 and 5.3, and by interacting particles, sections 5.4 and 5.5. Monte Carlo simulations of Lennard–Jones bosons show that position permutation loops cause the spike in heat capacity on the high temperature side of the transition. The non-local ground momentum state occupancy that macroscopically dominates on the low temperature side gives rise to superfluidity. The predicted critical velocity for superfluid flow in a pore is in reasonable agreement with measured values.

5.1 Introduction

This chapter analyzes the λ-transition in liquid helium, one of the more striking of the early quantum phenomenon (Keesom *et al* 1927, 1932). Signified by a sharp spike in the heat capacity at 2.17 K, it is associated with the onset of superfluidity. Keesom *et al* (1927) suggested that the second order nature of the λ-transition showed it to be a liquid–liquid phase transition, from helium I at high temperatures to helium II at low. London (1938) argued that it was an example of Bose–Einstein condensation, which general low temperature phenomenon marks the transition from majority energy excited state occupancy to majority energy ground state occupancy (Landsberg 1954). Tisza (1938) proposed that superconductivity below the λ-transition could be accounted for by the coexistence of two fluids, one comprising bosons in the energy excited states with normal viscosity, and the other comprising bosons in the energy ground state with zero viscosity.

Balibar (2014) has reviewed the history of low-temperature studies of liquid helium and the λ-transition. In particular, he discusses the ideal gas model that was originally used to relate the λ-transition to Bose–Einstein condensation by London (1938), and the two-fluid model of superconductivity proposed by Tisza (1938). Balibar (2014) quotes Landau (1941) in criticizing this model: [F London and]

doi:10.1088/978-0-7503-4055-7ch5

'L Tisza suggested that helium II should be considered as a degenerate ideal Bose gas... This point of view, however, cannot be considered as satisfactory... nothing would prevent atoms in a normal state from colliding with excited atoms, i.e., when moving through the liquid they would experience a friction and there would be no superfluidity at all. In this way the explanation advanced by Tisza not only has no foundation in his suggestions but is in direct contradiction to them'.

Section 5.2 gives the conventional London (1938) ideal gas model of Bose–Einstein condensation and the λ-transition. This treats the ground state separately and explicitly, and the excited states via an integral over the continuum. A feature of the present treatment is that it invokes a series of permutation loops, which provides a new perspective on the λ-transition and prepares the ground for the later analysis. That the permutation loop treatment ultimately gives the same final results as can be found in the usual textbooks should provide comfort in the approach.

In section 5.3, the ideal gas model is revisited using exact enumeration of the momentum states. This broadly confirms the explicit ground state plus continuum excited state approximation, but gives a more nuanced picture of Bose–Einstein condensation. The change in energy state occupancy is continuous across the λ-transition, with about 10% of the bosons being in the ground state at the transition. Also, it is found that it is the growth and decay of permutation loops that causes the spike in the heat capacity. Section 5.4 presents Monte Carlo simulations of a Lennard–Jones model of liquid helium. Wave function symmetrization effects are included up to the heptamer loop level. Along the Lennard–Jones saturation density line, which is significantly higher than for actual helium, the heat capacity is observed to sharply increase with decreasing temperature, apparently diverging at about 4–5 K. This resembles the λ-transition in liquid helium, albeit at a somewhat higher temperature. In the simulations the symmetrization loop contributions to the total heat capacity are dominant leading up to the transition.

In section 5.5, the role of the interaction potential is analyzed on the low-temperature side of the λ-transition. A certain factorization of the partition function together with the relatively negligible contribution from the potential energy explain the success of the ideal gas model for the heat capacity in this regime. The analysis also shows that momentum correlations *increase* the subsystem permutation loop entropy.

The conclusion, section 5.6, builds on the evidence gleaned in the preceding four sections to offer a physical explanation of the λ-transition and Bose–Einstein condensation more generally. It suggests that it is the non-local nature of momentum loops that is responsible for superfluidity. It also says that the loop entropy drives the stochastic collisional dynamics to enhance momentum correlations and to preserve the loops. This final point answers Landau's (1941) above quoted criticism of the London's (1938) ideal gas treatment of Bose–Einstein condensation, and Tisza's (1938) two-fluid superconductivity model.

5.2 Ideal gas approach to the λ-transition

This section presents the conventional ideal gas description of the λ-transition in the heat capacity of liquid He^4. It begins with the analysis of the quantum ideal gas from section 4.1, which is based on an integral over the momentum or energy continuum. Pathria (1972) argues that such an integral neglects the contributions from the ground state (because in three-dimensional space the volume element vanishes at zero energy), and that therefore one has to add explicitly the quantum ground state contribution to the continuum energy integral. The initial presentation follows Pathria (1972, section 7.1), except for the emphasis on the permutation loop contributions.

5.2.1 Loop forms of the grand potential and number

In order to show explicitly the essential assumptions that underpin the conventional approach, the analysis begins with the expression derived in this book for the loop grand potential for the ideal gas, equation (4.6), which transforms the sum over momentum eigenstates into an integral over momenta, $\sum_{\mathbf{n}^l} \Rightarrow (V^l/(2\pi\hbar)^{3l}) \int \mathrm{d}\mathbf{p}^l$.

We depart from that expression to follow conventional treatments (Pathria 1972, section 7.1): because in three dimensions $\mathrm{d}\mathbf{p}_j = 4\pi p_j^2\, \mathrm{d}p_j$, this gives zero weight to the ground state of zero momentum, $\mathbf{p}_j = \mathbf{0}$. Hence the transformation to the continuum represents the contribution from the excited states only, and the discrete ground state contribution has to be added explicitly. Therefore, beginning with equation (4.6) for bosons, and retaining the ground state explicitly while transforming the excited states to the continuum, the loop grand potential becomes

$$
\begin{aligned}
-\beta\Omega^{+,(l)} &= \frac{z^l}{l}\sum_{\mathbf{n}^l=\mathbf{0}}^{\infty} e^{-\beta\mathcal{H}^{(l)}_{\mathbf{n}^l}}\chi^{+,(l)}_{\mathbf{n}^l} \\
&= \frac{z^l}{l}e^{-\beta\mathcal{H}^{(l)}_0}\chi^{+,(l)}_0 + \frac{z^l}{l}\sum_{\mathbf{n}^l>0} e^{-\beta\mathcal{H}^{(l)}_{\mathbf{n}^l}}\chi^{+,(l)}_{\mathbf{n}^l} \\
&= \frac{z^l}{l} + \frac{z^l}{l}\frac{V^l}{(2\pi\hbar)^{3l}}\int \mathrm{d}\mathbf{p}^l\, e^{-\beta\mathcal{H}_{\mathbf{p}}}\chi^{+,(l)}_{\mathbf{p}}, \quad l \geqslant 1.
\end{aligned}
\tag{5.1}
$$

The explicit ground state contribution, the first term z^l/l, follows from the facts that the ground state energy vanishes, $\mathcal{H}^{(l)}_0 = 0$, and the symmetrization factor for the ground state is unity, $\chi^{+,(l)}_0 = 1$, equation (4.5). The integral may be extended to the origin because the integrand goes to zero, and so, it is argued, the ground state is not counted twice. This is equation (4.6) to which has been added the explicit contribution from the ground state. Hence the loop grand potential for bosons, equation (4.8), with the added ground state is

$$
-\beta\Omega^{+,(l)} = z^l l^{-1} + \Lambda^{-3}Vz^l l^{-5/2}, \quad l \geqslant 1, \tag{5.2}
$$

where the thermal wavelength is

$$
\Lambda \equiv \sqrt{\beta h^2/2\pi m}\,. \tag{5.3}
$$

Here and below $z = e^{\beta\mu}$ is the fugacity, where μ is the chemical potential and $\beta = 1/k_{\mathrm{B}}T$ is the inverse temperature. In the context of the following discussion, small fugacity means $z \to 0$, and large fugacity means $z \to 1^-$.

5.2.1.1 Digression: single-particle energy states

This expression for the boson loop grand potential including the ground state contribution was derived from the continuum results for the quantum ideal gas given in section 4.1. Alternatively it is worthwhile to derive the result from the expression for the loop grand potential as a sum over energy states for a general single-particle state system. Equation (3.58), repeated as equation (4.28), may be written for bosons with the ground state explicit

$$\begin{aligned} -\beta\Omega^{+,(l)} &= \frac{z^l}{l}\sum_{n=0}^{\infty} d_n e^{-\beta l\varepsilon_n} \\ &= \frac{z^l}{l}e^{-\beta l\varepsilon_0} + \frac{z^l}{l}\sum_{n>0}^{\infty} d_n e^{-\beta l\varepsilon_n} \\ &= \frac{z^l}{l} + \frac{z^l}{l}\frac{2\pi(2m)^{3/2}V}{h^3}\int_0^{\infty} \mathrm{d}\varepsilon\ \varepsilon^{1/2}e^{-\beta l\varepsilon} \\ &= z^l l^{-1} + \Lambda^{-3}Vz^l l^{-5/2}, \quad l \geqslant 1. \end{aligned} \tag{5.4}$$

The ε_n that appears here is the energy of a single-particle state, and d_n is the number of degenerate states. (The degeneracy varies from 1 to 48 in an erratic fashion.) For the ideal gas, the energy of a particle with momentum $\mathbf{p}$ is $\varepsilon = p^2/2m$, where m is its mass. The spacing of momentum states is $\Delta_p = 2\pi\hbar/L$ per particle per dimension, where L is the edge length of the subsystem (Messiah 1961). In three dimensions the subsystem volume is $V = L^3$. Hence the volume element of energy space per particle is $\mathrm{d}\mathbf{p} = 4\pi p^2\ \mathrm{d}p = 4\pi m\sqrt{2\,m\varepsilon}\ \mathrm{d}\varepsilon$. The third equality uses the fact that the ground state energy for the ideal gas is zero, $\varepsilon_0 = 0$. The quadrature for the final equality is given by Gradshteyn and Ryzhik (1980, equation (3.371)). One can see that this is the same result as was derived above from the continuum ideal gas analysis.

5.2.1.2 Number of bosons

The average number of bosons contributed by an l-loop is the fugacity derivative of the loop grand potential, $\bar{N}^{+,(l)} = z(\partial(-\beta\Omega^{+,(l)})/\partial z)_{T,V}$. This is evidently just l times the loop grand potential itself,

$$\bar{N}^{+,(l)} = z^l + \Lambda^{-3}Vz^l l^{-3/2}, \quad l \geqslant 1. \tag{5.5}$$

One can identify z^l as the average number of bosons contributed by an l-loop to the ground state. This is actually lz^l/l, where the l in the numerator is the number of bosons in the loop, and that in the denominator is the correction for over-counting, namely the number of cyclic permutations of the particles that leave the loop unchanged. The average number of bosons in the ground state in total is

$$\bar{N}_0^+ = \sum_{l=1}^{\infty} z^l = \frac{z}{1-z}. \tag{5.6}$$

The number in the ground state evidently diverges as $z \to 1^-$. One can invert this to express the fugacity as $z = \bar{N}_0^+/(1 + \bar{N}_0^+)$, which is strictly less than one.

The average total number of bosons is the sum of that in the ground state and that in the excited states, $\bar{N}^+ = \bar{N}_0^+ + \bar{N}_*^+$. For the average number of bosons in a loop of given size, the ratio of those in the excited states to those in the ground state is independent of the fugacity and decreases with increasing size, $\bar{N}_*^{+,(l)}/\bar{N}_0^{+,(l)} \sim l^{-3/2} \to 0,\ l \to \infty$. As $z \to 1^-$, larger loops contribute increasingly to the total number, and hence the total proportion of contributions from the ground state also increases, $\bar{N}_0^+/\bar{N}^+ \to 1,\ z \to 1^-$. One can conclude that as the fugacity approaches one from below, large symmetrization loops become filled in the ground state but not in the excited states. This is said to be the phenomenon of Bose–Einstein condensation.

It has just been shown that for particle number the ground state contribution becomes dominant as the fugacity approaches unity. In contrast for the grand potential the ground state contribution may be neglected, which can be seen from the total,

$$\begin{aligned} -\beta\Omega^+ &= -\beta \sum_{l=1}^{\infty} \Omega^{+,(l)} \\ &= -\ln(1 - z) + \Lambda^{-3} V \sum_{l=1}^{\infty} z^l l^{-5/2}. \end{aligned} \tag{5.7}$$

The excited state contribution here is extensive, being proportional to volume V. It is obvious that for small values of the fugacity the excited states dominate this. In the opposite limit, $z \to 1 - 1/\bar{N}_0^+$, the ground state contribution grows as the logarithm of system size, $-\beta\Omega_0^+ \sim \ln \bar{N}$, which again is relatively negligible. Hence for all values of the fugacity only the excited state contribution to the boson grand potential needs to be retained (Pathria 1972, section 7.1).

Perhaps the difference in the two thermodynamic properties can be understood by noting that the contribution with loop size falls off faster for the grand potential than for number, and hence large loops are relatively unimportant for the former. But as the fugacity approaches unity, large loops increasingly contribute, and they occupy the ground state. This explains why the ground state is relatively negligible for the grand potential but not for number.

The loop energy for these ideal bosons is

$$\bar{E}^{+,(l)} = \left(\frac{\partial(\beta\Omega^{+,(l)})}{\partial\beta} \right)_{V,z} = \frac{3}{2\beta} \Lambda^{-3} V z^l l^{-5/2}. \tag{5.8}$$

Obviously the ground state does not contribute to this. Since the total pressure is related to the total grand potential by $\beta\Omega^+ = -\beta p^+ V$, we may identify the contribution to the pressure from each permutation loop as $p^{+,(l)} = -\Omega^{+,(l)}/V$. Hence one sees that the pressure is proportional to the energy,

$$p^{+,(l)} = \frac{2E^{+,(l)}}{3V}, \text{ and } p^{+} = \frac{2E^{+}}{3V}. \tag{5.9}$$

This is the usual result for the ideal gas.

5.2.2 Total grand potential and number

We now work with the total grand potential and number, which are the sum of the loop contributions. To this end it is useful to define the Bose–Einstein integral (Pathria 1972, appendix D)

$$\begin{aligned} g_n(z) &= \frac{1}{\Gamma(n)} \int_0^{\infty} \mathrm{d}x \; x^{n-1} \frac{ze^{-x}}{1 - ze^{-x}} \\ &= \sum_{l=1}^{\infty} z^l l^{-n}. \end{aligned} \tag{5.10}$$

This is related to the Riemann zeta-function, $\zeta(n) = g_n(1)$. The symmetrization loop analysis provides a physical interpretation for the hitherto purely mathematical fugacity expansion of the Bose–Einstein integral. With this one sees that the total grand potential, neglecting the ground state contribution, is

$$-\beta\Omega^{+} = -\sum_{l=1}^{\infty} \beta\Omega^{+,(l)} = \Lambda^{-3} V g_{5/2}(z), \tag{5.11}$$

and, keeping the ground state, the total number is

$$\bar{N}^{+} = \sum_{l=1}^{\infty} \bar{N}^{+,(l)} = \bar{N}_0^{+} + \bar{N}_*^{+} = \bar{N}_0^{+} + \Lambda^{-3} V g_{3/2}(z). \tag{5.12}$$

In the limit $z \to 0$, where $g_{3/2}(z) \sim z$, then $\bar{N}_0^{+} \sim z$ and $\bar{N}_*^{+} \sim \Lambda^{-3} V z$. The second is a factor of volume larger than the first, and so for a macroscopic system one concludes that for small values of the fugacity most bosons are in excited states, and that $\bar{N}^{+} \approx \bar{N}_*^{+} \sim \Lambda^{-3} V z$, which is the classical result.

The Bose–Einstein integral $g_{3/2}(z)$ increases monotonically with z and has largest value at the terminus of its domain, $g_{3/2}(1) = \zeta(3/2) = 2.612$. It follows that the total number of excited bosons at a given temperature is bounded, $\bar{N}_*^{+}(z, V, T) \leqslant \Lambda(T)^{-3} V \zeta(3/2)$.

For large values of the fugacity, $z \lesssim 1$, the Bose–Einstein integral is a relatively slowly varying function of z, $g_{3/2}(z) \approx g_{3/2}(1) = \zeta(3/2)$. Hence as the fugacity is increased the actual number of bosons in the excited states is close to the upper bound at the specified temperature.

When the total number of bosons exceeds the maximum number that can be accommodated in the excited states at the specified temperature, $N^{+} > \Lambda^{-3} V \zeta(3/2)$, then the excess must go into the ground state. This is Bose–Einstein condensation. Rewriting this criterion for the temperature one can say that condensation occurs when

$$T < T_c \equiv \frac{h^2}{2\pi m k_B}\left(\frac{N^+}{V\zeta(3/2)}\right)^{2/3}. \tag{5.13}$$

The thermal wavelength at this condensation temperature is about 40% larger than the average spacing of neighbors in the liquid.

For low temperatures, $T \leqslant T_c$, the system can be regarded as a mixture of two phases:

- the gaseous phase, which consists of $\bar{N}_*^+ = (T/T_c)^{3/2}N$ bosons in excited states
- the condensed phase, which consists of $\bar{N}_0^+ = N^+ - \bar{N}_*^+$ bosons in the ground state.

For $T > T_c$, there is only a gaseous phase of bosons in excited states $\bar{N}_*^+ = N^+$ and $\bar{N}_0^+ = 0$. This corresponds to Keesom's (1927) helium I. Below the λ-transition these two coexisting phases that emerge from London's (1938) ideal gas model of Bose–Einstein condensation correspond to Tisza's (1938) two-fluid model of superconductivity, with bosons in the ground state forming the zero-viscosity liquid. Using different models on either side of the transition makes it appear sharper than would otherwise be the case. Figure 5.1 plots the fraction of the system in the ground and excited states according to this model.

The boson number density is $\rho(z, V, T) = \bar{N}^+(z, V, T)/V$. For temperatures greater than the condensation temperature, $T > T_c$, $\rho(z, V, T) = \Lambda(T)^{-3}g_{3/2}(z)$. In the high temperature limit this is $\rho(z, V, T) = \Lambda(T)^{-3}z$, $T \to \infty$, which is the classical result. For $T < T_c$, the ground state ground state plus continuum excited states ideal gas picture gives $\rho(z, V, T) = z/(1-z)V + \Lambda(T)^{-3}\zeta(3/2)$. One expects that for fixed N, $z \sim 1 - \mathcal{O}(V^{-1})$.

The pressure is proportional to the grand potential,

$$\beta p^+(z, V, T) = \frac{-\beta\Omega^+(z, V, T)}{V} = \Lambda(T)^{-3}g_{5/2}(z), \tag{5.14}$$

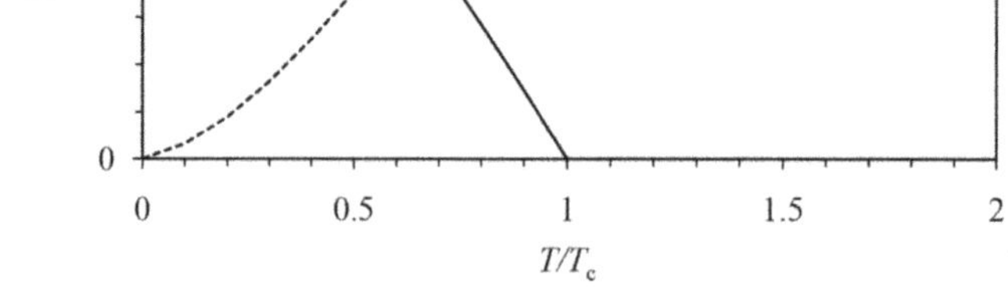

Figure 5.1. Ideal gas model of Bose–Einstein condensation. The graph shows the fraction of ideal bosons in the ground state (full curve) and in excited states (dashed curve) as a function of temperature.

as also derived above, equation (5.9). The present result derives from classical thermodynamics and is based on the facts that the pressure is the volume derivative of the grand potential, and that the grand potential is proportional to the volume. In the present case we have shown above that the ground state contribution to the grand potential, which is not proportional to the volume, can be neglected. The excited state contribution is proportional to the volume, and so the above result holds in the present case. One sees that the pressure depends upon only two intensive variables, even in the condensed regime.

It was shown in earlier chapters that wave function symmetrization created an effective attraction between pairs of bosons (cf section 2.3.1 and also equation (3.41)). Hence compared to the classical ideal gas, the ideal boson fluid should have a lower pressure at the same temperature and density. This can be seen to be the case in figure 4.1, which, accepting that it neglects the ground state contribution to the density, is only valid in the non-condensed regime.

5.2.3 Heat capacity

Now for the heat capacity at constant number and volume. One has at constant volume $\mathrm{d}\bar{N} = (\partial \bar{N}/\partial z)_{T,V}\,\mathrm{d}z + (\partial \bar{N}/\partial T)_{z,V}\,\mathrm{d}T$. Hence $(\partial z/\partial T)_{N,V} = -(\partial \bar{N}/\partial T)_{z,V} /(\partial \bar{N}/\partial z)_{T,V}$. Now

$$\left(\frac{\partial \bar{N}^+}{\partial z}\right)_{T,V} = \frac{1}{(1-z)^2} + \Lambda^{-3} V z^{-1} g_{1/2}(z), \tag{5.15}$$

since $g_n'(z) = z^{-1} g_{n-1}(z)$, and

$$\left(\frac{\partial \bar{N}^+}{\partial T}\right)_{z,V} = \frac{3}{2} T^{-1} \Lambda^{-3} V g_{3/2}(z). \tag{5.16}$$

From the average loop energy found above, the average energy is

$$\bar{E}^+(z, V, T) = \sum_{l=1}^{\infty} \bar{E}^{+,(l)}(z, V, T) = \frac{3}{2\beta} \Lambda^{-3} V g_{5/2}(z). \tag{5.17}$$

Putting these together the heat capacity per particle at constant number and volume is

$$\begin{aligned} \frac{C_V^+}{k_B N^+} = \frac{1}{k_B N^+}\left(\frac{\partial \bar{E}^+}{\partial T}\right)_{N,V} &= \frac{1}{k_B N^+}\left\{\left(\frac{\partial \bar{E}^+}{\partial T}\right)_{z,V} - \left(\frac{\partial \bar{E}^+}{\partial z}\right)_{T,V}\left(\frac{\partial \bar{N}^+}{\partial T}\right)_{z,V}\left(\frac{\partial \bar{N}^+}{\partial z}\right)_{T,V}^{-1}\right\} \\ &= \frac{1}{k_B N^+}\left\{\frac{15 k_B}{4}\Lambda^{-3} V g_{5/2}(z) - \right. \\ &\quad \left. \frac{3 k_B T}{2}\Lambda^{-3} V z^{-1} g_{3/2}(z) \frac{(3/2) T^{-1} \Lambda^{-3} V g_{3/2}(z)}{(1-z)^{-2} + \Lambda^{-3} V z^{-1} g_{1/2}(z)}\right\}. \end{aligned} \tag{5.18}$$

We now evaluate this for the ideal gas model above and below the transition temperature.

For $T \leqslant T_c$, $\bar{N}_*^+ = (T/T_c)^{3/2} N^+$ and $\bar{N}_0^+ = N^+(1 - (T/T_c)^{3/2})$. In this regime we can take $g_n(z) = g_n(1) = \zeta(n)$, and $(\partial \bar{N}^+/\partial z)_{T,V} = (1 - z)^{-2}$. This gives

$$\begin{aligned} \frac{C_V^+}{k_B N^+} &= \frac{1}{k_B N^+}\left\{\frac{15k_B}{4}\Lambda^{-3}V\zeta(5/2) - \frac{9k_B}{4}\Lambda^{-6}V^2\zeta(3/2)^2 z^{-1}(1 - z)^2\right\} \\ &\approx \frac{15}{4\rho\Lambda^3}\zeta(5/2), \quad T \leqslant T_c. \end{aligned} \tag{5.19}$$

The second term in the first equality is proportional to the fugacity times the square of the ratio of the number of bosons in excited states to the number in the ground state, which may be taken to be negligible for temperature less than the condensation temperature. (The neglect of this term is equivalent to neglecting the added explicit ground state contribution and retaining only the continuum energy integral. The term that remains says that only bosons in excited states contribute to the heat capacity.) Since in the ideal gas model of Bose–Einstein condensation $\rho\Lambda(T)^3 = (T/T_c)^{-3/2}\zeta(3/2)$, this says that at low temperatures the heat capacity is proportional to $T^{3/2}$.

At the condensation temperature, where $\rho\Lambda(T_c)^3 = \zeta(3/2)$ this is

$$\frac{C_V^+(T_c)}{k_B N^+} = \frac{15\zeta(5/2)}{4\zeta(3/2)} = 1.925. \tag{5.20}$$

For $T > T_c$, we can neglect the ground state contributions and take $N^+ = \bar{N}_*^+ = \Lambda^{-3}Vg_{3/2}(z)$, or $\rho\Lambda^3 = g_{3/2}(z)$, which implicitly gives $z(\rho, T)$. This gives

$$\begin{aligned} \frac{C_V^+}{k_B N^+} &= \frac{1}{k_B\Lambda^{-3}Vg_{3/2}(z)}\left\{\frac{15k_B}{4}\Lambda^{-3}Vg_{5/2}(z) - \frac{9k_B}{4}\frac{\Lambda^{-3}Vg_{3/2}(z)^2}{g_{1/2}(z)}\right\} \\ &= \frac{15g_{5/2}(z)}{4g_{3/2}(z)} - \frac{9g_{3/2}(z)}{4g_{1/2}(z)}, \quad T > T_c. \end{aligned} \tag{5.21}$$

As $z \to 0$, this gives $C_V^+/k_B N^+ \to (15/4) - (9/4) = 3/2$, which is the classical result.

As $z \to 1$, $g_{1/2}(z) \to \infty$, and the second term in can be neglected. The formula then gives $C_V^+/k_B N^+ = 15\zeta(5/2)/4\zeta(3/2)$, which equals the result found above at the condensation temperature. Hence the model predicts that the heat capacity is continuous at the condensation temperature. However it does give a second order discontinuity at the transition point.

For plotting purposes use the facts that $\rho\Lambda(T)^3 = g_{3/2}(z)$, $T \geqslant T_c$, which implicitly gives $z(\rho, T)$, $\rho\Lambda(T)^3/\rho\Lambda(T_c)^3 = (T_c/T)^{3/2}$ at constant density, and $\rho\Lambda(T_c)^3 = \zeta(3/2)$. Hence

$$\frac{C_V^+}{k_B N^+} = \frac{15}{4\rho\Lambda^3}\zeta(5/2) = \frac{15\zeta(5/2)}{4\zeta(3/2)}(T/T_c)^{3/2}, \quad T \leqslant T_c, \tag{5.22}$$

and

$$\frac{C_V^+}{k_B N^+} = \frac{15 g_{5/2}(z)}{4 g_{3/2}(z)} - \frac{9 g_{3/2}(z)}{4 g_{1/2}(z)}, \qquad T > T_c, \tag{5.23}$$

with $T(z)/T_c = [\zeta(3/2)/g_{3/2}(z)]^{2/3}$.

These are plotted in figure 5.2, where the second order discontinuity at $T = T_c$ is evident. It can be seen that the heat capacity asymptotes to the classical ideal gas result at high temperatures. Also it falls to zero as $T \to 0$, where almost all the bosons are in the ground state of zero energy, and those that are excited by increasing the temperature from absolute zero are a negligible fraction of the total. The famous λ shape of the curve is reasonably clear in the figure. The similarity in shape to the measured heat capacity of He^4 allows the laboratory data to be interpreted as signifying ground state condensation, which is ultimately due to the fully symmetrized wave function of bosons. Undoubtedly particle interactions play some role in the quantitative values of the heat capacity. For example, the transition in He^4 is measured at 2.17 K, and the ideal gas model gives $T_c = 3.13$ K (Pathria 1972, section 7.1). Also the peak value of the predicted heat capacity is a factor of three or more smaller than the measured values.

5.2.4 Critique of the ideal gas model

In many ways the qualitative and semi-quantitative success of the ideal gas model in describing the λ-transition in liquid helium is surprising. In the first place, it is unusual for an ideal gas model to have any quantitative accuracy for a liquid. After all the reason that a liquid condenses from a gas is because of the attractions between the particle, which interactions are neglected entirely in the ideal gas model.

Second, the assumptions that the ground state is unoccupied on the high temperature side of the transition, and that the continuum integral refers solely but entirely to the excited states, seems to be an extreme simplification that is responsible at least in part for the second order discontinuity and spike in the heat

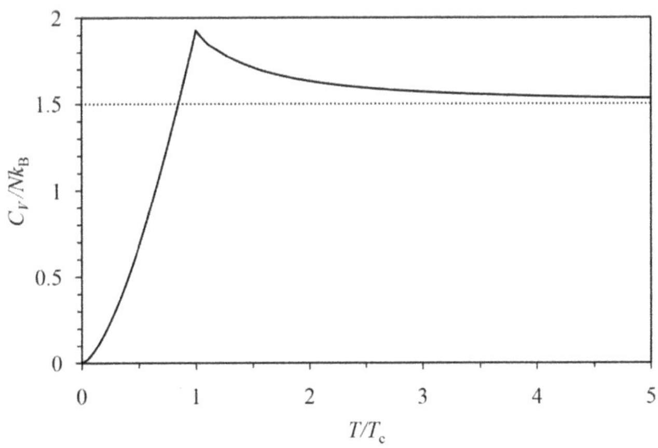

Figure 5.2. Specific heat capacity of ideal bosons as a function of temperature. The dotted line is the classical result.

capacity. But a more continuous transition would leave unexplained the molecular origin of the measured second order discontinuity in the heat capacity.

Third, it would be useful to have an explicit derivation showing that the sum over quantum states can be rigorously expressed as the ground state alone plus the integral over the continuum. The current argument is physically plausible but lacks mathematical rigor.

Given the success of the model, one could perhaps conclude that the inter-particle potential gives negligible contribution to the heat capacity at low temperatures (which conclusion is supported by the simulation data reported below). Further, it seems that the contributions from boson statistics dominate the λ-transition, and these can be obtained reliably with the ideal gas model. How to include interactions to correct the relatively small discrepancies in the transition temperature and in the value of the heat capacity is not so obvious.

5.3 Ideal gas: exact enumeration

One can further explore the ideal gas model of Bose–Einstein condensation and the λ-transition in liquid helium by the exact enumeration of the quantum states. This avoids the transformation to the continuum, and the need to use different functions and assumptions on either side of the transition. By avoiding such assumptions, it should offer reliable insight into the physical state of the system at the λ-transition.

For non-interacting particles the energy states are single-particle states. For the present ideal gas the single-particle energy eigenvalues are

$$\mathcal{H}_{\mathbf{n}_1}^{(1)} = \frac{\Delta_p^2}{2m}[n_x^2 + n_y^2 + n_z^2], \quad n_\alpha = 0, \pm 1, \pm 2, \ldots, \quad \Delta_p = 2\pi\hbar/L. \tag{5.24}$$

Here the volume of the system is $V = L^3$.

From chapter 3, the loop grand potential for bosons is given by equation (3.58),

$$\begin{aligned} -\beta\Omega^{+,(l)} &= \frac{z^l}{l}\sum_{\mathbf{n}_1} e^{-l\beta\mathcal{H}_{\mathbf{n}_1}^{(1)}} \\ &= \frac{z^l}{l}\left[\sum_{n_x=-\infty}^{\infty} e^{-\pi l(\Lambda/L)^2 n_x^2}\right]^3, \quad l \geqslant 1, \end{aligned} \tag{5.25}$$

where the thermal wavelength is $\Lambda = \sqrt{h^2\beta/2\pi m}$. The total grand potential is

$$-\beta\Omega^+ = -\sum_{n_x, n_y, n_z} \ln[1 - ze^{-\pi(\Lambda/L)^2 n^2}], \quad n^2 = n_x^2 + n_y^2 + n_z^2. \tag{5.26}$$

This is exact. The nature of its derivation in chapter 3 means that loop intersections are precluded.

One sees from these two equations the regime in which the excited states are occupied. The magnitude of the exponent for the loop grand potential in the first excited state, $n^2 = 1$, is

$$\pi l \frac{\Lambda^2}{L^2} = \frac{\pi l}{N^{2/3}} (\rho \Lambda^3)^{2/3}. \tag{5.27}$$

Near the λ-transition one has $\rho\Lambda^3 = \mathcal{O}(1)$, although it can be larger than this at lower temperatures. With this value one sees that loops $l \gtrsim N^{2/3}$ must be in the ground state, but smaller loops need not be. For the grand potential itself (set $l = 1$ in this formula), one sees that the exponent is negligible for a macroscopic system, and so the occupancy of the excited states can be expected to be significant at the λ-transition. This says that it is not until the thermal wavelength is of the same order as the system size, $\Lambda/L = \mathcal{O}(1)$, that the occupancy of the excited states becomes negligible.

The number of bosons in a loop of size l is

$$\bar{N}^{+,(l)} = z \frac{\partial(-\beta\Omega^{+,(l)})}{\partial z} = z^l \left[\sum_{n_x=-\infty}^{\infty} e^{-\pi l (\Lambda^2/L^2) n_x^2} \right]^3, \quad l \geqslant 1. \tag{5.28}$$

The total number is

$$\bar{N}^{+} = \sum_{n_x, n_y, n_z} \frac{z e^{-\pi(\Lambda^2/L^2)n^2}}{1 - z e^{-\pi(\Lambda^2/L^2)n^2}}. \tag{5.29}$$

The number of bosons in a ground state loop is $\bar{N}_0^{+,(l)} = z^l$, and the total number in the ground state is $\bar{N}_0^+ = z/(1-z)$.

The average energy is

$$\bar{E}^{+} = \frac{\partial(\beta\Omega^+)}{\partial\beta} = \sum_{n_x, n_y, n_z} \frac{z e^{-\pi(\Lambda^2/L^2)n^2} \pi n^2 \Lambda^2/\beta L^2}{1 - z e^{-\pi(\Lambda^2/L^2)n^2}}. \tag{5.30}$$

The heat capacity per particle at constant number and volume is

$$\begin{aligned} \frac{C_V^+}{k_B N^+} &= \frac{1}{k_B N^+} \left(\frac{\partial \bar{E}^+}{\partial T} \right)_{N,V} \\ &= \frac{-\beta^2}{N^+} \left\{ \left(\frac{\partial \bar{E}^+}{\partial \beta} \right)_{z,V} - \left(\frac{\partial \bar{E}^+}{\partial z} \right)_{T,V} \left(\frac{\partial \bar{N}^+}{\partial \beta} \right)_{z,V} \left(\frac{\partial \bar{N}^+}{\partial z} \right)_{T,V}^{-1} \right\}. \end{aligned} \tag{5.31}$$

For this one requires

$$\left(\frac{\partial \bar{E}^+}{\partial \beta} \right)_{z,V} = \sum_{n_x, n_y, n_z} \frac{-z e^{-\pi(\Lambda^2/L^2)n^2} [\pi n^2 \Lambda^2/\beta L^2]^2}{[1 - z e^{-\pi(\Lambda^2/L^2)n^2}]^2}, \tag{5.32}$$

$$\left(\frac{\partial \bar{E}^+}{\partial z} \right)_{T,V} = \sum_{n_x, n_y, n_z} \frac{e^{-\pi(\Lambda^2/L^2)n^2} \pi n^2 \Lambda^2/\beta L^2}{[1 - z e^{-\pi(\Lambda^2/L^2)n^2}]^2}, \tag{5.33}$$

$$\left(\frac{\partial \bar{N}^+}{\partial \beta}\right)_{z,V} = \sum_{n_x,n_y,n_z} \frac{-ze^{-\pi(\Lambda^2/L^2)n^2}\pi n^2\Lambda^2/\beta L^2}{[1 - ze^{-\pi(\Lambda^2/L^2)n^2}]^2}, \tag{5.34}$$

and

$$\left(\frac{\partial \bar{N}^+}{\partial z}\right)_{T,V} = \sum_{n_x,n_y,n_z} \frac{e^{-\pi(\Lambda^2/L^2)n^2}}{[1 - ze^{-\pi(\Lambda^2/L^2)n^2}]^2}. \tag{5.35}$$

5.3.1 Numerical results

Figure 5.3 shows the specific heat capacity at constant density for ideal bosons at a function of temperature near the λ-transition. The ground state plus continuum excited states model, equations (5.19) and (5.21) (full curve) is the same as in figure 5.2. The exact enumeration results use the formula given in this section, with the fugacity for a given number of bosons, density, and thermal wavelength being obtained by bisection methods using equation (5.29). The sum over states was terminated when the magnitude of the Maxwell–Boltzmann exponent exceeded 20.

It can be seen that there is relatively good agreement between the ground state plus continuum excited states model and the exact results for finite N. Indeed it appears that in the thermodynamic limit, the present exact results converge upon the earlier results. As an example of this, the present peak occurs at: $\{N, \rho\Lambda_c^3, C_{V,c}/Nk_B\} = \{500, 2.25, 1.948\}$, $\{1000, 2.34, 1.944\}$, $\{2000, 2.40, 1.940\}$ and $\{5000, 2.45, 1.937\}$. The ground state plus continuum excited states model gives $\rho\Lambda_c^3 = 2.612$ and $C_{V,c}/Nk_B = 1.925$.

Figure 5.4 shows the average number of ideal bosons in the energy ground state (solid curve) and in the first (long-dashed curve) and second (short-dashed curve)

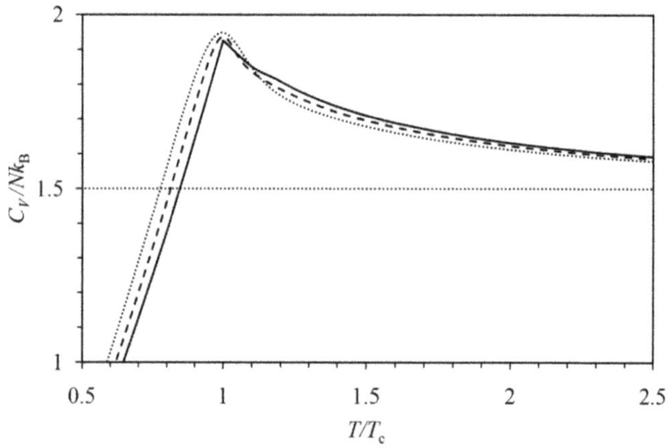

Figure 5.3. Specific heat capacity of ideal bosons as a function of temperature. The solid curve is the ground state plus continuum excited states model, section 5.2, and the remaining curves are exact results for fixed density $\rho = 0.3$ and $N = 500$ (dotted) and $N = 5000$ (short dashed). The condensation temperature T_c was defined by the location of the maximum for each curve. The dotted line is the classical result.

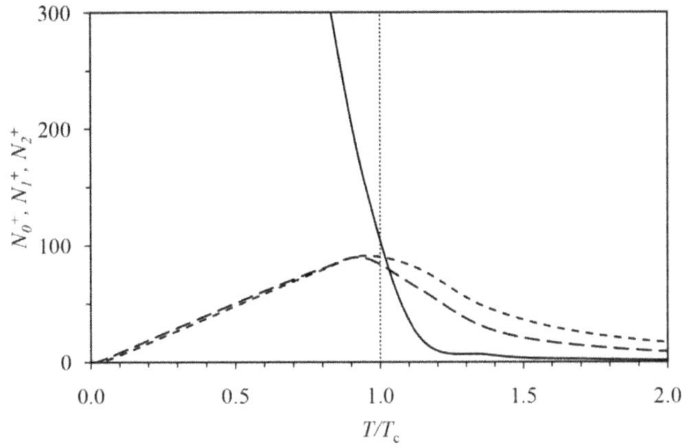

Figure 5.4. Average number of ideal bosons in the energy ground state (solid curve), in the first excited energy state (long-dashed curve), and in the second excited energy state (short-dashed curve). The total number is N^+=1000 and the density is $\rho = 0.3$, both fixed while the temperature and fugacity are varied. The dotted line is a guide to the eye.

excited energy states around the λ-transition. The number in the ground state increases monotonically with decreasing temperature. The number in the first excited energy state at high temperatures is low, but more than in the ground energy state. The number in the second excited state is higher still. Of course this is due to the degeneracy of the energy states being 6 and 12 respectively. The occupancy of a first excited momentum state is always less than that of the ground momentum (equals energy) state. Obviously most bosons are distributed amongst even higher excited energy states in this regime. The number density for the energy continuum would be peaked at a non-zero value that increased with increasing temperature.

On the low temperature side of the transition, the number of bosons in the first and second excited energy states reaches a peak before decreasing to zero at absolute zero. For the first excited energy state this peak occurs at $T/T_c = 0.957$ for $N^+ = 1000$, and 0.951 for $N^+ = 5000$. Below this peak the order of energy state occupancy is inverted from the order above the transition. Figure 5.5 shows the number of bosons as monomers and in the first three loops as a function of temperature. At high temperatures most bosons are monomers and permutation loops are practically non-existent. The number of monomers decreases monotonically with decreasing temperature. Obviously, this means that the number of bosons in permutation loops increases monotonically with decreasing temperature. The number of bosons in each of the loops shown increases with decreasing temperature and reaches a peak on the high temperature side of the transition. For dimers this peak occurs at $T/T_c = 1.44$ for $N^+ = 1000$, and 1.48 for $N^+ = 5000$, which is almost within the plotting point digitization error. The number of bosons in dimers exceeds those as trimers, which exceeds those as tetramers, for all temperatures shown. The gap between these decreases with decreasing temperature.

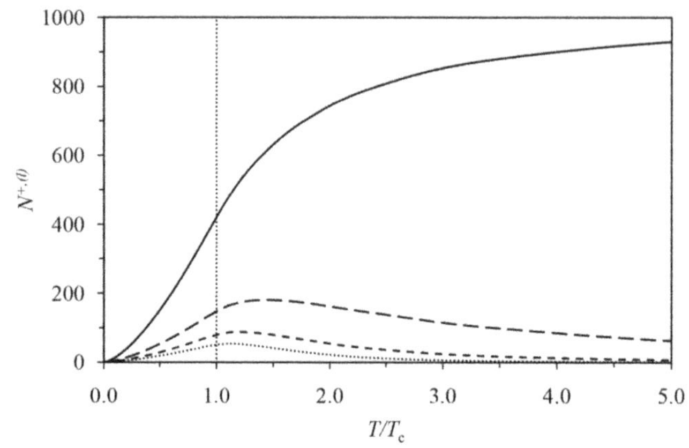

Figure 5.5. Average number of ideal bosons as monomers (solid curve), as dimers (long-dashed curve), as trimers (short-dashed curve), and as tetramers (dotted curve). The total number is $N^+ = 1000$ and the density is $\rho = 0.3$, both fixed while the temperature and fugacity are varied. The dotted line is a guide to the eye.

The decrease with decreasing temperature in the number of bosons in the first several loops on the low temperature side of the transition, together with the decrease in the number of monomers, suggests that larger loops are being populated at the expense of monomers and smaller loops (because the total number of bosons is fixed). But bosons in these larger loops must increasingly be in the ground state, because of the l in the Maxwell–Boltzmann exponent in equation (5.28). (It was pointed out in section 5.2.1 that large loops are mainly in the ground state, and that large loops increasingly dominate as $T \to 0$.) This is consistent with the increasing occupancy of the ground state shown in figure 5.4.

The total heat capacity is the sum of the heat capacity of the monomers and loops. The loop heat capacity varies with the excited state of the loop. Ground state bosons contribute nothing to the heat capacity. Hence one can see the reason why the heat capacity first increases and then decreases with decreasing temperature at the λ-transition: on the high temperature side the number of bosons in small loops is increasing, and a significant number of these bosons are in excited states. On the low temperature side the number of bosons in small loops decreases as they are cannibalized by large loops of bosons in the ground state.

5.4 The λ-transition for interacting bosons

The section develops a computer algorithm for simulating wave function symmetrization effects in general, and gives results for liquid helium near its λ-transition in particular.

The simulation algorithm is designed to produce classical equilibrium canonical averages, such as the average energy $\langle \mathcal{H} \rangle_{\text{cl}}$, as well as classical averages of quantum loop symmetrization functions, such as $\langle \overset{\circ}{\eta}{}_{\text{q}}^{\pm(l)} \rangle_{\text{cl}}$ and similar. As will be shown these can be combined to produce any desired quantity for the quantum system. The classical averages can also be called monomer averages. All of the following averages are classical and the subscript on the angular brackets is therefore dropped.

The commutation function, which is derived in chapter 7 and for which various series expansions and numerical examples are given in chapters 8–11, is set to unity in this and the following section. (For the ideal gas it is identically unity.) This is one limitation of the simulation algorithm. The reasons for neglecting (i.e., setting to unity) the commutation function are partly that it is difficult to calculate, and partly that it might indeed be unimportant (in view of the success of the ideal gas approach). In general the commutation function exponent at large separations is proportional to the gradient of the interaction potential, and it therefore goes to zero faster than the potential itself. Hence when the system is dominated by long range effects, as appears to be the case here, the commutation function can be neglected. It can also be mentioned that equation (7.23) below shows that the Maxwell–Boltzmann-weighted integral over phase space of the temperature derivative of the commutation function must vanish, which suggests that it is a reasonable approximation to set the temperature derivative throughout phase space to zero, which is equivalent to taking the commutation function to be a constant equal to its high temperature limit of unity. A final reason for setting it to unity is that the commutation function depends upon the particles' momenta, and in its absence the various momentum integrals can be performed analytically, which gives a great saving in computer time.

As will be discussed in the concluding section 5.6, another limitation of the following simulations is that they rely upon ideal solution theory: the permutation loops must be so dilute that their interaction is unlikely. The loop grand potential, whose derivatives give the loop series for the heat capacity calculated in the simulations, is derived by neglecting correlations between the permutation loops. It is not clear that these limitations are of any significance.

A third limitation is that the simulations invoke integrals over the momentum continuum. As discussed in section 5.2, these represent excited states only. Of course on the high temperature side of the transition this is the same as London's (1938) ideal gas model, which also neglects the ground state contribution to the excited state continuum, section 5.2.2.

5.4.1 Momentum integrals

5.4.1.1 Partition function and configuration integral

Consider a canonical equilibrium system with N particles, volume V, in thermal contact with a reservoir of temperature $T = 1/k_B\beta$. The classical Hamiltonian is the usual sum of kinetic and potential energy, $\mathcal{H}(\mathbf{\Gamma}) = \mathcal{K}(\mathbf{p}) + U(\mathbf{q})$, with the kinetic energy being $\mathcal{K}(\mathbf{p}) = p^2/2m = \sum_{j=1}^{N}\mathbf{p}_j \cdot \mathbf{p}_j/2m$. The classical or monomer partition function is

$$\begin{aligned} Z(N, V, T) &= \frac{1}{h^{3N}N!}\int \mathrm{d}\mathbf{p}\,\mathrm{d}\mathbf{q}\; e^{-\beta\mathcal{K}(\mathbf{p})}e^{-\beta U(\mathbf{q})} \\ &= \frac{1}{h^{3N}N!}(2\pi m/\beta)^{3N/2}\int \mathrm{d}\mathbf{q}\; e^{-\beta U(\mathbf{q})} \\ &\equiv \frac{\Lambda^{-3N}}{N!}Q(N, V, T). \end{aligned} \qquad (5.36)$$

The configuration integral is $Q \equiv \int \mathrm{d}\mathbf{q}\; e^{-\beta U(\mathbf{q})}$, and the thermal wavelength is $\Lambda = \sqrt{h^2\beta/2\pi m}$.

The classical average of the single loop symmetrization function for $l \geqslant 2$ is

$$\begin{aligned}\left\langle \overset{\circ}{\eta}{}_{\mathrm{q}}^{\pm(l)} \right\rangle &= \frac{1}{Zh^{3N}N!}\int \mathrm{d}\boldsymbol{\Gamma}\; e^{-\beta\mathcal{H}(\boldsymbol{\Gamma})}\overset{\circ}{\eta}{}_{\mathrm{q}}^{\pm(l)}(\boldsymbol{\Gamma}) \\ &= \frac{1}{Zh^{3N}N!}\frac{(\pm 1)^{l-1}N!}{(N-l)!l}\int \mathrm{d}\mathbf{p}\,\mathrm{d}\mathbf{q}\; e^{-\beta\mathcal{K}(\mathbf{p})}e^{-\beta U(\mathbf{q})} \\ &\quad\times e^{-\mathbf{q}_{12}\cdot\mathbf{p}_1/i\hbar}e^{-\mathbf{q}_{23}\cdot\mathbf{p}_2/i\hbar}\ldots e^{-\mathbf{q}_{l1}\cdot\mathbf{p}_l/i\hbar}.\end{aligned} \tag{5.37}$$

The momentum integral here is

$$\begin{aligned}&\int \mathrm{d}\mathbf{p}\; e^{-\beta\mathcal{K}(\mathbf{p})}e^{-\mathbf{q}_{12}\cdot\mathbf{p}_1/i\hbar}e^{-\mathbf{q}_{23}\cdot\mathbf{p}_2/i\hbar}\ldots e^{-\mathbf{q}_{l1}\cdot\mathbf{p}_l/i\hbar} \\ &= (2\pi m/\beta)^{3(N-l)/2}\int \mathrm{d}\mathbf{p}^l \prod_{j=1}^{l} e^{-\beta p_j^2/2m}\prod_{j=1}^{l} e^{-\mathbf{q}'_{j,j+1}\cdot\mathbf{p}_j/i\hbar} \\ &= h^{3N}\Lambda^{-3N}\prod_{j=1}^{l} e^{-\pi\Lambda^{-2}q'^2_{j,j+1}},\end{aligned} \tag{5.38}$$

where $\mathbf{q}'_{j+1} = \mathbf{q}_{j+1}$, $j = 1, 2, \ldots, l-1$ and $\mathbf{q}'_{l+1} = \mathbf{q}_1$. Hence for $l \geqslant 2$,

$$\begin{aligned}-\beta\Omega^{\pm(l)} &= \left\langle \overset{\circ}{\eta}{}_{\mathrm{q}}^{\pm(l)} \right\rangle \\ &= \frac{1}{Zh^{3N}N!}\frac{(\pm 1)^{l-1}N!}{(N-l)!l}h^{3N}\Lambda^{-3N}\int \mathrm{d}\mathbf{q}\; e^{-\beta U(\mathbf{q})}\prod_{j=1}^{l} e^{-\pi\Lambda^{-2}q'^2_{j,j+1}} \\ &= (\pm 1)^{l-1}\frac{N!}{(N-l)!l}\frac{\int \mathrm{d}\mathbf{q}\; e^{-\beta U(\mathbf{q})}\prod_{j=1}^{l} e^{-\pi\Lambda^{-2}q'^2_{j,j+1}}}{\int \mathrm{d}\mathbf{q}\; e^{-\beta U(\mathbf{q})}} \\ &= (\pm 1)^{l-1}\frac{N!}{(N-l)!l}\left\langle e^{-\pi\Lambda^{-2}q^2_{l,1}}\prod_{j=1}^{l-1} e^{-\pi\Lambda^{-2}q^2_{j,j+1}} \right\rangle \\ &= \langle G^{\pm(l)}(\mathbf{q}^l)\rangle.\end{aligned} \tag{5.39}$$

This is a classical configuration average (i.e., over positions). For an l-loop composed of $\mathbf{q}^l = \{\mathbf{q}_{j_1}, \mathbf{q}_{j_2}, \ldots, \mathbf{q}_{j_l}\}$, the sum of the square of the bond lengths is $\mathcal{L}_l(\mathbf{q}^l)^2 = q^2_{j_1,j_2} + q^2_{j_2,j_3} + \cdots + q^2_{j_l,j_1}$. With this, the loop Gaussian to be averaged is

$$\begin{aligned}G^{\pm(l)}(\mathbf{q}^l) &= (\pm 1)^{l-1}\frac{N!}{(N-l)!l}e^{-\pi\mathcal{L}_l(\{\mathbf{q}_1, \mathbf{q}_2, \ldots, \mathbf{q}_l\})^2/\Lambda^2} \\ &= (\pm 1)^{l-1}\sum_{i=1}^{N-l+1}\sum_{j_2,j_3,\ldots,j_l}^{N}{}^{(j_a>i,\, j_a\neq j_b)}\; e^{-\pi\mathcal{L}_l(\mathbf{q}^l)^2/\Lambda^2}.\end{aligned} \tag{5.40}$$

The equality of these two expressions is to be understood after the fluid averaging process. (The final expression is correct in all circumstances.) For what follows, it is important to note that the Gaussian loop factor depends upon temperature. Note that the clockwise loop $\{i, j_2, j_3, \ldots, j_l\}$ must be counted as distinct from the anti-clockwise version, $\{i, j_l, j_{l-1}, \ldots, j_2\}$ even though the two loop Gaussians have the same numerical value.

5.4.1.2 Average energy

The kinetic energy is included in the average monomer energy,

$$\bar{E}^{(1)} \equiv \langle \mathcal{H} \rangle = \frac{-\partial Z}{Z \partial \beta} = \frac{3Nk_B T}{2} + \langle U \rangle. \tag{5.41}$$

This is of course just the classical average energy.

The average energy for the loops $l \geqslant 2$ is

$$\begin{aligned} \bar{E}^{\pm(l)} &= \frac{\partial \beta \Omega^{\pm(l)}}{\partial \beta} \\ &= -\beta \Omega^{\pm(l)} \frac{1}{Q} \frac{\partial Q}{\partial \beta} + \frac{1}{Q} \int \mathrm{d}\mathbf{q}\; e^{-\beta U(\mathbf{q})} \left\{ U(\mathbf{q}) G^{\pm(l)}(\mathbf{q}^l) - \frac{\partial G^{\pm(l)}(\mathbf{q}^l)}{\partial \beta} \right\} \\ &= \left\langle \frac{-\pi \mathcal{L}_l^2}{\beta \Lambda^2} G^{\pm(l)} \right\rangle + \langle U\, G^{\pm(l)} \rangle - \langle U \rangle \langle G^{\pm(l)} \rangle. \end{aligned} \tag{5.42}$$

There is no kinetic energy contribution to these. The first term on the right hand side of the average loop energy, which is negative, might be called the ideal loop contribution to the average loop energy, because it is the only term that survives if $U = 0$. It is negative because the effective potential for bosons induced by wave function symmetrization is attractive, equation (3.42). It arises from the derivative of the loop Gaussian,

$$\begin{aligned} \frac{\partial G^{\pm(l)}(\mathbf{q}^l)}{\partial \beta} &= (\pm 1)^{l-1} \frac{N!}{(N-l)!\, l} \frac{\partial}{\partial \beta} e^{-\pi \mathcal{L}_l(\mathbf{q}^l)^2/\Lambda^2} \\ &= \frac{\pi \mathcal{L}_l(\mathbf{q}^l)^2}{\beta \Lambda^2} G^{\pm(l)}(\mathbf{q}^l). \end{aligned} \tag{5.43}$$

5.4.1.3 Heat capacity

The monomer contribution to the heat capacity is

$$C_V^{(1)} = \left(\frac{\partial \bar{E}^{(1)}}{\partial T} \right)_{N,V} = \frac{1}{k_B T^2} \langle [\mathcal{H} - \langle \mathcal{H} \rangle]^2 \rangle, \tag{5.44}$$

or

$$C_V^{(1)} = \frac{3Nk_B}{2} + \frac{1}{k_B T^2}[\langle U^2 \rangle - \langle U \rangle^2]. \tag{5.45}$$

This is the usual classical result.

The contributions to the heat capacity at constant number and volume from loops $l \geqslant 2$ is

$$\begin{aligned} C_V^{(l)} = \left(\frac{\partial \bar{E}^{\pm(l)}}{\partial T} \right)_{N,V} &= C_V^{(l),\mathrm{idx}} - \frac{1}{k_B T^2} \left\{ \frac{\partial \beta \Omega^{\pm(l)}}{\partial \beta} \langle U \rangle + \beta \Omega^{\pm(l)} \frac{\partial \bar{U}}{\partial \beta} - \frac{1}{Q} \frac{\partial Q}{\partial \beta} \langle U \; G^{\pm(l)} \rangle \right. \\ &\quad \left. - \langle U^2 \; G^{\pm(l)} \rangle + \left\langle U \; \frac{\partial G^{\pm(l)}}{\partial \beta} \right\rangle \right\} \\ &= C_V^{(l),\mathrm{idx}} - \frac{1}{k_B T^2} \left\{ \bar{E}^{\pm(l)} \langle U \rangle + \beta \Omega^{\pm(l)} [\langle U \rangle^2 - \langle U^2 \rangle] + \langle U \rangle \langle U \; G^{\pm(l)} \rangle \right. \\ &\quad \left. - \langle U^2 \; G^{\pm(l)} \rangle + \left\langle U \; \frac{\pi \mathcal{L}_l^2}{\beta \Lambda^2} G^{\pm(l)} \right\rangle \right\}. \end{aligned} \tag{5.46}$$

Gathering terms this may be written as

$$\begin{aligned} C_V^{(l)} = C_V^{(l),\mathrm{idx}} &+ \frac{1}{k_B T^2} \left\{ \left\langle \frac{\pi \mathcal{L}_l^2}{\beta \Lambda^2} \; G^{\pm(l)} \right\rangle \langle U \rangle - \left\langle U \; \frac{\pi \mathcal{L}_l^2}{\beta \Lambda^2} G^{\pm(l)} \right\rangle \right. \\ &+ \left. \left\langle [U - \langle U \rangle]^2 \left[G^{\pm(l)} - \langle G^{\pm(l)} \rangle \right] \right\rangle \right\}. \end{aligned} \tag{5.47}$$

The fluctuation correlation part explicit here need not be positive.

In the loop heat capacity, $C_V^{(l),\mathrm{idx}}$ is the ideal excess contribution. This is thus called because it comes from the ideal contribution to the loop energy. The strictly ideal part of it can be obtained from the following expression by setting $U = 0$. In full it is

$$\begin{aligned} C_V^{(l),\mathrm{idx}} &\equiv \left(\frac{\partial \left\langle \frac{-\pi \mathcal{L}_l^2}{\beta \Lambda^2} \; G^{\pm(l)} \right\rangle}{\partial T} \right)_{N,V} \\ &= \frac{-1}{k_B T^2} \left\{ \left\langle \frac{\pi \mathcal{L}_l^2}{\beta \Lambda^2} \; G^{\pm(l)} U \right\rangle - \left\langle \frac{\pi \mathcal{L}_l^2}{\beta \Lambda^2} \; G^{\pm(l)} \right\rangle \langle U \rangle - \left\langle \frac{\pi^2 \mathcal{L}_l^4}{\beta^2 \Lambda^4} \; G^{\pm(l)} \right\rangle \right. \\ &\quad \left. + \left\langle \frac{2\pi \mathcal{L}_l^2}{\beta^2 \Lambda^2} \; G^{\pm(l)} \right\rangle \right\}. \end{aligned} \tag{5.48}$$

This should be understood as an abbreviated notation. In more detail one has

$$C_V^{(l),\mathrm{idx}} \equiv \left(\frac{\partial \left\langle \frac{-\partial G^{\pm(l)}}{\partial \beta} \right\rangle}{\partial T} \right)_{N,V}$$
$$= \frac{-1}{k_B T^2} \left\{ \left\langle \frac{\partial G^{\pm(l)}}{\partial \beta} U \right\rangle - \left\langle \frac{\partial G^{\pm(l)}}{\partial \beta} \right\rangle \langle U \rangle - \left\langle \frac{\partial^2 G^{\pm(l)}}{\partial \beta^2} \right\rangle \right\}, \tag{5.49}$$

with

$$\frac{\partial^2 G^{\pm(l)}}{\partial \beta^2} = \frac{\partial}{\partial \beta} \left\{ (\pm 1)^{l-1} \frac{\pi}{\beta \Lambda^2} \sum_{i=1}^{N-l+1} \sum_{j_2, j_3, \ldots, j_l}^{N} {}^{(j_a > i,\, j_a \neq j_b)} e^{-\pi \mathcal{L}_l(\mathbf{q}^l)^2 / \Lambda^2} \mathcal{L}_l(\mathbf{q}^l)^2 \right\}$$
$$= \frac{-(\pm 1)^{l-1} 2\pi}{\beta^2 \Lambda^2} \sum_{i=1}^{N-l+1} \sum_{j_2, j_3, \ldots, j_l}^{N} {}^{(j_a > i,\, j_a \neq j_b)} e^{-\pi \mathcal{L}_l(\mathbf{q}^l)^2 / \Lambda^2} \mathcal{L}_l(\mathbf{q}^l)^2 \tag{5.50}$$
$$+ \frac{(\pm 1)^{l-1} \pi^2}{\beta^2 \Lambda^4} \sum_{i=1}^{N-l+1} \sum_{j_2, j_3, \ldots, j_l}^{N} {}^{(j_a > i,\, j_a \neq j_b)} e^{-\pi \mathcal{L}_l(\mathbf{q}^l)^2 / \Lambda^2} \mathcal{L}_l(\mathbf{q}^l)^4.$$

This second derivative of the Gaussian is the explicit non U or ideal gas contribution to the heat capacity. Of course one should not forget that its average involves the l-particle density, which gives the structure of the fluid and which depends strongly upon the potential and the temperature.

The simulation results obtained to date all show that the individual loop contributions to the heat capacity are positive. I have no analytic proof that this must always be true. Of course the total heat capacity over monomers and loops must be positive for the thermodynamic state to be stable.

5.4.2 Monte Carlo algorithm

The preceding analysis, which performs the momentum integrals analytically, casts the necessary statistical quantities in terms of classical canonical equilibrium averages, which are just Maxwell–Boltzmann weighted integrals in position space. These lend themselves directly to the usual Monte Carlo (Allen and Tildesley 1987) or stochastic molecular dynamics (Attard 2018) computer simulation procedures.

In the canonical equilibrium simulations the number N and volume V was fixed, and most of the results were obtained as a function of temperature along a path of fixed total density, $\rho_{\mathrm{tot}} = N/V$. According to the notions advanced in sections 5.2 and 5.3, this represents $N_* = N$ bosons in excited states and does not include those in the ground state.

In the results obtained here the Metropolis Monte Carlo algorithm was used. A cut-off $R_{\mathrm{cut}} = 3.5\sigma$ was used for the Lennard–Jones potential. A spatial neighbor table with cubic cells of size $\approx 0.7\sigma$ was used. This is about three times more efficient than the commonly used cells of size R_{cut}, since the small cells considerably reduce the neighborhood volume that needs to be searched, it being only about twice the size of the actual cut-off sphere. The neighbor tables mean that the computer time required for a given statistical accuracy scales sub-linearly with system size. The neighbor table also makes the evaluation of the loop functions more efficient, as is now detailed.

The novel aspect of the simulations is the evaluation of the loop Gaussians, $\langle G^{(l)}\rangle$, and related quantities such as $\langle UG^{(l)}\rangle$, $\langle \mathcal{L}_l^2 G^{(l)}\rangle$, etc. There are $N!/(N-l)!l$ distinct l-loops, and since the number of atoms in the simulation is $N \sim \mathcal{O}(10^3)$, it would be prohibitive to evaluate all of them for every configuration. Here a more efficient algorithm is described.

In order for the Gaussian to be non-zero, the loop has to be compact. Therefore one can confine the loops to be considered to a defined neighborhood $\mathcal{N}_i$ of each atom i. Let $j \in \mathcal{N}_i$ if and only if $q_{ij} \leqslant R_{\text{cut}}$ and $j > i$. (The condition $j > i$ allows distinct loops to be generated with the first atom i also having the lowest index, which means that the combinatorial factor of l in the denominator is no longer required.) It is not essential to use the same cut-off as for the potential, but it is convenient to do so because the same neighbor table can be used for both the potential and the loop calculations. With this approximation the expression for the loop Gaussian becomes

$$\begin{aligned} G^{(l)}(\mathbf{q}^N) &= \sum_{i=1}^{N-l+1} \sum_{j_2, j_3, \ldots, j_l}^{N} {}^{\text{all}>i,\ \text{all}\neq} e^{-\pi \mathcal{L}_l(\mathbf{q}_i,\, \mathbf{q}_{j_1},\, \ldots,\, \mathbf{q}_{j_l})^2/\Lambda^2} \\ &\approx \sum_{i=1}^{N-l+1} \sum_{j_2, j_3, \ldots, j_l}^{\mathcal{N}_i} {}^{\text{all}>i,\ \text{all}\neq} e^{-\pi \mathcal{L}_l(\mathbf{q}_i,\, \mathbf{q}_{j_1},\, \ldots,\, \mathbf{q}_{j_l})^2/\Lambda^2}. \end{aligned} \tag{5.51}$$

In the results reported below, the cut-off was $R_{\text{cut}} = 3.5\sigma$, and the thermal wavelength was $\Lambda \lesssim 2\sigma$, depending on the temperature. For a loop with one atom outside the neighborhood $\mathcal{L}_l^2 > 2R_{\text{cut}}^2$, the exponent is $\lesssim -2\pi \times 3.5^2/2^2 = -19$, and the Gaussian is $\lesssim 4 \times 10^{-9}$. This is a fairly conservative estimate. To the extent that this is small compared to unity, neglecting loops that extend beyond the neighborhood should not unduly underestimate the loop contribution to the various averages.

For the results reported below, the number of atoms in the defined neighborhood is $\mathcal{N}_i \sim \mathcal{O}(150)$ (for small i). The number of possible l-loops for each such atom is $\mathcal{N}_i!/(\mathcal{N}_i - l)!$, which is still much too large to evaluate exhaustively except for $l = 2$ and possibly $l = 3$. Instead a random sample $M_i < \mathcal{N}_i$ of the possible loops was evaluated, and the loop Gaussian was estimated as

$$G^{(l)}(\mathbf{q}^N) = \sum_{i=1}^{N-l+1} \frac{\mathcal{N}_i!}{(\mathcal{N}_i - l)!} \frac{1}{M_i} \sum_{a=1}^{M_i} {}^{(\mathbf{q}_a^{l-1} \in \mathcal{N}_i)} e^{-\pi \mathcal{L}_l(\mathbf{q}_i,\, \mathbf{q}_a^{l-1})^2/\Lambda^2}. \tag{5.52}$$

Usually, for each i, the number of trimers evaluated explicitly was $M_i = \mathcal{N}_i^2/4$, for tetramers it was $M_i = \mathcal{N}_i^3/27$, and for pentamers it was $M_i = \mathcal{N}_i^4/256$. In one test, for heptamers $M_i = \mathcal{N}_i^6/12^6$ was increased to $\mathcal{N}_i^6/9^6$, the computer time increased by a factor of 2.5, and the statistical error was reduced by a factor of 0.56. The inverse square root of the time factor was 0.63.

A further refinement was added to the loop Gaussian evaluation algorithm. Loops were generated using nested sums to randomly choose the next atom in the loop from amongst the neighbors of the first atom. This 'grows' a loop by adding bonds to the existing stem, in a tree-like arrangement. The current branch is pruned if the length of the stem exceeds a pre-set threshold, typically the sum of the bonds squared is less than 12 or 18 Λ^2. (In one test the smaller threshold ran about twice as fast as the larger one, and about five times as fast as no pruning.) Upon hitting the threshold, the current and all future loops based on this stem are given zero weight without actually calculating them, the counter is incremented by how many such future loops there would have been, and the nested sums are restarted at the next atom in line for the current loop position after the atom that caused the threshold to be exceeded. In this final form of the loop Gaussian evaluation algorithm, the number of trees actually generated is much less than those that would possibly be generated by the nested sums, which is much less than the number of possible loops of neighbors, which is much less than the number of possible loops of all atoms. Tests confirm the accuracy and the efficacy of this approach.

Finally, although the loop Gaussian evaluation algorithm is rather efficient compared to what it could have been, it can still be the rate determining step for the Monte Carlo algorithm. Under these circumstances it is important that the number of calls to the evaluation regime be restricted and as much information as possible be extracted from each call. This means that the position configurations for successive calls should be uncorrelated. Since the classical Metropolis Monte Carlo algorithm for generating a trajectory through position space is relatively cheap in terms of computer time, in the results reported below 100 attempted trial moves were made for each and every atom in the system between successive calls to the loop Gaussian evaluation subroutine.

5.4.3 Results for liquid helium

Results are now given for the classical canonical equilibrium Metropolis Monte Carlo simulations of liquid helium-4 at low temperatures. These include permutation loop contributions up to heptamers using the algorithm described above. The Lennard–Jones pair potential was used, $u(r) = 4\varepsilon[(\sigma/r)^{12} - (\sigma/r)^6]$, with parameters appropriate for helium: well-depth $\varepsilon = 10.22k_B$ J, core diameter $\sigma = 0.255\,6$ nm, and mass $m = 4.003 \times 1.660\,54 \times 10^{-27}$ kg (van Sciver 2012). Results were at overall number density $\rho_{tot}\sigma^3 = 0.3$ (mass density $\rho_{m,tot} = 119.4$ kg m^{-3}). The thermal wavelength at $k_B T/\varepsilon = 1.3$ ($T = 13.286\,0$ K) is $\Lambda = 0.936\,64\sigma$, and that at 0.4 (4.088 K) is $1.688\,6\sigma$.

The number of atoms used was $N = 500$, giving edge length of the cubic simulation cell $L = 11.86\sigma$, for overall number density $\rho_{tot}\sigma^3 = 0.3$. Periodic boundary conditions were applied and the minimum image convention was used. The pair potential was cut-off at $R_{cut} = 3.5\sigma$. This was used to define the neighborhood volume for the small cell spatially-based neighbor tables. The same neighbor tables was used for the permutation loop Gaussian calculations, as described above. The threshold for pruning the permutation loops was set at $12\Lambda^2$. A typical simulation attempted $0.4–1.6 \times 10^6$ moves per atom, with an acceptance rate of 30%–50%. It took from several hours to a day to complete a simulation at a single temperature on an adolescent desk-top personal computer.

The results reported below are along the line of liquid–vapor coexistence in the Lennard–Jones model. The density of the liquid is significantly higher than the specified overall density (table 5.1). The central density in the table was obtained for a sphere of radius σ about the center of mass of the simulation cell. The latter was obtained after a uniform shift of the atoms (including moving some out one face and in the opposite face of the central simulation cell) so as to minimize the second mass moment. This somewhat convoluted procedure was necessitated by the periodic boundary conditions. The higher the central density, the more inhomogeneous is the system. At high temperatures the liquid practically extended homogeneously throughout the system. The liquid region took the form of a diffuse spherical droplet for $T \gtrsim 11$ K, transitioning to a lens or oblate ellipsoid or cylinder for $T \lesssim 10$ K. These shapes, particularly the cylinder, are evidently affected by the cubic periodic boundary conditions. The results reported below are averaged across the full system, although obviously they are dominated by the contributions from the liquid.

The central densities using the Lennard–Jones parameters of helium in the simulations are much higher than the measured liquid density at saturation, which is 146.2 kg m^{-3} at 2.177 K and 100.8 kg m^{-3} at 5 K, with the critical point at 5.195 K (Donnelly and Barenghi 1998). It was confirmed that the simulations were performed for the fluid state by monitoring the diffusion of the atoms.

The high liquid density is undoubtedly an artifact of the model used, and it is worth listing its artificial aspects. One contribution to the high density of the

Table 5.1. Central number density in the Monte Carlo simulations. The overall number density is $\rho_{tot} = 0.3$ ($\rho_{m,tot} = 119.4$ kg m^{-3})† and $N = 500$.

$k_B T/\varepsilon$	$\rho\sigma^3$	$\rho\Lambda^3$	T (K)†	ρ_m (kg m^{-3})†	T (K)‡	ρ_m (kg m^{-3})‡
0.4	0.98	4.72	4.088	390.04	1.742	155.55
0.5	0.91	3.13	5.110	366.16	2.177	144.44
0.6	0.88	2.31	6.132	350.24	2.612	139.68
0.7	0.81	1.68	7.154	322.38	3.048	128.57
0.8	0.77	1.31	8.176	306.46	3.483	122.22
0.9	0.73	1.04	9.198	290.54	3.919	115.87
1	0.62	0.76	10.220	246.76	4.354	98.41
1.1	0.55	0.58	11.242	218.90	4.789	87.30
1.2	0.39	0.36	12.264	155.22	5.225	61.90
1.3	0.37	0.30	13.286	147.26	5.660	58.73

† $\varepsilon_{He} = 10.22\, k_B$ J and $\sigma_{He} = 0.255\,6$ nm.
‡ $0.43\varepsilon_{He}$ and $1.4\,\sigma_{He}$.

liquid is the Laplace pressure due to the surface tension of the highly curved interface of the droplets. (Small droplets are nevertheless stable in a finite-sized system of fixed number (Moody and Attard 2002).) Possibly a more important contribution is the approximate nature of the Lennard–Jones pair potential itself, which is likely inaccurate in the short-separation core-dominated region that becomes important at high densities, even though the parameters are derived by fitting scattering data (van Sciver 2012). Three-body potentials, such as the Axilrod–Teller triple dipole potential, and higher many-body potentials have been neglected, and these tend to be net repulsive (Barker and Klein 1973, Attard 1992). The commutation function (see chapter 7), which is neglected in the present simulations, is of unknown importance and effect at these low temperatures. Finally, the simulations are based on the momentum continuum and so they represent the contribution from bosons in excited states only. This effect is less important on the high temperature side of the λ-transition. In general for interacting bosons, it is unclear how to separate out the ground state bosons, or how to obtain their specific contribution, or how they effect the liquid–vapor equilibrium.

It would be possible to bring the simulated λ-transition temperature and density (and approximately the saturation density curve) into agreement with the measured values by making the replacements $\varepsilon \Rightarrow 0.43\varepsilon$, and $\sigma \Rightarrow 1.4\sigma$, as in the final two columns in table 5.1. The results reported below use the original Lennard–Jones parameters fitted for helium. Ultimately the discrepancy between the simulations and the measurements is due to the Lennard–Jones potential itself rather than the parameter values that it uses. The value of the dimensionless quantity $\rho\Lambda^3$ would not be changed by any such replacement.

In view of these approximations and artifacts, one should not expect the simulations to provide an accurate quantitative description of liquid helium-4. Instead, they should be regarded as an initial investigation of the effects of wave function symmetrization on the λ-transition, and as providing molecular-level insight into the physical mechanisms that underlay the associated phenomena.

Figure 5.6 shows the average potential energy per particle as a function of temperature at constant overall density, $\rho_{\text{tot}}\sigma^3 = 0.3$. See table 5.1 for the saturated liquid density at each temperature. One can see that the total potential energy decreases with decreasing temperature. The total potential energy is dominated by the monomer or classical contribution. The kinetic energy, $\langle \mathcal{K} \rangle = 3Nk_\text{B}T/2$, can simply be added to this term.

The loop contributions to the average energy are small. For $k_\text{B}T/\varepsilon > 0.5$ ($T > 5.11$ K), the contributions decrease in magnitude with increasing loop size l. For lower temperatures than this the opposite trend may be seen in the figure. This suggests that the permutation loop series converges at higher temperatures, and diverges at lower temperatures. At $k_\text{B}T/\varepsilon = 0.5$, for the total density $\rho_{\text{tot}}\sigma^3 = 0.3$, one has $\rho_{\text{tot}}\Lambda^3 = 1.03$, and for the central density $\rho\sigma^3 = 0.91$, one has $\rho\Lambda^3 = 3.1$.

The heat capacity at constant volume is shown in figure 5.7. Above about $T = 8$ K it is dominated by the monomer (classical) contribution. There is a local maximum

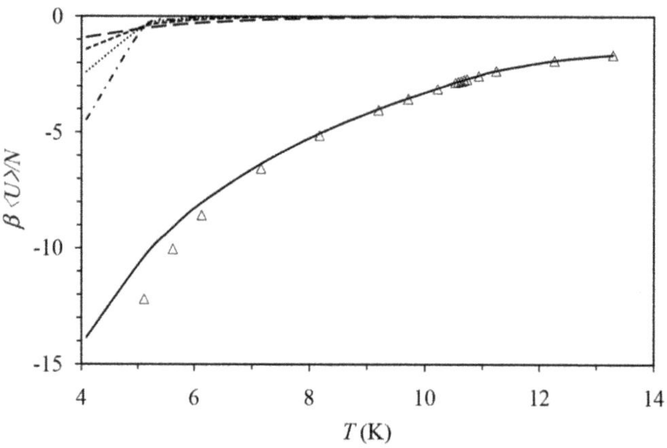

Figure 5.6. Average potential energy per particle for Lennard–Jones He^4 as a function of temperature at constant overall density $\rho_{tot}\sigma^3 = 0.3$. The contribution curves are monomer (classical, full), dimer (long dashed), trimer (short dashed), tetramer (dotted), and pentamer (dash-dotted). The symbols are the total of these. The statistical error is less than the size of the symbols.

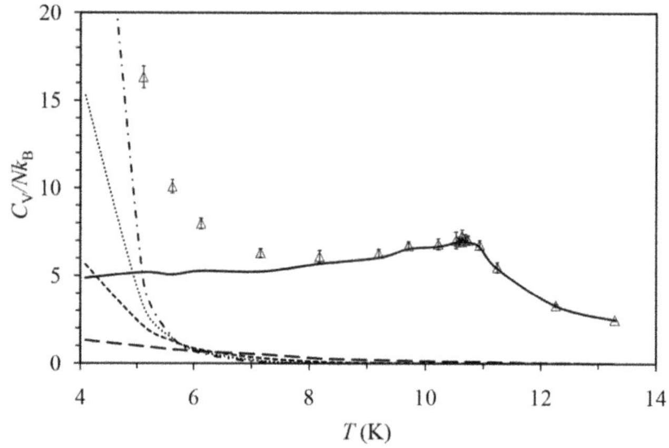

Figure 5.7. Constant volume heat capacity per particle for Lennard–Jones He^4 as a function of temperature at constant overall density $\rho_{tot}\sigma^3 = 0.3$. The curves and symbols are as in the preceding figure. Each arm of the error bar is twice the statistical error on the mean (i.e., 96% confidence level).

at about 10.6 K, that is relatively broad, and that is almost entirely classical in structure. These three points (the high temperature, the broad maximum, and the classical dominance) rule it out as a candidate for the λ-transition in liquid helium-4. This maximum appears to be associated with the transition from a diffuse spherical liquid droplet at higher temperatures to an oblate ellipsoid in the simulations at lower temperatures.

The classical contribution to the heat capacity monotonically decreases with decreasing temperature below this maximum. In contrast the total heat capacity begins to increase again at about 8 K, and continues to increase for the rest of the simulated regime. This is due to the increasingly significant contributions from the permutation loops, which increase with decreasing temperature.

The loop contributions to the heat capacity decrease with increasing loop size l for $k_{\rm B}T/\varepsilon > 0.6$ ($T > 6.132$ K). They show the opposite trend for lower temperatures than this, with the cross-over point clear in figure 5.7. Therefore for temperatures lower than about 6 K, the permutation loop series appears to be divergent, and therefore the results in the figure for the total heat capacity for this regime must be treated with caution. The individual loop contributions are reliable.

Based on the sharp increase in either the individual loop contributions or the total heat capacity for $T \lesssim 5$ K it seems reasonable to conclude that this is the high temperature side of the λ-transition for the Lennard–Jones model of helium-4. This is obviously a specific quantum effect, indeed an effect due entirely to wave function symmetrization, since the classical contribution has the opposite behavior to that measured. At this temperature and overall density, $\rho_{\rm tot}\Lambda^3 = 1.03$. (The central density is $\rho\sigma^3 = 0.91$, and $\rho\Lambda^3 = 3.1$.) The dominant contribution to the loop heat capacities $l \in [2, 7]$ in the simulations at this temperature comes from the non-U part, which is to say the temperature derivative of the thermal wavelength in the loop Gaussian, equation (5.50). This is the same contribution that occurs for the ideal gas, although of course here its average is affected by the structure in the Lennard–Jones system.

The discrepancy between the laboratory transition temperature of 2.17 K and the present estimate of 5 K can be attributed in part to the fact that the liquid density in the simulations is many times the saturated liquid density of helium-4 (table 5.1). This probably exacerbates the errors due to using the ideal solution approximation for the permutation loops (neglecting loop correlations), and the neglect of the ground state. The last two probably cause the loop contribution to the heat capacity to be increasingly overestimated with decreasing temperature.

Figure 5.8 shows the purely classical radial distribution function at three temperatures. At the highest temperature shown, there is a single peak at $r/\sigma = 1.10 \pm .02$ that represents nearest neighbors, and then more or less monotonic decay to unity. At the middle temperature the nearest-neighbor peak has grown, but remains at the same position, and there are signs of a secondary peak at $r/\sigma = 2.15 \pm .02$. It is noticeable that the radial distribution function is $g(r) = 1.24$ at the cut-off distance used for the potential. This is rather larger than the asymptotic limit of unity. At the lowest temperature there is noticeable widening and possible bifurcation of the secondary peak. The first four maxima are at $r/\sigma =$ 1.10, 1.94, 2.10, and 2.87. This seems to signify the beginning of the onset of crystalline structure in the liquid. It is noticeable that the oscillatory structure has not died out by the cut-off radius. Although the classical radial distribution function at this lowest temperature displays more structure than at higher temperatures, one would say that the differences are more quantitative than qualitative. This reinforces the point that the high value of the heat capacity at this temperature shown in figure 5.7 is a quantum effect that may be perturbed by the underlying classical structure but is in no way due to it.

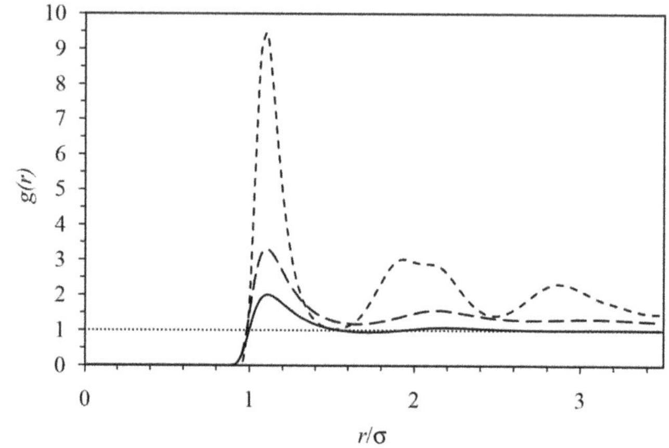

Figure 5.8. Classical radial distribution function for Lennard–Jones He^4 at $\rho_{tot}\sigma^3 = 0.3$. The temperatures are $k_BT/\varepsilon = 1.3$ (13.3 K, full curve), 1.0 (10.22 K, long-dashed curve), and 0.4 (4.09 K, short-dashed curve). The dotted line is a guide to the eye.

The growth in the height of the nearest-neighbor peak with decreasing temperature corresponds to an increasing adsorption excess. This causes the simulation time to markedly increase with decreasing temperature as more neighbors have to be accounted for within the cut-off of the loop calculations. The increase in adsorption excess causes the growth in the loop symmetrization contributions to the heat capacity. This fact underlies the Markov approximation for the various loop averages that is developed in section 6.3.

5.5 Interactions on the far side

The treatment of the λ-transition for interacting bosons with computer simulations using the algorithm detailed in the preceding section is currently limited to the high temperature side, due to the dilute loop approximation, and also to the neglect of the ground state, which become increasingly important below the transition. In seeking to quantify the effects of the interaction potential on the low temperature side of the λ-transition one needs a different approach.

In this section an analytic but approximate approach is pursued that is motivated on physical grounds. It explains mathematically the success of the ideal gas model detailed in sections 5.2 and 5.3 by demonstrating the insensitivity of the system to the interaction potential at low temperatures.

5.5.1 Factorization and the heat capacity

We proceed with discrete momentum states. The canonical equilibrium partition function for bosons, equation (3.6), neglecting the commutation function, and with $p_{j\alpha} = n_{j\alpha}\Delta_p$, $\Delta_p = 2\pi\hbar/L$, is

$$\begin{aligned} Z^{+}(N, V, T) &= \frac{1}{N!}\sum_{\mathbf{n}}\sum_{\hat{\mathrm{P}}}\langle\mathbf{n}|e^{-\beta\hat{\mathcal{H}}}|\hat{\mathrm{P}}\mathbf{n}\rangle \\ &= \frac{1}{N!}\sum_{\mathbf{n}}\sum_{\hat{\mathrm{P}}}\frac{1}{V^N}\int \mathrm{d}\mathbf{q}e^{-\beta\mathcal{H}(\mathbf{q},\,\mathbf{p}_\mathbf{n})}e^{\mathbf{q}\cdot\mathbf{p}_\mathbf{n}/i\hbar}e^{-\mathbf{q}\cdot\hat{\mathrm{P}}\mathbf{p}_\mathbf{n}/i\hbar} \\ &= \frac{1}{N!}\sum_{\hat{\mathrm{P}}}\frac{1}{V^N}\int \mathrm{d}\mathbf{q}e^{-\beta U(\mathbf{q})}\sum_{\mathbf{n}} e^{-\beta\Delta_p^2 n^2/2m}e^{\mathbf{q}\cdot(\mathbf{n}-\mathbf{n}')\Delta_p/i\hbar}, \\ \mathbf{n}' &\equiv \hat{\mathrm{P}}\mathbf{n}. \end{aligned} \tag{5.53}$$

Here $j = 1, 2, \ldots, N$, $\alpha = x, y, z$, and $\sum_{\mathbf{n}} = \prod_{j,\alpha}\sum_{n_{j\alpha}=-\infty}^{\infty}$. The second equality sets the commutation function to unity. (See chapter 7 for its definition; arguments for setting it to unity were given in the introduction to section 5.4.)

This section focuses on low temperatures, $\beta \to \infty$. In this case there is a limited number of accessible momentum states available, say $|n_{j\alpha}| \lesssim L/\Lambda$. (In subsection 5.6.1.3 below the number of accessible momentum states per particle at the λ-transition is shown to be on the order of $n_p = \Lambda^{-3}V \lesssim N$.) Under these circumstances the permutation loops are predominantly momentum loops in which all particles in a loop are in the same momentum state. In this case the momentum sum and the position integral factorize (because if $n_{j\alpha} - n_{j'\alpha} = 0$, then the Fourier factor is unity for all values of $\mathbf{q}$). The momentum factor that arises is

$$Z^{+,\,\mathrm{id}}(N, V, T) = \frac{1}{N!}\sum_{\hat{\mathrm{P}}}\prod_{j,\alpha}\sum_{n_{j\alpha}=-\infty}^{\infty} e^{-\beta\Delta_p^2 n_{j\alpha}^2/2m}\delta(n_{j\alpha} - n_{j'\alpha}). \tag{5.54}$$

All the wave function symmetrization effects are contained within this momentum factor. Because of the factorization, this is purely an ideal gas result, and it is directly connected with the grand canonical ideal gas results of section 5.2 and 5.3. As in that analysis, the sum over momentum states can be written as the integral over the momentum continuum, plus the explicit ground state contribution.

The remaining factor is the classical configuration integral, $Q^{\mathrm{cl}}(N, V, T) = \int \mathrm{d}\mathbf{q}\; e^{-\beta U(\mathbf{q})}$, and so one has

$$Z^{+}(N, V, T) = Z^{+,\,\mathrm{id}}(N, V, T)V^{-N}Q^{\mathrm{cl}}(N, V, T), \quad \beta \to \infty. \tag{5.55}$$

The average energy is just the logarithmic temperature derivative of this,

$$\begin{aligned} \left\langle E\right\rangle_{N,\,V,\,T}^{+} &= \frac{-V^{-N}}{Z^{+}(N, V, T)}\left\{Q^{\mathrm{cl}}(N, V, T)\frac{\partial Z^{+,\,\mathrm{id}}(N, V, T)}{\partial\beta}\right. \\ &\quad \left. + Z^{+,\,\mathrm{id}}(N, V, T)\frac{\partial Q^{\mathrm{cl}}(N, V, T)}{\partial\beta}\right\} \\ &= \left\langle E\right\rangle_{N,\,V,\,T}^{+,\mathrm{id}} + \left\langle U\right\rangle_{N,\,V,\,T}^{\mathrm{cl}}. \end{aligned} \tag{5.56}$$

This is the sum of the quantum average energy for ideal bosons plus the classical average potential energy of their interactions. Hence the heat capacity at constant density is similarly the sum of the quantum ideal part and the classical interaction part. The former is given by the ground state plus continuum excited states ideal gas heat capacity derived in subsection 5.2.3.

Approaching absolute zero, the potential energy is practically constant as the structure of the fluid is insensitive to temperature, $\langle U\rangle^{\mathrm{cl}}_{N,\,V,\,T} \to$ const. Hence its contribution to the heat capacity must be negligible, and to a good approximation the dominant energy contribution to the heat capacity at constant density is that from the ideal gas. In other words, at low temperatures approaching absolute zero, the present analysis of interacting bosons shows that the classical excess contributes negligibly to the heat capacity. The ideal gas expression, equation (5.18), dominates even for interacting bosons. One concludes that interactions make little difference to the ground state plus continuum excited states ideal gas model below the transition temperature.

It should be emphasized that the main approximation in the above analysis is to include only permutation loops consisting of bosons all in the same momentum state. This neglects permutation loops composed from bosons in different momentum states. These give the dominant contribution to the heat capacity in the vicinity of the λ-transition temperature, and so the present analysis cannot be expected to be accurate in that regime.

5.5.2 Permutation entropy

There is an alternative way to view the preceding factorization that brings out its explicit relationship to entropy and that sheds light on the nature of the system in the superfluid regime.

The microstates $\mathbf{n} = \{\mathbf{n}_1, \mathbf{n}_2, \ldots, \mathbf{n}_N\}$, which we take to be single-particle momentum microstates, $\mathbf{p}_{\mathbf{n}} = \Delta_p \mathbf{n}$, $\Delta_p = 2\pi\hbar/L$, $n_{j\alpha} = 0, \pm 1, \pm 2, \ldots$, have an internal subsystem entropy due to permutations,

$$S^{\mathrm{s},\pm}_{\mathbf{n}} = k_{\mathrm{B}} \ln \chi^{\pm}_{\mathbf{n}}. \tag{5.57}$$

Here $\chi^{\pm}_{\mathbf{n}}$ is the symmetrization factor, and its relation to entropy was mentioned in connection with equation (3.6). For bosons, with which we are here concerned, the symmetrization factor is given by equation (2.20),

$$\chi^{+}_{\mathbf{n}} = \prod_{\mathbf{a}} N_{\mathbf{a}}(\mathbf{n})!, \tag{5.58}$$

with the occupancy of the single-particle momentum state $\mathbf{a}$ being

$$N_{\mathbf{a}}(\mathbf{n}) = \sum_{j=1}^{N} \delta(\mathbf{n}_j - \mathbf{a}). \tag{5.59}$$

Here $\mathbf{a} = \{a_x, a_y, a_z\}$, $a_\alpha = 0, \pm 1, \pm 2, \ldots$.

Obviously we add this internal entropy to the canonical equilibrium reservoir entropy to get the total entropy of the subsystem microstate,

$$S_{\mathbf{n}}^{\pm} = S_{\mathbf{n}}^{\mathrm{s},\pm} + S_{\mathbf{n}}^{\mathrm{r}} = k_{\mathrm{B}} \ln \chi_{\mathbf{n}}^{\pm} - \frac{\mathcal{H}(\mathbf{q}, p_{\mathbf{n}})}{T}. \tag{5.60}$$

The specific contribution from the permutation entropy can be elucidated in terms of a trial momentum move,

$$\mathbf{p}_{\mathrm{a}} \to \mathbf{p}_{\mathrm{b}}, \quad \text{or} \quad \{N_{\mathrm{a}}, N_{\mathrm{b}}\} \to \{N_{\mathrm{a}} - 1, N_{\mathrm{b}} + 1\}. \tag{5.61}$$

Obviously $N_{\mathrm{a}} \geqslant 1$. The change in the internal subsystem entropy for this is

$$\Delta S_{\mathbf{n}}^{\mathrm{s},+}(\mathbf{a} \to \mathbf{b}) = k_{\mathrm{B}} \ln \frac{N_{\mathrm{b}} + 1}{N_{\mathrm{a}}}. \tag{5.62}$$

This is positive if the number of bosons in the destination state ends up being greater than the number in the initial state (the rich get richer). This what drives the bosons' momenta to become highly correlated at low temperatures, the formation of large momentum loops, and increased momentum state occupancy. (Single boson transitions and occupied state transitions are discussed in the context of superfluidity and the momentum gap in section 5.6.2.4 below.)

The change in reservoir entropy, $\Delta S_{\mathbf{n}}^{\mathrm{r}}(\mathbf{a} \to \mathbf{b}) = -\beta(p_{\mathrm{b}}^2 - p_{\mathrm{a}}^2)/2m$, should be added to this. This tends to drive the particles into the momentum ground state. One can independently change the particles positions, $\Delta S^{\mathrm{r}}(\mathbf{q} \to \mathbf{q}') = -\beta(U(\mathbf{q}') - U(\mathbf{q}))$. (These changes in momentum are conveniently but not necessarily done one particle at a time, and similarly for position.) The changes in entropy can be used in conjunction with, for example, the Metropolis algorithm, to generate a trajectory through (semi-discrete) phase space (which includes the ground momentum state) that ensures that the trajectory average equals the equilibrium average. They can also be used to formulate stochastic dissipative equations of motion (see chapters 4 and 5 of Attard 2015).

The canonical equilibrium partition function for bosons, equation (3.6), is

$$\begin{aligned} Z^{+}(N, V, T) &= \frac{1}{N!} \sum_{\mathbf{n}} \sum_{\hat{\mathrm{P}}} \langle \phi_{\mathbf{n}}(\mathbf{q}) | e^{-\beta \hat{\mathcal{H}}} | \phi_{\mathbf{n}}(\hat{\mathrm{P}}\mathbf{q}) \rangle \\ &= \frac{1}{N!V} \sum_{\mathbf{n}} \sum_{\hat{\mathrm{P}}} \int \mathrm{d}\mathbf{q}\; e^{\mathbf{q}\cdot\mathbf{p}_{\mathbf{n}}/i\hbar} e^{-\beta\hat{\mathcal{H}}} e^{\mathbf{q}\cdot\hat{\mathrm{P}}\mathbf{p}_{\mathbf{n}}/i\hbar} \\ &\approx \frac{1}{N!V} \sum_{\mathbf{n}} \sum_{\hat{\mathrm{P}}} \int \mathrm{d}\mathbf{q}\; e^{\mathbf{q}\cdot\mathbf{p}_{\mathbf{n}}/i\hbar} e^{-\beta\mathcal{H}(\mathbf{q},\mathbf{p}_{\mathbf{n}})} e^{\mathbf{q}\cdot\hat{\mathrm{P}}\mathbf{p}_{\mathbf{n}}/i\hbar} \\ &\approx \frac{1}{N!V} \sum_{\mathbf{n}} \int \mathrm{d}\mathbf{q}\; e^{-\beta\mathcal{H}(\mathbf{q},\mathbf{p}_{\mathbf{n}})} \chi_{\mathbf{n}}^{+} \\ &= \frac{1}{N!V} \sum_{\mathbf{n}} e^{-\beta\mathcal{K}_{\mathbf{n}}} e^{S_{\mathbf{n}}^{\mathrm{s},+}/k_{\mathrm{B}}} \int \mathrm{d}\mathbf{q}\; e^{-\beta U(\mathbf{q})}. \end{aligned} \tag{5.63}$$

The third equality neglects the commutation function. The fourth equality makes the same approximation as in the preceding subsection, namely it neglects permutations between bosons in different momentum states on the grounds that these give highly oscillatory Fourier factors. Also, the expectation value is dominated by regions beyond the range of the pair potentials and therefore orthogonality can be invoked. This means that we only need retain permutations amongst bosons in the same momentum state, the number of which is given by $\chi_{\mathbf{n}}^{+}$, and the logarithm of which number gives the internal subsystem entropy of the microstate. The final factor is just the configurational integral, $Q^{\rm cl}(N, V, T)$, and so this expression is the same as that given in the preceding subsection, equation (5.53), except that here the sum over permutations has been performed explicitly to give the subsystem internal microstate entropy.

The symmetrization factor differs from unity only at low temperatures when there are a restricted number of momentum states accessible so that there is high probability of some being occupied by multiple bosons. It is the entropy embodied in the symmetrization factor, and its increase with increasing momentum correlation, that is ultimately responsible for superfluidity below the λ-transition. Although the present results for discrete momentum states on the low temperature side of the λ-transition have been cast in terms of permutation loops, one could present them instead in terms momentum state occupancy.

5.5.3 Kinky loops

The preceding two subsections analyzed the low-temperature side of the λ-transition based on the idea that the permutation loops were dominated by those with the bosons in the same momentum state, and that only these loops need be taken into account. This naturally raises the question of testing the validity of that assumption by formulating corrections to it.

Obviously the first correction to the single-momentum state model is the case of permutation loops with two momentum states. Consider then an l-loop comprising a bosons in the momentum state $\mathbf{p}_\alpha$, and $l - a$ bosons in the momentum state $\mathbf{p}_\gamma$. (Strictly speaking, α and γ are three-dimensional integer vectors.) Since the symmetrization function for the loop consists of Fourier factors of the type $e^{-\mathbf{q}_{j_{k+1}}\cdot\mathbf{p}_{j_{k+1},j_k}/i\hbar}$, its oscillatory behavior is minimized by grouping together the bosons in the respective states, so that there are just two changes in momentum between neighbors (kinks) going around the loop.

As a specific example, choose two initial bosons from the fluid, the first at $\{\mathbf{q}_\alpha, \mathbf{p}_\alpha\}$ and the second at $\{\mathbf{q}_\gamma, \mathbf{p}_\gamma\}$. (Strictly speaking, the first subscript of each should be α_q and γ_q.) Choose another $a - 1$ bosons with momentum $\mathbf{p}_\alpha$ and another $l - a - 1$ bosons with momentum $\mathbf{p}_\gamma$. Form a two-kink permutation loop from these, with the first boson of the initial pair placed in the first position, and the second in the $a + 1$ st position.

The number of possible two-kink l-loops based on the original pair is quite large; it grows exponentially with the number of bosons in the system occupying the two given momentum states. If there are N_α bosons besides the first with momentum $\mathbf{p}_\alpha$, and N_γ bosons besides the second with momentum $\mathbf{p}_\gamma$, then the permutation loop entropy associated with this particular two-kink loop is

$$S^{\mathrm{s},\,+,\,(l)}(\alpha,\,\gamma;\,a) = k_{\mathrm{B}} \ln \frac{N_\alpha! N_\gamma!}{(N_\alpha - a + 1)!(N_\gamma - l + a + 1)!}. \tag{5.64}$$

One sees that this is maximized by the maximal loop, $S^{\mathrm{s},\,+,\,(N_\alpha+N_\gamma+2)}(\alpha,\,\gamma;\,N_\alpha + 1) = k_{\mathrm{B}} \ln(N_\alpha! N_\gamma!)$.

The symmetrization function for the given two-kink loop is

$$\prod_{j=1}^{l} e^{-\mathbf{q}_{k_{j+1}} \cdot \mathbf{p}_{k_{j+1},k_j}/\mathrm{i}\hbar} = e^{-\mathbf{q}_\alpha \cdot [\mathbf{p}_\alpha - \mathbf{p}_\gamma]/\mathrm{i}\hbar} e^{-\mathbf{q}_\gamma \cdot [\mathbf{p}_\gamma - \mathbf{p}_\alpha]/\mathrm{i}\hbar} = e^{-\mathbf{q}_{\alpha\gamma} \cdot \mathbf{p}_{\alpha\gamma}/\mathrm{i}\hbar}. \tag{5.65}$$

Note that this is independent of the positions of all the bosons except the initial pair. This non-localization property is ultimately the reason for the macroscopic number of two-kink loops. In the general case this result might provide a basis for simulating the primary correction to the ideal gas-configuration integral factorization approximation derived in the preceding two sections.

The oscillatory behavior of the symmetrization function can be further minimized by choosing that sub-set of all two-kink loops such that the initial pair of bosons have a difference in momentum state that is approximately perpendicular to their difference in position,

$$|\mathbf{q}_{\alpha\gamma} \cdot \mathbf{p}_{\alpha\gamma}| \lesssim \frac{\pi\hbar}{2}. \tag{5.66}$$

An example of such a seven-boson, two-kink loop is sketched in figure 5.9. In so far as this is non-localized for the initial pair, it seems that such two-kink loops are more likely than two-kink loops in general.

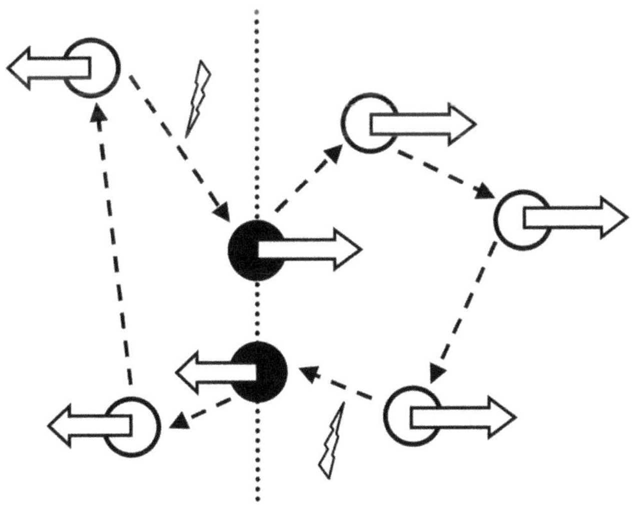

Figure 5.9. Sketch of a two-kink 7-loop. The circles represent bosons (original pair filled), the block arrows symbolize their momentum state, the dashed arrows depict pair transpositions, the lightning bolts denote kinks, and the dotted line portrays the z-axis. Except for the initial pair, the pictured positions have no meaning.

5.6 Permutation loops, the λ-transition, and superfluidity

This concluding section draws on the evidence to address two questions: what is the nature of the λ-transition, and what are the implications for superfluidity? The explanation for the λ-transition provides a detailed physical picture that differs a little from the usual understanding of Bose–Einstein condensation. It emerges that the non-local momentum ground state is ultimately responsible for superfluidity, which adds nuance to existing two-fluid theories while providing a molecular picture of the role of collisions and excited states.

5.6.1 The λ-transition

5.6.1.1 Simulation of liquid helium

The simulation results, section 5.4, were obtained by transforming to the momentum continuum and performing the momentum integrals explicitly. On the high temperature side of the λ-transition we do not expect invoking the momentum continuum to be unreliable or to be specifically responsible for the observed phenomena. Similarly the implicit neglect of the ground state contributions can be expected to have more significant consequences at low temperatures than at high. Simplifying assumptions such as using the Lennard–Jones pair potential and neglecting the commutation function, and their consequences such as the larger than measured saturation density, will have quantitative effects, but should not change the qualitative features of the simulations.

It was found that below about 11 K the monomer (classical) heat capacity decreases with decreasing temperature. Below about 8 K the loop heat capacity $l \geqslant 2$ increases with decreasing temperature. This increasingly sharp rise in the heat capacity is thus a specific quantum phenomenon due to wave function symmetrization. The computed increase in loop heat capacity correlates with the measured increase in heat capacity approaching T_λ^+ in the laboratory. The simulation data supports the widely held notion that the λ-transition is related to wave function symmetrization and boson state occupancy statistics, and as such it is a form of Bose–Einstein condensation.

According to the simulations, section 5.4, the permutation loop series for the heat capacity diverges approaching $T = 5$ K. One could make this temperature equal the measured T_λ by rescaling the Lennard–Jones parameter $\varepsilon_{\rm He}$, as in table 5.1, but we have not done this as it seems better to emphasize the approximate nature of the Lennard–Jones potential itself. In any case, the loop series in the simulations currently cannot be used reliably at lower temperatures. Using the central density from table 5.1 at the predicted transition temperature of $T = 5$ K gives the dimensionless parameter $\rho\Lambda^3 = 3.1$. (Using $T = 4$ K gives $\rho\Lambda^3 = 4.72$.) This is not far off the measured value for helium-4 at the λ-transition, $\rho\Lambda^3 = 4.57$, and a little closer than the ideal gas model prediction, $\rho\Lambda_{\rm c}^3 = 2.612$, section 5.2.

5.6.1.2 Simulated divergence of the heat capacity

The properties of liquid helium have been comprehensively tabulated (Donnelly and Barenghi 1998). The experimentally measured heat capacity has an extremely sharp maximum at the λ-transition temperature, with $\partial C_V/\partial T|_{T_\lambda^\pm} \to \mp\infty$. The maximum value is large, but probably not divergent. The heat capacity appears to be continuous across the transition, so that the discontinuity is second order.

The permutation loop Lennard–Jones simulations also give a sharp increase in the heat capacity approaching the λ-transition. The loop series is actually divergent.

There are several approximations in the simulations: the simple Lennard–Jones potential that gives a high saturation density and high transition temperature, table 5.1; the neglect of the ground state by the continuum momentum integral, section 5.2; the neglect of the commutation function, chapter 7; and the neglect of correlations between loops in expressing the grand potential as a series of loop grand potentials, section 7.2.2. These doubtless have a quantitative effect on the simulated heat capacity but it seems unlikely that they would qualitatively change the simulated divergence or its molecular origin.

The approximation in the simulations that possibly becomes most significant approaching the λ-transition is the implicit neglect of interactions and correlations between loops. In deriving the expression for the grand potential of an interacting system as a series of loop grand potentials, the average of products of loops is factored as the product of independent loop averages, equation (7.35). This is argued to be valid in the thermodynamic limit for compact loops, section 7.2.2. Similar to an ideal solution, the loops are taken to be dilute solutes in a dense fluid of monomers.

In the present case, approaching the λ-transition, this assumption possibly breaks down as the increasing number, size, and extent of the loops causes them to interact with each other. In these circumstances the loops can no longer be considered independent, and the series of loop grand potentials may become a poor approximation to the actual grand potential of the system.

The simulated divergence of the heat capacity is dominated by the non-potential (i.e., ideal) contributions, which is to say the average of the temperature derivative of the loop Gaussians, equation (5.50) (cf the discussion of figure 5.7). The loop Gaussians themselves are an ideal gas term, but their average involve integrals weighted by the particle correlation functions. The latter most certainly depend upon the interaction potential. This last point is important because it explains how the heat capacity for interacting bosons, both measured and simulated, can be so much greater than that for the ideal gas at the λ-transition. Hence the simulated divergence in the heat capacity is due to the change in structure of the fluid (increase in number, size, and extent of loops), which is missing in the ideal gas, sections 5.2 and 5.3.

In summary, what the simulations unambiguously show is that approaching the λ-transition from the high temperature side the number, size, and extent of permutation loops increase, as does their contribution to the heat capacity. Whether or not the simulations are quantitatively accurate for the rate of divergence in the heat capacity, the molecular picture of the λ-transition that emerges is that it is a type of percolation transition. The number and size of permutation loops surpasses a threshold density that allows their concatenation to form loops of macroscopic spatial extent.

5.6.1.3 Position and momentum symmetrization loops

A symmetrization loop, in either the discrete quantum picture or in the continuum phase space prior to momentum integration, consists of loops of Fourier factors that can be written either as $e^{-\mathbf{q}_{j_k,j_{k+1}}\cdot\mathbf{p}_{j_k}/i\hbar}$, or as $e^{-\mathbf{q}_{j_k}\cdot\mathbf{p}_{j_k,j_{k+1}}/i\hbar}$. Although both forms yield the same value when evaluated over the loop, they serve to explain the different regions of phase space where the permutation loops contribute. In order for the loop to which it belongs to be significant upon averaging, the exponent must be small in magnitude. Hence in practice there are two cases: either all $q_{j_k,j_{k+1}}$ are small (position loops) or else all $p_{j_k,j_{k+1}}$ are small or zero (momentum loops). This division is convenient as the two types are qualitatively different.

Position loops are localized in position space. They are the ones that give rise to the loop phenomena that have been identified on the high temperature side of the λ-transition in the simulations, section 5.4. Position loops are usually small in size l, and small in spatial extent. For high temperatures they are few in number. Approaching the transition from the high temperature side, $T \to T_\lambda^+$, their number, size, and extent increases.

Increasingly for $T < T_\lambda$, loops are localized in momentum space (momentum loops). They tend to be large in size l and large in spatial extent. (These assertions are justified below.) They are an example of the non-locality of quantum mechanics. The particles belonging to a momentum loop are highly correlated in momentum notwithstanding the fact that they are separated beyond the range of the interaction potential. With quantized momenta, all atoms in a loop can be in the same single-particle discrete momentum state, $\mathbf{p}_{j_k,j_{k+1}} = \mathbf{0}$ (cf section 5.5). Quantum leaps in momenta around the loop limit the loop's size and spatial extent in a way that having all the bosons in the same momentum state does not. Indeed to be definite, one can define a momentum loop to be one in which all bosons are in the same momentum state, and position loops to be everything else. (By this definition both the permutation loop simulations of section 5.4, and the kinky loops of section 5.5.3, invoke position loops.)

When there are many more accessible momentum states than bosons, most bosons are monomers. Any loop that they belong to is a position loop. When the number of accessible momentum states is comparable to or less than the number of bosons, a significant number, or even the majority, of bosons share a momentum eigenvalue with others.

In the quantized picture, the number of accessible single-particle momentum states is $n_p \approx (\sqrt{2\pi m k_B T}/\Delta_p)^3 = \Lambda^{-3}V$. This formula for n_p comes from the momentum eigenvalues and the exponent of the Maxwell–Boltzmann factor for a single boson. The factor of 2π is an order of magnitude constant that simplifies the following estimate. Since $\Delta_p = 2\pi\hbar/L$, the number of accessible single-particle momentum states is macroscopic even at low temperatures, $n_p \propto T^{3/2}V$. However, the number of bosons is also macroscopic, and the probability that none of the N bosons is in the same momentum state as any other is vanishingly small when $n_p = \mathcal{O}(N)$ (birthday paradox). This occurs at low temperatures when $\rho\Lambda^3 = \mathcal{O}(1)$. More exactly, with the above formula, $n_p = N \Leftrightarrow \rho\Lambda^3 = 1$. In the simulations of the

Lennard–Jones liquid, the λ-transition occurs at about $\rho\Lambda^3 = 3.1$ (central density). The measured value for helium-4 is $\rho\Lambda^3 = 4.57$, and that for the ideal gas is $\rho\Lambda_c^3 = 2.612$. These suggest that the present criterion, $\rho\Lambda^3 \gtrsim 1$ (equivalently, multiple occupancy of momentum states), is necessary but not sufficient for the λ-transition.

In any case, for temperatures near $\rho\Lambda^3 \approx 1$, there must be a significant number of momentum states occupied by more than one boson. This is the condition for the formation of momentum loops. As the temperature is further decreased, the number of accessible momentum states per boson decreases, and so the number of momentum loops that can have different momentum eigenvalues also decreases. It follows that the number of bosons with the same momentum eigenvalue increases. It will be shown below that the size l of the momentum loops also increases. Assuming few loops, or even one maximal loop for each distinct momentum eigenvalue, these mean that the number of momentum loops eventually decreases with decreasing temperature. This is consistent with the ideal gas results in figure 5.5.

The momentum state occupancy is sometimes a more convenient description than the individual loops that make up the permutations of the state. The two descriptions are ultimately equivalent. Bearing this in mind, we shall nevertheless continue to invoke momentum loops just to emphasize the distinction from position loops in which spatial order is important. Also, this is consistent with the formulation of the grand potential as a series of loop grand potentials, and it is exact for the ideal gas.

In summary, on the high temperature side of the λ-transition, the system is dominated by position loops, whose number, size l, and spatial extent all increase with decreasing temperature. Passing through the transition momentum loops become increasingly common; the occupancy of low-lying momentum states increases as the number of accessible momentum states decreases with decreasing temperature. Indeed it will be argued below that the transition marks the switch in dominance of the two types of loops.

5.6.1.4 Molecular contributions to the entropy of loops

Position loops do not strongly restrict momenta, although they definitely favor low-lying momentum states. Momentum loops do not strongly restrict positions. Strictly, momentum loops comprise bosons with the same momenta, which can only occur in the quantized case, and since they are non-local they can extend spatially throughout the whole system.

These differences in the two types of loops induce a difference in the configurational or molecular contribution to their entropy. (The combinatorial entropy of loops is treated below.) Momentum loops require highly correlated momenta, and they therefore have low momentum entropy. This is quite costly at high temperatures where there are many momentum states available. It is much less costly at low temperatures where there are few accessible momentum states.

In contrast the spatial structure of the liquid is relatively insensitive to temperature, and so the configuration cost of the short-range spatial order of position loops changes little with temperature. This is particularly true at temperatures below the λ-transition.

What favors momentum loops is that the atoms that comprise them can range throughout the volume of the system, because the differences in consecutive momenta can (or must) be precisely zero. Momentum loops are unconstrained by space. As mentioned, they are an example of quantum non-locality. In three-dimensional space, there are many more possible loops comprising far-separated consecutive atoms than close-separated ones because the number of possible partners increases quadratically with separation.

Position loops never have the same freedom in momentum space because the atoms never attain zero separation. And also, the accessible volume of momentum space decreases with decreasing temperature. Instead there is a threshold where the thermal wavelength extends beyond the nearest-neighbor separation in which the atoms in position loops can be composed from the available momentum states.

In this context it is worth reiterating that the permutation loop Monte Carlo simulations are based upon performing the continuum momentum integrals analytically, which gives the thermal wavelength. This treats the momenta as independent and, because of the momentum continuum, the bosons never occupy the same momentum state. It also neglects the contributions of bosons in the ground momentum state.

Consider a point in phase space in the discrete momentum picture prior to the momentum integration that gives the thermal wavelength. For a given positional configuration that could form a position l-loop, $\mathbf{q}^l = \{\mathbf{q}_{j_1}, \mathbf{q}_{j_2}, \dots, \mathbf{q}_{j_l}\}$, the permutation weight of the loop, which can be written in position loop form, $\prod_{k=1}^{l} e^{-\mathbf{p}_{j_k}\cdot\mathbf{q}_{j_k, j_{k+1}}/i\hbar}$, is maximized (i.e., is made less oscillatory) by low-lying values of the momenta. The momentum ground state, $\mathbf{p}_{j_k} = \mathbf{0}$, removes all restrictions on the bond $\mathbf{q}_{j_k, j_{k+1}}$, which obviously increases dramatically the number and extent of the viable loops. But even the excited states, $p_{j_k, x} = 2\pi n\hbar/L$, for small n allows large separations while still giving a small loop exponent. Since such small momenta means that averaging over slight changes in position gives a non-zero result, this case maximizes the combinatorial entropy contribution of the position loops (see below). Because it increases the loop configurational and combinatorial entropy, position loops induce condensation into low-lying momentum states. Due to the position correlation function, the positional configurations suitable for forming position loops are much more common for interacting bosons than for an ideal gas. Therefore, the occupancy of low-lying momentum states must be greater for interacting bosons than for an ideal gas.

This discussion of the configurational contribution to the loop entropy suggests that momentum loops become more common as the temperature is lowered, consistent with the ideal gas analysis, figure 5.5. That position loops are favored on the high temperature side of the λ-transition is confirmed by the data from the Monte Carlo simulations, section 5.4.

5.6.1.5 Combinatorial entropy of loops

The loop grand potential in the form $-\beta\Omega^{\pm,(l)}$ measures the entropy due to loops. The reader may have trouble accepting this, particularly in view of the fact that position loops require atoms localized to small spatial neighborhoods, and momentum loops require the atoms to have highly correlated momenta. To understand the loop entropy, one has to recall that the loop expansion begins with the monomer or classical term. This accounts for in its entirety the classical picture of entropy in terms of molecular configurations in phase space. The loop entropy is in addition to the classical entropy, and it is irrelevant that the loops themselves consist of ordered sets of molecules.

Loops are distinct arrangements of molecules, and such arrangements can be counted. Hence there is an entropy associated with them. The partition function, for example, equation (3.6), is the sum not just over energy states, but also over wave function symmetrization permutations. The logarithm of the partition function is the total entropy. In the case of bosons, all distinct permutation states add to the entropy. In the case of fermions, some permutations preclude otherwise acceptable states, which subtracts from the total entropy. As shown in equation (3.42) the attractive effective pair potential for bosons, and the repulsive one for fermions, come from the dimer loop. This effective potential is entropic in nature, as is signified by its temperature-dependence.

5.6.1.6 Dominance of large momentum loops

Large l momentum loops are favored over small ones, as is now shown. Let $N_{\mathbf{a}}$ be the number of bosons in the momentum state $\mathbf{a}$, $\sum_{\mathbf{a}} N_{\mathbf{a}} = N$. Let there be $m_{\mathbf{a}l}$ loops of size l in this momentum state, $\sum_l l m_{\mathbf{a}l} = N_{\mathbf{a}}$. The distribution of loops given $N_{\mathbf{a}}$ is determined by the entropy of the macrostate $\{m_{\mathbf{a}l}\}$,

$$S(\{m_{\mathbf{a}l}\}|N_{\mathbf{a}}) = k_{\mathrm{B}} \ln \left[\frac{N_{\mathbf{a}}!}{\prod_{l=1}^{N_{\mathbf{a}}} m_{\mathbf{a}l}!\, l^{m_{\mathbf{a}l}}} \right], \quad \sum_{l=1}^{N_{\mathbf{a}}} l m_{\mathbf{a}l} = N_{\mathbf{a}}. \tag{5.67}$$

Each of the l rigid rotations (i.e., cyclic permutations) of a loop of size l leaves the loop unchanged, which explains the factor of $l^{m_{\mathbf{a}l}}$. The factor of $m_{\mathbf{a}l}!$ represents permutations of loops of the same size.

This shows that large loops are favored over monomers and small loops, and that for a given number of bosons in the same momentum state, the entropy is maximized if they a form a single permutation loop. In particular, the entropy is zero if the $N_{\mathbf{a}}$ bosons are all monomers, $m_{\mathbf{a}l} = N_{\mathbf{a}}\delta_{l,1}$, and it is $k_{\mathrm{B}} \ln(N_{\mathbf{a}} - 1)!$ if there is a single maximal loop, $m_{\mathbf{a}l} = \delta_{l,n_{\mathbf{a}}}$.

The same conclusion was reached in section 5.5.2, where equation (5.62) shows that the internal entropy of the momentum microstates increases with transitions to more highly occupied momentum states. Section 5.5.3 analyzed a specific type of momentum loop, namely one with two kinks. There it was concluded that for a given original pair of bosons, the combinatorial loop entropy was maximized by the maximal two-kink loop. These results explain the prevalence of positive momentum correlations and of bosons in the same single-particle quantum state.

As mentioned above, instead of subdividing the permutations into momentum loops, it is often more convenient to focus on the occupancy of the momentum state as a whole. In the present example of $N_{\mathbf{a}}$ bosons occupying the single-particle momentum state **a**, there are $\chi_{\mathbf{a}}^{+}(N_{\mathbf{a}}) = N_{\mathbf{a}}!$ possible arrangements in total, which is not very much greater than the $(N_{\mathbf{a}} - 1)!$ arrangements if they all formed a single maximal momentum loop. The logarithm of this gives the internal entropy of the momentum state with this occupancy, $S_{\mathbf{a}}^{+,\mathrm{int}} = k_{\mathrm{B}} \ln N_{\mathbf{a}}!$.

In summary, as the temperature is decreased below the transition temperature, the number of accessible momentum states decreases, the number of atoms in a given accessible momentum state increases, and the size l of the momentum loops increases.

5.6.1.7 Loops, condensation, and the phase transition

Laboratory measurements show that the λ-transition signifies a phase change in liquid helium. The temperature $T = 2.17$ K marks a sharp maximum (possibly a divergence) in the heat capacity, and a discontinuous change in sign of its temperature derivative. It also marks a weak maximum in the density, as well as a sharp transition to the superfluid state (Donnelly and Barenghi 1998).

The above subsections outline the evidence for the increase in number, size, and extent of position loops approaching the λ-transition from the high temperature side. It was also pointed out that momentum loops are non-local, driven to maximal size by combinatorial entropy, and favored at low temperatures by position configuration entropy. It is plausible therefore that the λ-transition swaps the dominance of these two types of loops. This subsection sets out a mechanism by which this occurs, and explains how this is consistent with the experimental evidence summarized above for the λ-transition.

Above it was pointed out that the configurational entropy of position loops in phase space is increased by bosons in low-lying momentum states as they allow large separations between consecutive bosons without an excessively oscillatory permutation loop factor. Bosons in the ground momentum state are particularly efficacious in this regard. The combinatorial loop entropy is also increased due to the greater number, extent, and size of the loops that occupancy of the low-lying momentum states enables. It was concluded that position loops in an interacting system nucleate the occupancy of low-lying momentum states, including the ground momentum state, over and above what would occur in an ideal gas.

It is a moot point whether one says that position permutation loops cause bosons to condense into low-lying momentum states, or that bosons in low-lying momentum states enable more position permutation loops. In so far as this effect does not occur in an ideal gas, it is arguable that the position loops are the cause, and that the condensation is the result. Either way, the configurational and the combinatorial entropy of the position loops are increased as more bosons in low-lying momentum states enable more and more various position loops. One can say that the increase in number, size, and extent of position loops nucleates the condensation of bosons into low-lying momentum states, and vice versa. The catalytic mechanism appears to be the gain in permutation entropy by the non-local concatenation of compact position permutation chains into loops, each chain headed by a ground momentum state boson. This suggests that the λ-transition results from a form of positive feedback between the number of position loops and the number of bosons in low-lying momentum states. Once a critical concentration is reached both grow explosively to macroscopic size.

The divergence of the heat capacity in the simulations largely captures this effect. The low-lying excited momentum states are included in the continuum momentum integrals. In essence they give rise to the increase in the thermal wavelength with decreasing temperature. What the simulations don't include is the contribution from the bosons in the ground momentum state. In so far as these additionally enhance the formation of position loops, they might be expected to add to the divergence of the heat capacity.

At the λ-transition the simulations suggest that the position loops become macroscopic in number, size, and extent. Based on the above arguments one concludes that there must be an associated increase in the occupancies of the ground and low-lying momentum states. In fact, for $T \leqslant T_\lambda$, the number of bosons in the momentum ground state itself must be macroscopic, and the number of bosons in excited but low-lying momentum states must be microscopic. This is because the number of bosons N_ε in the energy state ε is proportional to $e^{-\beta N_\varepsilon \varepsilon}$, which can never be macroscopic for an excited state. (This argument conflates the ground momentum state with the ground energy state, which is strictly true only for the ideal gas.)

The situation for the heat capacity is different for the ideal gas, which in the momentum continuum (i.e., with thermal wavelength representing the excited states) has uniform positional correlations. Because the interaction potential induces positional correlations, the number, size, and extent of position loops is much greater for real bosons than for the ideal case. Hence the heat capacity for interacting bosons, both computed and measured, is much greater than that for the ideal gas.

One sees that ground momentum state condensation for interacting bosons is also different to that for the ideal gas, sections 5.5 and 5.3. For the present interacting bosons ground momentum state occupancy becomes macroscopic discontinuously at the λ-transition. For the exact ideal gas ground state occupancy increases continuously through the λ-transition, figure 5.4, depending on the size of the system. (For the two-fluid approximation to the ideal gas, section 5.2, the ground state occupancy artificially increases from zero at the λ-transition.) In so far as superfluidity is closely connected with ground momentum state occupancy (see below), the present arguments for interacting bosons give a sharp superfluid transition coincident with the λ-transition, consistent with experimental observation.

It seems difficult to reconcile this sudden increase in ground momentum state occupancy with the fact that the experimentally measured enthalpy is continuous at the λ-transition (Donnelly and Barenghi 1998). Of course, if the measured heat capacity were divergent at the λ-transition, then the energy and enthalpy would have to be discontinuous there. The bottom line is that it does not seem possible to account for the sharp superfluid transition that is actually observed without having a sharp condensation into the momentum ground state.

Returning to the heat capacity, position loops continue to exist and to contribute on the low temperature side of the λ-transition. This follows because the experimentally measured heat capacity continues to be much greater than that of the ideal gas, which is due to momentum loops alone, section 5.2 and 5.3. The analysis of section 5.5 says that when momentum loops dominate, the classical excess free energy should simply be added to the ideal gas loop free energy. But the classical excess heat capacity for the Lennard–Jones model at the λ-transition is $C_{\mathrm{V}}^{\mathrm{LJ,cl,ex}}/Nk_{\mathrm{B}} \approx 3.5$, figure 5.7, which on its own is not sufficient to raise the ideal gas result to the measured values close to the λ-transition. One must therefore conclude that the large values of the heat capacity measured for $T \lesssim T_\lambda$ are indicative of the continued presence of position permutation loops.

As the temperature is decreased below T_λ, the number of bosons in the ground momentum state must further increase. Also the number of accessible momentum states must decrease, leading to increased momentum correlations and increased occupancy of those low-lying momentum states that remain occupied. In these circumstances momentum loops increase, and the system increasingly looks like an ideal gas. These mean that the heat capacity, even though still dominated by position loops, must decrease with decreasing temperature for $T < T_\lambda$. This is consistent with the laboratory fact that $\partial C_{\mathrm{V}}/\partial T > 0$ in this regime (Donnelly and Barenghi 1998). Approaching absolute zero the measured heat capacity vanishes as $T^{3/2}$, just as in the ideal gas, which indicates that momentum loops and momentum state occupancy become dominant at very low temperatures.

Experimental data show that the liquid saturation density for helium-4 is a maximum at the λ-transition (Donnelly and Barenghi 1998). Usually in a simple liquid the density increases monotonically with decreasing temperature because the thermal energy makes the nearest-neighbor separation on average greater than the minimum in the pair potential, which effect decreases with decreasing temperature. The relatively sharp increase in density in liquid helium approaching the λ-transition from the high temperature side is consistent with the increase in position permutation loops, since these create an attractive effective potential between bosons, as was shown explicitly for the case of dimers in equation (3.42). The more gradual decrease in density with decreasing temperature on the low temperature side of the transition suggests that this extra attraction is diminishing. In part this appears due to the decline in position permutation loops themselves. Also the attractive force due to the Gaussian effective potential weakens once the thermal wavelength significantly exceeds the separation at which the pair potential is a minimum as further increase does little to alter the structure of the liquid. In any case it should be mentioned that the measured density appears to be continuous, but its temperature derivative appears discontinuous at T_λ (Donnelly and Barenghi 1998, Sachdeva and Nuss 2010).

5.6.1.8 Bose–Einstein condensation

If the λ-transition in liquid helium is to be regarded as a form of Bose–Einstein condensation, then these results suggest that the latter should *not* be narrowly defined as the point at which it becomes favorable for multiple bosons to occupy the single-particle momentum (or energy) state. In liquid helium above the λ-transition, but with $\rho\Lambda^3 \lesssim 1$, momentum states occupied by multiple bosons are certainly common. The formation of momentum loops and multiple occupancy of momentum states may sharply increase at the λ-transition and the momentum ground state may become macroscopically occupied due to nucleation by the position loops. But even for $T \lesssim T_\lambda^-$, after the transition has taken place, the greater than ideal heat capacity is evidence that positions loops remain significant. The ideal gas calculations of section 5.2 and 5.3, which are perhaps the canonical example of Bose–Einstein condensation, differ significantly from the λ-transition in that they give a heat capacity that has a relatively low maximum and a relatively weak second order discontinuity. For the reasons discussed in the above subsection, the sharp spike of the measured λ-transition seems to have proximal cause in the growth of position loops, and the condensation into momentum loops or momentum states appears to be a secondary effect.

Of course this may simply be a semantic argument. It would be fair to say that the position loops themselves are an example of the correlational attraction of bosons due to their symmetry properties. As such it would be quite reasonable to regard their growth to macroscopic size through the whole system at the λ-transition as a form of Bose–Einstein condensation. The concomitant growth in the occupancy of individual momentum states, and the macroscopic occupancy of the ground momentum state, which are close to the more conventional interpretation of Bose–Einstein condensation, is not the only phenomenon associated with the λ-transition.

Bose–Einstein condensates represent a low-temperature state of matter for bosons (Cornell *et al* 1999). Their nature is not addressed in detail here, but doubtless the picture of the λ-transition derived here also applies to such condensates. One cannot help but observe that the relatively small size of the condensates means that the quantization of momentum states is even more marked than for liquid helium, and so one would expect permutation loops to play a significant role in their properties. Also, the low density of the condensates compared to liquid helium make them an attractive proposition for the application of quantum statistical mechanics in classical phase space.

5.6.2 Superfluidity

5.6.2.1 Two-fluid theory

The two currently accepted theories for superfluidity are the ideal gas two-fluid theory of London (1938) and Tsiza (1938), and the phonon-roton two-fluid theory of Landau (1941). The ideal gas two-fluid theory asserts that the system consists of coexisting superfluid and normal fluid, with the superfluid being bosons in the ground energy state, which are said to have zero viscosity, and the normal fluid being bosons in excited states having the usual viscosity (Tsiza 1938). The superfluid is said to be able to flow through the normal fluid, just like the aether, and vice versa. The equilibrium thermodynamics of the two-component mixture is given by the ideal gas approach of London (1938), as detailed in section 5.2.

The phonon-roton theory of Landau (1941) is also a type of two-fluid scheme, with the superfluid again being bosons in the ground energy state, forming a quiescent background, and again said to have zero viscosity. But in this case the normal fluid consists of a gas of elementary energy excitations, which are related to quantized sound waves, which are termed phonons and rotons, and which are taken to be bosons. The empirical energy-momentum spectrum fitted from thermodynamic data by Landau (1941) also fits later neutron scattering measurements (Pathria 1972), although this casts no light on the nature of rotons or on their relevance to superfluidity. The theory is applicable far below the λ-transition close to absolute zero (Pathria 1972).

It is not clear to the present author how Landau's (1941) phonon-roton two-fluid model overcomes his own objection to Tisza's (1938) ideal gas two-fluid model: 'nothing would prevent atoms in a normal state from colliding with excited atoms, i.e., when moving through the liquid they would experience a friction and there would be no superfluidity at all' (Landau 1941). Feynman (1954), despite modifying and extending Landau's model, also expressed doubts: 'The division into a normal fluid and superfluid although yielding a simple model for understanding the final equations, appears artificial from a microscopic point of view. This opinion is shared by Landau and by Dingle.'

In any case the two models are more alike than is often recognized (Balibar 2017). The difference between the models is that London (1938) and Tsiza (1938) use an ideal gas approach, whereas Landau (1941) incorporates particle interactions at some level via quantized sound waves. One thing both have in common is the assertion that superfluidity sets in sharply at the λ-transition even though the fraction of superfluid component is zero in the model at the transition. Another thing in common is the etherial nature of the two-fluid model, with the superfluid component being ghost-like bosons in the energy ground state. A final common point is the absence of any explicit mechanism that explains exactly how it is that energy ground state bosons flow without viscosity. In favor of the models of Tisza (1938) and Landau (1941) are the detailed two-fluid hydrodynamic calculations that are consistent with a range of superfluid phenomena.

5.6.2.2 Momentum loop entropy in collisions

The present equilibrium analysis of the λ-transition in this chapter has so far not directly addressed the origin of superfluidity. This is more properly a topic for non-equilibrium thermodynamics and statistical mechanics, specifically in relation to hydrodynamics (Attard 2012, chapter 5 and section 9.6). Nevertheless given that the λ-transition in helium-4 also marks the onset of superfluidity, the present equilibrium account of the transition provides a new perspective on the nature of superfluidity.

The present work focuses on momentum states, as distinct from the energy states that form the basis of the two models discussed above. The fluid after the λ-transition was here characterized by the increasing importance of momentum permutation loops. It was argued that position loops nucleate momentum loops, and that therefore the explosive growth in number, size, and extent of position loops at the λ-transition induces a similar transition in momentum state occupancy, with the occupancy of the ground momentum state becoming macroscopic. All bosons in a momentum loop are in the same momentum state and the loop is non-local.

This means that there is no spatial relationship between the momentum-correlated bosons. We first explore the consequences of permutation loop entropy for collisions, and below we analyze the effects of non-locality.

In general the equations of motion of particles in a fluid, classical (Attard 2012, sections 7.4 and 11.1) or quantum (Attard 2015, sections 4.2, 5.1 and 5.2), are stochastic and dissipative. The stochastic terms tend to randomize the momenta or quantum state, and the dissipative terms increase the reservoir or subsystem entropy. In the case of classical shear flow it is the dissipative terms that drive the tendency to homogenize and to zero the momenta, which gives rise to classical viscous drag (Attard 2012, chapters 2 and 5 and section 9.6).

What is different to the classical case in the present problem is that below the λ-transition there is an additional contribution to the entropy, namely the permutation loop entropy. This favors correlated momentum states, which in the non-equilibrium case do not necessarily average to zero. For the present quantum systems, the entropy whose gradient gives the driving force for the stochastic dissipative Schrödinger equation and quantum microstate transitions (e.g., Attard 2015, equation (5.39)), should include the internal subsystem permutation entropy, which can be added to the reservoir entropy, as in the present equation (5.60). The general principle is that the dissipative term in the stochastic equations of motion supplies a force in the direction of increasing entropy precisely calibrated to the variance of the random forces that ultimately ensures the correct probability distribution.

The relevance of this to superfluidity is as follows. Below the λ-transition, for an equilibrium quantum system, forward and backward momentum states are equally likely, and hence the most likely single-particle momentum state is the zero momentum state, $\bar{\mathbf{n}} = \mathbf{0}$. In fact the number of bosons in this ground momentum state is macroscopic. As the most occupied state, it is the state with the most and the largest momentum permutation loops. It is therefore the momentum state with largest internal loop entropy. It is also the state of largest reservoir entropy. The dissipative terms in the equations of motion in this equilibrium case drive the system toward the zero momentum state, whereas the stochastic terms tend to drive the system randomly into excited momentum states. These plus the adiabatic term balance each other to maintain the equilibrium probability distribution.

For a non-equilibrium case, again quantum and again below the λ-transition, we consider an external field or pressure gradient that drives flow in the positive z-direction. In this case the most likely single-particle momentum state can be labeled $\bar{n}_z > 0$, with the transverse components zero. By definition this must be the most highly occupied state, and therefore it is the state with the most and the largest momentum permutation loops, and the largest internal loop entropy. The dissipative terms in the equations of motion in this non-equilibrium case drive the system toward the momentum state $\bar{n}_z$ (cf equation (5.62)), whereas the stochastic terms tend to drive the system randomly into nearby momentum states. These plus the adiabatic term balance each other to maintain the non-equilibrium probability distribution.

Ultimately what is different about the non-equilibrium fluid above and below the λ-transition is that above the transition the entropy is predominantly configurational, which is maximized by random momenta, and so in any flow the collisional driving force is toward momentum dissipation, which is manifest as non-zero shear viscosity. Below the λ-transition, the momentum loop entropy dominates, and this is maximized by correlated momenta. Hence in any flow the collisional driving force is toward momentum preservation, which means zero shear viscosity.

This answers Landau's (1941) objection that collisions should randomize the momenta and give rise to a non-zero viscosity. The momentum loop entropy causes boson correlations, which gives an entropic driving force that maintains their momenta, which fact Landau (1941) did not consider. (It appears that Landau (1941) did not accept London's (1938) idea that the λ-transition was a manifestation of Bose–Einstein condensation (Balibar 2014), so it is hardly surprising that he did not consider the consequences of correlations induced by multiple bosons occupying the same single-particle state.)

5.6.2.3 Non-local momentum correlations

In addition to the permutation loop entropy dominating the stochastic dissipative collisions, the non-local nature of the momentum correlations induced by the loops is essential for superfluidity. To elucidate this point we summarize the role of space above and below the λ-transition.

Recall the molecular origin of shear viscosity in a classical fluid. In shear flow, the inhomogeneous momentum flux is dissipated by molecular collisions. In effect faster moving fluid layers are slowed and adjacent slower layers are accelerated as momentum is transferred between them. This momentum dissipation has the effect of increasing the subsystem entropy, since the order represented by spatial variations in momentum flux is a state of low configurational entropy.

In general for a classical fluid, or for a quantum fluid above the λ-transition, the momentum correlations must be spatially localized, which gives rise to shear flow. For example, for Poiseuille flow, which is laminar flow in a pipe or channel under a pressure gradient, the flow is zero at the walls and increases continuously toward the center.

In the case of bosons below the λ-transition, the non-local momentum correlations prohibit the Poiseuille parabolic profile, since the high momentum state in the center of the channel would induce the same state in the bosons near the walls. Non-local momentum correlations imply momentum spatial homogeneity: the only mathematical solution for non-zero flow is plug flow in which the most likely momentum state of the bosons is uniform across the channel. This of course is also the classical solution for inviscid hydrodynamic flow.

5.6.2.4 Momentum gap

In understanding the nature of superfluidity a crucial piece of evidence is the existence of a critical velocity above which laboratory measurements show that superfluid flow through pores is destroyed. Pathria (1972, section 10.8) gives measured experimental values for the critical velocity of $v_c = 0.13$, 0.08, and 0.04 m s^{-1} for capillary diameter $D = 0.12$, 0.79, and 3.9 μm, respectively. The common explanation is that above a critical velocity the production and growth of excitations destroy the superfluid. The type of excitations generally invoked are associated with the rotons of Landau's (1941) theory, which are commonly interpreted as vortex rings.

In the context of the present approach to the λ-transition based upon momentum state occupancy, the fact that the critical velocity increases with decreasing pore diameter suggests instead that we should compare it to the velocity of the first excited transverse momentum state. For the rectangular geometry that we have focused on thus far this is $v_y = \Delta_p/m = 2\pi\hbar/mL$. (This would apply to a liquid film, where the vanishing of the wave function at the surface and at the liquid–vapor interface creates just such a transverse momentum gap.) In cylindrical geometry it is $v_r = 2j_{01}\hbar/mD$ (Blinder 2011), where $j_{01} = 2.4$ is the first zero of the zeroth order Bessel function. This calculated velocity is plotted in figure 5.10, together with the measured values and the vortex prediction of Kawatra and Pathria (1966).

Although it appears in figure 5.10 that the first excited transverse momentum state overestimates the critical velocity for small pore diameters, this may be misleading as other data indicate the opposite. For example, the reported value $v_c = 7$ m s^{-1} for $D = 20$ nm (Allum *et al* 1977) is about double the present prediction of 3.8 m s^{-1}. Hence the present prediction is within about a factor of two of the measured critical velocity over a range of a factor of 200 in pore diameters. Given the simplicity of the present theory, the absence of adjustable parameters, and the possibility of material-specific and statistical contributions to the wall collisions, the correlation of the present prediction with the measured values is satisfactory.

These results suggest that superfluid flow down a channel in the z-direction occurs in the transverse momentum ground state, $\bar{\mathbf{n}} = \{0, 0, \bar{n}_z\}$. This is the macroscopically-occupied, non-equilibrium, momentum ground state. Since the channel is usually much longer than it is wide, the spacing of momentum states in the flow direction is much less than those in the transverse direction. Such flow with all bosons in the transverse momentum ground state is presumably collisionless, which one expects for inviscid flow. It will be shown shortly that when the flow velocity

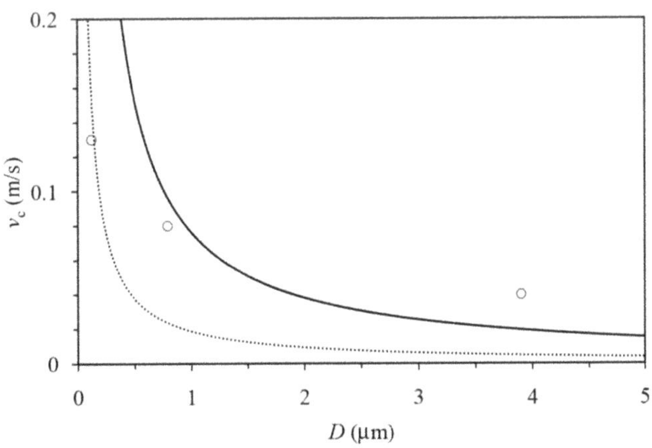

Figure 5.10. Velocities associated with superfluid flow of helium through a pore of diameter D. The circles are the measured critical flow velocity (Pathria 1972, section 10.8), the solid curve is the velocity of the first excited transverse momentum state $v_r = 2j_{01}\hbar/mD$ (cylindrical boundary conditions), and the dotted curve is the vortex result of Kawatra and Pathria (1966) $v_c = 1.18\hbar/mD$.

exceeds the velocity of the first excited transverse momentum state, interactions and collisions with the wall are able to excite the bosons out of the state $\bar{\mathbf{n}}$, thereby destroying the superfluid flow.

The measured critical velocities are about 1–3 orders of magnitude smaller than a component of the thermal velocity of an individual boson, which at $T = 1$ K is 45 m s^{-1}. But this measure is not entirely appropriate because bosons taking part in the superfluid flow occupy a single-momentum state, and so one cannot really consider them individually. Following the λ-transition, the overwhelming (i.e., macroscopic) number of bosons in the pore are in the non-equilibrium ground momentum state $\bar{\mathbf{n}} = \{0, 0, \bar{n}_z\}$, and a negligible fraction (i.e., microscopic) are in excited momentum states. The external potential or pressure gradient provides the driving force that effectively raises the equilibrium ground momentum state to $\bar{n}_z > 0$, the most likely state in the pore.

The stochastic dissipative equations of motion provide a quantitative estimate of the effect of a collision. Suppose that all the bosons in the pore initially occupy the most likely state $\bar{\mathbf{n}}$, $N_{\bar{\mathbf{n}}} \approx \rho V$, where ρ is the density and V is the volume of the pore. Let $\overline{v}_z = \bar{n}_z \Delta_{pz}/m$ be the flow velocity for this initial state. Consider the final state to be the first excited transverse momentum state and longitudinal ground state, $\mathbf{n}' = \{1, 0, 0\}$. We consider two possible collisions with the wall related to this final state. In the first case one particular boson transitions to the final state. In this case the changes in internal and reservoir (wall) entropy from the collision are (cf equation (5.60))

$$\Delta S^{\text{int}} = k_{\text{B}} \ln\left[\frac{(N_{\bar{\mathbf{n}}} - 1)!}{N_{\bar{\mathbf{n}}}!}\frac{1!}{0!}\right] = -k_{\text{B}} \ln N_{\bar{\mathbf{n}}}, \quad \text{and} \quad \Delta S^{\text{r}} = \frac{-m(v_r^2 - \overline{v}_z^2)}{2T}, \tag{5.68}$$

respectively. The first term is negative and much larger than the second because $N_{\bar{\mathbf{n}}}$ is macroscopic, whereas the second is of molecular size. This means that the dissipative part of the stochastic equations of motion suppress such individual boson collisions. This is a dynamic example of the tendency of bosons to occupy the same state, otherwise known as Bose–Einstein condensation.

The second type of collision to be considered is when all of the bosons in the pore are excited into the state $\mathbf{n}'$. This might be called an occupied state collision. In this case the respective changes in entropy are

$$\Delta S^{\text{int}} = k_{\text{B}} \ln\left[\frac{0!}{N_{\bar{\mathbf{n}}}!}\frac{N_{\bar{\mathbf{n}}}!}{0!}\right] = 0, \quad \text{and} \quad \Delta S^{\text{r}} = \frac{-mN_{\bar{\mathbf{n}}}(v_r^2 - \overline{v}_z^2)}{2T}. \tag{5.69}$$

For this occupied state collision the decrease in the internal entropy of the initial state cancels with the increase for the final state. In contrast, the change in reservoir entropy is now macroscopic. It is positive if and only if the flow velocity $\overline{v}_z$ exceeds the velocity of the first excited transverse momentum state v_r. Because it is macroscopic, its exponential is effectively a Heaviside step function so that the dissipative part of the stochastic equations of motion either forbid or else mandate such an occupied state collision and transition, depending on the relative size of the two velocities.

These calculations do not rule out bosons in the pore being in excited momentum states. In fact the classical probability of N_1 bosons being in the first excited transverse momentum state $\mathbf{n}'$ is proportional to $\exp -\beta N_1 m v_r^2/2$. For a typical pore the exponent is order unity when $N_1 = \mathcal{O}(10^5)$, which is negligible compared to the total number for say $L_z = 1$ mm, $N_0 = \mathcal{O}(10^{13})$. If the classical probability is applicable, which is not at all clear, then superfluid flow can occur even with a small minority of bosons in such excited states. It is not until a macroscopic number of bosons make the transition from the non-equilibrium ground state to the first excited transverse state, as in the second calculation, that superfluid flow is disrupted.

The first calculation shows that when the most likely state is occupied by macroscopic numbers of bosons, then excitation or transition by individual bosons is effectively forbidden. (This result holds even taking into account the fact that N_1 may be non-zero but microscopic.) The second calculation shows that the measured critical velocity is indeed the first excited transverse state velocity, $v_c \equiv v_r$, and that the necessary and sufficient condition for stable superfluid flow is that the flow velocity is less than this critical velocity.

This calculation combines all three of the above mechanisms for superfluidity. The non-local nature of momentum correlations is necessary for almost all of the bosons in the pore to be in a single-momentum state. The stochastic dissipative equations of motion are necessary to show how entropy drives occupied state collisions, and how the superfluid flow state $\bar{\mathbf{n}} = \{0, 0, \bar{n}_z\}$ is stable up until the flow velocity exceeds the velocity of the first excited transverse momentum state. And the transverse momentum gap is necessary for the critical velocity to be non-zero.

Superfluid flow occurs not just in microscopic pores but also in macroscopic bulk systems. For example, at the λ-transition turbulent boiling in a macroscopic sample is suppressed due to the high thermal conductivity of superfluid convective flow (Balibar 2014). In such a case a transverse momentum gap also exists, albeit much smaller than in a pore or thin film. This would place an upper bound on the velocity of the convective superfluid flow.

References

Allen M P and Tildesley D J 1987 *Computer Simulations of Liquids* (Oxford: Oxford University Press)

Allum D R, McClintock P V E and Phillips A 1977 The breakdown of superfluidity in liquid ^{4}He: an experimental test of Landau's theory *Phil. Trans. R. Soc.* **284** 179

Attard P 1992 Simulation results for a fluid with the Axilrod–Teller triple dipole potential *Phys. Rev.* A **45** 5649

Attard P 2012 *Non-Equilibrium Thermodynamics and Statistical Mechanics: Foundations and Applications* (Oxford: Oxford University Press)

Attard P 2015 *Quantum Statistical Mechanics: Equilibrium and Non-Equilibrium Theory from First Principles* (Bristol: IOP Publishing)

Attard P 2018 *Entropy Beyond the Second Law. Thermodynamics and Statistical Mechanics for Equilibrium, Non-Equilibrium, Classical, and Quantum Systems* (Bristol: IOP Publishing)

Balibar S 2014 Superfluidity: how quantum mechanics became visible *History of Artificial Cold, Scientific, Technological and Cultural Issues* ed K Gavroglu (Dordrecht: Springer) pp 93–117

Balibar S 2017 Laszlo Tisza and the two-fluid model of superfluidity *Comptes Rendus Physique* **18** 586 .

Barker J A and Klein M L 1973 Monte Carlo calculations for solid and liquid argon *Phys. Rev.* B **7** 4707

Blinder S M 2011 Quantum-mechanical particle in a cylinder *Demonstrations Project* (http://demonstrations.wolfram.com/QuantumMechanicalParticleInACylinder/Wolfram)

Cornell E A, Ensher J R and Wieman C E 1999 Experiments in dilute atomic Bose–Einstein condensation arXiv:cond-mat/9903109

Donnelly R J and Barenghi C F 1998 The observed properties of liquid helium at the saturated vapor pressure *J. Phys. Chem. Ref. Data* **27** 1217

Feynman R P 1954 Atomic theory of the two-fluid model of liquid helium *Phys. Rev.* **94** 262

Gradshteyn I S and Ryzhik I M 1980 *Tables of Integrals, Series, and Products* (San Diego, CA: Academic)

Kawatra M P and Pathria R K 1966 Quantized vortices in an imperfect Bose gas and the breakdown of superfluidity in liquid helium II *Phys. Rev.* **151** 132

Keesom W H and Wolfke M 1927 Two different liquid states of helium *Comm. Phys. Lab. Leiden* **190b** 22

Keesom W H, Clusius K and Keesom A P 1932 Ueber die specifische warme des fussigen heliums *Comm. Phys. Lab. Leiden* **219e, 221d**

Landau L D 1941 Theory of the superfluidity of helium II *Phys. Rev.* **60** 356

Landau L D 1941 Two-fluid model of liquid helium II *J. Phys. USSR* **5** 71

Landsberg P T 1954 On Bose–Einstein condensation *Proc. Camb. Phil. Soc.* **50** 65

London F 1938 The λ-phenomenon of liquid helium and the Bose–Einstein degeneracy *Nature* **141** 643

Messiah A 1961 *Quantum Mechanics* vols 1 and 2 (Amsterdam: North-Holland)

Moody M P and Attard P 2002 Homogeneous nucleation of droplets from a supersaturated vapour phase *J. Chem. Phys.* **117** 6705

Pathria R K 1972 *Statistical Mechanics* (Oxford: Pergamon)

Sachdeva D and Nuss M 2010 Density of liquid He^4 as a function of temperature (http://spa-mxpweb.spa.umn.edu/s10/Projects/S10_DensityofLiquidHe/)

Tisza L 1938 Transport phenomena in helium II *Nature* **141** 913

van Sciver S W 2012 *Helium Cryogenics* 2nd edn (New York: Springer)

IOP Publishing

Quantum Statistical Mechanics in Classical Phase Space

Phil Attard

Chapter 6

Further applications

Exact analytic results are obtained for the quantum mechanics of a one-dimensional harmonic crystal (section 6.2). These provide in other chapters benchmark tests for a model with interacting particles. Here they are applied to the textbook example of the vibrational heat capacity of solids (section 6.1). Contrary to conventional wisdom, approaching absolute zero the exact benchmarks confirm the exponential decay predicted by Einstein (1907) rather than the power law decay predicted by Debye (1912). A loop Markov superposition approximation is derived (section 6.3), which is useful for obtaining wave function symmetrization corrections for interacting particles. The final application is symmetrization for spin-position factorization (section 6.4).

6.1 Vibrational heat capacity of solids

In this section the heat capacity of solids is obtained by treating the lattice vibrations as simple harmonic oscillators. This elementary approach, originally due to Einstein (1907) and later modified by Debye (1912), can be found in most textbooks on statistical mechanics. Arguably as an independent single-particle energy state method, it would be better placed in chapter 5. The reason for presenting it here is connected with the way it treats—or doesn't treat—wave function symmetrization. This turns out to be directly relevant to the exact and not so elementary analysis of the one-dimensional harmonic crystal in section 6.2.

Consider a system of N atoms with positions $\mathbf{q} = \{\mathbf{q}_1, \mathbf{q}_2, \dots, \mathbf{q}_N\}$. Three-dimensional space is assumed, $\mathbf{q}_j = \{q_{jx}, q_{jy}, q_{jz}\}$. We suppose that the system is a crystalline solid, with $\bar{\mathbf{q}}$ denoting the lattice positions, which are the positions of lowest potential energy. Let $\mathbf{d} = \mathbf{q} - \bar{\mathbf{q}}$ be the displacement of the atoms from their lattice positions.

doi:10.1088/978-0-7503-4055-7ch6

The potential energy may be expanded to second order about its minimum,

$$U(\mathbf{q}) = U_0 + \frac{1}{2}\mathbf{dd}\colon \underline{\underline{U}}''_0. \qquad (6.1)$$

Here the minimum value of the potential is $U_0 = U(\bar{\mathbf{q}})$, and the $3N \times 3N$ matrix of second derivatives is likewise evaluated at the lattice positions, $U''_{0;j\alpha,k\gamma} = \partial^2 U(\mathbf{q})/\partial q_{j\alpha}\partial q_{k\gamma}\big|_{\mathbf{q}=\bar{\mathbf{q}}}$. These second order terms represent the harmonic component of the lattice vibrations (Pathria 1972, section 7.3). With this quadratic form for the potential energy, the Hamiltonian operator is

$$\hat{\mathcal{H}}(\mathbf{q}) = \frac{-\hbar^2}{2m}\nabla^2 + U(\mathbf{q}). \qquad (6.2)$$

The important point to make is that this Hamiltonian operator is not symmetric with respect to particle permutations,

$$\hat{\mathcal{H}}(\hat{\mathbf{P}}\mathbf{q}) \neq \hat{\mathcal{H}}(\mathbf{q}). \qquad (6.3)$$

This is because the potential energy depends upon the displacement of each particle from its own lattice position. If particle j and k are transposed, then the original displacements $\mathbf{q}_j - \bar{\mathbf{q}}_j$ and $\mathbf{q}_k - \bar{\mathbf{q}}_k$ become $\mathbf{q}_k - \bar{\mathbf{q}}_j$ and $\mathbf{q}_j - \bar{\mathbf{q}}_k$, which are quite different. Hence the potential energy is not invariant with respect to particle permutation, and similarly the energy operator.

One concludes from this that the particles are not identical. This in turn means that the wave function has no particular symmetry with respect to particle permutations. It will be recalled that it is a fundamental requirement of quantum mechanics that the wave function be fully symmetric or fully anti-symmetric with respect to permutations of identical bosons or of identical fermions, respectively (Merzbacher 1970, Messiah 1961). Since the present lattice system does not comprise identical particles, no symmetry requirements are imposed on its wave function.

Because we are expanding about a minimum in the potential, the symmetric second derivative matrix is positive definite. Hence an orthogonal transformation diagonalizes it with positive eigenvalues,

$$\mathbf{dd}\colon \underline{\underline{U}}''_0 = (\underline{\underline{X}}^{\mathrm{T}}\mathbf{d})^{\mathrm{T}}\underline{\underline{X}}^{\mathrm{T}}\,\underline{\underline{U}}''_0\,\underline{\underline{X}}(\underline{\underline{X}}^{\mathrm{T}}\mathbf{d}) = \mathbf{q}'^{\mathrm{T}}\underline{\underline{\omega}}\mathbf{q}'. \qquad (6.4)$$

The diagonal matrix has elements $\omega_{n,n'} = \omega_n\delta_{n,n'}$, and the mode amplitudes are $\mathbf{q}' \equiv \underline{\underline{X}}^{\mathrm{T}}\mathbf{d}$. Because the transformation is orthogonal, $\underline{\underline{X}}\,\underline{\underline{X}}^{\mathrm{T}} = \underline{\underline{I}}$, the functional form of the classical kinetic energy is unchanged

$$\mathcal{K}(\mathbf{p}) = \frac{1}{2m}\mathbf{p}\cdot\mathbf{p} = \frac{m}{2}\dot{\mathbf{q}}\cdot\dot{\mathbf{q}} = \frac{m}{2}\dot{\mathbf{d}}\cdot\dot{\mathbf{d}} = \frac{m}{2}\dot{\mathbf{q}}'\cdot\dot{\mathbf{q}}' = \frac{1}{2m}\mathbf{p}'\cdot\mathbf{p}'. \qquad (6.5)$$

Hence the classical Hamiltonian takes the form of independent particles or modes,

$$\mathcal{H}(\mathbf{q}, \mathbf{p}) = U_0 + \sum_{j=1}^{3N}\left\{\frac{1}{2m}p_j'^2 + \frac{\omega_n}{2}q_j'^2\right\}. \qquad (6.6)$$

The fact that the modes have different frequencies means that they are not identical. Since there is one mode of each type, it would be meaningless to describe them as bosons or as fermions.

The lattice vibrations are the normal modes of the crystal. These dynamic distortions of the crystal are just sound waves. Since they are quantized, they can be regarded as particles and are called phonons. The solid may be considered as a set of $3N$ independent harmonic oscillators, with characteristic frequencies determined by the second derivatives of the interparticle potential. The energy eigenvalues are

$$E_{\mathbf{n}} = U_0 + \sum_{j=1}^{3N}\left[n_j + \frac{1}{2}\right]\hbar\omega_j. \tag{6.7}$$

Here $n_j = 0, 1, 2, \ldots$ is the energy state of phonon (oscillator) of type j.

For a single harmonic oscillator of frequency ω able to exchange energy with a reservoir of temperature $T = 1/\beta k_{\mathrm{B}}$, the average energy is

$$\begin{aligned}\bar{E}(T) &= \frac{\sum_n \epsilon_n e^{-\beta\epsilon_n}}{\sum_n e^{-\beta\epsilon_n}} \\ &= \frac{\sum_{n=0}^{\infty}[n + 1/2]\hbar\omega e^{-\beta[n+1/2]\hbar\omega}}{\sum_{n=0}^{\infty} e^{-\beta[n+1/2]\hbar\omega}} \\ &= \frac{\hbar\omega}{2} + \frac{\hbar\omega e^{-\beta\hbar\omega}}{1 - e^{-\beta\hbar\omega}}\end{aligned} \tag{6.8}$$

Hence for the present model of $3N$-independent oscillators, this gives the average energy of the crystal,

$$\bar{E}(N, T) = U_0 + \frac{\hbar}{2}\sum_{j=1}^{3N}\omega_j + \sum_{j=1}^{3N}\frac{\hbar\omega_j e^{-\beta\hbar\omega_j}}{1 - e^{-\beta\hbar\omega_j}}. \tag{6.9}$$

The first two terms give the ground state or binding energy of the crystal. This must be negative. The temperature-dependent part gives the heat capacity

$$\begin{aligned}C_{\mathrm{V}} &= \frac{\partial\bar{E}(N, T)}{\partial T} \\ &= k_{\mathrm{B}}\sum_{j=1}^{3N}\frac{(\beta\hbar\omega_j)^2 e^{-\beta\hbar\omega_j}}{(1 - e^{-\beta\hbar\omega_j})^2}.\end{aligned} \tag{6.10}$$

Einstein (1907) estimated the heat capacity of a crystal assuming that all modes had the same frequency, $\omega_j = \omega_{\mathrm{E}}$, in which case the formula reduces to

$$C_{\mathrm{V,E}} = 3Nk_{\mathrm{B}}\frac{(\beta\hbar\omega_{\mathrm{E}})^2 e^{-\beta\hbar\omega_{\mathrm{E}}}}{(1 - e^{-\beta\hbar\omega_{\mathrm{E}}})^2}. \tag{6.11}$$

In view of this one can define the Einstein temperature, $\Theta_{\mathrm{E}} \equiv \hbar\omega_{\mathrm{E}}/k_{\mathrm{B}}$. For temperatures much greater than this, $T \gg \Theta_{\mathrm{E}}$, $C_{\mathrm{V,E}} \sim 3Nk_{\mathrm{B}}$, which is the classical limit. Conversely at low temperatures, $T \ll \Theta_{\mathrm{E}}$, $C_{\mathrm{V,E}} \sim 3Nk_{\mathrm{B}}(\Theta_{\mathrm{E}}/T)^2 e^{-\Theta_{\mathrm{E}}/T}$, which is to say that the heat capacity vanishes exponentially.

Debye (1912) modified Einstein's assumption to allow for the continuous distribution of modes up to a cut-off frequency ω_{D}. The latter is determined by the density of modes, $g(\omega)$, and the total number of modes, $3N$, such that (Pathria 1972)

$$3N = \int_0^{\omega_{\mathrm{D}}} \mathrm{d}\omega\; g(\omega) = \int_0^{\omega_{\mathrm{D}}} \mathrm{d}\omega\; V\left\{\frac{\omega^2}{2\pi^2 c_{\mathrm{L}}^2} + 2\frac{\omega^2}{2\pi^2 c_{\mathrm{T}}^2}\right\}. \tag{6.12}$$

Here c_{L} and c_{T} are the velocity of propagation of longitudinal and transverse modes, respectively. This gives the Debye frequency

$$\omega_{\mathrm{D}}^3 = 18\pi^2\frac{N}{V}\left\{c_{\mathrm{L}}^{-2} + 2c_{\mathrm{T}}^{-2}\right\}^{-1}. \tag{6.13}$$

The corresponding Debye temperature is $\Theta_{\mathrm{D}} \equiv \hbar\omega_{\mathrm{D}}/k_{\mathrm{B}}$. The heat capacity in the Debye approximation is $C_{\mathrm{V,D}} = 3Nk_{\mathrm{B}}D(x_0)$, with (Pathria 1972)

$$D(x_0) = \frac{3}{x_0^3}\int_0^{x_0} \mathrm{d}x\; \frac{x^4 e^{-x}}{(1 - e^{-x})^2}, \quad x_0 \equiv \frac{\hbar\omega_{\mathrm{D}}}{k_{\mathrm{B}}T} = \frac{\Theta_{\mathrm{D}}}{T}. \tag{6.14}$$

For high temperatures, $T \gg \Theta_{\mathrm{D}}$, the heat capacity asymptotes as $C_{\mathrm{V,D}} = 3Nk_{\mathrm{B}}[1 - (\Theta_{\mathrm{D}}/T)^2/20 + \cdots]$. To leading order this is the classical result. In the opposite limit of low temperatures (Pathria 1972),

$$C_{\mathrm{V,D}} \sim Nk_{\mathrm{B}}\frac{12\pi^4}{5}\left(\frac{T}{\Theta_{\mathrm{D}}}\right)^3, \quad T \ll \Theta_{\mathrm{D}}. \tag{6.15}$$

This says that the heat capacity vanishes as the cube of temperature approaching absolute zero, which is slower than the exponential dependence predicted by Einstein.

Note that the Debye approach implicitly assumes that that the speed of sound is independent of temperature, which implies that the Debye temperature is similarly constant. The speed of sound can be written as the reciprocal of the square root of the adiabatic compressibility and the mass density. This is equivalent to assuming that the mode frequency is linear in the wave number, which may be true for small wave numbers. Depending on the system and the temperature regime, these may or may not be reasonable assumptions.

Figure 6.1 compares the Einstein and Debye predictions for the heat capacity due to lattice vibrations. At high temperatures both asymptote to the classical result. As

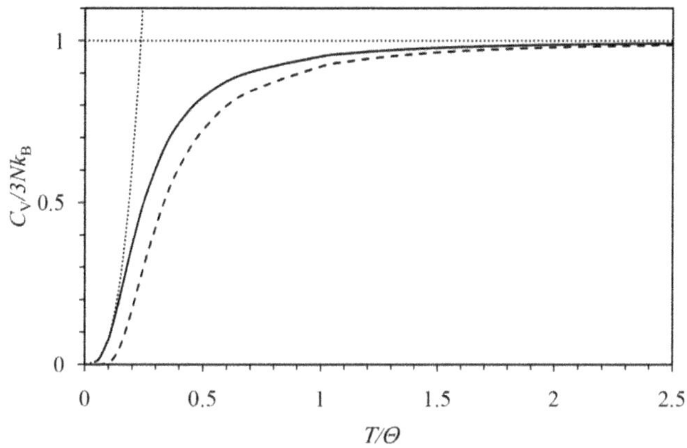

Figure 6.1. Heat capacity due to lattice vibrations. The full curve is the Debye model and the dashed curve the Einstein model. The dotted line is the high temperature asymptote and the dotted curve is the low temperature T^3 asymptote to the Debye model.

the temperature is decreased, the heat capacity is reduced and approaches zero at absolute zero. The heat capacity in Einstein's model vanishes faster than that in Debye's. The T^3 behavior predicted by the latter is said to be in agreement with laboratory measurements (Pathria 1972). (But see the comparison with exact results in figure 6.5 below.)

6.2 One-dimensional harmonic crystal

In this section is analyzed a one-dimensional harmonic crystal in which the particles are attached by linear springs to each other and to lattice sites. A feature of the analysis, which follows that of Attard (2019), is that the energy eigenfunctions and eigenvalues are obtained exactly. This makes it one of the few quantum mechanical many-particle systems for which exact analytic results are known. The normal modes given below were originally obtained in the context of classical statistical mechanics (Attard 2002). Numerical results from this model will be used in section 9.6.3 and 10.5.1.

In the following analysis a misconception in the original quantum treatment (Attard 2019) is corrected: now it is pointed out that the lack of particle permutation symmetry of the model potential means that the energy eigenfunctions do not obey any symmetrization rule. This same point was made above in the analysis of lattice vibrations, section 6.1. Because the wave function is not symmetrized, arguably the phonons, which are the quantized vibrations, should not be called bosons.

6.2.1 Model

Consider a one-dimensional harmonic crystal in which the particles are attached by linear springs to each other and to lattice sites. Let the coordinate of particle j be q_j, and let its lattice position (i.e., the position of the minimum of the external harmonic potential acting on it) be $\bar{q}_j = j\Delta_q$. The lattice spacing is also the relaxed interparticle spring length. There are fixed 'wall' particles at $q_0 = 0$ and $q_{N+1} = (N+1)\Delta_q$. Let $d_j \equiv q_j - \bar{q}_j$ be the displacement from the lattice position; for the wall particles, $d_0 = d_{N+1} = 0$. The system has overall number density $\rho = \Delta_q^{-1}$.

In this model, there is an external harmonic potential of spring constant κ acting on each particle centered at its lattice site. The interparticle spring has strength λ and relaxed length Δ_q. With these the potential energy is

$$\begin{aligned} U(\mathbf{q}) &= \frac{\kappa}{2}\sum_{j=1}^{N}[q_j - \bar{q}_j]^2 + \frac{\lambda}{2}\sum_{j=0}^{N}[q_{j+1} - q_j - \Delta_q]^2 \\ &= \frac{\kappa}{2}\sum_{j=1}^{N} d_j^2 + \frac{\lambda}{2}\sum_{j=0}^{N}[d_{j+1} - d_j]^2 \\ &= \frac{-\lambda}{2}\underline{\underline{K}}^{(N)}\colon \mathbf{dd}. \end{aligned} \tag{6.16}$$

Here $\underline{\underline{K}}^{(N)}$ is an $N \times N$ tridiagonal matrix with elements

$$K_{jk} = \begin{cases} K, & j = k \\ 1, & j = k \pm 1, \\ 0, & \text{otherwise}, \end{cases} \tag{6.17}$$

where $K \equiv -2 - \kappa/\lambda$.

It should be emphasized that the lattice positions are in order, $\bar{q}_j < \bar{q}_{j+1}$, $j = 1, 2, \dots, N$. However for the particle positions themselves there is no constraint on their order save that encouraged by the potential itself. The fact that each particle is tied to its respective lattice position and to its labeled neighbors means that they can be distinguished one from another. That is, they are not identical particles. The fact that the present potential is not invariant with respect to a permutation of the particle positions is just another way of saying that the system is not composed from identical particles. This point was also made in the discussion of the vibrational heat capacity of solids, section 6.1.

As has been mentioned many times in this book, the wave function of identical particles must be either fully symmetric (bosons) or full anti-symmetric (fermions) (Messiah 1961, Merzbacher 1970). Because the particles are not identical to each other, the wave function of the present model of the harmonic crystal does not need to obey any particular symmetry requirements.

6.2.2 Eigenvalues and eigenvectors

The eigenvalues of the potential energy matrix are required, and these may be obtained from the characteristic equation,

$$S_N(K-\mu) \equiv |\underline{\underline{K}}^{(N)} - \mu \mathrm{I}|$$
$$= \begin{vmatrix} K-\mu & 1 & 0 & \dots & 0 \\ 1 & K-\mu & 1 & 0 & \dots \\ 0 & 1 & K-\mu & 1 & \dots \\ \vdots & & & \ddots & \vdots \\ 0 & \dots & 0 & 1 & K-\mu \end{vmatrix}$$
$$= (K-\mu)|\underline{\underline{K}}^{(N-1)} - \mu \mathrm{I}|$$
$$- \begin{vmatrix} 1 & 1 & 0 & \dots & 0 \\ 0 & K-\mu & 1 & 0 & \dots \\ 0 & 1 & K-\mu & 1 & \dots \\ \vdots & & & \ddots & \vdots \\ 0 & \dots & 0 & 1 & K-\mu \end{vmatrix}$$
$$= (K-\mu)S_{N-1}(K-\mu) - S_{N-2}(K-\mu). \tag{6.18}$$

This is just the recursion relation for the Tchebyshev polynomials of the second kind, which are denoted $S_n(x) = U_n(x/2)$ by Abramowitz and Stegun (1970, equation (22.7.6)). They give $S_N(2\cos\theta) = U_N(\cos\theta) = \sin((N+1)\theta)/\sin\theta$, which evidently vanishes when $\theta_n = \pm n\pi/(N+1)$, $n = 1, 2, \dots, N$. Hence the characteristic equation vanishes when $K - \mu_n = 2\cos\theta_n$, or

$$\mu_n = K + 2\cos\frac{n\pi}{N+1}, \quad n = 1, 2, \dots, N. \tag{6.19}$$

These are the eigenvalues, and since $K \equiv -2 - \kappa/\lambda$, they are negative.

The corresponding eigenvectors, $\mathbf{u}_n$, have elements

$$u_{n,j} = \sqrt{\frac{2}{N+1}} \sin\frac{jn\pi}{N+1}, \quad j = 1, 2, \dots, N. \tag{6.20}$$

It may be shown that these form an orthonormal set.

Figure 6.2 shows the positions of the particles in the eigenvectors of the potential energy matrix for the harmonic crystal consisting of four particles. For the lowest frequency mode, the particles are all displaced in the same direction (in phase), with the central two particles having twice the amplitude of the outer two. For the second lowest frequency, the middle pair of particles are out of phase. For the second highest frequency, both outer pairs of particles are out of phase. And finally, for the highest frequency, all three consecutive pairs of particles are out of phase.

6.2.3 Normal modes

The matrix of eigenvectors,

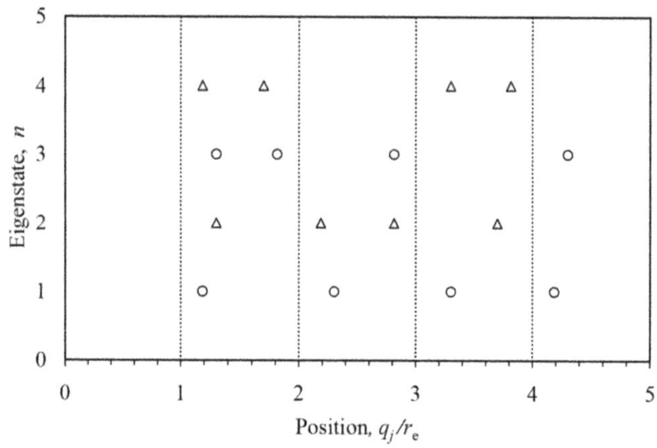

Figure 6.2. Eigenvectors of the potential energy matrix, as described by positions of the $N = 4$ particles ($\lambda = \kappa = m\omega_{LJ}^2$). The corresponding eigenvalues increase in magnitude from bottom to top. The dotted lines indicate the lattice positions $\bar{q}_j$. The displacements are all scaled by the same amount.

$$\underline{\underline{X}} \equiv \{\mathbf{u}_1, \mathbf{u}_2, \ldots, \mathbf{u}_N\} = \begin{pmatrix} u_{1,1} & u_{2,1} & \cdots & u_{N,1} \\ u_{1,2} & u_{2,2} & \cdots & u_{N,2} \\ \vdots & & \ddots & \vdots \\ u_{1,N} & u_{2,N} & \cdots & u_{N,N} \end{pmatrix}, \tag{6.21}$$

is orthogonal, $\underline{\underline{X}}^{\mathrm{T}}\underline{\underline{X}} = \underline{\underline{X}}\,\underline{\underline{X}}^{\mathrm{T}} = \mathrm{I}$. (It is also symmetric, $\underline{\underline{X}}^{\mathrm{T}} = \underline{\underline{X}}$.) With this the potential energy may be written

$$\begin{aligned} U(\mathbf{q}) &= \frac{-\lambda}{2}\underline{\underline{K}}:\mathbf{dd} \\ &= \frac{-\lambda}{2}(\underline{\underline{X}}^{\mathrm{T}}\mathbf{d})^{\mathrm{T}}\underline{\underline{X}}^{\mathrm{T}}\,\underline{\underline{K}}\,\underline{\underline{X}}(\underline{\underline{X}}^{\mathrm{T}}\mathbf{d}) \\ &= \frac{-\lambda}{2}\mathbf{q}'^{\mathrm{T}}\underline{\underline{D}}\mathbf{q}'. \end{aligned} \tag{6.22}$$

Here $\underline{\underline{D}} \equiv \underline{\underline{X}}^{\mathrm{T}}\underline{\underline{K}}\,\underline{\underline{X}}$ is diagonal, $D_{n,n'} = \mu_n \delta_{n,n'}$, and the mode amplitudes are defined as

$$\mathbf{q}' \equiv \underline{\underline{X}}^{\mathrm{T}}\mathbf{d}. \tag{6.23}$$

The amplitudes are a function of the ordered particle positions $\tilde{\mathbf{q}}(\mathbf{q})$.

Suppose that the particles have mass m, and that the momentum of the jth particle is $p_j = m\dot{q}_j$. Hence the classical kinetic energy is

$$\mathcal{K}(\mathbf{p}) = \frac{1}{2m}\mathbf{p}\cdot\mathbf{p} = \frac{1}{2m}\mathbf{p}'\cdot\mathbf{p}', \tag{6.24}$$

where $\mathbf{p}' \equiv \underline{\underline{X}}^{\mathrm{T}}\mathbf{p}$. Accordingly the classical Hamiltonian is

$$\mathcal{H}(\Gamma) = \mathcal{K}(\mathbf{p}) + U(\mathbf{q}) = \frac{1}{2m}\mathbf{p}' \cdot \mathbf{p}' - \frac{\lambda}{2}\underline{\underline{D}}: \mathbf{q}'\mathbf{q}'. \tag{6.25}$$

Evidently, the normal modes represent independent harmonic oscillators.

6.2.4 Quantum mechanics

The Hamiltonian operator in the normal mode representation is

$$\begin{aligned} \hat{\mathcal{H}} &= \frac{1}{2m}\hat{\mathbf{p}}' \cdot \hat{\mathbf{p}}' - \frac{\lambda}{2}\underline{\underline{D}}: \mathbf{q}'\mathbf{q}' \\ &= \frac{1}{2m}\sum_{n=1}^{N} \hat{p}_n'^2 - \frac{\lambda}{2}\sum_{n=1}^{N} \mu_n q_n'^2 \\ &= \sum_{n=1}^{N} \frac{\hbar\omega_n}{2}\left\{\hat{P}_n^2 + Q_n^2\right\}, \end{aligned} \tag{6.26}$$

where $\hat{P}_n \equiv \hat{p}_n'/\sqrt{m\hbar\omega_n}$, $Q_n \equiv q_n'\sqrt{m\omega_n/\hbar}$, and $m\omega_n^2 \equiv -\lambda\mu_n$. Since $\mu_n < 0$ and $\lambda > 0$, the frequencies are real.

For each mode, the energy eigenvalues are (Messiah 1961, Merzbacher 1970)

$$E_{n,l_n} = (l_n + 0.5)\hbar\omega_n, \quad l_n = 0, 1, 2, \ldots, \tag{6.27}$$

and the corresponding energy eigenfunctions are the Hermite functions (Messiah 1961, Merzbacher 1970),

$$\phi_{n,l_n}(Q_n) \equiv \frac{1}{\sqrt{2^{l_n} l_n! \sqrt{\pi}}} e^{-Q_n^2/2} \mathrm{H}_{l_n}(Q_n), \tag{6.28}$$

where $\mathrm{H}_{l_n}(Q)$ is the Hermite polynomial of degree l_n. These are normalized over the mode amplitudes,

$$\int_{-\infty}^{\infty} \mathrm{d}Q_n\, \phi_{n,l_n}(Q_n)^*\, \phi_{n,l_n}(Q_n) = 1. \tag{6.29}$$

The eigenfunctions of the system have product form,

$$\phi_\ell(\mathbf{Q}) = \prod_{n=1}^{N} \phi_{n,l_n}(Q_n), \tag{6.30}$$

and the energy in such an eigenstate is of course $E_\ell = \sum_{n=1}^{N} E_{n,l_n}$.

Because the transformation to modes $\mathbf{q} \Rightarrow \mathbf{Q}$ is orthogonal, these eigenfunctions are orthonormalized in the sense that

$$\int \mathrm{d}\mathbf{Q}\, \phi_{\ell'}(\mathbf{Q})^*\, \phi_\ell(\mathbf{Q}) = C \int \mathrm{d}\mathbf{q}\, \phi_{\ell'}(\mathbf{Q})^*\, \phi_\ell(\mathbf{Q}) = \delta_{\ell',\ell}. \tag{6.31}$$

Here $C \equiv \prod_{n=1}^{N} \sqrt{m\omega_n/\hbar}$ is a constant independent of the state of the modes, and so it has no more than trivial effects.

As mentioned above, this particular model potential for the harmonic crystal distinguishes the particles one from another (i.e., they are not identical), and so the energy eigenfunctions have no particular symmetry requirements and it is not possible to analyze the system in terms of bosons or fermions.

The reason for giving the energy eigenfunctions is that they facilitate the calculation of such quantities as the density profile of the crystal and the commutation function. For other quantities one can use the known results for the simple harmonic oscillator together with the fact that the modes are independent. In particular, the average energy of the harmonic crystal is

$$\bar{E}(T) = \sum_{n=1}^{N} \left\{ \frac{\hbar\omega_n}{2} + \frac{\hbar\omega_n e^{-\beta\hbar\omega_n}}{1 - e^{-\beta\hbar\omega_n}} \right\}, \tag{6.32}$$

and the heat capacity is

$$C_V(T) = k_B \sum_{n=1}^{N} \frac{(\beta\hbar\omega_n)^2 e^{-\beta\hbar\omega_n}}{(1 - e^{-\beta\hbar\omega_n})^2}. \tag{6.33}$$

The latter goes to zero exponentially as the temperature approaches absolute zero, in agreement with the prediction of Einstein (1907).

6.2.5 Numerical results

Figure 6.3 shows the quantized energy levels for the harmonic crystal for several sets of parameters. Initially the energy of each level increases rapidly with level number. But for large level numbers, the rate of change of energy with energy level is sub-linear. Or to put it another way, the density of energy states increases with increasing energy. For $l \gtrsim 10$, the data is well fitted by $E_l \propto l^{1/N}$, as can be seen in the inset to the figure.

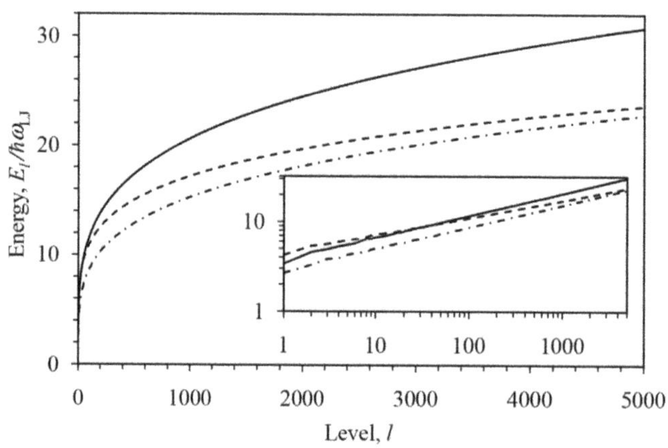

Figure 6.3. The first 5000 quantum energy levels for $N = 4$, $\lambda = \kappa = m\omega_{LJ}^2$ (solid curve); for $N = 4$, $\lambda = m\omega_{LJ}^2$, $\kappa = 0$, (dash–dotted curve); and for $N = 5$, $\lambda = \kappa = m\omega_{LJ}^2$ (dashed curve). In all cases, $\Delta_q = r_e$. Inset: Log–log plot.

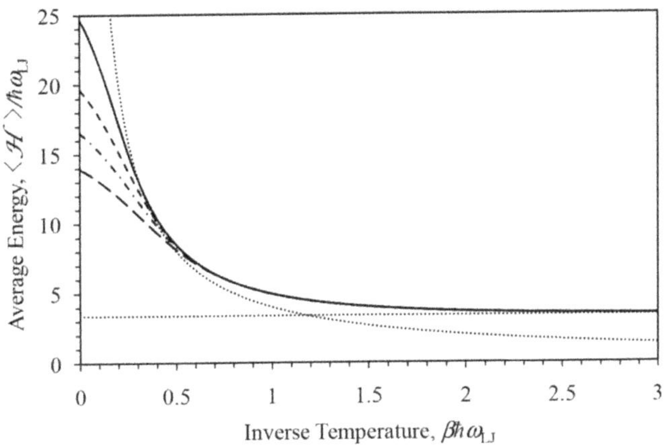

Figure 6.4. Average energy calculated with various numbers of levels (N = 4, $\Delta_q = r_e$, and $\lambda = \kappa = m\omega_{LJ}^2$). From bottom to top, l_{max}= 500 (long dashed), 1000 (dash–dotted), 2000 (short dashed), and 5000 (full). The dotted curve is the classical result, $\langle E \rangle_{cl} = N/\beta$. The dotted line is the energy of the ground state, $E_1 = \sum_{n=1}^{N} \hbar\omega_n/2$.

Figure 6.4 shows the average energy for a canonical equilibrium system as a function of inverse temperature. The data tests the dependence of the average on the number of levels used, l_{max}. One can see that for $\beta\hbar\omega_{LJ} \gtrsim 0.8$, the results for all $l_{max} \geqslant 500$ are indistinguishable. For $\beta\hbar\omega_{LJ} \lesssim 0.5$, there is a discernable difference between l_{max}= 2000 and 5000, and this difference increases with increasing temperature (decreasing inverse temperature).

The figure also shows the energy of the ground state, $E_1 = \sum_{n=1}^{N} \hbar\omega_n/2$. It can be seen that for these parameters, excited states make negligible contribution when the temperature is lower than $\beta\hbar\omega_{LJ} \gtrsim 2$.

The present crystal has $2N$ harmonic modes in the classical Hamiltonian (N in the potential energy, and N in the kinetic energy). Hence by the equipartition theorem, the average energy is $\langle \mathcal{H} \rangle_{cl} = N/\beta$, which is the dotted curve in figure 6.4. One can see that this lies increasingly below the quantum results as the temperature decreases (inverse temperature increases), and it lies increasingly above the quantum results as the temperature increases. In the present case there is a region, $0.3 \lesssim \beta\hbar\omega_{LJ} \lesssim 0.4$, in which the quantum results for l_{max} = 5000 coincide with the classical result. The data suggests what one knows to be true: the classical result must be the limiting result at high temperatures, but an increasing number of energy levels contribute to the average as the temperature increases. Hence for fixed l_{max}, there is always a temperature above which the quantum results become inaccurate.

One can draw the important conclusion from this figure that starting from the exact quantum approach is a very inefficient way of obtain the classical result, which is the exact high temperature limit.

Figure 6.5 shows the heat capacity of the crystal for the same case as the preceding figure. Two versions of the exact results are shown: the direct simple harmonic oscillator expression as the sum of the modes, equation (6.33), and the brute force approach that identifies and sums the first l_{max}= 5000 energy levels, using

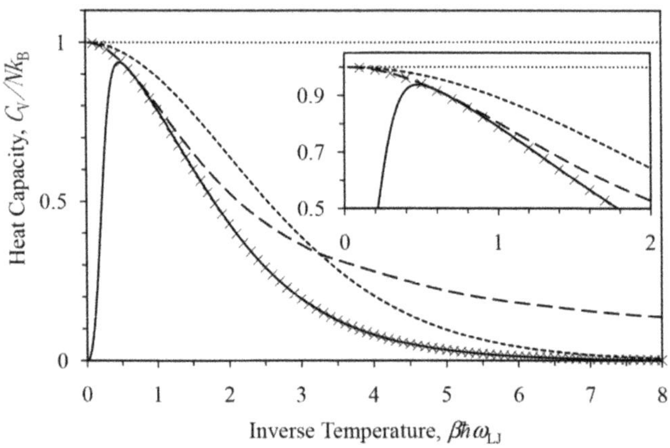

Figure 6.5. Heat capacity per particle of the harmonic crystal ($N = 4$, $\Delta_q = r_e$, and $\lambda = \kappa = m\omega_{LJ}^2$). The crosses are the exact analytic result, equation (6.33), the solid curve is the exact result with the first l_{max}= 5000 energy levels, the short-dashed curve is the Einstein model with $\omega_E = \omega_1$ ($k_B\Theta_E/\hbar\omega_{LJ} = 1.2$), and the long-dashed curve is the Debye model with $k_B\Theta_D/\hbar\omega_{LJ} = 3$. The dotted line is the classical asymptote. Inset: Magnified view at high temperatures.

the finite difference of the average energy from figure 6.4. It can be seen that the heat capacity goes to zero at zero temperature. The analytic exact result goes smoothly over to the classical asymptote at high temperatures. In contrast the brute force exact approach reaches a peak of about $C_V/Nk_B = 0.94$ at about $\beta\hbar\omega_{LJ} = 0.5$, and drops below the classical asymptote $C_V/Nk_B = 1$ for higher temperatures. This is an artifact of using a limited number of energy levels. As observed in connection with the preceding figure, the exact quantum approach is not recommended for obtaining the classical result. Figure 6.5 also tests the Debye model for the lattice vibrational heat capacity. The fitted Debye temperature was $k_B\Theta_D/\hbar\omega_{LJ} = 3$, which gives a reasonable fit for $\beta\hbar\omega_{LJ} \lesssim 0.8$. Although it correctly goes over to the classical limit at high temperatures, the Debye model with this parameter value does not fit the exact data at low temperatures. The Debye model takes the velocity of sound to be a constant, which neglects its dispersion with frequency and its temperature dependence.

The mode density of the harmonic crystal at $n = 1$, upon expanding the trigonometric functions for large N, is $g(\omega) = 2(N+1)^2\sqrt{m\kappa}/3\pi^2\lambda$, which gives $k_B\Theta_D/\hbar\omega_{LJ} = (3\pi^2N/2(N+1)^2)\lambda/\omega_{LJ}\sqrt{m\kappa} = 2.37$ for the present case, $N = 4$ and $\lambda = \kappa = m\omega_{LJ}^2$. The discrepancy between the high temperature fit and the zero temperature calculation underscores the limitation of the Debye model in neglecting the frequency and temperature dependence of the sound velocity.

It should be mentioned that the low temperature value, $k_B\Theta_D/\hbar\omega_{LJ} = 2.37$, did not give a good fit of the Debye model to the exact calculations at low temperatures. Further, equation (6.33) shows that the exact heat capacity must vanish exponentially approaching absolute zero, $\propto e^{-\beta\hbar\omega_1}$, as $T \to 0$. Interestingly enough this agrees with the prediction of the Einstein model, undoubtedly because at low temperatures

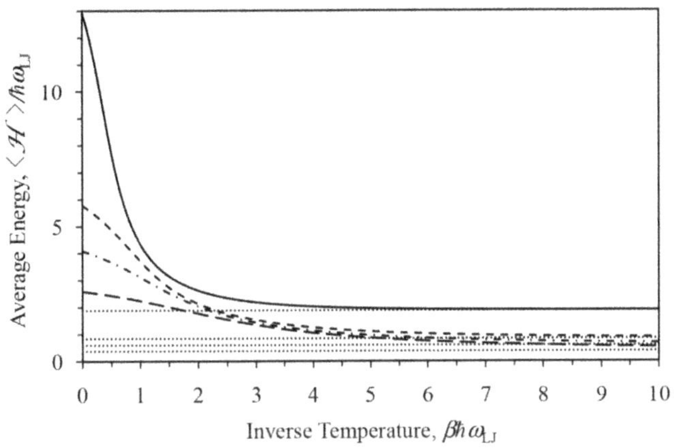

Figure 6.6. Average energy for various values of the interparticle spring constant λ ($N = 4$, $\Delta_q = r_e$, $\kappa = 0$, and $l_{max} = 5000$). From top to bottom, $\lambda / m\omega_{LJ}^2 = 0.5$ (solid curve), 0.1 (short-dashed curve), 0.05 (dash–dotted curve), and 0.02 (long-dashed curve). The dotted lines are the respective ground states.

only the lowest frequency mode is occupied, and Einstein (1907) assumes a single frequency for his model. In contrast Debye (1912) predicts for a one-dimensional system that the heat capacity should go to zero linearly with temperature as $T \to 0$, which is clearly incorrect. It can be seen in figure 6.5 that the model of Einstein becomes less accurate than that of Debye at intermediate temperatures as the higher order modes are activated. Figure 6.6 shows the average energy for several values of the interparticle spring constant λ. In this figure the lattice spring constant has been set to zero, $\kappa = 0$. The wall particles fixed at $q_0 = 0$ and $q_{N+1} = (N + 1)\Delta_q$ interact via λ with their moveable neighbors and serve to localize the harmonic crystal. In general at a given temperature, the average energy increases with increasing interparticle spring constant. In the low temperature limit, $\beta \to \infty$, the average energy is the ground state energy, $E_0 = \sum_{n=1}^{N} \hbar\omega_n / 2$. It can be seen in figure 6.6 that for $\beta\hbar\omega_{LJ} \gtrsim 3$–9 ($\lambda / m\omega_{LJ}^2 = 0.5$–0.02) the system may be considered to be in the ground state. The system reaches the ground state at higher temperatures for larger interparticle spring constants.

Figure 6.7 shows the density profiles corresponding to the same cases as the preceding figure at the temperature $\beta\hbar\omega_{LJ} = 2$. As mentioned, although the lattice spring constant has been set to zero, $\kappa = 0$, the fixed wall particles at $q_0 = 0$ and $q_{N+1} = (N + 1)\Delta_q$ prevent the moveable atoms from wandering because they interact via λ with their moveable neighbors. The profiles are normalized to integrate to $N = 4$. In general the peaks and troughs in the profiles indicate that the particles are mainly localized to their respective lattice positions. At the highest interparticle spring constant shown, the density is zero between lattice points, which means that there is little overlap between the particles. Conversely, at the lowest coupling shown, the density peaks are much broader and there is a high probability of finding a particle between the lattice positions.

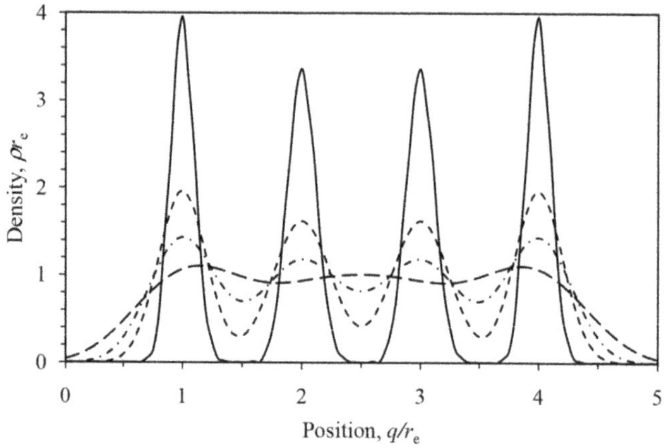

Figure 6.7. Density profiles for various values of the interparticle spring constant λ ($N = 4$, $\Delta_q = r_e$, $\kappa = 0$, $l_{max} = 5000$, and $\beta\hbar\omega_{LJ} = 2$). From top to bottom at the peaks, $\lambda/m\omega_{LJ}^2 = 0.5$ (solid curve), 0.1 (short-dashed curve), 0.05 (dash–dotted curve), and 0.02 (long-dashed curve).

6.3 Loop Markov superposition approximation

In chapter 5, the λ-transition in liquid helium was explored using Monte Carlo simulations, section 5.4. That method obtained successive terms in the series of permutation loop grand potentials. It is obvious that in such an approach there will always be a limit on how large in size l are the loops that can be included. In this section a practical approximation is developed that allows the infinite loop series to be resummed and calculated with no more than a one-dimensional Fourier transform and integration.

The Monte Carlo results in section 5.4 relied upon certain expressions for statistical averages in which the momentum integration was performed explicitly. This situation arises quite generally when the commutation function is neglected. (This function is introduced in the formal transformation to classical phase space in chapter 7.) Hence the approximation now given has wide applicability beyond liquid helium and Bose–Einstein condensation. The simulation results of section 5.4 serve as benchmarks against which the present approximate approach can be tested.

In the first instance we shall focus on the loop grand potential, $-\beta\Omega^{\pm,(l)} = (\pm 1)^{l-1}\langle G^{(l)}\rangle$. The loop Gaussian is $G^{(l)}(\mathbf{q}^l) = (N!/(N-l)!l)e^{-\pi\mathcal{L}_l(\mathbf{q}^l)/\Lambda^2}$ for averaging, with the sum of the squares of the bonds being $\mathcal{L}_l(\mathbf{q}^l)^2 = q_{12}^2 + q_{23}^2 + \cdots + q_{l1}^2$, and the thermal wavelength being $\Lambda = \sqrt{\beta h^2/2\pi m}$.

The present canonical equilibrium average of the loop Gaussian can be written in terms of the canonical equilibrium l-particle density. This is normalized such that $\int d\mathbf{q}^l\, \rho_N^{(l)}(\mathbf{q}^l) = N!/(N-l)!$ (Attard 2002, equation (7.72)). One has

$$-\beta\Omega^{\pm,(l)} = (\pm 1)^{l-1}\langle G^{(l)}\rangle = \frac{(\pm 1)^{l-1}(N-l)!}{N!}\int d\mathbf{q}^l\, \rho_N^{(l)}(\mathbf{q}^l)\, G^{(l)}(\mathbf{q}^l)$$
$$= \frac{(\pm 1)^{l-1}\rho^l}{l}\int d\mathbf{q}^l\, g_N^{(l)}(\mathbf{q}^l)\, e^{-\pi\mathcal{L}_l^2(\mathbf{q}^l)/\Lambda^2}. \tag{6.34}$$

Here ρ is the one-particle number density, a homogeneous system being assumed. This expression is formally exact (accepting the neglect of the commutation function and the validity of the loop grand potential).

We now introduce the Markov superposition approximation for the l-particle density. This approximation was originally developed for one-dimensional applications (Attard *et al* 1997), although it can readily be generalized to higher dimensions (Attard 1999, Attard 2018). Since the permutation loop is essentially a one-dimensional object in three-dimensional space, one can expect the Markov approach to be particularly easy to apply. Previous tests for highly correlated systems have shown that for the l-particle density the Markov superposition approximation is much more reliable than the Kirkwood superposition approximation (Attard *et al* 1997, Attard 1999, Attard 2018). The Markov superposition approximation for a loop of size $l \geqslant 2$ takes the l-loop distribution function to be the product of pair distribution functions consecutive around the loop,

$$g_N^{(l)}(\mathbf{q}^l) = g^{(2)}(q_{12})g^{(2)}(q_{23})\ldots g^{(2)}(q_{l1}). \tag{6.35}$$

A homogeneous system is again assumed, and the pair distribution function is taken to be only a function of separation. This is not quite the original Markov superposition approximation and it might more accurately be called a Markov–Kirkwood variant, as is briefly discussed in the tests below.

With this the loop Markov superposition approximation for the loop grand potential is

$$-\beta\Omega^{\pm,(l)} = \frac{(\pm 1)^{l-1}\rho^l}{l}\int \mathrm{d}\mathbf{q}^l\, e^{-\pi \mathcal{L}_l^2(\mathbf{q}^l)/\Lambda^2} g^{(2)}(q_{12})g^{(2)}(q_{23})\ldots g^{(2)}(q_{l1}), \quad l \geqslant 3. \tag{6.36}$$

The dimer contribution is

$$\begin{aligned} -\beta\Omega^{\pm,(2)} &= \frac{\pm\rho^2}{2}\int \mathrm{d}\mathbf{q}_1\,\mathrm{d}\mathbf{q}_2\, e^{-2\pi q_{12}^2/\Lambda^2} g^{(2)}(q_{12}) \\ &= \frac{\pm\rho^2 V}{2}\int \mathrm{d}\mathbf{q}_1\, e^{-2\pi q_1^2/\Lambda^2} g^{(2)}(q_1), \end{aligned} \tag{6.37}$$

which is exact.

Using the fact that in three-dimensional space

$$\int \mathrm{d}\mathbf{k}_1\, e^{-i\mathbf{k}_1\cdot\mathbf{q}_1} = (2\pi)^3\delta(\mathbf{q}_1), \tag{6.38}$$

the Fourier transform pair is defined to be

$$\hat{f}(k) = \int \mathrm{d}\mathbf{q}\, e^{-i\mathbf{k}\cdot\mathbf{q}} f(q), \quad \text{and} \quad f(q) = \frac{1}{(2\pi)^3}\int \mathrm{d}\mathbf{k}\, e^{i\mathbf{k}\cdot\mathbf{q}}\hat{f}(k). \tag{6.39}$$

For the present radial function the Fourier transform is one-dimensional, $\hat{f}(k) = 4\pi\int_0^\infty \mathrm{d}q\, q^2 f(q)\sin(kq)/kq$.

Define $F(q) \equiv e^{-\pi q^2/\Lambda^2} g^{(2)}(q)$ to be the Gaussian-weighted radial distribution function. This is a well-behaved short-ranged function whose Fourier transform is straightforward to obtain numerically. Then with the loop Markov superposition approximation, and using homogeneity,

$$\begin{aligned} -\beta\Omega^{\pm,(l)} &= \frac{(\pm 1)^{l-1}\rho^l}{l} \int \mathrm{d}\mathbf{q}^l \, F(q_{12})F(q_{23})\ldots F(q_{l1}) \\ &= \frac{(\pm 1)^{l-1}\rho^l}{l} V \int \mathrm{d}\mathbf{q}^{l-1} \, F(q_{12})F(q_{23})\ldots F(q_{l-1})F(q_1) \\ &= \frac{(\pm 1)^{l-1}\rho^l V}{l(2\pi)^{3l}} \int \mathrm{d}\mathbf{k}^l \int \mathrm{d}\mathbf{q}^{l-1} \, e^{i\mathbf{k}_1\cdot\mathbf{q}_{12}} e^{i\mathbf{k}_2\cdot\mathbf{q}_{23}} \ldots e^{i\mathbf{k}_{l-1}\cdot\mathbf{q}_{l-1}} e^{-i\mathbf{k}_l\cdot\mathbf{q}_1} \\ &\quad \times \hat{F}(k_1)\hat{F}(k_2)\ldots\hat{F}(k_l) \\ &= \frac{(\pm 1)^{l-1}\rho^l V}{(2\pi)^3 l} \int \mathrm{d}\mathbf{k}_1 \, \hat{F}(k_1)^l, \quad l \geqslant 3. \end{aligned} \tag{6.40}$$

This expression for the loop grand potential is just a one-dimensional integral of the Fourier transform of the Gaussian-weighted radial distribution function.

Readers from a certain era will recognize these as the ring diagrams, which underlie the hypernetted chain closure approximation (see Attard 2002, equation (10.78)). The function $F(q)$ plays the role of the total correlation function. The factor of 1/2 that appears in the hypernetted chain case corrects for double counting of the clockwise and mirrored anti-clockwise rings, which are identical. In the present case one must always traverse the loop clockwise, and so the anti-clockwise version is counted as distinct even though it is numerically equal to the clockwise version. For example, the 3-loop $1 \to 2 \to 3 \to 1$ is distinct from $1 \to 3 \to 2 \to 1$ even though their Gaussians are numerically equal.

The series of loop grand potentials gives the symmetrization contribution to the grand potential,

$$\begin{aligned} -\beta \sum_{l=2}^{\infty} \Omega^{\pm,(l)} &= \sum_{l=2}^{\infty} (\pm 1)^{l-1} \langle G^{(l)} \rangle \\ &= -\beta\Omega^{\pm,(2)} + \sum_{l=3}^{\infty} \frac{\pm(\pm\rho)^l V}{(2\pi)^3 l} \int \mathrm{d}\mathbf{k}_1 \, \hat{F}(k_1)^l \\ &= \frac{\mp V}{(2\pi)^3} \int \mathrm{d}\mathbf{k}_1 \left\{ \ln\left[1 \mp \rho\hat{F}(k_1)\right] \pm \rho\hat{F}(k_1) + \frac{\rho^2}{2}\hat{F}(k_1)^2 \right\} \\ &\quad \pm \frac{\rho^2 V}{2} \int \mathrm{d}\mathbf{q}_1 \, e^{-2\pi q_1^2/\Lambda^2} g^{(2)}(q_1). \end{aligned} \tag{6.41}$$

This is a surprisingly simple closed form expression for the total contribution to the grand potential from symmetrization loops. It requires only the classical canonical equilibrium radial distribution function as input, which is relatively trivial to obtain by computer simulation, integral equation approaches, or even x-ray or neutron scattering measurements.

In order for the argument of the logarithm to be positive, for bosons one requires $\rho\hat{F}(k_1) < 1$. Assuming that the Fourier transform is a maximum at $k_1 = 0$, this condition is

$$4\pi\rho \int_0^\infty \mathrm{d}q\; q^2 e^{-\pi q^2/\Lambda^2} g^{(2)}(q) < 1. \tag{6.42}$$

(For fermions, the integral must be >-1, which it certainly is.) Presumably, within the loop Markov superposition approximation, this defines the spinodal boundary of a thermodynamic phase.

6.3.1 Temperature derivatives

The temperature derivative (at constant number density) is

$$\frac{\partial(-\beta\Omega^{\pm,(l)})}{\partial\beta} = \frac{(\pm 1)^{l-1}\rho^l V}{(2\pi)^3} \int \mathrm{d}\mathbf{k}_1\; \hat{F}(k_1)^{l-1}\hat{\dot{F}}(k_1), \quad l \geqslant 2, \tag{6.43}$$

with

$$\dot{F}(q) \equiv \frac{\partial F(q)}{\partial\beta} = e^{-\pi q^2/\Lambda^2}\dot{g}^{(2)}(q) + \frac{\pi q^2}{\beta\Lambda^2} e^{-\pi q^2/\Lambda^2} g^{(2)}(q). \tag{6.44}$$

With

$$\rho_N^{(2)}(\mathbf{q}_N, \mathbf{q}_{N-1}) = \frac{N(N-1)}{Q(N, V, T)} \int_V \mathrm{d}\mathbf{q}^{N-2}\; e^{-\beta U(\mathbf{q}^N)} \tag{6.45}$$

one has

$$\begin{aligned} g^{(2)}(q) &= \frac{V^2}{Q(N, V, T)} \int_V \mathrm{d}\mathbf{q}^{N-2}\; e^{-\beta U(\mathbf{q}^{N-2}, q\hat{\mathbf{z}}, \mathbf{0})} \\ &= V^2 \left\langle \delta(\mathbf{q}_N - \mathbf{0})\, \delta(\mathbf{q}_{N-1} - q\hat{\mathbf{z}}) \right\rangle. \end{aligned} \tag{6.46}$$

Hence

$$\begin{aligned} \dot{g}^{(2)}(q) &= \langle U \rangle\, g^{(2)}(q) - \frac{V^2}{Q} \int_V \mathrm{d}\mathbf{q}^{N-2}\; e^{-\beta U(\mathbf{q}^{N-2}, q\hat{\mathbf{z}}, \mathbf{0})} U(\mathbf{q}^{N-2}, q\hat{\mathbf{z}}, \mathbf{0}) \\ &= \langle U \rangle\, g^{(2)}(q) - V^2 \left\langle U\, \delta(\mathbf{q}_N - \mathbf{0})\, \delta(\mathbf{q}_{N-1} - q\hat{\mathbf{z}}) \right\rangle. \end{aligned} \tag{6.47}$$

Alternatively, in total one has

$$\begin{aligned} \frac{\partial}{\partial\beta} \sum_{l=2}^{\infty} (-\beta\Omega^{\pm,(l)}) &= \frac{\pm V}{(2\pi)^3} \frac{\partial}{\partial\beta} \int \mathrm{d}\mathbf{k}_1\; \left\{ \ln\left[1 \mp \rho\hat{F}(k_1)\right] \pm \rho\hat{F}(k_1) \right\} \\ &= \frac{V}{(2\pi)^3} \int \mathrm{d}\mathbf{k}_1 \left\{ \rho\hat{\dot{F}}(k_1) - \frac{\rho\hat{\dot{F}}(k_1)}{1 \mp \rho\hat{F}(k_1)} \right\}. \end{aligned} \tag{6.48}$$

The second derivative is

$$\frac{\partial^2}{\partial\beta^2}\sum_{l=2}^{\infty}(-\beta\Omega^{(l)}) = \frac{V}{(2\pi)^3}\int \mathrm{d}\mathbf{k}_1\left\{\rho\hat{\ddot{F}}(k_1) - \frac{\rho\hat{\ddot{F}}(k_1)}{1 \mp \rho\hat{F}(k_1)} \mp \frac{\rho^2\hat{\dot{F}}(k_1)^2}{[1 \mp \rho\hat{F}(k_1)]^2}\right\}. \quad (6.49)$$

One has

$$\begin{aligned}\ddot{F}(q) &\equiv \frac{\partial^2 F(q)}{\partial\beta^2}\\ &= e^{-\pi q^2/\Lambda^2}\ddot{g}^{(2)}(q) + \frac{2\pi q^2}{\beta\Lambda^2}e^{-\pi q^2/\Lambda^2}\dot{g}^{(2)}(q)\\ &\quad+\left[\frac{\pi^2 q^4}{\beta^2\Lambda^4} - \frac{2\pi q^2}{\beta^2\Lambda^2}\right]e^{-\pi q^2/\Lambda^2}g^{(2)}(q),\end{aligned} \quad (6.50)$$

and

$$\begin{aligned}\ddot{g}^{(2)}(q) &= \frac{\partial}{\partial\beta}\left\{\langle U\rangle\, g^{(2)}(q) - V^2\left\langle U\,\delta(\mathbf{q}_N - \mathbf{0})\,\delta(\mathbf{q}_{N-1} - q\hat{\mathbf{z}})\right\rangle\right\}\\ &= \langle U\rangle\,\dot{g}^{(2)}(q) + \langle U\rangle^2\, g^{(2)}(q) - \langle U^2\rangle\, g^{(2)}(q)\\ &\quad+ V^2\left\langle U^2\,\delta(\mathbf{q}_N - \mathbf{0})\,\delta(\mathbf{q}_{N-1} - q\hat{\mathbf{z}})\right\rangle\\ &\quad- V^2\langle U\rangle\left\langle U\,\delta(\mathbf{q}_N - \mathbf{0})\,\delta(\mathbf{q}_{N-1} - q\hat{\mathbf{z}})\right\rangle\\ &= \langle U\rangle^2\, g^{(2)}(q) - V^2\langle U\rangle\left\langle U\,\delta(\mathbf{q}_N - \mathbf{0})\,\delta(\mathbf{q}_{N-1} - q\hat{\mathbf{z}})\right\rangle\\ &\quad+ \langle U\rangle^2\, g^{(2)}(q) - \langle U^2\rangle\, g^{(2)}(q)\\ &\quad+ V^2\left\langle U^2\,\delta(\mathbf{q}_N - \mathbf{0})\,\delta(\mathbf{q}_{N-1} - q\hat{\mathbf{z}})\right\rangle\\ &\quad- V^2\langle U\rangle\left\langle U\,\delta(\mathbf{q}_N - \mathbf{0})\,\delta(\mathbf{q}_{N-1} - q\hat{\mathbf{z}})\right\rangle\\ &= 2\langle U\rangle^2\, g^{(2)}(q) - 2V^2\langle U\rangle\left\langle U\,\delta(\mathbf{q}_N - \mathbf{0})\,\delta(\mathbf{q}_{N-1} - q\hat{\mathbf{z}})\right\rangle\\ &\quad- \langle U^2\rangle\, g^{(2)}(q) + V^2\left\langle U^2\,\delta(\mathbf{q}_N - \mathbf{0})\,\delta(\mathbf{q}_{N-1} - q\hat{\mathbf{z}})\right\rangle.\end{aligned} \quad (6.51)$$

6.3.2 Average bond length in a loop

The root mean square bond length for an l-loop is $\bar{b}_l = \sqrt{\langle\mathcal{L}_l^2 G^{(l)}\rangle/\langle G^{(l)}\rangle\, l}$. Define $F_2(q) \equiv q^2 e^{-\pi q^2/\Lambda^2}g^{(2)}(q) = q^2 F(q)$. One has

$$\begin{aligned}\langle\mathcal{L}_l^2 G^{(l)}\rangle &= \frac{l\rho^l}{l}\int \mathrm{d}\mathbf{q}^l\; e^{-\pi q_{12}^2/\Lambda^2}e^{-\pi q_{23}^2/\Lambda^2}\ldots e^{-\pi q_{l1}^2/\Lambda^2}\\ &\quad\times g^{(2)}(q_{12})g^{(2)}(q_{23})\ldots g^{(2)}(q_{l1})\; q_{12}^2\\ &= \rho^l\int \mathrm{d}\mathbf{q}^l\; F_2(q_{12})F(q_{23})\ldots F(q_{l1})\\ &= \frac{\rho^l V}{(2\pi)^3}\int \mathrm{d}\mathbf{k}_1\; \hat{F}_2(k_1)\hat{F}(k_1)^{l-1}, \quad l \geqslant 3.\end{aligned} \quad (6.52)$$

The l in the numerator of the first equality comes from the fact that there are l bonds $q_{j,j+1}$ that are equivalent to q_{12} upon averaging. Hence

$$\bar{b}_l^2 = \frac{\int \mathrm{d}\mathbf{k}_1\ \hat{F}_2(k_1)\hat{F}(k_1)^{l-1}}{\int \mathrm{d}\mathbf{k}_1\ \hat{F}(k_1)^l}, \quad l \geqslant 3. \tag{6.53}$$

The mean square dimer bond length is

$$\bar{b}_2^2 = \frac{\int \mathrm{d}\mathbf{q}_1\ e^{-2\pi q_1^2/\Lambda^2} g^{(2)}(q_1) q_1^2}{\int \mathrm{d}\mathbf{q}_1\ e^{-2\pi q_1^2/\Lambda^2} g^{(2)}(q_1)}. \tag{6.54}$$

6.3.3 Tests

The loop Markov superposition approximation is tested against simulation results for liquid helium near the computed λ-transition in table 6.1. As detailed in table 5.1, in the simulations the central density at this state point is approximately $\rho\sigma^3 = 0.62$ ($\rho_m = 246.76$ kg m^{-3}). The saturation mass density at 5 K is 100.8 kg m^{-3}, and that at 2.177 K is 146.2 kg m^{-3} (http://www.engineeringtoolbox.com). The thermal wavelength at the temperature of the data in the table is $\Lambda = 1.04\sigma$.

It can be seen that loop grand potential, the average of the loop Gaussian, given by the simulations decreases with increasing loop size. This trend also holds for the loop Markov superposition approximation. The latter is quantitatively accurate for dimers and trimers, but less so for tetramers where it is about a factor of two too large. It is about 50% too large for pentamers. The total Markov loop Gaussian is $\langle G \rangle = 0.801\,85$. Overall one would say that given the simplicity of the theory the performance of the loop Markov superposition approximation in predicting the average value of the loop Gaussian is quite satisfying.

Table 6.1. Boson loop grand potential, $\beta\Omega^{+,\,l} = \langle G^{(l)} \rangle$, and root mean square bond length $\bar{b}_l$ given by Monte Carlo simulations and by the loop Markov approximation for helium-4 at $k_B T/\varepsilon = 1.05$ ($T = 10.73$ K) and $\rho_{tot}\sigma^3 = 0.3$ ($\rho_m = 119.4$ kg m^{-3}).

	$\beta\Omega^{+,\,l}$		$\bar{b}_l$	
l	Monte Carlo	Markov	Monte Carlo	Markov
2	0.803 4(26)	0.787 4	1.071 1	1.071 0
3	0.016 130(97)	0.017 61	1.128 5	1.126 6
4	$6.824(63) \times 10^{-4}$	1.2×10^{-3}	1.134 8	1.126 8
5	$4.271(61) \times 10^{-5}$	6.9×10^{-5}	1.134 7	1.130 6
6		4.7×10^{-6}		1.131 9
7		3.4×10^{-7}		1.133 3

The stability integral, equation (6.42), has value 0.103 at this particular thermodynamic state, as obtained from the simulated radial distribution function. As can be seen from the loop grand potential data in the table, the loop series appears to be convergent. At the same overall density that integral has value 1.02 for $k_B T/\varepsilon = 0.6$ (6.132 K). At higher temperatures than this, the integral is less than one, and the loop heat capacity decreases with increasing loop size. At this temperature and below, the integral is greater than one, and the loop heat capacity increases with increasing loop size. This suggests that the stability criteria is relatively accurate.

It can also be seen in table 6.1 that the simulated root mean square loop bond length is relatively insensitive to the loop size. The simulations show that it increases weakly with increasing loop size. For comparison, the first peak of the simulated radial distribution function occurs at 1.11 ± 0.02. The atoms in a dimers are separated on average by less than this, but consecutive atoms in trimers and above are on average separated by more than this. Again the performance of the loop Markov superposition approximation is very good, since it predicts the same trend with loop size, and is within 1% of the Monte Carlo results.

The Markov approximation is tested at a lower temperature in table 6.2. The central density at this state point is $\rho\sigma^3 = 0.88$ ($\rho_m = 350.2$ kg m^{-3}). It can be seen that its performance for the grand potential in this case is not as good. The Markov results for the loop grand potential would agree better with the Monte Carlo results if they were divided by $l - 1$.

The bond lengths given by the loop Markov superposition approximation in table 6.2 are again in good agreement with the Monte Carlo results. This is not a very sensitive test. The bonds are slightly longer at this lower temperature than those that occur at the higher temperature of the previous table. This appears to be due to the fact that the classical penetration into the core region is greater at higher temperatures than at lower. (Incidentally, this point suggests that symmetrization effects will be larger when the commutation function is included as it gives greater quantum penetration into the core than the classical approach allows.)

In table 6.2, it can be seen that the loop grand potentials are large, but decreasing with increasing loop number. The Monte Carlo loop heat capacities for this thermodynamic state point, $k_B T/\varepsilon = 0.6$, $\rho_{tot}\sigma^3 = 0.3$, are approximately constant

Table 6.2. Same as preceding table but for $k_B T/\varepsilon = 0.6$ ($T = 6.132$ K).

	$\beta\Omega^{+,l}$		$\bar{b}_l$	
l	Monte Carlo	Markov	Monte Carlo	Markov
2	29.40(41)	30.04	1.116 41(94)	1.114 8
3	8.15(24)	16.50	1.180 6(12)	1.173 5
4	3.33(17)	11.09	1.195 1(12)	1.170 8
5	1.72(13)	6.49	1.201 1(11)	1.179 5
6		4.50		1.182 8
7		3.31		1.186 5

with loop size. The integral in equation (6.42) is 1.056. This state point seems to be on the border between convergence and divergence of the loop series. Probably convergence depends upon the loop quantity that is being obtained.

The probable reason for the overestimate of the loop grand potential evident in table 6.2 is that the version of the Markov approximation given above is actually a hybrid of the original and the Kirkwood superposition approximation. It is actually equal to the Kirkwood superposition approximation for $l = 3$. In a highly correlated system the Kirkwood superposition approximation is known to overestimate the correlations (Attard *et al* 1997, Attard 1999, Attard 2018). The original Markov superposition approximation applied to a loop is

$$g_N^{(l)}(\mathbf{q}^l) = g^{(2)}(q_{12})g^{(2)}(q_{23})\ldots g^{(2)}(q_{l-1,l}). \tag{6.55}$$

This contains $l - 1$ factors. Here there is nothing to prevent overlap of the first and last atoms in the loop, which is the region where the Gaussian between them is largest. It may be better to use instead

$$g_N^{(l)}(\mathbf{q}^l) = g^{(2)}(q_{12})g^{(2)}(q_{23})\ldots g^{(2)}(q_{l-1,l})\Theta(q_{l,1} - \sigma). \tag{6.56}$$

Here $\Theta(x)$ is the Heaviside step function. These alternatives will reduce the estimate of the loop grand potential and may well perform better in more highly coupled systems.

Table 6.3 compares results at the lower temperature $T = 5.11$ K for the Markov–Kirkwood superposition approximation, Markov 1, equation (6.35), and the Markov superposition approximation, Markov 2, equation (6.56). It can be seen that the latter performs much better for larger loops. The table also gives Monte Carlo results for the heat capacity. Whereas the series for the loop grand potential is borderline convergent, the loop series for the heat capacity is clearly divergent at this temperature, at least for the terms shown. The integral in equation (6.42) is 1.67.

On the basis of the present numerical results the loop Markov approximation could be described as semi-quantitative in accuracy. It certainly appears to give the

Table 6.3. Boson loop grand potential and heat capacity per particle given by Monte Carlo simulations and by the loop Markov approximation for helium at $k_BT/\varepsilon = 0.5$ ($T = 5.11$ K) and $\rho_{tot}\sigma^3 = 0.3$ ($\rho_m = 119.4$ kg m^{-3}).

	$\beta\Omega^{+,l}$			$C_V^{+,(l)}/Nk_B$
l	Monte Carlo	Markov 1	Markov 2	Monte Carlo
1				3.603(86)
2	64.477(49)	65.9	65.9	0.855(95)
3	31.450(49)	67.4	16.1	1.927(96)
4	21.239(54)	71.8	17.1	2.88(14)
5	19.060(78)	68.9	16.9	4.49(23)
6	17.92(14)	77.6	19.2	6.03(53)
7	19.83(25)	93.4	23.4	9.7 (10)

correct physical picture of the behavior of permutation loops in liquid helium near the λ-transition.

Higher order Markov approximations for the l-particle distribution function exist (Attard 2018, section 5.6.3). More accurate results might be obtained using, for example, products of the pair and triplet distribution functions.

6.4 Symmetrization for spin-position factorization

6.4.1 Exact form

The set of commuting dynamical variables for one particle j may be taken to be $\mathbf{x}_j = \{\mathbf{q}_j, \sigma_j\}$, where $\sigma_j \in \{-S, -S+1, \dots, S\}$ is the z-component of the spin of particle j (Messiah 1961, section 14.1, Merzbacher 1970, section 20.5). Note that here σ is *not* a spin operator or a Pauli spin matrix. Label the $2S+1$ spin eigenstates of particle j by $s_j \in \{-S, -S+1, \dots, S\}$, and the spin basis function by $\alpha_{s_j}(\sigma_j) = \delta_{s_j,\sigma_j}$. Note that this is *not* a spinor. For N particles, $\boldsymbol{\sigma} \equiv \{\sigma_1, \sigma_2, \dots, \sigma_N\}$, and similarly for $\mathbf{s}$ and $\mathbf{q}$, and the basis functions for spin space are $\alpha_{\mathbf{s}}(\boldsymbol{\sigma}) = \delta_{\mathbf{s},\boldsymbol{\sigma}} = \prod_{j=1}^{N} \delta_{s_j,\sigma_j}$. The states $\mathbf{n}$ may be single or multi-particle states.

An unsymmetrized wave function $\psi(\mathbf{x})$ in general has symmetrized form

$$\psi^{\pm}(\mathbf{x}) \equiv \frac{1}{\sqrt{\chi^{\pm} N!}} \sum_{\hat{\mathrm{P}}} (\pm 1)^p \psi(\hat{\mathrm{P}}\mathbf{x}), \tag{6.57}$$

with the symmetrization or overlap factor being

$$\chi^{\pm} \equiv \sum_{\hat{\mathrm{P}}} (\pm 1)^p \langle \psi(\hat{\mathrm{P}}\mathbf{x}) | \psi(\mathbf{x}) \rangle. \tag{6.58}$$

Alternatively, one can expand the wave function in terms of spin-position basis functions,

$$\psi(\mathbf{x}) = \sum_{\mathbf{s},\mathbf{n}} \langle \alpha_{\mathbf{s}} \phi_{\mathbf{n}} | \psi \rangle \, \Phi_{\mathbf{n},\mathbf{s}}(\mathbf{x}), \quad \Phi_{\mathbf{n},\mathbf{s}}(\mathbf{x}) \equiv \alpha_{\mathbf{s}}(\boldsymbol{\sigma}) \phi_{\mathbf{n}}(\mathbf{q}). \tag{6.59}$$

Here and below the $\alpha_{\mathbf{s}}$ will be called the spin basis functions, and the $\phi_{\mathbf{n}}$ will be called the position basis functions. This nomenclature favors brevity over precision; better might be, for example, the basis functions for spin and position space, respectively. Instead of position one could use the momentum representation. It will often prove useful to choose the $\phi_{\mathbf{n}}$ to be energy eigenfunctions. The $\alpha_{\mathbf{s}}(\boldsymbol{\sigma})$ and the $\phi_{\mathbf{n}}(\mathbf{q})$ form a complete orthonormal set.

The symmetrization of any wave function in spin-position space can be accomplished by using symmetrized basis functions in its expansion. The latter are given by

$$\Phi^{\pm}_{\mathbf{n},\mathbf{s}}(\mathbf{x}) = \frac{1}{\sqrt{\chi^{\pm}_{\mathbf{n},\mathbf{s}} N!}} \sum_{\hat{\mathrm{P}}} (\pm 1)^p \alpha_{\mathbf{s}}(\hat{\mathrm{P}}\boldsymbol{\sigma}) \phi_{\mathbf{n}}(\hat{\mathrm{P}}\mathbf{q}). \tag{6.60}$$

This is exact.

6.4.2 Approximate form

In place of this exact symmetrization, one can invoke an approximation that relies upon the factorization of the spin-position basis function into the sum of products of symmetrized position basis functions and symmetrized spin basis functions, namely

$$\Phi^{\pm}_{\mathbf{n},\,\mathbf{s}}(\mathbf{x}) = \begin{cases} \dfrac{1}{\sqrt{\tilde{\chi}^{+}_{\mathbf{n},\,\mathbf{s}}}}\left[\tilde{\alpha}^{+}_{\mathbf{s}}(\boldsymbol{\sigma})\tilde{\phi}^{+}_{\mathbf{n}}(\mathbf{q}) + \tilde{\alpha}^{-}_{\mathbf{s}}(\boldsymbol{\sigma})\tilde{\phi}^{-}_{\mathbf{n}}(\mathbf{q})\right] \\ \dfrac{1}{\sqrt{\tilde{\chi}^{-}_{\mathbf{n},\,\mathbf{s}}}}\left[\tilde{\alpha}^{+}_{\mathbf{s}}(\boldsymbol{\sigma})\tilde{\phi}^{-}_{\mathbf{n}}(\mathbf{q}) + \tilde{\alpha}^{-}_{\mathbf{s}}(\boldsymbol{\sigma})\tilde{\phi}^{+}_{\mathbf{n}}(\mathbf{q})\right] \end{cases}. \tag{6.61}$$

The merits or otherwise of this approximation are discussed at the end of section 6.4.3 below. Although the left-hand side is normalized by the overall factor of $\sqrt{\tilde{\chi}^{\pm}_{\mathbf{n},\,\mathbf{s}}}$, the individual factors on the right-hand side are not normalized. This is essential to the correct formulation of the ansatz and is emphasized by the tilde. The un-normalized symmetrized basis functions are

$$\tilde{\phi}^{\pm}_{\mathbf{n}}(\mathbf{q}) \equiv \frac{1}{\sqrt{N!}} \sum_{\hat{\mathrm{P}}} (\pm 1)^{p} \phi_{\mathbf{n}}(\hat{\mathrm{P}}\mathbf{q}), \tag{6.62}$$

and

$$\tilde{\alpha}^{\pm}_{\mathbf{s}}(\boldsymbol{\sigma}) \equiv \frac{1}{\sqrt{N!}} \sum_{\hat{\mathrm{P}}} (\pm 1)^{p} \alpha_{\mathbf{s}}(\hat{\mathrm{P}}\boldsymbol{\sigma}). \tag{6.63}$$

The $\alpha_{\mathbf{s}}(\boldsymbol{\sigma})$ and the $\phi_{\mathbf{n}}(\mathbf{q})$ are normalized. The $\sqrt{N!}$ here is an immaterial constant that is convenient but not essential. Respective symmetrization or overlap factors for use below may be defined in terms of these,

$$\chi^{\pm}_{\mathbf{n}} = \langle \tilde{\phi}^{\pm}_{\mathbf{n}} | \tilde{\phi}^{\pm}_{\mathbf{n}} \rangle, \quad \text{and} \quad \chi^{\pm}_{\mathbf{s}} = \langle \tilde{\alpha}^{\pm}_{\mathbf{s}} | \tilde{\alpha}^{\pm}_{\mathbf{s}} \rangle \tag{6.64}$$

It is essential that these symmetrization factors are not used to normalize the $\tilde{\phi}^{\pm}_{\mathbf{n}}(\mathbf{q})$ and the $\tilde{\alpha}^{\pm}_{\mathbf{s}}(\boldsymbol{\sigma})$ individually. The reason for this is that without individual normalization, all terms in the approximation for $\Phi^{\pm}_{\mathbf{n},\,\mathbf{s}}(\mathbf{x})$ have equal weight when written as permutation sums. If the individual factors were normalized, then the symmetric factors would have a different weight to the anti-symmetric factors, and since different products have two, one, or zero of each of these, individual terms in the permutation sum would have different weights.

This point also explains why the two products in each line of the approximation are simply added together symmetrically. In principle, the two products could be superposed with a relative phase factor and with a relative probability factor. The reason they aren't is that ultimately the expression is meant to approximate a permutation sum in which all terms have equal weight (apart from the $(-1)^p$ for fermions). These two points will be taken up in the next subsubsection 6.4.3, and with the concrete example of two particles following that.

The approximation is *sufficient* to ensure the symmetrization of the spin-position basis functions,

$$\Phi^{\pm}_{\mathbf{n},\,\mathbf{s}}(\hat{\mathrm{P}}\mathbf{x}) = (\pm 1)^{p} \Phi^{\pm}_{\mathbf{n},\,\mathbf{s}}(\mathbf{x}), \tag{6.65}$$

as can be confirmed by inspection.

The overall normalization factor for the symmetrized basis function $\Phi^{\pm}_{\mathbf{n},\,\mathbf{s}}(\mathbf{x})$ for the approximation is

$$\begin{aligned}\tilde{\chi}^{\pm}_{\mathbf{n},\,\mathbf{s}} &\equiv \begin{cases}\langle\tilde{\alpha}_{\mathbf{s}}^{+}|\tilde{\alpha}_{\mathbf{s}}^{+}\rangle\,\langle\tilde{\phi}_{\mathbf{n}}^{+}|\tilde{\phi}_{\mathbf{n}}^{+}\rangle + \langle\tilde{\alpha}_{\mathbf{s}}^{-}|\tilde{\alpha}_{\mathbf{s}}^{-}\rangle\,\langle\tilde{\phi}_{\mathbf{n}}^{-}|\tilde{\phi}_{\mathbf{n}}^{-}\rangle\\ \langle\tilde{\alpha}_{\mathbf{s}}^{+}|\tilde{\alpha}_{\mathbf{s}}^{+}\rangle\,\langle\tilde{\phi}_{\mathbf{n}}^{-}|\tilde{\phi}_{\mathbf{n}}^{-}\rangle + \langle\tilde{\alpha}_{\mathbf{s}}^{-}|\tilde{\alpha}_{\mathbf{s}}^{-}\rangle\,\langle\tilde{\phi}_{\mathbf{n}}^{+}|\tilde{\phi}_{\mathbf{n}}^{+}\rangle\end{cases}\\ &= \begin{cases}\chi_{\mathbf{s}}^{+}\chi_{\mathbf{n}}^{+} + \chi_{\mathbf{s}}^{-}\chi_{\mathbf{n}}^{-}\\ \chi_{\mathbf{s}}^{+}\chi_{\mathbf{n}}^{-} + \chi_{\mathbf{s}}^{-}\chi_{\mathbf{n}}^{+}.\end{cases}\end{aligned} \tag{6.66}$$

It should be noted that in certain states the Fermi exclusion principle means that $\tilde{\alpha}_{\mathbf{s}}^{-}(\boldsymbol{\sigma})$ or $\tilde{\phi}_{\mathbf{n}}^{-}(\mathbf{q})$ vanish. In such states $\chi_{\mathbf{s}}^{-}$ and $\chi_{\mathbf{n}}^{-}$ also respectively vanish.

6.4.3 Comparison of exact and approximate forms

One can label the $N!$ permutations of the dynamical variables $\boldsymbol{\sigma}_P$ and $\mathbf{q}_P$, $P = 1, 2, \ldots, N!$, in such a way that the permutation has the same parity as its label. The exact result is

$$\Phi^{\pm}_{\mathbf{n},\,\mathbf{s}}(\mathbf{x}) = \frac{1}{\sqrt{N!\chi^{\pm}_{\mathbf{n},\,\mathbf{s}}}}\sum_{P=1}^{N!}(\pm1)^{P}\alpha_{\mathbf{s}}(\boldsymbol{\sigma}_P)\phi_{\mathbf{n}}(\mathbf{q}_P). \tag{6.67}$$

The approximate expression may be written

$$\begin{aligned}\Phi^{\pm}_{\mathbf{n},\,\mathbf{s}}(\mathbf{x}) &= \frac{1}{\sqrt{\tilde{\chi}^{\pm}_{\mathbf{n},\,\mathbf{s}}}}\left[\tilde{\alpha}_{\mathbf{s}}^{+}(\boldsymbol{\sigma})\tilde{\phi}_{\mathbf{n}}^{\pm}(\mathbf{q}) + \tilde{\alpha}_{\mathbf{s}}^{-}(\boldsymbol{\sigma})\tilde{\phi}_{\mathbf{n}}^{\mp}(\mathbf{q})\right]\\ &= \frac{1}{N!\sqrt{\tilde{\chi}^{\pm}_{\mathbf{n},\,\mathbf{s}}}}\sum_{P',P''=1}^{N!}[(\pm1)^{P''} + (-1)^{P'}(\mp1)^{P''}]\alpha_{\mathbf{s}}(\boldsymbol{\sigma}_{P'})\phi_{\mathbf{n}}(\mathbf{q}_{P''})\\ &= \frac{2}{N!\sqrt{\tilde{\chi}^{\pm}_{\mathbf{n},\,\mathbf{s}}}}\sum_{P=1}^{N!}(\pm1)^{P}\alpha_{\mathbf{s}}(\boldsymbol{\sigma}_P)\phi_{\mathbf{n}}(\mathbf{q}_P)\\ &\quad+\frac{1}{N!\sqrt{\tilde{\chi}^{\pm}_{\mathbf{n},\,\mathbf{s}}}}\sum_{P',P''}^{(P'\neq P'')}(\pm1)^{P''}[1 + (-1)^{P''+P'}]\alpha_{\mathbf{s}}(\boldsymbol{\sigma}_{P'})\phi_{\mathbf{n}}(\mathbf{q}_{P''}).\end{aligned} \tag{6.68}$$

In the final equality, the single sum is over terms where the permutation of the spins is the same as that of the positions. This is the same as the exact result. (Evidently, if this were the only contribution, then the normalization constants would be related as $\tilde{\chi}^{\pm}_{\mathbf{n},\,\mathbf{s}} = 4\chi^{\pm}_{\mathbf{n},\,\mathbf{s}}/N!$.) The double sum, where the permutation of the spins differs from that of the particles, does not appear in the exact result and is unphysical in the sense that it disassociates the spin and position of each particle. It may be noted that only permutations of spin and position with the same parity make a non-zero contribution to this double sum. In the case of $N = 2$ no such terms exist, so the double sum is zero, and the approximation is exact in this case (see also next).

Notice that the agreement of part of the approximation with the exact formulation depends upon using the un-normalized symmetrized basis functions, $\tilde{\alpha}_{\mathrm{s}}^{\pm}(\boldsymbol{\sigma})$ and $\tilde{\phi}_{\mathrm{n}}^{\pm}(\mathbf{q})$, and upon superposing the two terms without any phase factor or probability weight. As mentioned, the justification for this is that this procedure weights all terms in the permutation sum equally.

In the light of this analysis of the symmetrized basis functions, it is worth discussing whether or not the approximation equation (6.61) has any advantages over the exact form, equation (6.60). One could argue that the sum of products in the approximate form has a transparent interpretation that lends itself to the physical interpretation of symmetrization effects, and of physical phenomena such as Bose–Einstein condensation and Fermi exclusion. Indeed, the long-standing electronic orbital theory for optical spectra is predicated on this sum of products form, (which is exact for $N = 2$; see also next). It could also be argued that there could be computational advantages to obtaining and storing the symmetrized spinless functions, and at a later stage combining them with the symmetrized spin functions for different values of S. Finally, it might be argued that exploring the properties of the symmetrized spinless functions leads directly to an understanding of the spatial (or momentum) localization of symmetrization effects, which is often missed in the formal treatment of symmetrization. Such localization shows that in many important terrestrial cases the nearest neighbor dimers give the dominant contribution, and for such pairs the approximation is exact.

The numerical results in this book using fully symmetrized and fully anti-symmetrized spinless wave functions implicitly assume that such arguments carry some weight.

6.4.4 Simple example, $N = 2$

For two particles, $N = 2$, the comparison of the exact formulation for symmetrization with the approximate form can be performed rather directly. With $\mathbf{x}' \equiv \hat{\mathrm{P}}_{12}\mathbf{x}$, the exact result is

$$\Phi_{\mathrm{n,\,s}}^{\pm}(\mathbf{x}) = \frac{1}{\sqrt{2\chi_{\mathrm{n,\,s}}^{\pm}}}\left[\alpha_{\mathrm{s}}(\boldsymbol{\sigma})\phi_{\mathrm{n}}(\mathbf{q}) \pm \alpha_{\mathrm{s}}(\boldsymbol{\sigma}')\phi_{\mathrm{n}}(\mathbf{q}')\right]. \tag{6.69}$$

The approximation gives

$$\begin{aligned}\Phi_{\mathrm{n,\,s}}^{\pm}(\mathbf{x}) &= \frac{1}{\sqrt{\tilde{\chi}_{\mathrm{n,\,s}}^{\pm}}}\left[\tilde{\alpha}_{\mathrm{s}}^{+}(\boldsymbol{\sigma})\tilde{\phi}_{\mathrm{n}}^{\pm}(\mathbf{q}) + \tilde{\alpha}_{\mathrm{s}}^{-}(\boldsymbol{\sigma})\tilde{\phi}_{\mathrm{n}}^{\mp}(\mathbf{q})\right] \\ &= \frac{1}{2\sqrt{\tilde{\chi}_{\mathrm{n,\,s}}^{\pm}}}\left[\{\alpha_{\mathrm{s}}(\boldsymbol{\sigma}) + \alpha_{\mathrm{s}}(\boldsymbol{\sigma}')\}\{\phi_{\mathrm{n}}(\mathbf{q}) \pm \phi_{\mathrm{n}}(\mathbf{q}')\}\right. \\ &\quad \left. + \{\alpha_{\mathrm{s}}(\boldsymbol{\sigma}) - \alpha_{\mathrm{s}}(\boldsymbol{\sigma}')\}\{\phi_{\mathrm{n}}(\mathbf{q}) \mp \phi_{\mathrm{n}}(\mathbf{q}')\}\right] \\ &= \frac{1}{\sqrt{\tilde{\chi}_{\mathrm{n,\,s}}^{\pm}}}\left[\alpha_{\mathrm{s}}(\boldsymbol{\sigma})\phi_{\mathrm{n}}(\mathbf{q}) \pm \alpha_{\mathrm{s}}(\boldsymbol{\sigma}')\phi_{\mathrm{n}}(\mathbf{q}')\right].\end{aligned} \tag{6.70}$$

This is the same as the exact form. (Evidently in this case $\tilde{\chi}^{\pm}_{\mathbf{n},\,\mathbf{s}} = 2\chi^{\pm}_{\mathbf{n},\,\mathbf{s}}$.)

One can illustrate the approximation further by making direct contact with conventional electronic orbital theory for the simple case of two fermions, $N = 2$, $S = 1/2$. In this case the symmetrized, un-normalized spin basis functions are

$$\begin{aligned}\tilde{\alpha}^{\pm}_{\mathbf{s}}(\boldsymbol{\sigma}) &\equiv \frac{1}{\sqrt{2}}\{\alpha_{s_1}(\sigma_1)\alpha_{s_2}(\sigma_2) \pm \alpha_{s_1}(\sigma_2)\alpha_{s_2}(\sigma_1)\} \\ &= \frac{1}{\sqrt{2}}\{\delta_{s_1,\sigma_1}\delta_{s_2,\sigma_2} \pm \delta_{s_1,\sigma_2}\delta_{s_2,\sigma_1}\},\end{aligned} \tag{6.71}$$

and the symmetrization or overlap factor is

$$\chi^{\pm}_{\mathbf{s}} = \sum_{\hat{\mathrm{P}}}(\pm 1)^p\langle\hat{\mathrm{P}}\mathbf{s}|\mathbf{s}\rangle = 1 \pm \delta_{s_1,s_2}. \tag{6.72}$$

This may be re-written to show the same and different state occupancies explicitly,

$$\begin{aligned}\tilde{\alpha}^{\pm}_{\mathbf{s}}(\boldsymbol{\sigma}) &= \frac{\delta_{s_1,s_2}}{\sqrt{2}}[\delta_{\mathbf{s},\sigma} \pm \delta_{\mathbf{s},\sigma}] + \frac{\bar{\delta}_{s_1,s_2}}{\sqrt{2}}[\bar{\delta}_{\sigma_1,\sigma_2}(\delta_{s_1,\sigma_1} \pm \delta_{s_1,\sigma_2})] \\ &= \frac{\delta_{s_1,s_2}}{\sqrt{2}}\begin{cases}2\delta_{\mathbf{s},\sigma} \\ 0\end{cases} + \frac{\bar{\delta}_{s_1,s_2}}{\sqrt{2}}\bar{\delta}_{\sigma_1,\sigma_2}[\delta_{s_1,\sigma_1} \pm \bar{\delta}_{s_1,\sigma_1}] \\ &= \frac{\delta_{s_1,s_2}}{\sqrt{2}}\begin{cases}2\delta_{\mathbf{s},\sigma} \\ 0\end{cases} + \frac{\bar{\delta}_{s_1,s_2}}{\sqrt{2}}\begin{cases}\bar{\delta}_{\sigma_1,\sigma_2} \\ \bar{\delta}_{\sigma_1,\sigma_2}[\delta_{s_1,\sigma_1} - \bar{\delta}_{s_1,\sigma_1}].\end{cases}\end{aligned} \tag{6.73}$$

Here and throughout, the complementary Kronecker delta is $\bar{\delta}_{j,k} \equiv 1 - \delta_{j,k}$.

For the basis functions in position one can consider single-particle states, $\phi_{\mathbf{n}}(\mathbf{q}) = \phi_{\mathbf{n}_1}(\mathbf{q}_1)\phi_{\mathbf{n}_2}(\mathbf{q}_2)$. In the case of electronic orbitals, typically $\mathbf{n}_j = \{n_j, l_j, m_j\}$. Then

$$\chi^{\pm}_{\mathbf{n}} = \sum_{\hat{\mathrm{P}}}(\pm 1)^p\langle\hat{\mathrm{P}}\mathbf{n}|\mathbf{n}\rangle = 1 \pm \delta_{\mathbf{n}_1,\mathbf{n}_2}, \tag{6.74}$$

and

$$\begin{aligned}\tilde{\phi}^{\pm}_{\mathbf{n}}(\mathbf{q}) &= \frac{1}{\sqrt{2}}\left[\phi_{\mathbf{n}_1}(\mathbf{q}_1)\phi_{\mathbf{n}_2}(\mathbf{q}_2) \pm \phi_{\mathbf{n}_1}(\mathbf{q}_2)\phi_{\mathbf{n}_2}(\mathbf{q}_1)\right] \\ &= \frac{\delta_{\mathbf{n}_1,\mathbf{n}_2}}{\sqrt{2}}\begin{cases}2\phi_{\mathbf{n}_1}(\mathbf{q}_1)\phi_{\mathbf{n}_2}(\mathbf{q}_2) \\ 0\end{cases} \\ &\quad + \frac{\bar{\delta}_{\mathbf{n}_1,\mathbf{n}_2}}{\sqrt{2}}\left[\phi_{\mathbf{n}_1}(\mathbf{q}_1)\phi_{\mathbf{n}_2}(\mathbf{q}_2) \pm \phi_{\mathbf{n}_1}(\mathbf{q}_2)\phi_{\mathbf{n}_2}(\mathbf{q}_1)\right].\end{aligned} \tag{6.75}$$

On the first line of the right-hand side of the final equality, the upper line $2\phi_{\mathbf{n}_1}(\mathbf{q}_1)\phi_{\mathbf{n}_2}(\mathbf{q}_2)$ is for bosons (the superscript + on the left-hand side), and the lower line 0 is for fermions (the superscript −). Shortly use will be made of a result that follows from this, namely

$$\tilde{\phi}_{\mathbf{n}}^{+}(\mathbf{q}) \pm \tilde{\phi}_{\mathbf{n}}^{-}(\mathbf{q}) = \delta_{\mathbf{n}_1,\mathbf{n}_2}\sqrt{2}\,\phi_{\mathbf{n}_1}(\mathbf{q}_1)\phi_{\mathbf{n}_2}(\mathbf{q}_2) + \bar{\delta}_{\mathbf{n}_1,\mathbf{n}_2}\sqrt{2}\begin{cases}\phi_{\mathbf{n}_1}(\mathbf{q}_1)\phi_{\mathbf{n}_2}(\mathbf{q}_2)\\ \phi_{\mathbf{n}_1}(\mathbf{q}_2)\phi_{\mathbf{n}_2}(\mathbf{q}_1).\end{cases} \tag{6.76}$$

The left-hand side is the sum and difference of the basis functions for bosons and for fermions.

The overall normalization factor is

$$\begin{aligned}\tilde{\chi}_{\mathbf{n},\,\mathbf{s}}^{\pm} &= \begin{cases}(1+\delta_{s_1,s_2})(1+\delta_{\mathbf{n}_1,\mathbf{n}_2}) + (1-\delta_{s_1,s_2})(1-\delta_{\mathbf{n}_1,\mathbf{n}_2})\\ (1+\delta_{s_1,s_2})(1-\delta_{\mathbf{n}_1,\mathbf{n}_2}) + (1-\delta_{s_1,s_2})(1+\delta_{\mathbf{n}_1,\mathbf{n}_2})\end{cases}\\ &= 2(1 \pm \delta_{s_1,s_2}\delta_{\mathbf{n}_1,\mathbf{n}_2}).\end{aligned} \tag{6.77}$$

For fermions, this vanishes if both are in the same state (i.e., both spin state and energy state). Since the normalization factor appears in the denominator of $\Phi_{\mathbf{n},\,\mathbf{s}}^{\pm}(\mathbf{x})$ as the square root, the symmetrized spin-position basis function, $\Phi_{\mathbf{n},\,\mathbf{s}}^{-}(\mathbf{x})$, also vanishes in this case.

Putting these together, the anti-symmetrized two fermion wave function is

$$\begin{aligned}
&\psi_{\mathbf{n},\,\mathbf{s}}^{-}(\mathbf{q},\,\boldsymbol{\sigma})\\
&= \frac{1}{\sqrt{\tilde{\chi}_{\mathbf{n},\,\mathbf{s}}^{-}}}\left[\alpha_{\mathbf{s}}^{+}(\boldsymbol{\sigma})\tilde{\phi}_{\mathbf{n}}^{-}(\mathbf{q}) + \alpha_{\mathbf{s}}^{-}(\boldsymbol{\sigma})\tilde{\phi}_{\mathbf{n}}^{+}(\mathbf{q})\right]\\
&= \frac{\sqrt{2}}{\sqrt{\tilde{\chi}_{\mathbf{n},\,\mathbf{s}}^{-}}}\delta_{s_1,s_2}\delta_{\mathbf{s},\boldsymbol{\sigma}}\tilde{\phi}_{\mathbf{n}}^{-}(\mathbf{q}) + \frac{\bar{\delta}_{s_1,s_2}}{\sqrt{\tilde{\chi}_{\mathbf{n},\,\mathbf{s}}^{-}}}\frac{1}{\sqrt{2}}\Big[\delta_{s_1,\sigma_1}\delta_{s_2,\sigma_2}\left\{\tilde{\phi}_{\mathbf{n}}^{-}(\mathbf{q}) + \tilde{\phi}_{\mathbf{n}}^{+}(\mathbf{q})\right\}\\
&+ \delta_{s_1,\sigma_2}\delta_{s_2,\sigma_1}\left\{\tilde{\phi}_{\mathbf{n}}^{-}(\mathbf{q}) - \tilde{\phi}_{\mathbf{n}}^{+}(\mathbf{q})\right\}\Big]\\
&= \frac{\sqrt{2}}{\sqrt{\tilde{\chi}_{\mathbf{n},\,\mathbf{s}}^{-}}}\delta_{s_1,s_2}\delta_{\mathbf{s},\boldsymbol{\sigma}}\tilde{\phi}_{\mathbf{n}}^{-}(\mathbf{q}) + \frac{\bar{\delta}_{s_1,s_2}}{\sqrt{2\tilde{\chi}_{\mathbf{n},\,\mathbf{s}}^{-}}}\delta_{s_1,\sigma_1}\delta_{s_2,\sigma_2}\left\{\delta_{\mathbf{n}_1,\mathbf{n}_2}\sqrt{2}\phi_{\mathbf{n}_1}(\mathbf{q}_1)\phi_{\mathbf{n}_2}(\mathbf{q}_2)\right.\\
&\left.+ \bar{\delta}_{\mathbf{n}_1,\mathbf{n}_2}\sqrt{2}\phi_{\mathbf{n}_1}(\mathbf{q}_1)\phi_{\mathbf{n}_2}(\mathbf{q}_2)\right\}\\
&- \frac{\bar{\delta}_{s_1,s_2}}{\sqrt{2\tilde{\chi}_{\mathbf{n},\,\mathbf{s}}^{-}}}\delta_{s_1,\sigma_2}\delta_{s_2,\sigma_1}\left\{\delta_{\mathbf{n}_1,\mathbf{n}_2}\sqrt{2}\phi_{\mathbf{n}_1}(\mathbf{q}_1)\phi_{\mathbf{n}_2}(\mathbf{q}_2)\right.\\
&\left.+ \bar{\delta}_{\mathbf{n}_1,\mathbf{n}_2}\sqrt{2}\phi_{\mathbf{n}_1}(\mathbf{q}_2)\phi_{\mathbf{n}_2}(\mathbf{q}_1)\right\}\\
&= \delta_{s_1,s_2}\delta_{\mathbf{s},\boldsymbol{\sigma}}\frac{\bar{\delta}_{\mathbf{n}_1,\mathbf{n}_2}}{\sqrt{2}}\left[\phi_{\mathbf{n}_1}(\mathbf{q}_1)\phi_{\mathbf{n}_2}(\mathbf{q}_2) - \phi_{\mathbf{n}_1}(\mathbf{q}_2)\phi_{\mathbf{n}_2}(\mathbf{q}_1)\right]\\
&+ \frac{\bar{\delta}_{s_1,s_2}}{\sqrt{2}}\delta_{s_1,\sigma_1}\delta_{s_2,\sigma_2}\phi_{\mathbf{n}_1}(\mathbf{q}_1)\phi_{\mathbf{n}_2}(\mathbf{q}_2)\\
&- \frac{\bar{\delta}_{s_1,s_2}}{\sqrt{2}}\delta_{s_1,\sigma_2}\delta_{s_2,\sigma_1}\left\{\delta_{\mathbf{n}_1,\mathbf{n}_2}\phi_{\mathbf{n}_1}(\mathbf{q}_1)\phi_{\mathbf{n}_2}(\mathbf{q}_2) + \bar{\delta}_{\mathbf{n}_1,\mathbf{n}_2}\phi_{\mathbf{n}_1}(\mathbf{q}_2)\phi_{\mathbf{n}_2}(\mathbf{q}_1)\right\}.
\end{aligned} \tag{6.78}$$

For $s_1 = s_2$ this is

$$\psi^-_{\mathbf{n},\,\mathbf{s}}(\mathbf{q},\,\boldsymbol{\sigma}) = \frac{1}{\sqrt{2}}\delta_{\mathbf{s},\boldsymbol{\sigma}}\bar{\delta}_{\mathbf{n}_1,\mathbf{n}_2}\left\{\phi_{\mathbf{n}_1}(\mathbf{q}_1)\phi_{\mathbf{n}_2}(\mathbf{q}_2) - \phi_{\mathbf{n}_1}(\mathbf{q}_2)\phi_{\mathbf{n}_2}(\mathbf{q}_1)\right\}. \tag{6.79}$$

This vanishes if $\mathbf{n}_1 = \mathbf{n}_2$, which is just the Fermi exclusion principle.

For $\mathbf{n}_1 = \mathbf{n}_2$ one has

$$\psi^-_{\mathbf{n},\,\mathbf{s}}(\mathbf{q},\,\boldsymbol{\sigma}) = \frac{\bar{\delta}_{s_1,s_2}}{\sqrt{2}}\phi_{\mathbf{n}_1}(\mathbf{q}_1)\phi_{\mathbf{n}_2}(\mathbf{q}_2)[\delta_{s_1,\sigma_1}\delta_{s_2,\sigma_2} - \delta_{s_1,\sigma_2}\delta_{s_2,\sigma_1}]. \tag{6.80}$$

This vanishes if $s_1 = s_2$, which is again the Fermi exclusion principle. The term in brackets is equivalent to the so-called singlet state.

For $\mathbf{n}_1 \neq \mathbf{n}_2$ one has

$$\begin{aligned}\psi^-_{\mathbf{n},\,\mathbf{s}}(\mathbf{q},\,\boldsymbol{\sigma}) = \delta_{s_1,s_2}\delta_{\mathbf{s},\boldsymbol{\sigma}}\tilde{\phi}^-_{\mathbf{n}}(\mathbf{q}) &+ \frac{\bar{\delta}_{s_1,s_2}}{\sqrt{2}}\delta_{s_1,\sigma_1}\delta_{s_2,\sigma_2}\phi_{\mathbf{n}_1}(\mathbf{q}_1)\phi_{\mathbf{n}_2}(\mathbf{q}_2)\\ &- \frac{\bar{\delta}_{s_1,s_2}}{\sqrt{2}}\delta_{s_1,\sigma_2}\delta_{s_2,\sigma_1}\phi_{\mathbf{n}_1}(\mathbf{q}_2)\phi_{\mathbf{n}_2}(\mathbf{q}_1).\end{aligned} \tag{6.81}$$

These terms are equivalent to the so-called triplet state (the first term represents the ++ and $^{--}$ states, and the second and third terms are a superposition of the +− and −+ state).

References

Abramowitz M and Stegun I A 1970 *Handbook of Mathematical Functions* 9th edn (New York: Dover)

Attard P, Jepps O G and Marčelja S 1997 Information content of signals using correlation function expansions of the entropy *Phys. Rev.* E **56** 4052

Attard P 1999 Markov superposition expansion for the entropy and correlation functions in two and three dimensions *Statistical Physics on the Eve of the Twenty-First Century* ed M T Batchelor and L T Wille (Singapore: World Scientific)

Attard P 2002 *Thermodynamics and Statistical Mechanics: Equilibrium by Entropy Maximisation* (London: Academic)

Attard P 2018 *Entropy Beyond the Second Law. Thermodynamics and Statistical Mechanics for Equilibrium, Non-Equilibrium, Classical, and Quantum Systems* (Bristol: IOP Publishing)

Attard P 2019 Fermionic phonons: exact analytic results and quantum statistical mechanics for a one dimensional harmonic crystal arXiv:1903.06866

Debye P 1912 Zur theorie der spezifischen wärme (The theory of specific heat) *Ann. Phys.* **344** 789

Einstein A 1907 Plancksche theorie der strahlung und die theorie der spezifischen wärme (Planck's theory of radiation and the theory of specific heat) *Ann. Phys.* **22** 180

Merzbacher E 1970 *Quantum Mechanics* 2nd edn (New York: Wiley)

Messiah A 1961 *Quantum Mechanics* vols 1 and 2 (Amsterdam: North-Holland)

Pathria R K 1972 *Statistical Mechanics* (Oxford: Pergamon)

Chapter 7

Phase space formalism for the partition function and averages

It is shown that the formally exact transformation of the quantum partition function to classical phase space gives rise to commutation and symmetrization functions. The former accounts for the non-commutativity of the position and momentum operators in the Maxwell–Boltzmann operator. The latter generalizes particle state occupancy rules to the continuum that is classical phase space.

The same commutation function can be used for the statistical average of any pure function of the position operator or of the momentum operator, or the sum thereof. In the dilute loop approximation, the symmetrization function can be written as a series of products of disconnected permutation loops, which series can be exponentially resummed. The average energy can be written as a series of single-loop fluctuation averages, and similarly for other thermodynamic derivatives and averages.

7.1 Partition function in classical phase space

The aim of this chapter is to formulate without approximation quantum statistical mechanics in classical phase space. To this end expressions for the grand canonical equilibrium partition function and statistical averages are derived. One of the major issues that will recur in later chapters is accounting for the non-commutativity of position and momentum operators in the transformation to classical phase space, a point in which represents the simultaneous specification of the particles' positions and momenta. A second issue is the accounting for the consequences of wave function symmetrization for bosons and fermions in the continuum of states that is classical phase space.

In section 3.3, it was shown that the symmetrization factor correctly accounted for particle statistics and state occupancy so that a sum over distinct allowed states could be written as a sum over all states with it as weighting factor. At the end of

doi:10.1088/978-0-7503-4055-7ch7

that section, equation (3.29) showed that one did not have to invoke explicitly the symmetrization factor. Instead one could include directly in the sum the symmetrized permutations of the basis functions

$$\begin{aligned}\mathrm{TR}'\hat{f} &= \frac{1}{N!}\sum_{\mathbf{n}} \chi_{\mathbf{n}}^{\pm} \langle \phi_{\mathbf{n}}^{\pm} | \hat{f} | \phi_{\mathbf{n}}^{\pm} \rangle \\ &= \frac{1}{(N!)^2}\sum_{\mathbf{n}} \sum_{\hat{\mathrm{P}}', \hat{\mathrm{P}}''} (\pm 1)^{p'+p''} \langle \phi_{\mathbf{n}}(\hat{\mathrm{P}}'\mathbf{r}) | \hat{f}(\mathbf{r}) | \phi_{\mathbf{n}}(\hat{\mathrm{P}}''\mathbf{r}) \rangle \\ &= \frac{1}{N!}\sum_{\mathbf{n}} \sum_{\hat{\mathrm{P}}} (\pm 1)^{p} \langle \phi_{\mathbf{n}}(\hat{\mathrm{P}}\mathbf{r}'') | \hat{f}(\mathbf{r}'') | \phi_{\mathbf{n}}(\mathbf{r}'') \rangle .\end{aligned} \tag{7.1}$$

Here $\phi_{\mathbf{n}}$ is any complete orthonormal set of unsymmetrized basis functions. The final equality assumes that the operator is symmetric, $\hat{f}(\hat{\mathrm{P}}\mathbf{r}) = \hat{f}(\mathbf{r})$, so that after a change in the dummy position integration variable, $\hat{\mathrm{P}}''\mathbf{r} \Rightarrow \mathbf{r}''$, and dummy summation permutation operator, $\hat{\mathrm{P}}'\hat{\mathrm{P}}''^{-1} \Rightarrow \hat{\mathrm{P}}$, the summand is independent of $\hat{\mathrm{P}}''$ and the sum may be replaced by a factor of $N!$. This is a fecund way to view the trace as it manifests the fundamental axiom that the wave function must be symmetrized without resorting to the lemma concerning the restricted occupancy of single-particle states. However it must be observed that in order to get correctly the factors of $1/(N!)^2$ and $1/N!$ in the second and third equality it is necessary to derive the result from the first equality involving the symmetrization factor. It has already been proven that this factor enables the trace to be written as an unrestricted sum over all states, sections 2.3.2 and 3.3. If one simply takes a guess at the form of the second equality, one is liable to get this wrong (Kirkwood 1933, equation (10); see section 8.2.2 below). In any case this result provides the starting point for transforming the partition function to classical phase space.

7.1.1 Grand partition function

Consider N spinless particles in d dimensions confined to a hypercube of volume $V = L^d$. The position representation will be invoked, $\mathbf{r} = \{\mathbf{r}_1, \mathbf{r}_2, \ldots, \mathbf{r}_N\}$, with $\mathbf{r}_j = \{r_{jx}, r_{jy}, \ldots, r_{jd}\}$. The unsymmetrized position and momentum eigenfunctions are respectively (Messiah 1961)

$$|\mathbf{q}\rangle = \delta(\mathbf{r} - \mathbf{q}), \quad \text{and} \quad |\mathbf{p}\rangle = \frac{e^{-\mathbf{p}\cdot\mathbf{r}/i\hbar}}{V^{N/2}}. \tag{7.2}$$

The position eigenvalues belong to the continuum. The momentum eigenvalues are discrete with spacing $\Delta_{\mathrm{p}} = 2\pi\hbar/L$ per particle per dimension (Messiah 1961); almost immediately the continuum limit for this will be taken. Each set of eigenfunctions is complete.

These will now be used to transform the grand partition function to classical phase space, after which each stage of the derivation will be discussed in detail. The grand partition function is

$$
\begin{aligned}
\Xi^{\pm} &= \mathrm{TR}'\left\{e^{-\beta\hat{\mathcal{H}}}\right\} \\
&= \sum_{N=0}^{\infty} \frac{z^N}{N!} \sum_{\hat{\mathrm{P}}} (\pm 1)^p \sum_{\mathbf{p}} \langle \hat{\mathrm{P}}\mathbf{p} | e^{-\beta\hat{\mathcal{H}}} | \mathbf{p} \rangle \\
&= \sum_{N=0}^{\infty} \frac{z^N}{N!} \sum_{\hat{\mathrm{P}}} (\pm 1)^p \sum_{\mathbf{p}} \int \mathrm{d}\mathbf{q}\, \langle \hat{\mathrm{P}}\mathbf{p} | \mathbf{q} \rangle \, \langle \mathbf{q} | e^{-\beta\hat{\mathcal{H}}} | \mathbf{p} \rangle \\
&= \sum_{N=0}^{\infty} \frac{z^N}{h^{dN} N!} \sum_{\hat{\mathrm{P}}} (\pm 1)^p \int \mathrm{d}\boldsymbol{\Gamma}\, \frac{\langle \mathbf{q} | e^{-\beta\hat{\mathcal{H}}} | \mathbf{p} \rangle}{\langle \mathbf{q} | \mathbf{p} \rangle} \frac{\langle \hat{\mathrm{P}}\mathbf{p} | \mathbf{q} \rangle}{\langle \mathbf{p} | \mathbf{q} \rangle} \\
&\equiv \sum_{N=0}^{\infty} \frac{z^N}{h^{dN} N!} \int \mathrm{d}\boldsymbol{\Gamma}\, e^{-\beta\mathcal{H}(\boldsymbol{\Gamma})} \omega_{\mathrm{p}}(\boldsymbol{\Gamma}) \eta_{\mathrm{q}}^{\pm}(\boldsymbol{\Gamma}).
\end{aligned}
\tag{7.3}
$$

Here $z = e^{\beta\mu}$ is the fugacity, $\beta = 1/k_{\mathrm{B}}T$ is the inverse temperature, h is Planck's constant, and $\boldsymbol{\Gamma} = \{\mathbf{p}, \mathbf{q}\}$ is a point in classical phase space. The grand partition function is related to the canonical partition function as $\Xi^{\pm}(\mu, V, T) = \sum_{N=0}^{\infty} z^N Z^{\pm}(N, V, T)$.

The first equality here has the appearance of von Neumann trace form for the partition function (Messiah 1961, Merzbacher 1970, Pathria 1972). It arises from the collapse of the total wave function due to entanglement of the subsystem with the reservoir upon energy (and particle) exchange (see chapter 12, and also Attard 2015, Attard 2018a). The prime on the trace signifies that it is over unique states, which is to say that each distinct state can only appear once, or if more than once each occurrence must have appropriately reduced weight. Not all workers avert to this requirement.

The second equality writes the trace as a sum over momentum states. It also symmetrizes the eigenfunctions by summing over all particle permutations $\hat{\mathrm{P}}$, with the upper (plus) sign for bosons, the lower (minus) sign for fermions, according to the parity p of the permutation. The factor of $N!$ in the denominator corrects for the double counting of the states due to this symmetrization, equations (3.29) and (3.52). Kirkwood (1933) gives the equivalent of this second equality as his equation (10) for $Z(N, V, T)$, although he erroneously neglects a factor of $N!$ in the denominator (cf the transition from the second to the third equality in equation (7.1); see also section 8.2.2 below). The present formulation of particle statistics is formally exact, and carries over to the continuum where it would otherwise be impossible to enforce particle state occupancy rules.

The third equality inserts the completeness condition $\int \mathrm{d}\mathbf{q}\, |\mathbf{q}\rangle \langle \mathbf{q}| = \delta(\mathbf{r}' - \mathbf{r}'')$, equation (3.20). This draws a distinction between the position representation coordinate $\mathbf{r}$ and the position eigenvalue that appears in phase space $\mathbf{q}$. This distinction, and $\mathbf{q}$ itself, is not present in the second equality, but is important because it will allow commutation effects and symmetrization effects to be encapsulated in separate phase functions, and it will be essential to the local state expansion in chapter 10. The position basis functions are not symmetrized as they were in equation (3.36) because the basis functions for the whole position space are required here. Introducing the completeness condition to the left of the Maxwell–Boltzmann operator as here produces an asymmetry in position and momentum that is discussed shortly.

The fourth equality transforms to the momentum continuum. The factor from the momentum volume element, $\Delta_{\mathrm{p}}^{-dN}$, combines with the factor of $V^{-N} = \langle \mathbf{q}|\mathbf{p}\rangle \langle \mathbf{p}|\mathbf{q}\rangle$ that cancels the denominator introduced here to give the prefactor h^{-dN}. This is now an integral over classical phase space.

The fifth equality writes the phase space integral in terms of the commutation function ω_{p}, and symmetrization function $\eta_{\mathrm{q}}^{\pm}$, both of which will be defined explicitly next. If at the third equality the completeness condition had instead been inserted to the right of the Maxwell–Boltzmann operator one would have ended up with position and momentum swapped in these subscripts,

$$\Xi^{\pm} = \sum_{N=0}^{\infty} \frac{z^N}{h^{dN} N!} \int \mathrm{d}\boldsymbol{\Gamma}\, e^{-\beta\mathcal{H}(\boldsymbol{\Gamma})} \omega_{\mathrm{q}}(\boldsymbol{\Gamma}) \eta_{\mathrm{p}}^{\pm}(\boldsymbol{\Gamma}). \tag{7.4}$$

Obviously the partition function must remain unchanged.

The symmetrization functions are defined as

$$\eta_{\mathrm{q}}^{\pm}(\mathbf{p}, \mathbf{q}) \equiv \frac{1}{\langle \mathbf{p}|\mathbf{q}\rangle} \sum_{\hat{\mathrm{P}}} (\pm 1)^p \langle \hat{\mathrm{P}}\mathbf{p}|\mathbf{q}\rangle, \tag{7.5}$$

and

$$\eta_{\mathrm{p}}^{\pm}(\mathbf{p}, \mathbf{q}) \equiv \frac{1}{\langle \mathbf{q}|\mathbf{p}\rangle} \sum_{\hat{\mathrm{P}}} (\pm 1)^p \langle \hat{\mathrm{P}}\mathbf{q}|\mathbf{p}\rangle. \tag{7.6}$$

Since $\langle \hat{\mathrm{P}}\mathbf{p}|\mathbf{q}\rangle = \langle \mathbf{p}|\hat{\mathrm{P}}^{-1}\mathbf{q}\rangle$, it is evident that these are just the complex conjugate of each other, $\eta_{\mathrm{p}}^{\pm}(\mathbf{p}, \mathbf{q}) = \eta_{\mathrm{q}}^{\pm}(\mathbf{p}, \mathbf{q})^*$. Also, since $|\mathbf{p}\rangle \equiv e^{-\mathbf{p}\cdot\mathbf{r}/i\hbar}/V^{N/2}$, $\eta_{\mathrm{p}}^{\pm}(\mathbf{p}, \mathbf{q})^* = \eta_{\mathrm{p}}^{\pm}(-\mathbf{p}, \mathbf{q})$. The symmetrization function is related to, but is different from, the symmetrization or overlap factor, $\chi_{\mathbf{n}}^{\pm}$, introduced in the preceding chapters. The former is a sort of asymmetric version of the latter. The symmetrization function will be discussed in greater detail below, section 7.2.

The commutation function introduced here is very closely related to a function introduced by Wigner (1932) and analyzed by Kirkwood (1933). Although the application differs, it has essentially the same defining equation,

$$e^{-\beta\mathcal{H}(\mathbf{p}, \mathbf{q})} \omega_{\mathrm{p}}(\mathbf{p}, \mathbf{q}) = \frac{\langle \mathbf{q}|e^{-\beta\hat{\mathcal{H}}}|\mathbf{p}\rangle}{\langle \mathbf{q}|\mathbf{p}\rangle}, \tag{7.7}$$

and

$$e^{-\beta\mathcal{H}(\mathbf{p}, \mathbf{q})} \omega_{\mathrm{q}}(\mathbf{p}, \mathbf{q}) = \frac{\langle \mathbf{p}|e^{-\beta\hat{\mathcal{H}}}|\mathbf{q}\rangle}{\langle \mathbf{p}|\mathbf{q}\rangle}. \tag{7.8}$$

For the complex conjugates one has $\omega_p(\mathbf{p}, \mathbf{q}) = \omega_q(\mathbf{p}, \mathbf{q})^*$, and also $\omega_p(\mathbf{p}, \mathbf{q})^* = \omega_p(-\mathbf{p}, \mathbf{q})$. High temperature expansions for the commutation function have been given (Wigner 1932, Kirkwood 1933, Attard 2018b) and will be treated in chapter 8. Other expansions for the commutation function will be derived in chapters 9, 10 and 11.

The commutation function in phase space can also be written as a series of energy eigenfunctions and eigenvalues, $\hat{\mathcal{H}}|\mathbf{n}\rangle = E_\mathbf{n}|\mathbf{n}\rangle$. Using the completeness properties of these one obtains (Attard 2018b)

$$\begin{aligned} e^{-\beta\mathcal{H}(\mathbf{p},\mathbf{q})}\omega_p(\mathbf{p}, \mathbf{q}) &= \frac{\langle\mathbf{q}|e^{-\beta\hat{\mathcal{H}}}|\mathbf{p}\rangle}{\langle\mathbf{q}|\mathbf{p}\rangle} \\ &= \frac{1}{\langle\mathbf{q}|\mathbf{p}\rangle}\sum_\mathbf{n}\langle\mathbf{q}|e^{-\beta\hat{\mathcal{H}}}|\mathbf{n}\rangle\,\langle\mathbf{n}|\mathbf{p}\rangle \\ &= \frac{1}{\langle\mathbf{q}|\mathbf{p}\rangle}\sum_\mathbf{n} e^{-\beta E_\mathbf{n}}\,\langle\mathbf{q}|\mathbf{n}\rangle\,\langle\mathbf{n}|\mathbf{p}\rangle, \end{aligned} \tag{7.9}$$

and analogously for ω_q. There is reason to believe that this formulation in terms of energy eigenvalues and eigenfunctions is useful in computational applications, as will be demonstrated in chapters 10 and 11.

It is clear that the grand partition function is real, because the imaginary parts of ω and of $\eta^\pm$ are odd in momentum. Furthermore, as discussed for equation (7.4), the partition function is unchanged by the replacement $\omega_p\eta_q^\pm \Rightarrow \omega_q\eta_p^\pm$. Since these are the complex conjugate of each other, this proves that the partition function is real. On the occasion below we shall omit the subscripts on these, in which case it is implicit that whichever is used for ω, the other must be used for $\eta^\pm$.

The integrand of the grand partition function, $e^{-\beta\mathcal{H}(\Gamma)}\omega_p(\Gamma)\eta_q^\pm(\Gamma)$, can be interpreted as the weight density of a point in classical phase space. In classical statistical mechanics and in classical probability theory it is essential that the weight density be real and non-negative. But in quantum statistical mechanics there is no such requirement, and in general the quantum weight density of phase space is complex. Any problems associated with ω and $\eta^\pm$ being complex are practical and computational rather than fundamental and principled.

Interestingly enough, when integrated over momenta the consequent position weight density is real and non-negative, and when integrated over positions the consequent momentum weight density is real and non-negative. Proof: The Maxwell–Boltzmann operator is Hermitian, and integrating over one is equivalent to the diagonal expectation value or matrix element of the operator with respect to the other label. Integrating over the positions gives the momentum weight density as proportional to $\langle\mathbf{p}^\pm|e^{-\beta\hat{\mathcal{H}}}|\mathbf{p}^\pm\rangle = \sum_\mathbf{n} e^{-\beta E_\mathbf{n}}|\langle\mathbf{n}|\mathbf{p}^\pm\rangle|^2$, which is a sum of non-negative, real terms. Similarly, integrating over the momenta gives the position weight density as proportional to $\langle\mathbf{q}|e^{-\beta\hat{\mathcal{H}}}|\mathbf{q}\rangle = \sum_\mathbf{n} e^{-\beta E_\mathbf{n}}|\langle\mathbf{n}|\mathbf{q}\rangle|^2$, which again is a sum of non-negative, real terms.

7.1.2 Statistical averages

The grand canonical average of an operator can be transformed to phase space in a derivation identical to that used above for the grand partition function. One obtains

$$\begin{aligned}\langle \hat{A}\rangle^{\pm}_{z,\beta,V} &= \frac{1}{\Xi^{\pm}}\mathrm{TR}'\left\{e^{-\beta\hat{\mathcal{H}}}\hat{A}\right\}\\ &= \frac{1}{\Xi^{\pm}}\sum_{N=0}^{\infty}\frac{z^N}{h^{dN}N!}\int \mathrm{d}\mathbf{\Gamma}\; e^{-\beta\mathcal{H}(\mathbf{\Gamma})}A(\mathbf{\Gamma})\omega_{\mathrm{A,p}}(\mathbf{\Gamma})\eta_{\mathrm{q}}^{\pm}(\mathbf{\Gamma}).\end{aligned} \tag{7.10}$$

There are in fact six equivalent versions of this average, corresponding to the 3! permutations of the order of the three operators $e^{-\beta\hat{\mathcal{H}}}$, $|\mathbf{q}\rangle\langle\mathbf{q}|$, and $\hat{A}$. (One could probably make this twelve by swapping position and momentum for the initial trace.) We shall examine only four in detail: two with $|\mathbf{q}\rangle\langle\mathbf{q}|$ after (as here) or before (swap the subscripts p and q on the commutation and symmetrization functions) $e^{-\beta\hat{\mathcal{H}}}\hat{A}$, times two with $\hat{A}$ after (as here) or before (to be indicated by a tilde) $e^{-\beta\hat{\mathcal{H}}}$.

The commutation functions for the average are defined by

$$e^{-\beta\mathcal{H}(\mathbf{p},\mathbf{q})}A(\mathbf{p},\mathbf{q})\omega_{\mathrm{A,p}}(\mathbf{p},\mathbf{q}) = \frac{\langle\mathbf{q}|e^{-\beta\hat{\mathcal{H}}}\hat{A}|\mathbf{p}\rangle}{\langle\mathbf{q}|\mathbf{p}\rangle}, \tag{7.11}$$

and

$$e^{-\beta\mathcal{H}(\mathbf{p},\mathbf{q})}A(\mathbf{p},\mathbf{q})\omega_{\mathrm{A,q}}(\mathbf{p},\mathbf{q}) = \frac{\langle\mathbf{p}|e^{-\beta\hat{\mathcal{H}}}\hat{A}|\mathbf{q}\rangle}{\langle\mathbf{p}|\mathbf{q}\rangle}. \tag{7.12}$$

Swapping the order of the operators gives $\tilde{\omega}$,

$$e^{-\beta\mathcal{H}(\mathbf{p},\mathbf{q})}A(\mathbf{p},\mathbf{q})\tilde{\omega}_{\mathrm{A,p}}(\mathbf{p},\mathbf{q}) = \frac{\langle\mathbf{q}|\hat{A}e^{-\beta\hat{\mathcal{H}}}|\mathbf{p}\rangle}{\langle\mathbf{q}|\mathbf{p}\rangle}, \tag{7.13}$$

and

$$e^{-\beta\mathcal{H}(\mathbf{p},\mathbf{q})}A(\mathbf{p},\mathbf{q})\tilde{\omega}_{\mathrm{A,q}}(\mathbf{p},\mathbf{q}) = \frac{\langle\mathbf{p}|\hat{A}e^{-\beta\hat{\mathcal{H}}}|\mathbf{q}\rangle}{\langle\mathbf{p}|\mathbf{q}\rangle}. \tag{7.14}$$

In the simplest case the operator being averaged is an ordinary function of the momentum and position operators, $\hat{A} = A(\hat{\mathbf{p}},\hat{\mathbf{q}})$, which leads directly to the phase space function $A(\mathbf{p},\mathbf{q})$ appearing as here. More generally, one should probably use the right-hand side to define the product $(A\omega_{\mathrm{A,p}})(\mathbf{p},\mathbf{q})$, etc, as a single phase space function.

Assume that the operator is Hermitian, $\hat{A} = \hat{A}^{\dagger}$, that the phase function is real, $A(\mathbf{p},\mathbf{q}) = A(\mathbf{p},\mathbf{q})^{*}$, and that the average is real. Taking the complex conjugate of the defining equation yields

$$\begin{aligned}e^{-\beta\mathcal{H}(\mathbf{p},\mathbf{q})}A(\mathbf{p},\mathbf{q})\omega_{\mathrm{A,p}}(\mathbf{p},\mathbf{q})^{*} &= \frac{\langle\mathbf{p}|\hat{A}e^{-\beta\hat{\mathcal{H}}}|\mathbf{q}\rangle}{\langle\mathbf{p}|\mathbf{q}\rangle}\\ &\equiv e^{-\beta\mathcal{H}(\mathbf{p},\mathbf{q})}A(\mathbf{p},\mathbf{q})\tilde{\omega}_{\mathrm{A,q}}(\mathbf{p},\mathbf{q}).\end{aligned} \tag{7.15}$$

Of course if $\hat{A} = A(\hat{\mathcal{H}})$, it commutes with the Maxwell–Boltzmann operator. In such a case $\tilde{\omega}_{\mathrm{A,q}} = \omega_{\mathrm{A,q}}$, and $\tilde{\omega}_{\mathrm{A,p}} = \omega_{\mathrm{A,p}}$.

In general

$$\omega_{\mathrm{A,q}}(\mathbf{\Gamma}) \neq \omega_{\mathrm{q}}(\mathbf{\Gamma}) \quad \text{and} \quad \omega_{\mathrm{A,p}}(\mathbf{\Gamma}) \neq \omega_{\mathrm{p}}(\mathbf{\Gamma}). \tag{7.16}$$

Unlike classical statistical mechanics, for quantum statistical mechanics the phase space weight depends upon the function being averaged. However in the case of 'pure' functions, from the definitions one can show that

$$\begin{aligned} \text{if } \hat{A} = A(\hat{\mathbf{p}}) \text{ then } & \omega_{\mathrm{A,p}}(\mathbf{\Gamma}) = \omega_{\mathrm{p}}(\mathbf{\Gamma}) \\ \text{and } & \tilde{\omega}_{\mathrm{A,q}}(\mathbf{\Gamma}) = \omega_{\mathrm{q}}(\mathbf{\Gamma}). \end{aligned} \tag{7.17}$$

And also that

$$\begin{aligned} \text{if } \hat{A} = A(\hat{\mathbf{q}}) \text{ then } & \omega_{\mathrm{A,q}}(\mathbf{\Gamma}) = \omega_{\mathrm{q}}(\mathbf{\Gamma}) \\ \text{and } & \tilde{\omega}_{\mathrm{A,p}}(\mathbf{\Gamma}) = \omega_{\mathrm{p}}(\mathbf{\Gamma}). \end{aligned} \tag{7.18}$$

One can see that any linear combination of pure functions of the position and momentum operators has an average that can be arranged as a linear combination of averages with the original commutation function, ω_{p} or ω_{q}. This is now illustrated explicitly for the energy operator.

7.1.3 Energy

The first expression for the average energy takes $\hat{A} = \hat{\mathcal{H}}$ and uses $\omega_{\mathcal{H},\mathrm{p}}$ or $\omega_{\mathcal{H},\mathrm{q}}$, giving

$$\begin{aligned} \langle \hat{\mathcal{H}} \rangle^{\pm}_{z,\beta,V} &= \frac{1}{\Xi^{\pm}} \sum_{N=0}^{\infty} \frac{z^N}{h^{dN} N!} \int \mathrm{d}\mathbf{\Gamma}\; e^{-\beta\mathcal{H}(\mathbf{\Gamma})} \mathcal{H}(\mathbf{\Gamma}) \omega_{\mathcal{H},\mathrm{p}}(\mathbf{\Gamma}) \eta^{\pm}_{\mathrm{q}}(\mathbf{\Gamma}) \\ &= \frac{1}{\Xi^{\pm}} \sum_{N=0}^{\infty} \frac{z^N}{h^{dN} N!} \int \mathrm{d}\mathbf{\Gamma}\; e^{-\beta\mathcal{H}(\mathbf{\Gamma})} \mathcal{H}(\mathbf{\Gamma}) \omega_{\mathcal{H},\mathrm{q}}(\mathbf{\Gamma}) \eta^{\pm}_{\mathrm{p}}(\mathbf{\Gamma}). \end{aligned} \tag{7.19}$$

Obviously because the energy operator commutes with the Maxwell Boltzmann operator, $\tilde{\omega}_{\mathcal{H},\mathrm{q}} = \omega_{\mathcal{H},\mathrm{q}}$ and $\tilde{\omega}_{\mathcal{H},\mathrm{p}} = \omega_{\mathcal{H},\mathrm{p}}$.

The Hamiltonian is the sum of the kinetic and potential energies, which is indeed a linear combination of pure functions,

$$\hat{\mathcal{H}} = \mathcal{K}(\hat{\mathbf{p}}) + U(\hat{\mathbf{q}}). \tag{7.20}$$

Hence one can also write

$$\begin{aligned} \langle \hat{\mathcal{H}} \rangle^{\pm}_{z,\beta,V} &= \langle \hat{\mathcal{K}} \rangle^{\pm}_{z,\beta,V} + \langle \hat{U} \rangle^{\pm}_{z,\beta,V} \\ &= \frac{1}{\Xi^{\pm}} \sum_{N=0}^{\infty} \frac{z^N}{h^{dN} N!} \int \mathrm{d}\mathbf{\Gamma}\; e^{-\beta\mathcal{H}(\mathbf{\Gamma})} \left\{ \mathcal{K}(\mathbf{p}) \omega_{\mathrm{p}}(\mathbf{\Gamma}) \eta^{\pm}_{\mathrm{q}}(\mathbf{\Gamma}) + U(\mathbf{q}) \omega_{\mathrm{q}}(\mathbf{\Gamma}) \eta^{\pm}_{\mathrm{p}}(\mathbf{\Gamma}) \right\} \\ &= \frac{1}{\Xi^{\pm}} \sum_{N=0}^{\infty} \frac{z^N}{h^{dN} N!} \int \mathrm{d}\mathbf{\Gamma}\; e^{-\beta\mathcal{H}(\mathbf{\Gamma})} \mathcal{H}(\mathbf{\Gamma}) \omega_{\mathrm{p}}(\mathbf{\Gamma}) \eta^{\pm}_{\mathrm{q}}(\mathbf{\Gamma}) \\ &= \frac{1}{\Xi^{\pm}} \sum_{N=0}^{\infty} \frac{z^N}{h^{dN} N!} \int \mathrm{d}\mathbf{\Gamma}\; e^{-\beta\mathcal{H}(\mathbf{\Gamma})} \mathcal{H}(\mathbf{\Gamma}) \omega_{\mathrm{q}}(\mathbf{\Gamma}) \eta^{\pm}_{\mathrm{p}}(\mathbf{\Gamma}). \end{aligned} \tag{7.21}$$

The penultimate equality follows by taking the complex conjugate of the average potential energy in the second equality. The final equality follows by taking the complex conjugate of the average kinetic energy in the second equality. One sees from these that one can obtain the average energy using the same commutation function as for the partition function.

Differentiating with respect to $-\beta$ the defining equation for ω_p, equation (7.7), one obtains

$$\mathcal{H}(\mathbf{\Gamma})e^{-\beta\mathcal{H}(\mathbf{\Gamma})}\omega_\mathrm{p}(\mathbf{\Gamma}) - e^{-\beta\mathcal{H}(\mathbf{\Gamma})}\frac{\partial\omega_\mathrm{p}(\mathbf{\Gamma})}{\partial\beta} = \frac{\langle\mathbf{q}|\hat{\mathcal{H}}e^{-\beta\hat{\mathcal{H}}}|\mathbf{p}\rangle}{\langle\mathbf{q}|\mathbf{p}\rangle} = e^{-\beta\mathcal{H}(\mathbf{\Gamma})}\mathcal{H}(\mathbf{\Gamma})\omega_{\mathcal{H},\mathrm{p}}(\mathbf{\Gamma}), \quad (7.22)$$

since $\tilde{\omega}_{\mathcal{H},\mathrm{p}} = \omega_{\mathcal{H},\mathrm{p}}$. Hence $\omega_{\mathcal{H},\mathrm{p}} = \omega_\mathrm{p} - \mathcal{H}^{-1}\partial\omega_\mathrm{p}/\partial\beta$. Inserting the left-hand side into equation (7.19), and comparing this with equation (7.21) one sees that in order for the two expressions for the average energy to be equal one must have

$$\sum_{N=0}^{\infty}\frac{z^N}{h^{dN}N!}\int \mathrm{d}\mathbf{\Gamma}\; e^{-\beta\mathcal{H}(\mathbf{\Gamma})}\frac{\partial\omega_\mathrm{p}(\mathbf{\Gamma})}{\partial\beta}\eta_\mathrm{q}^{\pm}(\mathbf{\Gamma}) = 0. \quad (7.23)$$

An analogous result holds for $\partial\omega_\mathrm{q}(\mathbf{\Gamma})/\partial\beta$.

7.1.3.1 A useful result

It is now proven that this vanishes individually for each particle number N and for each permutation $\hat{\mathrm{P}}$. This turns out to be necessary for thermodynamic consistency, as will be explained below.

With $|\mathbf{p}\rangle = V^{-N/2}e^{-\mathbf{p}\cdot\mathbf{r}/i\hbar}$, one has that $\hat{\mathcal{H}}|\mathbf{p}\rangle = \mathcal{H}(\mathbf{p}, \mathbf{r})|\mathbf{p}\rangle$. Also since the permutations are between identical particles, $\mathcal{H}(\mathbf{p}, \mathbf{q}) = \mathcal{H}(\hat{\mathrm{P}}\mathbf{p}, \mathbf{q}) = \mathcal{H}(\mathbf{p}, \hat{\mathrm{P}}\mathbf{q})$, for any permutation $\hat{\mathrm{P}}$. (This assumes that there are no velocity dependent forces, which is the usual case. Lorentz forces need thought.) Hence

$$\mathcal{H}(\mathbf{p}, \mathbf{q})\langle\hat{\mathrm{P}}\mathbf{p}|\mathbf{q}\rangle = \langle\hat{\mathrm{P}}\mathbf{p}|\hat{\mathcal{H}}|\mathbf{q}\rangle. \quad (7.24)$$

With this one has

$$\begin{aligned}
&\sum_{N=0}^{\infty}\frac{z^N}{h^{dN}N!}\int \mathrm{d}\mathbf{\Gamma}\; e^{-\beta\mathcal{H}(\mathbf{\Gamma})}\frac{\partial\omega_\mathrm{p}(\mathbf{\Gamma})}{\partial\beta}\eta_\mathrm{q}^{\pm}(\mathbf{\Gamma})\\
&= \sum_{N=0}^{\infty}\frac{z^N}{h^{dN}N!}\int \mathrm{d}\mathbf{\Gamma}\; e^{-\beta\mathcal{H}(\mathbf{\Gamma})}\eta_\mathrm{q}^{\pm}(\mathbf{\Gamma})\left\{\mathcal{H}(\mathbf{\Gamma})\omega_{\mathcal{H},\mathrm{p}}(\mathbf{\Gamma}) - \mathcal{H}(\mathbf{\Gamma})\omega_\mathrm{p}(\mathbf{\Gamma})\right\}\\
&= \sum_{N=0}^{\infty}\frac{z^N V^N}{h^{dN}N!}\sum_{\hat{\mathrm{P}}}(\pm 1)^p\int \mathrm{d}\mathbf{\Gamma}\;\langle\hat{\mathrm{P}}\mathbf{p}|\mathbf{q}\rangle\left\{\langle\mathbf{q}|e^{\beta\hat{\mathcal{H}}}\hat{\mathcal{H}}|\mathbf{p}\rangle - \mathcal{H}(\mathbf{p}, \mathbf{q})\langle\mathbf{q}|e^{\beta\hat{\mathcal{H}}}|\mathbf{p}\rangle\right\}\\
&= \sum_{N=0}^{\infty}\frac{z^N V^N}{h^{dN}N!}\sum_{\hat{\mathrm{P}}}(\pm 1)^p\int \mathrm{d}\mathbf{\Gamma}\left\{\langle\hat{\mathrm{P}}\mathbf{p}|\mathbf{q}\rangle\,\langle\mathbf{q}|e^{\beta\hat{\mathcal{H}}}\hat{\mathcal{H}}|\mathbf{p}\rangle - \langle\hat{\mathrm{P}}\mathbf{p}|\hat{\mathcal{H}}|\mathbf{q}\rangle\,\langle\mathbf{q}|e^{\beta\hat{\mathcal{H}}}|\mathbf{p}\rangle\right\}\\
&= \sum_{N=0}^{\infty}\frac{z^N}{h^{dN}N!}\sum_{\hat{\mathrm{P}}}(\pm 1)^p\int \mathrm{d}\mathbf{p}\;\left\{\langle\hat{\mathrm{P}}\mathbf{p}|e^{-\beta\hat{\mathcal{H}}}\hat{\mathcal{H}}|\mathbf{p}\rangle - \langle\hat{\mathrm{P}}\mathbf{p}|\hat{\mathcal{H}}e^{-\beta\hat{\mathcal{H}}}|\mathbf{p}\rangle\right\}\\
&= 0.
\end{aligned} \quad (7.25)$$

The term in braces vanishes for each N and for each permutation $\hat{\mathrm{P}}$, as was to be shown. This result will prove important shortly. It also vanishes for each momentum configuration $\mathbf{p}$, which will not be required.

7.2 Loop expansion, grand potential and average energy

7.2.1 Symmetrization loops

As discussed in section 3.4.2, any particular particle permutation operator can be factored into loop permutation operators, where a loop is a closed connected sequence of successive pair transpositions. Hence the sum over all permutation operators can be written as the sum over all possible factors of loop permutations, equation (3.44)

$$\sum_{\hat{\mathrm{P}}}(\pm 1)^p\, \hat{\mathrm{P}} = \hat{\mathrm{I}} \pm \sum_{i,j}{}' \hat{\mathrm{P}}_{ij} + \sum_{i,j,k}{}' \hat{\mathrm{P}}_{ij}\hat{\mathrm{P}}_{jk} + \sum_{i,j,k,l}{}' \hat{\mathrm{P}}_{ij}\hat{\mathrm{P}}_{kl} \pm \ldots \tag{7.26}$$

Here $\hat{\mathrm{P}}_{jk}$ is the transpose of particles j and k. The prime on the sums restrict them to unique loops, with each index being different. The first term is just the identity. The second term is a dimer loop, the third term is a trimer loop, and the fourth term shown is the product of two different dimers.

With this, the symmetrization function, $\eta_{\mathrm{q}}^{\pm}(\mathbf{\Gamma}) = \sum_{\hat{\mathrm{P}}}(\pm 1)^p \langle \hat{\mathrm{P}}\mathbf{p}|\mathbf{q}\rangle / \langle \mathbf{p}|\mathbf{q}\rangle$, is the sum of the expectation values of these loops. The monomer symmetrization function is obviously unity, $\eta_{\mathrm{q}}^{(1)} \equiv \langle \mathbf{p}|\mathbf{q}\rangle / \langle \mathbf{p}|\mathbf{q}\rangle = 1$.

The dimer symmetrization function in the microstate $\mathbf{\Gamma}$ for particles j and k is

$$\begin{aligned} \eta_{\mathrm{q};jk}^{\pm(2)} &= \frac{\pm\langle \hat{\mathrm{P}}_{jk}\mathbf{p}|\mathbf{q}\rangle}{\langle \mathbf{p}|\mathbf{q}\rangle} \\ &= \frac{\pm\langle \mathbf{p}_k|\mathbf{q}_j\rangle\langle \mathbf{p}_j|\mathbf{q}_k\rangle}{\langle \mathbf{p}_j|\mathbf{q}_j\rangle\langle \mathbf{p}_k|\mathbf{q}_k\rangle} \\ &= \pm\, e^{(\mathbf{q}_k-\mathbf{q}_j)\cdot\mathbf{p}_j/i\hbar} e^{(\mathbf{q}_j-\mathbf{q}_k)\cdot\mathbf{p}_k/i\hbar}. \end{aligned} \tag{7.27}$$

Recall that $|\mathbf{p}\rangle \equiv e^{-\mathbf{p}\cdot\mathbf{r}/i\hbar}/V^{N/2}$. Note that since the basis functions are the product of single-particle functions, the expectation value factorizes leaving only the permuted particles to contribute (cf section 3.4.3).

Similarly the trimer symmetrization function for particles j, k, and l is

$$\begin{aligned} \eta_{\mathrm{q};jkl}^{\pm(3)} &= \frac{\langle \hat{\mathrm{P}}_{jk}\hat{\mathrm{P}}_{kl}\mathbf{p}|\mathbf{q}\rangle}{\langle \mathbf{p}|\mathbf{q}\rangle} \\ &= \frac{\langle \mathbf{p}_k|\mathbf{q}_j\rangle\langle \mathbf{p}_j|\mathbf{q}_l\rangle\langle \mathbf{p}_l|\mathbf{q}_k\rangle}{\langle \mathbf{p}_j|\mathbf{q}_j\rangle\langle \mathbf{p}_k|\mathbf{q}_k\rangle\langle \mathbf{p}_l|\mathbf{q}_l\rangle} \\ &= e^{(\mathbf{q}_j-\mathbf{q}_k)\cdot\mathbf{p}_k/i\hbar} e^{(\mathbf{q}_k-\mathbf{q}_l)\cdot\mathbf{p}_l/i\hbar} e^{(\mathbf{q}_l-\mathbf{q}_j)\cdot\mathbf{p}_j/i\hbar}. \end{aligned} \tag{7.28}$$

In general the l-loop symmetrization function is

$$\eta^{\pm(l)}_{\mathrm{q};1\ldots l} = (\pm 1)^{l-1} e^{\mathbf{q}_{1l}\cdot\mathbf{p}_l/i\hbar} \prod_{j=1}^{l-1} e^{\mathbf{q}_{j+1,j}\cdot\mathbf{p}_j/i\hbar}, \tag{7.29}$$

where $\mathbf{q}_{jk} \equiv \mathbf{q}_j - \mathbf{q}_k$. Recall that $\eta_\mathrm{p} = \eta_\mathrm{q}^*$.

The product of Fourier exponentials that occurs here makes each specific l-loop symmetrization function highly oscillatory unless successive particles around the loop are close together in either momentum or position space. This means that the only non-zero contributions to phase space integrals come from such compact loops, since otherwise their oscillations would average to zero. (See equation (7.38) below for an explicit example.)

With these loop symmetrization functions, the symmetrization function itself can be written as a series of loop products,

$$\eta^\pm_\mathrm{q}(\mathbf{\Gamma}) = 1 + \sum_{ij}{}' \eta^{\pm(2)}_{\mathrm{q};ij} + \sum_{ijk}{}' \eta^{\pm(3)}_{\mathrm{q};ijk} + \sum_{ijkl}{}' \eta^{\pm(2)}_{q;ij}\eta^{\pm(2)}_{\mathrm{q};kl} + \cdots \tag{7.30}$$

Here the superscript is the order of the loop, and the subscripts are the atoms involved in the loop. The prime signifies that the sum is over unique loops (i.e., each configuration of particles in loops occurs once only) with each index different (i.e., no particle may belong to more than one loop).

Figure 7.1 is a diagrammatic representation of the loop expansion of the symmetrization function. The filled circles represent a sum over particle labels, which must obey certain constraints, and the bonds represent a Fourier pair factor. The full power of such diagrams becomes apparent upon integration over phase space. These are the analog of the Mayer f-bond diagrams that have proved useful in classical statistical mechanics (Hansen and McDonald 1986, Attard 2002).

In this one can identify the terms that consist solely of a single loop, and define the single-loop symmetrization function,

$$\begin{aligned}\overset{\circ}{\eta}{}^\pm_\mathrm{q}(\mathbf{\Gamma}) &\equiv \sum_{jk}{}' \eta^{\pm(2)}_{\mathrm{q};jk} + \sum_{jkm}{}' \eta^{\pm(3)}_{\mathrm{q};jkm} + \sum_{jkmn}{}' \eta^{\pm(4)}_{\mathrm{q};jkmn} + \cdots \\ &\equiv \sum_{l=2}^{\infty} \overset{\circ}{\eta}{}^{\pm(l)}_\mathrm{q}(\mathbf{\Gamma}).\end{aligned} \tag{7.31}$$

This gives the single l-loop symmetrization function $\overset{\circ}{\eta}{}^{\pm(l)}_\mathrm{q}(\mathbf{\Gamma})$ as the sum over the $N!/(N-l)!l$ distinct arrangements of the l particle labels. This is sketched in diagrammatic terms in figure 7.2. In taking the average of each single-loop

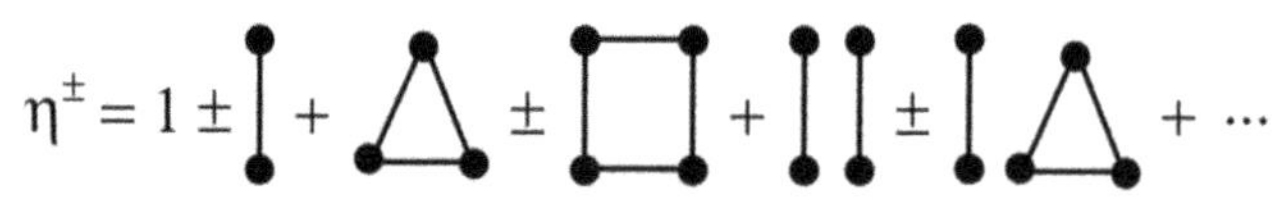

Figure 7.1. Diagrammatic representation of the loop expansion of the symmetrization function, equation (7.30).

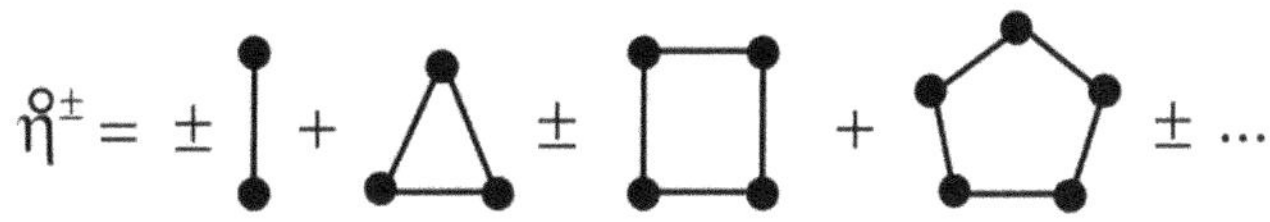

Figure 7.2. Diagrammatic representation of the single loop version of the symmetrization function, equation (7.31).

symmetrization function, since the particles are identical, all the terms in the sum over particles gives the same average value. Therefore, what will be shown next to be the loop grand potential can be obtained by evaluating any one arrangement, say particles 1, 2, … , l,

$$\begin{aligned} -\beta\Omega^{\pm(l)}_{\omega_{\rm p}} &\equiv \left\langle \overset{\circ}{\eta}{}^{\pm(l)}_{\rm q}(\mathbf{\Gamma}) \right\rangle_{\omega_{\rm p}} \\ &= \left\langle \sum_{j_1 \ldots j_l}{}' \eta^{\pm(l)}_{{\rm q};j_1 \ldots j_l} \right\rangle_{\omega_{\rm p}} \\ &= \left\langle \frac{N!}{(N-l)!l} \eta^{\pm(l)}_{{\rm q};1\ldots l} \right\rangle_{\omega_{\rm p}}. \end{aligned} \tag{7.32}$$

Here and below we often attach subscripts to the grand partition function, grand potential, and statistical average to indicate the quantum phase functions that are being used in the integrand. The full quantum grand potential is written $\Omega_{\omega_{\rm p}\eta^{\pm}_{\rm q}} \equiv \Omega^{\pm}$, and the classical grand potential, which has $\omega_{\rm p}(\mathbf{\Gamma}) = \eta^{\pm}_{\rm q}(\mathbf{\Gamma}) = 1$, is $\Omega_{\rm cl}$.

7.2.2 Grand potential

The grand potential is the thermodynamic free energy of an open equilibrium system (i.e., one in which the subsystem exchanges energy and particles with a reservoir). In general the grand potential is the logarithm of the grand partition function,

$$\Omega^{\pm}(\mu, V, T) = -k_{\rm B}T \ln \Xi^{\pm}(\mu, V, T). \tag{7.33}$$

This is of course just the usual relationship linking thermodynamic free energies to statistical mechanics.

In the present section an expression for the grand potential is derived that invokes a factorized form for integrals involving the symmetrization function. This factorization ansatz is exact for non-interacting systems, section 3.4.3, and it gives the known results for the quantum ideal gas, section 4.1, and for a system of non-interacting quantum harmonic oscillators, section 4.2. The ansatz neglects correlations between loops in the interacting case, and in this case it is expected to be exact for dilute loops, as is discussed in subsection 7.2.2.1 below. The thermodynamic limit is the large volume limit taken at constant fugacity or density. For interacting bosons the factorization ansatz appears to break down at low temperatures, section 5.4.

In fact the factorization ansatz is not essential for the theory, since one can just take the grand potential from the logarithm of the partition function, the latter having been evaluated using the full series for the symmetrization function, and similarly for any average. Undoubtedly, however, it would be somewhat tedious to calculate the series in full, and the exponential that results from the factorization is convenient and much more rapidly converging.

The monomer grand partition function comes from setting $\eta^{\pm}=1$,

$$\Xi^{\pm}_{\omega_{\rm p}} = \sum_{N=0}^{\infty} \frac{z^N}{h^{dN} N!} \int \mathrm{d}\mathbf{p}\, \mathrm{d}\mathbf{q}\; e^{-\beta\mathcal{H}(\mathbf{p},\,\mathbf{q})}\omega_{\rm p}(\mathbf{p},\,\mathbf{q}). \tag{7.34}$$

Recall that the imaginary part of $\omega_{\rm p}$ is odd in momentum, and that $\omega_{\rm p}^* = \omega_{\rm q}$. Hence the subscript p or q (and also the superscript ±) is redundant in the monomer case. Nevertheless it seems best to signify explicitly which of these commutation functions is being used.

The ratio of the full, $\Xi^{\pm} \equiv \Xi_{\omega_{\rm p}\eta_{\rm q}^{\pm}}$, to the monomer, $\Xi_{\omega_{\rm p}}$, grand partition function is the monomer average of the symmetrization function,

$$\begin{aligned}
\frac{\Xi_{\omega_{\rm p}\eta_{\rm q}^{\pm}}}{\Xi_{\omega_{\rm p}}} &= \left\langle \eta_{\rm q}^{\pm} \right\rangle_{\omega_{\rm p}} \\
&= 1 + \left\langle \sum_{ij}{}' \eta_{{\rm q};ij}^{\pm(2)} \right\rangle_{\omega_{\rm p}} + \left\langle \sum_{ijk}{}' \eta_{{\rm q};ijk}^{\pm(3)} \right\rangle_{\omega_{\rm p}} + \left\langle \sum_{ijkl}{}' \eta_{{\rm q};ij}^{\pm(2)} \eta_{{\rm q};kl}^{\pm(2)} \right\rangle_{\omega_{\rm p}} + \cdots \\
&= 1 + \left\langle \frac{N}{(N-2)!2} \eta_{{\rm q};12}^{\pm(2)} \right\rangle_{\omega_{\rm p}} + \left\langle \frac{N!}{(N-3)!3} \eta_{{\rm q};123}^{\pm(3)} \right\rangle_{\omega_{\rm p}} \\
&\quad + \frac{1}{2} \left\langle \frac{N!}{(N-2)!2} \eta_{{\rm q};12}^{\pm(2)} \right\rangle_{\omega_{\rm p}}^2 + \cdots \\
&= \sum_{\{m_l\}} \frac{1}{m_l!} \prod_{l=2}^{\infty} \left\langle \frac{N!}{(N-l)!l} \eta_{{\rm q};1\ldots l}^{\pm(l)} \right\rangle_{\omega_{\rm p}}^{m_l} \\
&= \prod_{l=2}^{\infty} \sum_{m_l=0}^{\infty} \frac{1}{m_l!} \left\langle \frac{N!}{(N-l)!l} \eta_{{\rm q};1\ldots l}^{\pm(l)} \right\rangle_{\omega_{\rm p}}^{m_l} \\
&\equiv \prod_{l=2}^{\infty} e^{-\beta\Omega_{\omega_{\rm p}}^{\pm(l)}}.
\end{aligned} \tag{7.35}$$

The third and following equalities write the average of the product as the product of the averages. For example, the double dimer term has been written as $\langle (N!/8(N-4)!)\eta_{{\rm q};12}^{\pm(2)}\eta_{{\rm q};34}^{\pm(2)} \rangle_{\omega_{\rm p}} = \langle (N!/2(N-2)!)\eta_{{\rm q};12}^{\pm(2)} \rangle_{\omega_{\rm p}}^2/2$, and similarly for other products. This is valid in the thermodynamic limit, since the product of the average of two loops scales as V^2, whereas the correlated interaction of two loops scales as V, and similarly for all the other products. As mentioned above, only compact loops contribute to the integral; see the discussion on the excess dimer correlation in subsection 7.2.2.1, as well as section 5.6 for a counter-example. The combinatorial factor accounts for the number of unique loops in each term.

The grand potential is given by the logarithm of the partition function $-\beta\Omega^{\pm} = \ln \Xi^{\pm}$. The monomer grand potential is given by $-\beta\Omega^{\pm}_{\omega_{\rm p}} = \ln \Xi^{\pm}_{\omega_{\rm p}}$. For the monomer term the superscript $\pm$ is redundant and will usually be dropped.

The difference between the full grand potential and the monomer grand potential is just the series of loop grand potentials,

$$\begin{aligned} -\beta[\Omega_{\omega_{\rm p}\eta_{\rm q}^{\pm}} - \Omega_{\omega_{\rm p}}] &= \ln \frac{\Xi_{\omega_{\rm p}\eta_{\rm q}^{\pm}}}{\Xi_{\omega_{\rm p}}} \\ &= \sum_{l=2}^{\infty} \left\langle \frac{N!}{(N-l)!l} \eta_{{\rm q};1\ldots l}^{\pm(l)} \right\rangle_{\omega_{\rm p}} \\ &\equiv -\beta \sum_{l=2}^{\infty} \Omega_{\omega_{\rm p}}^{\pm(l)}. \end{aligned} \tag{7.36}$$

Here $\eta_{\rm q}$ is paired with $\omega_{\rm p}$; one could alternatively pair $\eta_{\rm p}$ and $\omega_{\rm q}$. These give the same, real, value for the grand potential.

This factorization of the grand partition function and series for the grand potential is in essence based on classical concepts of short-ranged correlations. It presupposes that the commutation function is in some sense a localized function.

7.2.2.1 Excess dimer correlation

Above, the product of two dimers was written as

$$\left\langle \sum_{jkmn}{}' \eta_{{\rm q};jk}^{\pm(2)} \eta_{{\rm q};mn}^{\pm(2)} \right\rangle_{\omega_{\rm p}} = \frac{1}{2} \left\langle \frac{N!}{(N-2)!2} \eta_{{\rm q};jk}^{\pm(2)} \right\rangle_{\omega_{\rm p}}^2. \tag{7.37}$$

The prime on the sum indicates that each configuration of loops occurs once only, and that no two indeces are equal. On the right-hand side the indeces are arbitrary apart from $j \neq k$. The right-hand side is an approximation because it treats the two sums on the left-hand side as independent.

With $\eta_{{\rm q};jk}^{\pm(2)} = e^{-{\bf p}_{jk}\cdot{\bf q}_{jk}/i\hbar}$, in the case of a classical average (i.e., $\omega_{\rm p} = 1$), one can perform the momentum integrals by completing the square,

$$\begin{aligned} \left\langle \eta_{{\rm q};jk}^{\pm(2)} \right\rangle_{\rm cl} &= \frac{\pm Z_{\rm cl}^{-1}}{h^{3N} N!} \int {\rm d}\boldsymbol{\Gamma}\; e^{-\beta\mathcal{H}(\boldsymbol{\Gamma})} e^{-{\bf p}_{jk}\cdot{\bf q}_{jk}/i\hbar} \\ &= \frac{\pm Q_{\rm cl}^{-1}}{V^N N!} \int {\rm d}{\bf q}\; e^{-\beta U({\bf q})} e^{-mq_{jk}^2/2\beta\hbar^2} e^{-mq_{kj}^2/2\beta\hbar^2} \\ &= \frac{\pm Q_{\rm cl}^{-1}}{V^N N!} \int {\rm d}{\bf q}\; e^{-\beta U({\bf q})} e^{-\pi q_{jk}^2/\Lambda^2} e^{-\pi q_{kj}^2/\Lambda^2}, \end{aligned} \tag{7.38}$$

where the thermal wavelength is $\Lambda \equiv \sqrt{2\pi\hbar^2\beta/m}$. The integrand is zero except for separations less than the thermal wavelength, $q_{jk} \lesssim \Lambda$. This means that effectively particle j is tied to particle k, which means that a volume integral is lost, so that $\left\langle \eta_{\mathrm{q};jk}^{\pm(2)} \right\rangle_{\mathrm{cl}} \sim \mathcal{O}(V^{-1})$.

It is clear from this that $\langle \sum'_{jk} \eta_{\mathrm{q};jk}^{\pm(2)} \rangle_{\omega_{\mathrm{p}}}$ involves an integral over position space of the two particle density $\rho^{(2)}$. Recall that in classical statistical mechanics, $\rho^{(2)}$ at large separations goes asymptotically as the square of the singlet densities. Similarly, $\left\langle \sum'_{jkmn} \eta_{\mathrm{q};jk}^{\pm(2)} \eta_{\mathrm{q};mn}^{\pm(2)} \right\rangle_{\omega_{\mathrm{p}}}$ involves an integral of the four particle density $\rho^{(4)}$, which for large separations between the pairs goes like $\rho^{(4)}(\mathbf{q}_j, \mathbf{q}_k, \mathbf{q}_m, \mathbf{q}_n) \sim \rho^{(2)}(q_{jk})\rho^{(2)}(q_{mn})$. Hence in the thermodynamic limit, the relative error in the factorization of the dimer product vanishes

$$\frac{\left\langle \sum'_{jkmn} \eta_{\mathrm{q};jk}^{\pm(2)} \eta_{\mathrm{q};mn}^{\pm(2)} \right\rangle_{\omega_{\mathrm{p}}} - \frac{1}{2}\left\langle \sum'_{jk} \eta_{\mathrm{q};jk}^{\pm(2)} \right\rangle_{\omega_{\mathrm{p}}}^2}{\frac{1}{2}\left\langle \sum'_{jk} \eta_{\mathrm{q};jk}^{\pm(2)} \right\rangle_{\omega_{\mathrm{p}}}^2} \to 0 \quad V \to \infty, \ z, T = \mathrm{const.} \tag{7.39}$$

The denominator is $\mathcal{O}(V^{-2})$. The numerator is $\mathcal{O}(V^{-3})$, because three volume integrals are lost when all four particles are correlated, which is required for the difference to be non-zero. An analogous argument holds for the arbitrary product of arbitrary loops. This justifies the factorization of the symmetrization function in the thermodynamic limit.

7.2.3 Energy

The most likely energy can be written as the temperature derivative of the grand potential. As above the latter can be written as a series of loop derivatives. One has

$$\bar{E}_{\omega_{\mathrm{p}}\eta_{\mathrm{q}}^{\pm}} = \frac{\partial \beta \Omega^{\pm}_{\omega_{\mathrm{p}}\eta_{\mathrm{q}}}}{\partial \beta} = \sum_{l=1}^{\infty} \bar{E}_{\omega_{\mathrm{p}}}^{\pm(l)}. \tag{7.40}$$

The monomer term is

$$\begin{aligned}
\bar{E}_{\omega_{\mathrm{p}}} &= \frac{\partial \beta \Omega_{\omega_{\mathrm{p}}}}{\partial \beta} = \frac{-\partial \ln \Xi_{\omega_{\mathrm{p}}}}{\partial \beta} \\
&= \frac{1}{\Xi_{\omega_{\mathrm{p}}}} \sum_{N=0}^{\infty} \frac{z^N}{h^{dN} N!} \int \mathrm{d}\boldsymbol{\Gamma}\ e^{-\beta \mathcal{H}(\boldsymbol{\Gamma})} \left\{ \mathcal{H}(\boldsymbol{\Gamma}) \omega_{\mathrm{p}}(\boldsymbol{\Gamma}) - \frac{\partial \omega_{\mathrm{p}}(\boldsymbol{\Gamma})}{\partial \beta} \right\} \\
&= \langle \mathcal{H} \rangle_{\omega_{\mathrm{p}}} \\
&= \langle \mathcal{H} \rangle_{\omega_{\mathcal{H},\mathrm{p}}}.
\end{aligned} \tag{7.41}$$

In and about equation (7.22) above, it was shown that $\omega_{\mathcal{H},\mathrm{p}} = \omega_{\mathrm{p}} - \mathcal{H}^{-1}\partial\omega_{\mathrm{p}}/\partial\beta$. It was also shown in equation (7.25) that the integral of the second part must vanish *for each* N and for each permutation $\hat{\mathrm{P}}$. One can therefore interchange $\omega_{\mathcal{H},\mathrm{p}}$ and ω_{p} in the present monomer integrals, including $\Xi_{\omega_p} = \Xi_{\omega_{\mathcal{H},p}}$. This says that $\bar{E}_{\omega_p} = \bar{E}_{\omega_{\mathcal{H},p}}$, so that the commutation function in the subscript is redundant.

For the loops $l \geqslant 2$ one has

$$\begin{aligned}
\bar{E}^{\pm(l)}_{\omega_p} &= \frac{\partial \beta \Omega^{\pm(l)}_{\omega_p}}{\partial \beta} \\
&= \frac{-\partial}{\partial \beta} \left\langle \frac{N!}{(N-l)!l} \eta^{\pm(l)}_{q,\,1\ldots l} \right\rangle_{\omega_p} \\
&= -\beta \Omega^{\pm(l)}_{\omega_p} \frac{1}{\Xi_{\omega_p}} \frac{\partial \Xi_{\omega_p}}{\partial \beta} + \frac{1}{\Xi_{\omega_p}} \sum_{N=l}^{\infty} \frac{z^N h^{-dN}}{(N-l)!l} \int \mathrm{d}\boldsymbol{\Gamma}\; e^{-\beta \mathcal{H}(\boldsymbol{\Gamma})} \\
&\quad \times \left\{ \mathcal{H}(\boldsymbol{\Gamma}) \omega_p(\boldsymbol{\Gamma}) - \frac{\partial \omega_p(\boldsymbol{\Gamma})}{\partial \beta} \right\} \eta^{\pm(l)}_{q,\,1\ldots l}(\boldsymbol{\Gamma}) \\
&= \beta \Omega^{\pm(l)}_{\omega_p} \bar{E}_{\omega_p} + \left\langle \mathcal{H}\, \mathring{\eta}^{\pm(l)}_{q} \right\rangle_{\omega_p} \\
&= \beta \Omega^{\pm(l)}_{\omega_p} \bar{E}_{\omega_{\mathcal{H},p}} + \left\langle \mathcal{H}\, \mathring{\eta}^{\pm(l)}_{q} \right\rangle_{\omega_{\mathcal{H},p}}.
\end{aligned} \tag{7.42}$$

Recall that $\mathring{\eta}^{\pm(l)}_{q}(\boldsymbol{\Gamma}) = \sum'_{j_1 \ldots j_l} \eta^{\pm(l)}_{q;j_1\ldots j_l}$. Again, equation (7.25) has been applied here, showing that one can equally use $\omega_{\mathcal{H},p}$ or ω_p. This expression has the form of the average of a fluctuation,

$$\bar{E}^{\pm(l)}_{\omega_p} = \left\langle \left[\mathcal{H} - \bar{E}_{\omega_p}\right]\left[\mathring{\eta}^{\pm(l)}_{q} - \langle \mathring{\eta}^{\pm(l)}_{q} \rangle_{\omega_p}\right] \right\rangle_{\omega_p}, \tag{7.43}$$

where $l \geqslant 2$, and where either commutation function may be used. This result will be derived by a different route in section 7.3. below.

7.2.3.1 Heat capacity

The heat capacity at constant volume and fugacity is

$$C_V^{\pm} = \frac{\partial \bar{E}^{\pm}}{\partial T} = -k_B \beta^2 \frac{\partial \bar{E}^{\pm}}{\partial \beta} = -k_B \beta^2 \sum_{l=1}^{\infty} \frac{\partial \bar{E}^{\pm(l)}}{\partial \beta}. \tag{7.44}$$

The monomer contribution comes from (ω_p is not constant)

$$\begin{aligned}
\frac{\partial \bar{E}^{(1)}}{\partial \beta} &= \frac{\partial}{\partial \beta} \frac{1}{\Xi_{\omega_p}} \sum_{N=0}^{\infty} \frac{z^N}{h^{dN} N!} \int \mathrm{d}\boldsymbol{\Gamma}\; e^{-\beta \mathcal{H}(\boldsymbol{\Gamma})} \mathcal{H}(\boldsymbol{\Gamma}) \omega_p(\boldsymbol{\Gamma}) \\
&= \left\langle \mathcal{H} - \frac{\partial \ln \omega_p}{\partial \beta} \right\rangle_{\omega_p} \bar{E}^{(1)} - \left\langle \mathcal{H}^2 - \mathcal{H} \frac{\partial \ln \omega_p}{\partial \beta} \right\rangle_{\omega_p} \\
&= (\bar{E}^{(1)})^2 - \left\langle \mathcal{H}^2 - \mathcal{H} \frac{\partial \ln \omega_p}{\partial \beta} \right\rangle_{\omega_p}.
\end{aligned} \tag{7.45}$$

The loop contribution $l \geqslant 2$ comes from the derivative of equation (7.42), with $\omega_{\mathcal{H},\mathrm{p}} \Rightarrow \omega_{\mathrm{p}}$,

$$\begin{aligned}\frac{\partial \bar{E}^{\pm(l)}}{\partial \beta} &= \frac{\partial}{\partial \beta}\left\{\beta\Omega_l^{\mp}\,\bar{E}^{\pm(1)} + \frac{1}{\Xi^{\pm}_{\omega_p}}\sum_{N=l}^{\infty}\frac{z^N h^{-dN}}{(N-l)!l}\int \mathrm{d}\boldsymbol{\Gamma}\; e^{-\beta\mathcal{H}(\boldsymbol{\Gamma})}\mathcal{H}(\boldsymbol{\Gamma})\omega_{\mathrm{p}}(\boldsymbol{\Gamma})\eta^{\pm(l)}_{\mathrm{q};1\ldots l}(\boldsymbol{\Gamma})\right\}\\ &= \bar{E}^{\pm(l)}\bar{E}^{(1)} + \beta\Omega_l^{\mp}\frac{\partial\bar{E}^{(1)}}{\partial\beta} + \bar{E}^{(1)}\left[\bar{E}^{\pm(l)} - \beta\Omega_l^{\mp}\bar{E}^{(1)}\right] - \left\langle \mathcal{H}^2 - \mathcal{H}\frac{\partial \ln\omega_{\mathrm{p}}}{\partial\beta}\right\rangle_{\omega_{\mathrm{p}}\mathring{\eta}^{\pm(l)}_{\mathrm{q}}}.\end{aligned} \tag{7.46}$$

7.2.4 Average number

Making the usual assumption that most likely values equal average values, in classical thermodynamics and statistical mechanics the average number is (Attard 2002)

$$\langle N\rangle_{\mu,V,T} = \frac{-\partial\Omega}{\partial\mu}. \tag{7.47}$$

For quantum systems, the right-hand side is a sum of monomer and loop potentials, so that one has

$$\langle N\rangle_{\mu,V,T} = \sum_{l=1}^{\infty}\langle N\rangle^{\pm(l)} \equiv \sum_{l=1}^{\infty}\frac{-\partial\Omega^{\pm(l)}_{\omega_{\mathrm{p}}}}{\partial\mu}. \tag{7.48}$$

The monomer term is $\langle N\rangle^{(1)} \equiv \langle N\rangle_{\omega_{\mathrm{p}}}$, or

$$\langle N\rangle_{\omega_{\mathrm{p}}} = \frac{-\partial\Omega_1}{\partial\mu} = \frac{k_{\mathrm{B}}T}{\Xi_{\omega_{\mathrm{p}}}}\frac{\partial\Xi_{\omega_{\mathrm{p}}}}{\partial\mu} = \frac{1}{\Xi_{\omega_{\mathrm{p}}}}\sum_{N=1}^{\infty}\frac{Nz^N}{N!h^{3N}}\int \mathrm{d}\boldsymbol{\Gamma}\; e^{-\beta\mathcal{H}(\boldsymbol{\Gamma})}\omega_{\mathrm{p}}(\boldsymbol{\Gamma}). \tag{7.49}$$

The symmetrization quantum corrections to the average number are given by the terms $l \geqslant 2$,

$$\begin{aligned}\langle N\rangle^{\pm(l)} &= \frac{-\partial\Omega^{\pm(l)}_{\omega_{\mathrm{p}}}}{\partial\mu}\\ &= \frac{\Omega^{\pm(l)}_{\omega_{\mathrm{p}}}}{\Xi_{\omega_{\mathrm{p}}}}\frac{\partial\Xi_{\omega_{\mathrm{p}}}}{\partial\mu} + \frac{1}{\Xi_{\omega_{\mathrm{p}}}}\sum_{N=l}^{\infty}\frac{Nz^N}{l(N-l)!h^{3N}}\int \mathrm{d}\boldsymbol{\Gamma} e^{-\beta\mathcal{H}(\boldsymbol{\Gamma})}\omega_{\mathrm{p}}(\boldsymbol{\Gamma})\eta^{\pm(l)}_{\mathrm{q};1\ldots l}\\ &= \beta\Omega^{\pm(l)}_{\omega_{\mathrm{p}}}\langle N\rangle_{\omega_{\mathrm{p}}} + \left\langle \mathring{\eta}^{\pm(l)}(\boldsymbol{\Gamma})N\right\rangle_{\omega_{\mathrm{p}}}\\ &= \left\langle \mathring{\eta}^{\pm(l)}(\boldsymbol{\Gamma})\left[N - \langle N\rangle_{\omega_{\mathrm{p}}}\right]\right\rangle_{\omega_{\mathrm{p}}}\\ &= \left\langle \Delta\mathring{\eta}^{\pm(l)}(\boldsymbol{\Gamma})\,\Delta N\right\rangle_{\omega_{\mathrm{p}}}.\end{aligned} \tag{7.50}$$

In the third equality both terms are $\mathcal{O}(N^2)$. The fourth equality shows the cancelation between them, with the residue quantum correction presumably $\mathcal{O}(N)$. Interestingly enough, this thermodynamic derivative has the linearized form of the factorized average that now follows.

7.2.5 Factorization of averages

The above results for the average energy and for the loop grand potential relied upon the factorization of the symmetrization function,

$$\langle \eta^{\pm} \rangle \approx \prod_{\ell=2}^{\infty} \langle e^{\mathring{\eta}^{\pm(\ell)}} \rangle. \tag{7.51}$$

Using the same idea, the average of an operator that is a linear combination of 'pure' operators can be written

$$\begin{aligned} \langle \hat{A} \rangle_{\mu,V,T} &= \langle A \rangle_{\omega_p \eta_q^{\pm}} \\ &= \frac{\left\langle \eta_q^{\pm} \right\rangle_{A\omega_p}}{\left\langle \eta_q^{\pm} \right\rangle_{\omega_p}} \langle A \rangle_{\omega_p} \\ &= \langle A \rangle_{\omega_p} \prod_{\ell=2}^{\infty} \exp\left[\left\langle \mathring{\eta}_q^{\pm(\ell)} \right\rangle_{A\omega_p} - \left\langle \mathring{\eta}_q^{\pm(\ell)} \right\rangle_{\omega_p} \right] \\ &= \langle A \rangle_{\omega_p} \exp \sum_{\ell=2}^{\infty} \frac{1}{\langle A \rangle_{\omega_p}} \left\{ \left\langle \mathring{\eta}_q^{\pm(\ell)} A \right\rangle_{\omega_p} - \left\langle \mathring{\eta}_q^{\pm(\ell)} \right\rangle_{\omega_p} \langle A \rangle_{\omega_p} \right\} \\ &\approx \langle A \rangle_{\omega_p} + \sum_{\ell=2}^{\infty} \left\langle \Delta \mathring{\eta}_q^{\pm(\ell)} \Delta A \right\rangle_{\omega_p}. \end{aligned} \tag{7.52}$$

The fluctuations are $\Delta \mathring{\eta}_q^{\pm(\ell)} \equiv \mathring{\eta}_q^{\pm(\ell)} - \left\langle \mathring{\eta}_q^{\pm(\ell)} \right\rangle_{\omega_p}$ and $\Delta A \equiv A - \langle A \rangle_{\omega_p}$. In so far as fluctuations are relatively negligible, the linearization for the final approximation should be accurate.

This result shows how quantum corrections due to symmetrization effects appear successively as terms from single loops of increasing size. The result relies upon the localization inherent in the permutation loop expansion.

7.2.6 Explicit comparison for the ideal gas

For single-particle states, equation (3.57) writes the grand partition function as the product of loop exponentials, $\Xi^{\pm}(\mu, V, T) = \Xi_1 \prod_{l=2}^{\infty} e^{Z_l^{\pm}}$, with the exponents being the loop grand potentials, equation (3.58),

$$-\beta \Omega^{\pm(l)} \equiv Z_l^{\pm} = \frac{z^l}{l} \sum_{\mathbf{n}_1,\ldots,\mathbf{n}_l} e^{-\beta \mathcal{H}^{(l)}_{\mathbf{n}_1\ldots\mathbf{n}_l}} \chi^{\pm,(l)}_{\mathbf{n}_1\ldots\mathbf{n}_l}. \tag{7.53}$$

In comparison, for the general case, equation (7.35) similarly factorizes the grand partition function, but now the loop grand potential is an average, equation (7.36). For interacting particles the loop grand potential is given as

$$\begin{aligned} -\beta\Omega^{\pm(l)}_{\omega_{\rm p}} &\equiv \left\langle \mathring{\eta}^{\pm(l)}_{\rm q}(\boldsymbol{\Gamma}) \right\rangle_{\omega_{\rm p}} \\ &= \frac{1}{\Xi_{\omega_{\rm p},1}} \sum_{N=l}^{\infty} \frac{z^N}{N!h^{3N}} \int \mathrm{d}\boldsymbol{\Gamma} e^{-\beta\mathcal{H}(\boldsymbol{\Gamma})} \omega_{\rm p}(\boldsymbol{\Gamma}) \mathring{\eta}^{\pm(l)}_{\rm q}(\boldsymbol{\Gamma}) \\ &= \frac{1}{\Xi_{\omega_{\rm p},1}} \sum_{N=l}^{\infty} \frac{z^N}{N!h^{3N}} \frac{N!}{(N-l)!l} \int \mathrm{d}\boldsymbol{\Gamma} e^{-\beta\mathcal{H}(\boldsymbol{\Gamma})} \omega_{\rm p}(\boldsymbol{\Gamma}) \eta^{\pm(l)}_{\mathrm{q};1\ldots l}. \end{aligned} \tag{7.54}$$

For the case of an ideal gas, $U = 0$, $\mathcal{H}(\boldsymbol{\Gamma}) = \mathcal{K}(\mathbf{p})$, and $\omega_{\rm p}(\boldsymbol{\Gamma}) = 1$. The integrals of the $N - l$ particles that are not part of the loop factorize, and the dummy summation variable may be relabeled, $N' = N - l$, so that these sum to the classical grand partition function, which cancels with the denominator to give

$$-\beta\Omega^{\pm(l),\mathrm{id}}_{\omega_{\rm p}} = \frac{z^l}{lh^{3l}} \int \mathrm{d}\boldsymbol{\Gamma}^l e^{-\beta\mathcal{K}(\boldsymbol{\Gamma}^l)}\, \eta^{\pm(l)}_{\mathrm{q};1\ldots l}. \tag{7.55}$$

One can see by inspection that this is the direct phase space analog of the loop grand potential expressed as a sum over single-particle states above.

It is axiomatic in thermodynamics that the average value of a quantity is independent of the type of thermodynamic system (closed, canonical, grand canonical, isobaric, etc) that is used to obtain it (at the equivalent thermodynamic state point). It is of interest to test this explicitly for the ideal gas by obtaining the loop grand potential using canonical rather than grand canonical averages. One then obtains

$$\begin{aligned} -\beta\Omega^{\pm(l),\mathrm{id}}_{\omega_{\rm p}} &= \frac{1}{Z^{\mathrm{id}}(N, V, T)} \frac{1}{N!h^{3N}} \frac{N!}{(N-l)!l} \int \mathrm{d}\boldsymbol{\Gamma} e^{-\beta\mathcal{K}(\boldsymbol{\Gamma})} \eta^{\pm(l)}_{\mathrm{q};1\ldots l} \\ &= \frac{N!}{\Lambda^{-3N} V^N} \frac{\Lambda^{-3(N-l)} V^{N-l}}{(N-l)!h^{3l}l} \int \mathrm{d}\boldsymbol{\Gamma}^l e^{-\beta\mathcal{K}(\boldsymbol{\Gamma}^l)} \eta^{\pm(l)}_{\mathrm{q};1\ldots l} \\ &\approx \frac{N^l \Lambda^{3l} V^{-l}}{lh^{3l}} \int \mathrm{d}\boldsymbol{\Gamma}^l e^{-\beta\mathcal{K}(\boldsymbol{\Gamma}^l)} \eta^{\pm(l)}_{\mathrm{q};1\ldots l} \\ &= \frac{z^l}{lh^{3l}} \int \mathrm{d}\boldsymbol{\Gamma}^l e^{-\beta\mathcal{K}(\boldsymbol{\Gamma}^l)} \eta^{\pm(l)}_{\mathrm{q};1\ldots l}. \end{aligned} \tag{7.56}$$

In the penultimate equality has been written $N!/(N-l)! \approx N^l$. The final equality follows since for the classical ideal gas, $\rho = \Lambda^{-3} z$. Since the averages are classical averages (in this case $\omega_{\rm p} = 1$), one must use the classical relationship between fugacity and number to compare the grand canonical and the canonical average. It is consoling that the two averages, at least in this case, are equal.

7.3 Multi-particle density

Consider a position configuration of n particles, $\mathbf{Q}^n = \{\mathbf{Q}_1, \mathbf{Q}_2, \ldots, \mathbf{Q}_n\}$. The distinct n-particle density quantum operator for this is

$$\hat{\rho}^{(n)}(\mathbf{Q}^n; \mathbf{r}) = \sum_{k_1,\ldots,k_n}^{N}{}'' \prod_{j=1}^{n} \delta(\mathbf{Q}_j - \mathbf{r}_{k_j}). \tag{7.57}$$

The sum is over the particle indeces, and the double prime indicates that in any term no two indeces are equal. If there are N particles in the system, then there are $N!/(N-n)!$ terms in the sum. This says any particle is at $\mathbf{Q}_1$, any different particle is at $\mathbf{Q}_2$, etc. The density operator is evidently unchanged by a permutation of the $\mathbf{Q}_j$ or of the particle labels.

Since this is only a function of the position operator, the commutation function ω_q or ω_p may be used for its average. The average n-particle density is

$$\begin{aligned}
&\rho^{(n)}_{\omega_\mathrm{p}\eta^\pm_\mathrm{q}}(\mathbf{Q}^n) \\
&\quad\equiv \langle \hat{\rho}^{(n)}(\mathbf{Q}^n; \mathbf{q}) \rangle_{\omega_\mathrm{p}\eta^\pm_\mathrm{q}} \\
&\quad= \frac{1}{\Xi_{\omega_\mathrm{p}\eta^\pm_\mathrm{q}}} \sum_{N=0}^{\infty} \frac{z^N}{h^{dN} N!} \int \mathrm{d}\mathbf{\Gamma}\, e^{-\beta\mathcal{H}(\mathbf{\Gamma})} \omega_\mathrm{p}(\mathbf{\Gamma}) \eta^\pm_\mathrm{q}(\mathbf{\Gamma}) \sum_{k_1,\ldots,k_n}^{N}{}'' \prod_{j=1}^{n} \delta(\mathbf{Q}_j - \mathbf{q}_{k_j}).
\end{aligned} \tag{7.58}$$

It is worth mentioning that one does not actually have to define the quantum operator density, $\hat{\rho}^{(n)}(\mathbf{Q}^n; \mathbf{q})$, but that instead one can introduce the phase function $\rho^{(n)}_{\omega_\mathrm{p}}(\mathbf{Q}^n)$ directly via the phase space integral. More generally one can define the n-particle density for positions and momenta in phase space as

$$\rho^{(n)}(\gamma^n; \mathbf{\Gamma}) = \sum_{k_1,\ldots,k_n}^{N}{}'' \prod_{j=1}^{n} \delta(\gamma_j - \mathbf{\Gamma}_{k_j}), \tag{7.59}$$

and its average as

$$\begin{aligned}
\rho^{(n)}_{\omega_\mathrm{p}\eta^\pm_\mathrm{q}}(\gamma^n) \equiv{}& \frac{1}{\Xi_{\omega_\mathrm{p}\eta^\pm_\mathrm{q}}} \sum_{N=0}^{\infty} \frac{z^N}{h^{dN} N!} \int \mathrm{d}\mathbf{\Gamma}\, e^{-\beta\mathcal{H}(\mathbf{\Gamma})} \omega_\mathrm{p}(\mathbf{\Gamma}) \eta^\pm_\mathrm{q}(\mathbf{\Gamma}) \\
&\times \sum_{k_1,\ldots,k_n}^{N}{}'' \prod_{j=1}^{n} \delta(\gamma_j - \mathbf{\Gamma}_{k_j}).
\end{aligned} \tag{7.60}$$

We shall use this below for the average energy.

The n-particle density for positions is evidently normalized as

$$\int \mathrm{d}\mathbf{Q}^n\, \rho^{(n)}_{\omega_\mathrm{p}\eta^\pm_\mathrm{q}}(\mathbf{Q}^n) = \left\langle \frac{N!}{(N-n)!} \right\rangle_{z,V,T}, \tag{7.61}$$

which is in agreement with the classical convention (Pathria 1972, Attard 2002). One also has the reduction condition

$$\int \mathrm{d}\mathbf{Q}_n\, \rho^{(n)}_{\omega_p\eta_q^\pm}(\mathbf{Q}^n) = \left\langle \frac{N!}{(N-n)!} \right\rangle_{z,V,T} \left\langle \frac{N!}{(N-n+1)!} \right\rangle^{-1}_{z,\,V,\,T} \rho^{(n-1)}_{\omega_p\eta_q^\pm}(\mathbf{Q}^{n-1}) \\ \approx (\bar{N}-n+1)\, \rho^{(n-1)}_{\omega_p\eta_q^\pm}(\mathbf{Q}^{n-1}). \tag{7.62}$$

This n-particle density can be used for the average of any operator that is a function of the position operator. For example, suppose that the potential energy is the sum of pair potentials,

$$U(\mathbf{q}) = \frac{1}{2}\sum_{j,k}'' u(\mathbf{q}_j,\, \mathbf{q}_k) = \sum_{j,k}' u(\mathbf{q}_j,\, \mathbf{q}_k). \tag{7.63}$$

The double prime again indicates that $j \neq k$, and the single prime indicates that the sum is over unique pairs, $j < k$. The average potential energy in this case is

$$\langle \hat{U} \rangle_{z,V,T} = \frac{1}{2}\int \mathrm{d}\mathbf{Q}_1\, \mathrm{d}\mathbf{Q}_2\, \rho^{(2)}_{\omega_p\eta_q^\pm}(\mathbf{Q}_1,\, \mathbf{Q}_2)\, u(\mathbf{Q}_1,\, \mathbf{Q}_2), \tag{7.64}$$

as can be confirmed by substituting in the definition of the pair density and interchanging the order of integrations. The factor of one half again corrects for double counting of each particle pair.

Define the density fluctuation

$$\Delta\hat{\rho}^{(n)}(\mathbf{Q}^n;\, \mathbf{q}) \equiv \hat{\rho}^{(n)}(\mathbf{Q}^n;\, \mathbf{q}) - \langle \hat{\rho}^{(n)}(\mathbf{Q}^n;\, \mathbf{q}) \rangle_{\omega_p} \\ \equiv \hat{\rho}^{(n)}(\mathbf{Q}^n;\, \mathbf{q}) - \rho^{(n)}_{\omega_p}(\mathbf{Q}^n). \tag{7.65}$$

Here the monomer average, $\eta_q^\pm(\boldsymbol{\Gamma}) = 1$, is used for the fluctuation, $\rho^{(n)}_{\omega_p}(\mathbf{Q}^n) \equiv \langle \hat{\rho}^{(n)}(\mathbf{Q}^n;\, \mathbf{q}) \rangle_{\omega_p}$. Similarly one has the symmetrization function,

$$\Delta\eta_q^\pm(\boldsymbol{\Gamma}) \equiv \eta_q^\pm(\boldsymbol{\Gamma}) - \left\langle \eta_q^\pm(\boldsymbol{\Gamma}) \right\rangle_{\omega_p}. \tag{7.66}$$

With these the average density may be written

$$\begin{aligned} \rho^{(n)}_{\omega_p\eta_q^\pm}(\mathbf{Q}^n) &\equiv \langle \hat{\rho}^{(n)}(\mathbf{Q}^n;\, \mathbf{q}) \rangle_{\omega_p\eta_q^\pm} \\ &= \frac{\Xi_{\omega_p}}{\Xi_{\omega_p\eta_q^\pm}} \left\langle \hat{\rho}^{(n)}(\mathbf{Q}^n;\, \mathbf{q})\, \eta_q^\pm(\boldsymbol{\Gamma}) \right\rangle_{\omega_p} \\ &= \left\langle \eta_q^\pm(\boldsymbol{\Gamma}) \right\rangle^{-1}_{\omega_p} \Big\{ \langle \hat{\rho}^{(n)}(\mathbf{Q}^n;\, \mathbf{q}) \rangle_{\omega_p} \left\langle \eta_q^\pm(\boldsymbol{\Gamma}) \right\rangle_{\omega_p} \\ &\quad + \left\langle \Delta\hat{\rho}^{(n)}(\mathbf{Q}^n;\, \mathbf{q}) \Delta\eta_q^\pm(\boldsymbol{\Gamma}) \right\rangle_{\omega_p} \Big\} \\ &= \rho^{(n)}_{\omega_p}(\mathbf{Q}^n) + \left\langle \Delta\hat{\rho}^{(n)}(\mathbf{Q}^n;\, \mathbf{q})\, \Delta\mathring{\eta}_q^\pm(\boldsymbol{\Gamma}) \right\rangle_{\omega_p}. \end{aligned} \tag{7.67}$$

This holds in the thermodynamic limit, $V \to \infty$, z, $T=$ const. Recall that $\mathring{\eta}^{\pm}_{\mathrm{q}}$ is the series of single-loop symmetrization functions defined in equation (7.31). From equation (7.36), the prefactor in the second equality is $\Xi_{\omega_\mathrm{p}}/\Xi_{\omega_\mathrm{p}\eta^{\pm}_{\mathrm{q}}} = \langle \eta^{\pm}_{\mathrm{q}} \rangle^{-1}_{\omega_\mathrm{p}} = e^{-\langle \mathring{\eta}^{\pm}_{\mathrm{q}} \rangle_{\omega_\mathrm{p}}}$. Also, since $\langle \Delta\hat{\rho}^{(n)} \rangle_{\omega_\mathrm{p}} = 0$, one can write successively

$$\begin{aligned}
\langle \Delta\hat{\rho}^{(n)} \Delta\eta^{\pm}_{\mathrm{q}} \rangle_{\omega_\mathrm{p}} &= \langle \Delta\hat{\rho}^{(n)} \eta^{\pm}_{\mathrm{q}} \rangle_{\omega_\mathrm{p}} \\
&= \langle \Delta\hat{\rho}^{(n)} [\eta^{\pm}_{\mathrm{q}} - 1] \rangle_{\omega_\mathrm{p}} \\
&= \langle \Delta\hat{\rho}^{(n)} \mathring{\eta}^{\pm}_{\mathrm{q}} \rangle_{\omega_\mathrm{p}} e^{\langle \mathring{\eta}^{\pm}_{\mathrm{q}} \rangle_{\omega_\mathrm{p}}} \\
&= \langle \Delta\hat{\rho}^{(n)} \Delta\mathring{\eta}^{\pm}_{\mathrm{q}} \rangle_{\omega_\mathrm{p}} \, e^{\langle \mathring{\eta}^{\pm}_{\mathrm{q}} \rangle_{\omega_\mathrm{p}}}.
\end{aligned} \tag{7.68}$$

The final equality in equation (7.67) is justified by the passage from the second to the third equality here. The average of the density fluctuation without loops vanishes, $\langle \Delta\hat{\rho}^{(n)}(\mathbf{Q}^n) \rangle_{\omega_\mathrm{p}} = 0$. Hence the only configurations that contribute to the average in the second equality in equation (7.68) are those with at least one loop in the vicinity of $\mathbf{Q}^n$. This means that in any product of loops in the symmetrization function $\eta^{\pm}_{\mathrm{q}}(\Gamma)$, one loop must be tied to $\mathbf{Q}^n$, and the rest are free to wander throughout the volume. Hence just as in the factorization of the grand partition function in section 7.2.2, the average of the density fluctuation times such a product can be written as the product of the average of the density fluctuation times the tied loop, times the product of the averages of each of the free loops. The symmetry number reflects the number and type of the free loops, because these are equivalent and distinct from the tied loop. Hence just as in section 7.2.2, they sum to the exponential, $e^{\langle \mathring{\eta}^{\pm}_{\mathrm{q}} \rangle_{\omega_\mathrm{p}}}$, which gives the third equality in equation (7.68). The final equality in that equation, combined with the prefactor $\langle \eta^{\pm}_{\mathrm{q}} \rangle^{-1}_{\omega_\mathrm{p}} = e^{-\langle \mathring{\eta}^{\pm}_{\mathrm{q}} \rangle_{\omega_\mathrm{p}}}$ from the penultimate equality in equation (7.67), gives the final equality in equation (7.67).

The result equation (7.67) for the multi-particle position density holds analogously for the multi-particle phase density, equation (7.60). It follows that the average of any phase function can be written in terms that involve only the single-loop symmetrization function, as is now shown for the energy.

7.3.1 Average energy factorized

Suppose that the classical Hamiltonian consists of many-body potentials $\mathcal{H}(\mathbf{\Gamma}) = \sum_n \mathcal{H}^{(n)}(\mathbf{\Gamma})$, with

$$\mathcal{H}^{(n)}(\mathbf{\Gamma}) = \frac{1}{n!} \sum_{k_1,\ldots,k_n}^{N}{}'' \; \mathcal{H}^{(n)}(\mathbf{\Gamma}_{k_1}, \ldots, \mathbf{\Gamma}_{k_n}). \tag{7.69}$$

The kinetic energy is included in the one-body term.

In terms of the position and momentum n-body density, equation (7.60), the average Hamiltonian energy may be written

$$\begin{aligned}
\langle\hat{\mathcal{H}}\rangle^{\pm}_{z,\,V,\,T} &= \sum_n \frac{1}{n!}\int \mathrm{d}\gamma^n\, \rho^{(n)}(\gamma^n)\, \mathcal{H}^{(n)}(\gamma^n) \\
&= \sum_n \frac{1}{n!}\int \mathrm{d}\gamma^n\, \mathcal{H}^{(n)}(\gamma^n)\Big\{\rho^{(n)}_{\omega_{\rm p}}(\gamma^n) + \Big\langle \Delta\rho^{(n)}(\gamma^n;\,\boldsymbol{\Gamma})\Delta\mathring{\eta}^{\pm}_{\rm q}(\boldsymbol{\Gamma})\Big\rangle_{\omega_{\rm p}}\Big\} \\
&= \langle\mathcal{H}(\boldsymbol{\Gamma})\rangle_{\omega_{\rm p}} + \Big\langle \Delta\mathcal{H}(\boldsymbol{\Gamma})\Delta\mathring{\eta}^{\pm}_{\rm q}(\boldsymbol{\Gamma})\Big\rangle_{\omega_{\rm p}} \\
&= \langle\mathcal{H}(\boldsymbol{\Gamma})\rangle_{\omega_{\rm p}} + \sum_{l=2}^{\infty}\Big\langle \Delta\mathcal{H}(\boldsymbol{\Gamma})\Delta\mathring{\eta}^{\pm(l)}_{\rm q}(\boldsymbol{\Gamma})\Big\rangle_{\omega_{\rm p}}.
\end{aligned} \tag{7.70}$$

This agrees with the expression obtained from the temperature derivative of the grand potential, section 7.2.3, where the loop contributions were given by equation (7.43),

$$\bar{E}^{\pm(l)}_{\omega_{\rm p}} = \Big\langle\Big[\mathcal{H} - \bar{E}_{\omega_{\rm p}}\Big]\Big[\mathring{\eta}^{\pm(l)}_{\rm q} - \langle\mathring{\eta}^{\pm(l)}_{\rm q}\rangle_{\omega_{\rm p}}\Big]\Big\rangle_{\omega_{\rm p}}, \tag{7.71}$$

for $l \geqslant 2$. It also agrees with equation (7.52).

7.4 Virial pressure

The thermodynamic pressure is the volume derivative of the grand potential (Pathria 1972, Attard 2002)

$$\bar{p}^{\pm} = \frac{-\partial\Omega^{\pm}}{\partial V} = \sum_{l=1}^{\infty}\frac{-\partial\Omega^{\pm}_l}{\partial V}. \tag{7.72}$$

With the subsystem occupying a hypercube of volume, $V = L^d$, one can perform the usual trick of scaling the position coordinates by the edge length L, $\mathbf{q}' = \mathbf{q}/L$, so that $\mathrm{d}\mathbf{q} = \mathbf{q}L^{-1}\mathrm{d}L$. Recall that the spacing between momentum states is $\Delta_p = 2\pi\hbar/L$, and so $\mathrm{d}\mathbf{p} = -\mathbf{p}L^{-1}\mathrm{d}L$. These give the change in a phase function as

$$\begin{aligned}
\mathrm{d}f(\boldsymbol{\Gamma}) &= \nabla_{\rm p} f\cdot \mathrm{d}\mathbf{p} + \nabla_{\rm q} f\cdot \mathrm{d}\mathbf{q} \\
&= \big\{\mathbf{q}\cdot\nabla_{\rm q} f - \mathbf{p}\cdot\nabla_{\rm p} f\big\}L^{-1}\mathrm{d}L.
\end{aligned} \tag{7.73}$$

In terms of the partition function, the pressure is

$$\begin{aligned}
\beta\bar{p}^{\pm} &= \frac{\partial \ln \Xi^{\pm}}{\partial V} \\
&= \frac{L}{dV\Xi^{\pm}}\sum_{N=0}^{\infty}\frac{z^N}{h^{dN}N!}\int \mathrm{d}\boldsymbol{\Gamma}\,\frac{\mathrm{d}}{\mathrm{d}L}\big\{e^{-\beta\mathcal{H}(\boldsymbol{\Gamma})}\omega_{\rm p}(\boldsymbol{\Gamma})\eta^{\pm}_{\rm q}(\boldsymbol{\Gamma})\big\}.
\end{aligned} \tag{7.74}$$

One has

$$\begin{aligned} \frac{\mathrm{d} V^{N/2}\langle \hat{\mathbf{P}}\mathbf{p}|\mathbf{q}\rangle}{\mathrm{d}L} &= \frac{\mathrm{d}}{\mathrm{d}L} e^{(\hat{\mathbf{P}}\mathbf{p})\cdot\mathbf{q}/i\hbar} \\ &= \frac{1}{i\hbar}\left\{\mathbf{q}\cdot(\hat{\mathbf{P}}\mathbf{p}) - (\hat{\mathbf{P}}\mathbf{p})\cdot\mathbf{q}\right\} e^{(\hat{\mathbf{P}}\mathbf{p})\cdot\mathbf{q}/i\hbar} \\ &= 0. \end{aligned} \tag{7.75}$$

It follows that $\mathrm{d}\eta_{\mathrm{q}}^{\pm}(\mathbf{\Gamma})/\mathrm{d}L = 0$, and that therefore

$$\begin{aligned} &L\frac{\mathrm{d}}{\mathrm{d}L}\left\{e^{-\beta\mathcal{H}(\mathbf{\Gamma})}\omega_{\mathrm{p}}(\mathbf{\Gamma})\eta_{\mathrm{q}}^{\pm}(\mathbf{\Gamma})\right\} \\ &= \eta_{\mathrm{q}}^{\pm}(\mathbf{\Gamma})\left\{\mathbf{q}\cdot\nabla_{\mathrm{q}} - \mathbf{p}\cdot\nabla_{\mathrm{p}}\right\}\left\{e^{-\beta\mathcal{H}(\mathbf{\Gamma})}\omega_{\mathrm{p}}(\mathbf{\Gamma})\right\}. \end{aligned} \tag{7.76}$$

The commutation function is given by

$$\begin{aligned} e^{-\beta\mathcal{H}(\mathbf{\Gamma})}\omega_{\mathrm{p}}(\mathbf{\Gamma}) &= \frac{\langle \mathbf{q}|e^{-\beta\hat{\mathcal{H}}}|\mathbf{p}\rangle}{\langle \mathbf{q}|\mathbf{p}\rangle} \\ &= e^{\mathbf{p}\cdot\mathbf{q}/i\hbar}\int \mathrm{d}\mathbf{r}\ \delta(\mathbf{r}-\mathbf{q})e^{-\beta\hat{\mathcal{H}}(\mathbf{r})}e^{-\mathbf{p}\cdot\mathbf{r}/i\hbar}. \end{aligned} \tag{7.77}$$

The prefactor is a constant with respect to L, and so one has

$$\begin{aligned} L\frac{\mathrm{d}}{\mathrm{d}L}e^{-\beta\mathcal{H}(\mathbf{\Gamma})}\omega_{\mathrm{p}}(\mathbf{\Gamma}) &= \mathbf{q}\cdot\nabla_{\mathrm{q}}\left\{e^{-\beta\mathcal{H}(\mathbf{\Gamma})}\omega_{\mathrm{p}}(\mathbf{\Gamma})\right\} - \mathbf{p}\cdot\nabla_{\mathrm{p}}\left\{e^{-\beta\mathcal{H}(\mathbf{\Gamma})}\omega_{\mathrm{p}}(\mathbf{\Gamma})\right\} \\ &= e^{\mathbf{p}\cdot\mathbf{q}/i\hbar}\int \mathrm{d}\mathbf{r}\ \left\{\mathbf{q}\cdot\delta'(\mathbf{r}-\mathbf{q})e^{-\beta\hat{\mathcal{H}}(\mathbf{r})}e^{-\mathbf{p}\cdot\mathbf{r}/i\hbar}\right. \\ &\quad \left. + (i\hbar)^{-1}\delta(\mathbf{r}-\mathbf{q})e^{-\beta\hat{\mathcal{H}}(\mathbf{r})}e^{-\mathbf{p}\cdot\mathbf{r}/i\hbar}\mathbf{p}\cdot\mathbf{r}\right\} \\ &= -e^{\mathbf{p}\cdot\mathbf{q}/i\hbar}\int \mathrm{d}\mathbf{r}\ \left\{\delta(\mathbf{r}-\mathbf{q})\mathbf{q}\cdot\nabla_{\mathbf{r}}[e^{-\beta\hat{\mathcal{H}}(\mathbf{r})}e^{-\mathbf{p}\cdot\mathbf{r}/i\hbar}]\right. \\ &\quad \left. - (i\hbar)^{-1}\delta(\mathbf{r}-\mathbf{q})e^{-\beta\hat{\mathcal{H}}(\mathbf{r})}e^{-\mathbf{p}\cdot\mathbf{r}/i\hbar}\mathbf{p}\cdot\mathbf{r}\right\} \\ &= \frac{\langle \mathbf{q}|e^{-\beta\hat{\mathcal{H}}}\mathbf{p}\cdot\mathbf{r}|\mathbf{p}\rangle}{i\hbar\langle \mathbf{q}|\mathbf{p}\rangle} - \frac{\langle \mathbf{q}|\mathbf{q}\cdot\nabla_{\mathbf{r}}e^{-\beta\hat{\mathcal{H}}}|\mathbf{p}\rangle}{\langle \mathbf{q}|\mathbf{p}\rangle} \\ &\equiv e^{-\beta\mathcal{H}(\mathbf{\Gamma})}\omega_{\mathrm{p}}(\mathbf{\Gamma})\mathcal{V}_{\mathrm{p}}(\mathbf{\Gamma}). \end{aligned} \tag{7.78}$$

The final equality defines what might be called the quantum phase space virial. Note that the $\mathbf{p}$ that appears explicitly in the middle of the first term to the right of the penultimate equality is the momentum variable, not the momentum operator, and that the gradient operator $\nabla_{\mathbf{r}}$ that appears explicitly in the second term acts on everything to its right.

With this the pressure is given by

$$dV\beta\bar{p}^{\pm} = \frac{1}{\Xi^{\pm}}\sum_{N=0}^{\infty}\frac{z^N}{h^{dN}N!}\int \mathrm{d}\mathbf{\Gamma}\ e^{-\beta\mathcal{H}(\mathbf{\Gamma})}\omega_{\mathrm{p}}(\mathbf{\Gamma})\mathcal{V}_{\mathrm{p}}(\mathbf{\Gamma})\eta_{\mathrm{q}}^{\pm}(\mathbf{\Gamma}). \tag{7.79}$$

or, equivalently, from the derivative of the series of loop grand potentials, this can also be written

$$\begin{aligned} dV\beta\bar{p}^{\pm} &= \left\langle \mathcal{V}_{\mathrm{p}}(\boldsymbol{\Gamma})\right\rangle_{\omega_{\mathrm{p}}} + \left\langle \Delta\mathcal{V}_{\mathrm{p}}(\boldsymbol{\Gamma})\, \Delta\mathring{\eta}_{\mathrm{q}}^{\pm}(\boldsymbol{\Gamma})\right\rangle_{\omega_{\mathrm{p}}} \\ &= \left\langle \mathcal{V}_{\mathrm{p}}(\boldsymbol{\Gamma})\right\rangle_{\omega_{\mathrm{p}}} + \sum_{l=2}^{\infty}\left\langle \Delta\mathcal{V}_{\mathrm{p}}(\boldsymbol{\Gamma})\, \Delta\mathring{\eta}_{\mathrm{q}}^{\pm(l)}(\boldsymbol{\Gamma})\right\rangle_{\omega_{\mathrm{p}}}. \end{aligned} \tag{7.80}$$

The gives the pressure as the sum of single-loop averages of the virial fluctuation and the single-loop symmetrization function fluctuation. Such a sum over single-loops is of the same form as the general expression for an average derived in equation (7.52).

In practice possibly the simplest way to obtain the quantum phase space virial will turn out to be from the right-hand side of the first equality in the defining equation (7.78)

$$\begin{aligned} e^{-\beta\mathcal{H}(\boldsymbol{\Gamma})}\omega_{\mathrm{p}}(\boldsymbol{\Gamma})\mathcal{V}_{\mathrm{p}}(\boldsymbol{\Gamma}) &= \mathbf{q}\cdot\nabla_{q}\left\{e^{-\beta\mathcal{H}(\boldsymbol{\Gamma})}\omega_{\mathrm{p}}(\boldsymbol{\Gamma})\right\} - \mathbf{p}\cdot\nabla_{\mathrm{p}}\left\{e^{-\beta\mathcal{H}(\boldsymbol{\Gamma})}\omega_{\mathrm{p}}(\boldsymbol{\Gamma})\right\} \\ &= -\beta\omega_{\mathrm{p}}(\boldsymbol{\Gamma})e^{-\beta\mathcal{H}(\boldsymbol{\Gamma})}\left\{\mathbf{q}\cdot\nabla_{\mathrm{q}}U(\mathbf{q}) - \mathbf{p}\cdot\nabla_{\mathrm{p}}\mathcal{K}(\mathbf{p})\right\} \\ &\quad + e^{-\beta\mathcal{H}(\boldsymbol{\Gamma})}\left\{\mathbf{q}\cdot\nabla_{\mathrm{q}}\omega_{\mathrm{p}}(\boldsymbol{\Gamma}) - \mathbf{p}\cdot\nabla_{\mathrm{p}}\omega_{\mathrm{p}}(\boldsymbol{\Gamma})\right\}. \end{aligned} \tag{7.81}$$

This is useful when the 'bare' commutation function ω_{p} can be otherwise obtained and differentiated.

In the classical limit of the present expression, $\omega_{\mathrm{p}}(\boldsymbol{\Gamma}) = \eta_{\mathrm{q}}^{\pm}(\boldsymbol{\Gamma}) = 1$, and assuming only an internal potential, this reduces to

$$\begin{aligned} e^{-\beta\mathcal{H}(\boldsymbol{\Gamma})}\mathcal{V}_{\mathrm{p}}^{\mathrm{cl}}(\boldsymbol{\Gamma}) &= \left\{\mathbf{q}\cdot\nabla_{\mathrm{q}}e^{-\beta\mathcal{H}(\boldsymbol{\Gamma})} - \mathbf{p}\cdot\nabla_{\mathrm{p}}e^{-\beta\mathcal{H}(\boldsymbol{\Gamma})}\right\}e^{-\beta\mathcal{H}(\boldsymbol{\Gamma})} \\ &= \left\{-\beta\mathbf{q}\cdot\nabla_{\mathrm{q}}U^{\mathrm{int}}(\mathbf{q}) + \frac{\beta p^2}{m}\right\}e^{-\beta\mathcal{H}(\boldsymbol{\Gamma})} \\ &= \left\{\beta\mathcal{V}^{\mathrm{cl,\,int}}(\mathbf{q}) + \frac{\beta p^2}{m}\right\}e^{-\beta\mathcal{H}(\boldsymbol{\Gamma})}. \end{aligned} \tag{7.82}$$

Therefore, in the classical limit $\omega_{\mathrm{p}}(\boldsymbol{\Gamma}) = \eta_{\mathrm{q}}^{\pm}(\boldsymbol{\Gamma}) = 1$, the pressure is given by

$$\begin{aligned} \beta\bar{p}_{\mathrm{cl}} &= \frac{1}{dV\Xi_{\mathrm{cl}}}\sum_{N=0}^{\infty}\frac{z^N}{h^{dN}N!}\int \mathrm{d}\boldsymbol{\Gamma}\; e^{-\beta\mathcal{H}(\boldsymbol{\Gamma})}\mathcal{V}_{\mathrm{p}}^{\mathrm{cl}}(\boldsymbol{\Gamma}) \\ &= \frac{1}{dV}\left\langle \beta\mathcal{V}^{\mathrm{cl,\,int}}(\mathbf{q}) + \frac{\beta p^2}{m}\right\rangle_{z,T} \\ &= \frac{1}{dV}\langle dN\rangle_{z,T} + \frac{\beta}{dV}\langle\mathcal{V}^{\mathrm{cl,\,int}}(\mathbf{q})\rangle_{z,T}. \end{aligned} \tag{7.83}$$

This agrees with the known classical expression for the virial pressure (Pathria 1972, Hansen and McDonald 1986, Attard 2002).

References

Attard P 2002 *Thermodynamics and Statistical Mechanics: Equilibrium by Entropy Maximisation* (London: Academic)

Attard P 2015 *Quantum Statistical Mechanics: Equilibrium and Non-Equilibrium Theory from First Principles* (Bristol: IOP Publishing)

Attard P 2018a Entropy Beyond the Second Law *Thermodynamics and Statistical Mechanics for Equilibrium, Non-Equilibrium, Classical, and Quantum Systems* (Bristol: IOP Publishing)

Attard P 2018b Algorithm for quantum many-particle systems by transformation to classical phase space with test results for quantum harmonic oscillators arXiv:1811.02032

Hansen J P and McDonald I R 1986 *Theory of Simple Liquids* (London: Academic)

Kirkwood J G 1933 Quantum statistics of almost classical particles *Phys. Rev.* **44** 31

Merzbacher E 1970 *Quantum Mechanics* 2nd edn (New York: Wiley)

Messiah A 1961 *Quantum Mechanics* vols 1 and 2 (Amsterdam: North-Holland)

Pathria R K 1972 *Statistical Mechanics* (Oxford: Pergamon)

Wigner E 1932 On the quantum correction for thermodynamic equilibrium *Phys. Rev.* **40** 749

IOP Publishing

Quantum Statistical Mechanics in Classical Phase Space

Phil Attard

Chapter 8

High temperature expansions for the commutation function

The temperature derivative of the defining equation for the commutation function yields recursion relations for the expansion coefficients in powers of Planck's constant or of inverse temperature. The commutation function is also written as a series of fluctuations of the phase space expectation value of the energy operator. The related expansions contain higher derivatives of the potential, and the terms grow rapidly in number and complexity. Exfact results for the simple harmonic oscillator indicate that the commutation function is largest at the potential minimum, and test the rate of convergence of the various expansions.

The necessity of counting unique quantum states once only is emphasized. This corrects the primary quantum correction to classical thermodynamics given by Kirkwood (1933), and confirms the second order correction given by Wigner (1932).

8.1 Preliminary definitions

The aim of this chapter is to derive high temperature expansions that are suited for computing the commutation function. It will be recalled that this function arose in the transformation to classical phase space derived in chapter 7. It accounts for the non-commutativity of the position and momentum operators, since if they did commute it would be unity.

In sections 8.2, 8.3, and 8.4 expansions for the commutation function are derived using a temperature-derivative procedure based on that originally given by Kirkwood (1933). Differences in detail arise from different definitions. (Also an error in Kirkwood's result for the primary quantum correction is corrected.) The present expansions are either high temperature, with terms involving successive powers of the inverse temperature, $\beta = 1/k_{\mathrm{B}}T$, k_{B} being Boltzmann's constant and T the absolute temperature. Or else they are in powers of Planck's constant h, with $\hbar = h/2\pi$. In section 8.5, the commutation function is written as a series of fluctuations in the phase space expectation value of the energy operator, and a recursion relation is developed

doi:10.1088/978-0-7503-4055-7ch8

for the coefficients. This provides an interesting physical interpretation of the high temperature expansions derived by the temperature-derivative method. Section 8.6.2 presents numerical results of the expansions applied to the simple harmonic oscillator.

8.1.1 History

The defining equation for what I have called the commutation function, equation (7.7), originally appeared in essence in Wigner's (1932) analysis of the leading order high temperature quantum correction to classical thermodynamics. Wigner's particular formulation of a type of integral wave function in a pseudo-phase space given in that paper is *not* related to the present formulation of quantum statistical mechanics in classical phase space, chapter 7.

Kirkwood (1933) showed that a temperature derivative of Wigner's defining equation gave an easier route to obtaining the primary high temperature quantum correction, and in addition he obtained the leading order correction for wave function symmetrization, again valid in the high temperature limit. (Actually his expression for the partition function (Kirkwood 1933, equation (10)) is missing a factor of $N!$, which is an error that appears to arise from the manner of its derivation.) Kirkwood's (1933) derivation gave a classical phase space form for the partition function that is rather close to the formulation given in chapter 7. The main difference is that the present formulation separates non-commutation and symmetrization effects into two separate phase functions, whereas Kirkwood kept the two as a single function. The present commutation function can be compared to Kirkwood's function with the identity permutation.

8.1.2 Definition

The present commutation function was defined in equation (7.7),

$$e^{-\beta\mathcal{H}(\mathbf{p},\,\mathbf{q})}\omega_{\mathrm{p}}(\mathbf{p},\,\mathbf{q}) = \frac{\langle\mathbf{q}|e^{-\beta\hat{\mathcal{H}}}|\mathbf{p}\rangle}{\langle\mathbf{q}|\mathbf{p}\rangle} = e^{\mathbf{q}\cdot\mathbf{p}/i\hbar}e^{-\beta\hat{\mathcal{H}}(\mathbf{q})}e^{-\mathbf{q}\cdot\mathbf{p}/i\hbar}, \tag{8.1}$$

and

$$e^{-\beta\mathcal{H}(\mathbf{p},\,\mathbf{q})}\omega_{\mathrm{q}}(\mathbf{p},\,\mathbf{q}) = \frac{\langle\mathbf{p}|e^{-\beta\hat{\mathcal{H}}}|\mathbf{q}\rangle}{\langle\mathbf{p}|\mathbf{q}\rangle} = e^{-\mathbf{q}\cdot\mathbf{p}/i\hbar}e^{-\beta\hat{\mathcal{H}}(\mathbf{q})}e^{\mathbf{q}\cdot\mathbf{p}/i\hbar}. \tag{8.2}$$

For the complex conjugates one has $\omega_{\mathrm{p}}(\mathbf{p},\,\mathbf{q}) = \omega_{\mathrm{q}}(\mathbf{p},\,\mathbf{q})^* = \omega_{\mathrm{p}}(-\mathbf{p},\,\mathbf{q})^*$.

The classical Hamiltonian is taken to be the sum of the kinetic energy and the potential energy, $\mathcal{H}(\mathbf{p},\,\mathbf{q}) = \mathcal{K}(\mathbf{p}) + U(\mathbf{q})$. The kinetic energy is

$$\mathcal{K}(\mathbf{p}) = \frac{p^2}{2m} = \frac{1}{2m}\sum_{j=1}^{N}\sum_{\alpha=x}^{d} p_{j\alpha}^2, \tag{8.3}$$

where m is the particle mass, which is assumed the same for all particles. The potential energy consists of one-body, two-body, etc, terms,

$$\begin{aligned} U(\mathbf{q}) &= U^{(1)}(\mathbf{q}) + U^{(2)}(\mathbf{q}) + U^{(3)}(\mathbf{q}) + \cdots \\ &= \sum_{j=1}^{N} u^{(1)}(\mathbf{q}_j) + \sum_{j<k}^{N} u^{(2)}(\mathbf{q}_j, \mathbf{q}_k) + \sum_{j<k<\ell}^{N} u^{(3)}(\mathbf{q}_j, \mathbf{q}_k, \mathbf{q}_\ell) + \cdots \end{aligned} \tag{8.4}$$

Commonly only a central pair potential is considered, $u^{(2)}(\mathbf{q}_j, \mathbf{q}_k) = u^{(2)}(q_{jk})$. Again all particles are assumed identical.

8.1.3 Classical limit

In the classical limit quantum effects, including non-commutativity, must be negligible. Hence the phase space weight must go over to the classical Maxwell–Boltzmann factor, which means that the commutation function becomes unity,

$$\omega_{\rm p}(\mathbf{p}, \mathbf{q}) = \omega_{\rm q}(\mathbf{p}, \mathbf{q}) = 1, \quad \beta = 0, \quad \text{or} \quad \hbar = 0. \tag{8.5}$$

8.1.4 Extensivity

In thermodynamics the concept of extensivity is extremely important (Attard 2002). An extensive function scales with the size of the system. System size can mean volume, or particle number. Some examples of extensive thermodynamic functions are the energy (both potential and kinetic), the free energy (Helmholtz, Gibbs, grand potential), the entropy (both subsystem and total). The reason why the logarithm of the partition function arises in statistical mechanics is that it is thereby made extensive and can be equated to the free energy. On a personal note, the present author's theory for non-equilibrium thermodynamics was initially driven by the fact that the first energy moment is an extensive variable (Onsager 1931, Attard 2012).

In the present book we have already met extensivity in the resummation of the symmetrization function to give the exponential of single loop functions and the loop grand potential, equations (3.58) and (7.36). This concept of exponential resummation of the symmetrization function originally arose when the author noticed that continuing the series of permutation corrections to the classical free energy given by Pathria (1972, section 5.5) would not be extensive.

Like the relation between the free energy and the partition function, one can define a version of the commutation function that is just the logarithm of that defined above,

$$W_{\rm p}(\mathbf{p}, \mathbf{q}) \equiv \ln \omega_{\rm p}(\mathbf{p}, \mathbf{q}), \quad \text{or} \quad \omega_{\rm p}(\mathbf{p}, \mathbf{q}) \equiv e^{W_{\rm p}(\mathbf{p}, \mathbf{q})}. \tag{8.6}$$

Based on the structure of the defining equation for the commutation function, there is reason to believe that W is extensive. In slightly loose nomenclature, we shall refer to both ω and W as the commutation function.

Of course the relation between ω and W is so trivial that one may question whether it is worth defining a separate function. Two points can be made in response. When mathematical operations are carried out the equations take on different forms, and it is possible that the structure of one version is simpler, or more conducive to physical interpretation, or more amenable to approximation or calculation, than the other. Second, when one introduces an approximation, such as truncating an infinite series after a finite number of terms, then the two versions are no longer mathematically equivalent, and one may prove more accurate and reliable than the other. As an example, from the strict mathematical point of view there is no need to introduce free energy into thermodynamics since it is just the logarithm of the partition function, and any equation in one could be transformed into an equation in the other. But experience shows that the thermodynamic formula and expansions take on a simpler forms when written in terms of free energy rather than in terms of the partition function.

8.2 Expansion 1

8.2.1 Derivation of expansion

Recall that the commutation function is defined by $e^{-\beta\mathcal{H}(\mathbf{p},\,\mathbf{q})}\omega_p(\mathbf{p},\,\mathbf{q};\,\beta)\langle\mathbf{q}|\mathbf{p}\rangle = \langle\mathbf{q}|e^{-\beta\hat{\mathcal{H}}}|\mathbf{p}\rangle$, the Hamiltonian operator is $\hat{\mathcal{H}} = (-\hbar^2/2m)\nabla_q^2 + U(\mathbf{q})$, and the momentum eigenfunction is $\langle\mathbf{q}|\mathbf{p}\rangle = e^{-\mathbf{p}\cdot\mathbf{q}/i\hbar}/V^{N/2}$. Following Kirkwood (1933) the definition of the commutation function may be differentiated with respect to inverse temperature. Suppressing the arguments the derivative of the left-hand side is

$$\frac{\partial}{\partial\beta}\left\{e^{-\beta\mathcal{H}}\omega_{\mathrm{p}}e^{-\mathbf{p}\cdot\mathbf{q}/i\hbar}\right\} = \left\{\frac{\partial\omega_{\mathrm{p}}}{\partial\beta} - \mathcal{H}\omega_{\mathrm{p}}\right\}e^{-\beta\mathcal{H}}e^{-\mathbf{p}\cdot\mathbf{q}/i\hbar}. \tag{8.7}$$

That of the right-hand side is

$$\begin{aligned}
&\frac{\partial}{\partial\beta}\langle\mathbf{q}|e^{-\beta\hat{\mathcal{H}}}|\mathbf{p}\rangle \\
&= -\hat{\mathcal{H}}\left\{e^{-\beta\mathcal{H}}\omega_{\mathrm{p}}e^{-\mathbf{p}\cdot\mathbf{q}/i\hbar}\right\} \\
&= \left\{-\mathcal{H}[e^{-\beta\mathcal{H}}\omega_{\mathrm{p}}] + \frac{\hbar^2}{2m}\nabla^2[e^{-\beta\mathcal{H}}\omega_{\mathrm{p}}] + 2\frac{i\hbar}{2m}\mathbf{p}\cdot\nabla[e^{-\beta\mathcal{H}}\omega_{\mathrm{p}}]\right\}e^{-\mathbf{p}\cdot\mathbf{q}/i\hbar}.
\end{aligned} \tag{8.8}$$

Equating these and rearranging gives

$$\frac{\partial\omega_{\mathrm{p}}}{\partial\beta} = \frac{i\hbar}{m}e^{\beta U}\mathbf{p}\cdot\nabla\left\{e^{-\beta U}\omega_{\mathrm{p}}\right\} + \frac{\hbar^2}{2m}e^{\beta U}\nabla^2\left\{e^{-\beta U}\omega_{\mathrm{p}}\right\}. \tag{8.9}$$

By inspection, $\omega_{\mathrm{p}}(\beta = 0) = 1$, and so integrating this gives

$$\begin{aligned}
\omega_{\mathrm{p}} = 1 &+ \frac{i\hbar}{m}\int_0^\beta \mathrm{d}\beta'\ e^{\beta' U}\mathbf{p}\cdot\nabla\left\{e^{-\beta' U}\omega_{\mathrm{p}}(\beta')\right\} \\
&+ \frac{\hbar^2}{2m}\int_0^\beta \mathrm{d}\beta'\ e^{\beta' U}\nabla^2\left\{e^{-\beta' U}\omega_{\mathrm{p}}(\beta')\right\}.
\end{aligned} \tag{8.10}$$

This is essentially identical to an expression given by Kirkwood (1933) for the case of the identity permutation. (In the present approach wave function symmetrization is treated differently to Kirkwood.)

This provides the basis for an expansion in powers of $\hbar$,

$$\omega_{\rm p} = \sum_{n=0}^{\infty} \omega_{{\rm p},n} \hbar^n. \tag{8.11}$$

The term $\omega_{\rm p,0} = 1$ is the classical term, and the terms $\omega_{{\rm p},n}$, $n \geqslant 1$ may be called the quantum corrections due to non-commutativity. They arise from the action of the kinetic energy operator on the potential energy and they involve gradients of the potential energy.

With $\omega_{\rm p,0} = 1$, successive substitution gives the first order correction as

$$\begin{aligned} \omega_{\rm p,1} &= \frac{\rm i}{m} \int_0^\beta {\rm d}\beta' \, e^{\beta' U} \mathbf{p} \cdot \nabla \left\{ e^{-\beta' U} \omega_{\rm p,0}(\beta') \right\} \\ &= \frac{\rm i}{m} \int_0^\beta {\rm d}\beta' \, (-\beta') \mathbf{p} \cdot \nabla U \\ &= \frac{-{\rm i}\beta^2}{2m} \mathbf{p} \cdot \nabla U, \end{aligned} \tag{8.12}$$

and the second order correction as

$$\begin{aligned} \omega_{\rm p,2} &= \frac{\rm i}{m} \int_0^\beta {\rm d}\beta' \, e^{\beta' U} \mathbf{p} \cdot \nabla \left\{ e^{-\beta' U} \omega_{\rm p,1}(\beta') \right\} \\ &\quad + \frac{1}{2m} \int_0^\beta {\rm d}\beta' \, e^{\beta' U} \nabla^2 \left\{ e^{-\beta' U} \omega_{\rm p,0}(\beta') \right\} \\ &= \frac{\rm i}{m} \int_0^\beta {\rm d}\beta' \left[(-\beta')(\mathbf{p} \cdot \nabla U) \frac{-{\rm i}\beta'^2}{2m} \mathbf{p} \cdot \nabla U + \frac{-{\rm i}\beta'^2}{2m} (\mathbf{p} \cdot \nabla)^2 U \right] \\ &\quad + \frac{1}{2m} \int_0^\beta {\rm d}\beta' \, \left\{ -\beta' \nabla^2 U + \beta'^2 \nabla U \cdot \nabla U \right\} \\ &= \frac{-\beta^4}{8m^2} (\mathbf{p} \cdot \nabla U)^2 + \frac{\beta^3}{6m^2} (\mathbf{p} \cdot \nabla)^2 U - \frac{\beta^2}{4m} \nabla^2 U + \frac{\beta^3}{6m} \nabla U \cdot \nabla U. \end{aligned} \tag{8.13}$$

These are essentially the same as Kirkwood (1933, equation (16)). In general

$$\begin{aligned} \omega_{{\rm p},n} &= \frac{\rm i}{m} \int_0^\beta {\rm d}\beta' \, e^{\beta' U} \mathbf{p} \cdot \nabla \left\{ e^{-\beta' U} \omega_{{\rm p},n-1}(\beta') \right\} \\ &\quad + \frac{1}{2m} \int_0^\beta {\rm d}\beta' \, e^{\beta' U} \nabla^2 \left\{ e^{-\beta' U} \omega_{{\rm p},n-2}(\beta') \right\}. \end{aligned} \tag{8.14}$$

This is essentially the same as Kirkwood (1933, equation (17)).

It is shown below that $\omega_{\rm p,1}$ plus $\omega_{\rm p,2}$ (and neglecting symmetrization corrections) give a configuration position probability density, section 8.2.3, and an average kinetic energy, section 8.2.4, that are the same as that given by Wigner (1932) and by Kirkwood (1933).

8.2.2 An error in Kirkwood

Kirkwood (1933) derived the first two (i.e., primary and secondary) quantum corrections to classical statistical mechanics. It turns out that there is an error in his analysis. Under normal circumstances one might ignore this, giving the benefit of doubt to an otherwise reliable scientist. However, because Kirkwood is an authority of the first rank, because the result is historically significant, because the error appears to be of conceptual rather than arithmetical or typographical nature, and because the erroneous result continues to contaminate the literature (e.g., Allen and Tildesley 1987), it seems important to explicitly address the error.

Kirkwood (1933, equation (10)) gives for the canonical equilibrium partition function (my notation)

$$Z(N, V, T) = \frac{1}{h^{3N}N!}\sum_{\hat{\mathrm{P}}', \hat{\mathrm{P}}''}(\pm 1)^{p'+p''}\int \mathrm{d}\mathbf{p}\,\mathrm{d}\mathbf{q}\; e^{(\hat{\mathrm{P}}'\mathbf{p})\cdot\mathbf{q}/i\hbar}e^{-\beta\hat{\mathcal{H}}(\mathbf{r})}e^{-(\hat{\mathrm{P}}''\mathbf{p})\cdot\mathbf{q}/i\hbar}. \tag{8.15}$$

One can compare this to equation (7.1), which transforms a sum over unique allowed states, to a sum over all states weighted with the symmetrization factor, to a sum over states and permutations. My equation provides the exact way to count unique states, since it is derived from the symmetrization factor, which was shown to count particle occupancy statistics correctly, and it provides the basis for the partition function, equation (7.3). Compared to my result, one sees that this expression by Kirkwood is missing a factor of $N!$ in the denominator. Since Kirkwood doesn't publish a derivation of his result, and since a single factor of $N!$ occurs in classical statistical mechanics, and since a related error is shown below to occur in his quantum correction explicitly, it seems likely that this is a conceptual rather than typographical or algebraic error. My guess is that Kirkwood did not advert to the need to count distinct states once and once only.

Kirkwood (1933) includes the particle permutator in what I call the commutation function ω_p (again my notation). He gives $\omega_\mathrm{p} = \sum_l \hbar^l \omega_{\mathrm{p},l}$, with $\omega_{\mathrm{p},0} = 1$,

$$\omega_{\mathrm{p},1} = \frac{-\mathrm{i}\beta^2}{2m}\hat{\mathrm{P}}\sum_{k=1}^{N}\mathbf{p}_k\cdot\nabla_k U. \tag{8.16}$$

This, his equation (16), in the case of the identity permutation is the same as equation (8.12). In the context of pair transposition Kirkwood invokes $\omega_1(jk)$ and $\omega_1(kj)$, without explicit definition, but which I interpret to mean

$$\omega_{\mathrm{p},1}(jk) = \frac{-\mathrm{i}\beta^2}{2m}\mathbf{p}_j\cdot\nabla_k U, \quad \text{and} \quad \omega_{\mathrm{p},1}(kj) = \frac{-i\beta^2}{2m}\mathbf{p}_k\cdot\nabla_j U. \tag{8.17}$$

His first quantum correction (Kirkwood 1933, equation (20)) is

$$\begin{aligned} Z_1^{\pm} = {} & \frac{\pm 1}{h^{3N}}\frac{1}{2}\sum_{k\neq j}\int \mathrm{d}\mathbf{p}^N\,\mathrm{d}\mathbf{q}^N\; e^{-\beta\mathcal{H}(\mathbf{q},\,\mathbf{p})} \\ & \times\left\{e^{-\mathbf{p}_{jk}\cdot\mathbf{q}_{jk}/i\hbar}\hbar\omega_{\mathrm{p},1}(jk) + e^{\mathbf{p}_{jk}\cdot\mathbf{q}_{jk}/i\hbar}\hbar\omega_{\mathrm{p},1}(kj)\right\}. \end{aligned} \tag{8.18}$$

I gave the linear order term for the commutation function (without including the particle permutator) as equation (8.12) above. It agrees with Kirkwood (1933, equation (16)) for the identity permutation. In my approach, the leading quantum correction due to symmetrization comes from the single dimer contribution to the symmetrization function, which is

$$\begin{aligned}\overset{\circ}{\eta}{}_q^{\pm(2)}(\mathbf{q},\mathbf{p}) &= \pm\, e^{-\mathbf{q}\cdot\mathbf{p}/i\hbar}\sum_{j<k} e^{(\hat{\mathrm{P}}_{jk}\mathbf{q})\cdot\mathbf{p}/i\hbar} \\ &= \pm\sum_{j<k} e^{-\mathbf{q}_j\cdot\mathbf{p}_j/i\hbar}e^{-\mathbf{q}_k\cdot\mathbf{p}_k/i\hbar}e^{\mathbf{q}_k\cdot\mathbf{p}_j/i\hbar}e^{\mathbf{q}_j\cdot\mathbf{p}_k/i\hbar}.\end{aligned} \tag{8.19}$$

The combination of $\omega_{p,1}$ and $\overset{\circ}{\eta}{}_q^{\pm(2)}$ yields for the primary quantum correction to the partition function

$$\begin{aligned}Z_{1,1}^{\pm(2)} &= \frac{\pm 1}{h^{3N}N!}\int \mathrm{d}\mathbf{q}\,\mathrm{d}\mathbf{p}\; e^{-\beta\mathcal{H}(\mathbf{q},\mathbf{p})}\frac{-i\hbar\beta^2}{2m}\mathbf{p}\cdot(\nabla U)\sum_{j<k} e^{-\mathbf{q}_{jk}\cdot\mathbf{p}_j/i\hbar}e^{-\mathbf{q}_{kj}\cdot\mathbf{p}_k/i\hbar} \\ &= \frac{\pm 1}{h^{3N}N!}\frac{-i\hbar\beta^2}{2m}\frac{1}{2}\sum_{j,k}^{(j\neq k)}\int \mathrm{d}\mathbf{q}\,\mathrm{d}\mathbf{p}\; e^{-\beta U(\mathbf{q})}e^{-\beta p^2/2m} \\ &\quad\times[\mathbf{p}_j\cdot\nabla_j U + \mathbf{p}_k\cdot\nabla_k U]e^{-\mathbf{q}_{jk}\cdot\mathbf{p}_j/i\hbar}e^{-\mathbf{q}_{kj}\cdot\mathbf{p}_k/i\hbar}.\end{aligned} \tag{8.20}$$

Compared to this, one can see that Kirkwood (1933, equation (20)), equation (8.18) above, has doubled up on this term. It seems to me that Kirkwood has arrived at his result by treating the pair transpositions $\hat{\mathrm{P}}_{jk}$ and $\hat{\mathrm{P}}_{kj}$ as distinct, whereas I treat them as one and the same state. In my judgment he double counts this term.

One can perform the momentum integrals to obtain from my result

$$\begin{aligned}Z_{1,1}^{\pm(2)} &= \frac{\pm 1}{h^{3N}N!}(2\pi m k_{\mathrm{B}}T)^{3N/2}\frac{-i\hbar\beta^2}{4m}\sum_{j,k}^{(j\neq k)}\int \mathrm{d}\mathbf{q}\; e^{-\beta U(\mathbf{q})}e^{-mq_{kj}^2/2\beta\hbar^2} \\ &\quad\times\left[\frac{-m}{i\hbar\beta}\mathbf{q}_{jk}\cdot\nabla_j U + \frac{-m}{i\hbar\beta}\mathbf{q}_{kj}\cdot\nabla_k U\right] \\ &= \frac{\pm(2\pi m k_{\mathrm{B}}T)^{3N/2}}{h^{3N}N!}\frac{\beta}{4}\sum_{j,k}^{(j\neq k)}\int \mathrm{d}\mathbf{q}\; e^{-\beta U(\mathbf{q})}e^{-mq_{kj}^2/2\beta\hbar^2} \\ &\quad\times\mathbf{q}_{jk}\cdot\left[\nabla_j U - \nabla_k U\right].\end{aligned} \tag{8.21}$$

The factor of $1/2N!$ that appears here is missing from the leading order correction given by Kirkwood (1933, equation (21)).

Interestingly enough Kirkwood (1933) recognizes that his correction is a factor of 2 larger than the supposedly same correction derived by Uhlenbeck and Gropper (1932). Kirkwood says that the latter is in error due to a trivial oversight. In my opinion it is Kirkwood's expression that is the one in error. (I have not checked Uhlenbeck and Gropper (1932).)

8.2.3 Position configuration weight density

Recall that in general for free energies, partition functions, and statistical averages we use a subscript of the form $f_{\omega_\mathrm{p}\eta_q^\pm}$ to denote the particular approximation that we are currently analyzing. In the full quantum case both factors in the subscript are often suppressed. The case $\omega_\mathrm{p} = \eta_q^\pm = 1$ is the classical case, and we usually denote this as f_cl. (The quantity $Z_{1,1}^{\pm(2)}$ introduced in equation (8.21) above does not follow either convention; instead it results from the leading order terms $\omega_{\mathrm{p},1}$ and $\mathring{\eta}_q^{\pm(2)}$.)

Wigner's (1932) formulation of the quantum states that represent phase space and the consequent phase space quasi-probability function neglect the effects of wave function symmetrization. This may be compared to setting $\eta^\pm = 1$ in the present approach, which is just the monomer contribution. The monomer grand potential is $\Omega_{\omega_\mathrm{p}}^\pm = -k_\mathrm{B}T \ln \Xi_{\omega_\mathrm{p}}^\pm$, with the monomer grand partition function being given by equation (7.34)

$$\begin{aligned}\Xi_{\omega_\mathrm{p}}^\pm &= \sum_N \frac{z^N}{N!h^{3N}} \int \mathrm{d}\boldsymbol{\Gamma}\, e^{-\beta\mathcal{H}(\boldsymbol{\Gamma})}\omega_\mathrm{p}(\boldsymbol{\Gamma}) \\ &= \Xi_\mathrm{cl}\left\langle\omega_\mathrm{p}\right\rangle_\mathrm{cl},\end{aligned} \tag{8.22}$$

where a classical average and the classical grand partition function appear in the final equality.

Hence the grand potential in this approximation is

$$\begin{aligned}\Omega_{\omega_\mathrm{p}}^\pm &= \Omega_\mathrm{cl} - k_\mathrm{B}T \ln\left\langle\omega_\mathrm{p}\right\rangle_\mathrm{cl} \\ &= \Omega_\mathrm{cl} - k_\mathrm{B}T\left\langle\hbar^2\omega_{\mathrm{p},2}\right\rangle_\mathrm{cl} + \mathcal{O}(\hbar^4).\end{aligned} \tag{8.23}$$

The final equality follows because the first quantum correction due to non-commutativity, equation (8.12), is proportional to the momentum $\mathbf{p}$, which averages to zero in an equilibrium system, $\left\langle\omega_{\mathrm{p},1}\right\rangle_\mathrm{cl} = 0$.

The second order term is given by equation (8.13),

$$\omega_{\mathrm{p},2} = \frac{-\beta^4}{8m^2}(\mathbf{p}\cdot\nabla U)^2 + \frac{\beta^3}{6m^2}(\mathbf{p}\cdot\nabla)^2 U - \frac{\beta^2}{4m}\nabla^2 U + \frac{\beta^3}{6m}\nabla U\cdot\nabla U. \tag{8.24}$$

The classical equilibrium phase space weight is an even function of the momenta, and, by the equipartition theorem or directly, $\langle\mathbf{pp}\rangle_\mathrm{cl} = mk_\mathrm{B}T\,\mathrm{I}$. Hence to quadratic order

$$\begin{aligned}\left\langle\omega_\mathrm{p}\right\rangle_\mathrm{cl} &= 1 + \hbar^2\left\langle\omega_{\mathrm{p},2}\right\rangle_\mathrm{cl} + \mathcal{O}(\hbar^4) \\ &= 1 - \hbar^2\left\langle\frac{\beta^2}{4m}\nabla^2 U - \frac{\beta^3}{6m}\nabla U\cdot\nabla U + \frac{\beta^3}{8m}\nabla U\cdot\nabla U - \frac{\beta^2}{6m}\nabla^2 U\right\rangle_\mathrm{cl} \\ &= 1 - \hbar^2\left\langle\frac{\beta^2}{12m}\nabla^2 U - \frac{\beta^3}{24m}\nabla U\cdot\nabla U\right\rangle_\mathrm{cl} \\ &= 1 - \frac{\hbar^2\beta^2}{24m}\langle\nabla^2 U\rangle_\mathrm{cl}.\end{aligned} \tag{8.25}$$

The final equality uses the fact that $\langle\nabla U\cdot\nabla U\rangle_\mathrm{cl} = \beta^{-1}\langle\nabla^2 U\rangle_\mathrm{cl}$, which may be derived by an integration by parts, assuming that either $e^{-\beta U(\mathbf{q})}$ or ∇U vanish on the boundary of the system. (Or else that boundary contributions are relatively negligible in the thermodynamic limit.) Numerical confirmation of this result can be found in section 8.4.1.1 below. This gives the first quantum correction to the grand potential due to the non-commutativity of the position and momentum operators. With this the second order quantum correction to the classical grand potential is given by

$$-\beta[\Omega_{\omega_{\rm p,2}} - \Omega_{\rm cl}] = \frac{-\hbar^2\beta^2}{24m}\langle\nabla^2 U\rangle_{\rm cl}. \tag{8.26}$$

The statistical average of an operator that is an ordinary function of the position coordinates can be taken with the ordinary commutation function, equation (7.18). Taking only the monomer term in such a case gives

$$\langle O(\mathbf{r})\rangle_{\mu,V,T} \approx \langle O\rangle_{\omega_{\rm p}} = \frac{\left\langle \omega_{\rm p} O\right\rangle_{\rm cl}}{\left\langle \omega_{\rm p}\right\rangle_{\rm cl}}. \tag{8.27}$$

It follows that since the operator is a function only of the position

$$\begin{aligned}&\left\langle \omega_{\rm p}(\boldsymbol{\Gamma})O(\mathbf{q})\right\rangle_{\rm cl}\\ &\quad= \langle O\rangle_{\rm cl} - \hbar^2\left\langle\left[\frac{\beta^2}{12m}\nabla^2 U - \frac{\beta^3}{24m}\nabla U\cdot\nabla U\right]O\right\rangle_{\rm cl}.\end{aligned} \tag{8.28}$$

Hence the leading non-commutativity correction to an average of a function of the position is

$$\begin{aligned}\langle O(\mathbf{r})\rangle_{\mu,V,T} &= \langle O(\mathbf{q})\rangle_{\omega_{\rm p}}\\ &= \frac{\left\langle \omega_{\rm p} O(\mathbf{q})\right\rangle_{\rm cl}}{\left\langle \omega_{\rm p}\right\rangle_{\rm cl}}\\ &= \langle O(\mathbf{q})\rangle_{\rm cl} - \hbar^2\left\langle\left[\frac{\beta^2}{12m}\nabla^2 U - \frac{\beta^3}{24m}\nabla U\cdot\nabla U\right]\right.\\ &\qquad\left.\times[O(\mathbf{q}) - \langle O(\mathbf{q})\rangle_{\rm cl}]\right\rangle_{\rm cl} + \mathcal{O}(\hbar^4).\end{aligned} \tag{8.29}$$

From this one can identify the probability of a configuration position in classical phase space, neglecting wave function symmetrization, to quadratic order in $\hbar$

$$\wp(\mathbf{q}) \propto e^{-\beta U(\mathbf{q})}\left\{1 - \frac{\hbar^2\beta^2}{12m}\nabla^2 U + \frac{\hbar^2\beta^3}{24m}\nabla U\cdot\nabla U\right\}. \tag{8.30}$$

This is identical to Wigner (1932, equation (28)). Kirkwood (1933) states that his configuration position probability density is also identical to Wigner's.

One has to distinguish three quantities: the configuration position probability density, the configuration momentum probability density, and the phase space probability density. As just mentioned, the present theory agrees with Wigner and Kirkwood for the configuration position probability density (in the absence of wave function symmetrization). As will be shown next, it gives the same result for the average kinetic energy, which suggests that it also agrees with Wigner and Kirkwood for the configuration momentum probability density (in the absence of wave function symmetrization). However the present theory does not give a phase space probability density, at least not one that is independent of the operator being averaged in the most general case for an arbitrary function of position and momentum operators. This is in contrast to Wigner (1932) and Kirkwood (1933) who each claim to give a phase space quasi-probability density. Unfortunately, those two densities do not agree with each other.

8.2.4 Average kinetic energy

Now as an example of an operator that is a function of the momentum operator, consider the kinetic energy operator, $\hat{\mathcal{K}} = (-\hbar^2/2m)\nabla^2$. Since the kinetic energy operator in a pure operator (i.e., an ordinary function of the momentum operator alone), we can use $\omega_{\rm p}$ as the commutation function, equation (7.17).

The first three terms in an expansion of the commutation function in powers of $\hbar$ were given above: the zeroth order term $\omega_{\rm p,0} = 1$, the first order term, equation (8.12),

$$\omega_{\rm p,1} = \frac{-i\beta^2}{2m}\mathbf{p}\cdot\nabla U, \tag{8.31}$$

and the second order term, equation (8.13),

$$\omega_{\rm p,2} = \frac{-\beta^4}{8m^2}(\mathbf{p}\cdot\nabla U)^2 + \frac{\beta^3}{6m^2}(\mathbf{p}\cdot\nabla)^2 U - \frac{\beta^2}{4m}\nabla^2 U + \frac{\beta^3}{6m}\nabla U\cdot\nabla U. \tag{8.32}$$

We shall retain only the single dimer term in the symmetrization function,

$$\mathring{\eta}_{\rm q}^{\pm(2)}(\mathbf{q},\mathbf{p}) = \pm\sum_{j<k} e^{-\mathbf{q}_{jk}\cdot\mathbf{p}_j/i\hbar} e^{-\mathbf{q}_{kj}\cdot\mathbf{p}_k/i\hbar}. \tag{8.33}$$

The average kinetic energy expanded to leading order is

$$\begin{aligned}\langle\mathcal{K}\rangle_{\mu,V,T} &= \frac{\left\langle\mathcal{K}(\mathbf{p})\omega_{\rm p}(\mathbf{\Gamma})\eta_{\rm p}^{\pm}(\mathbf{\Gamma})\right\rangle_{\rm cl}}{\left\langle\omega_{\rm p}(\mathbf{\Gamma})\eta_{\rm p}^{\pm}(\mathbf{\Gamma})\right\rangle_{\rm cl}}\\ &= \frac{\left\langle\mathcal{K}\left[1+\hbar\omega_{\rm p,1}+\hbar^2\omega_{\rm p,2}+\mathcal{O}(\hbar^3)\right]\left[1+\mathring{\eta}_{\rm q}^{\pm(2)}\right]\right\rangle_{\rm cl}}{\left\langle\left[1+\hbar\omega_{\rm p,1}+\hbar^2\omega_{\rm p,2}+\mathcal{O}(\hbar^3)\right]\left[1+\mathring{\eta}_{\rm q}^{\pm(2)}\right]\right\rangle_{\rm cl}}\\ &= \frac{\langle\mathcal{K}\rangle_{\rm cl}+\hbar\left\langle\mathcal{K}\omega_{\rm p,1}\mathring{\eta}_{\rm q}^{\pm(2)}\right\rangle_{\rm cl}+\hbar^2\left\langle\mathcal{K}\omega_{\rm p,2}\right\rangle_{\rm cl}+\cdots}{1+\hbar\left\langle\omega_{\rm p,1}\mathring{\eta}_{\rm q}^{\pm(2)}\right\rangle_{\rm cl}+\hbar^2\left\langle\omega_{\rm p,2}\right\rangle_{\rm cl}+\cdots}.\end{aligned} \tag{8.34}$$

8.2.4.1 Some momentum integrals

We shall require the following momentum integrals:

$$\int \mathrm{d}\mathbf{p}_\ell \; e^{-\beta p_\ell^2/2m} = (2\pi m k_\mathrm{B} T)^{3/2}, \tag{8.35}$$

$$\int \mathrm{d}\mathbf{p}_k \; e^{-\beta p_k^2/2m} e^{-\mathbf{q}_{kj}\cdot\mathbf{p}_k/i\hbar} = (2\pi m k_\mathrm{B} T)^{3/2} e^{-m q_{kj}^2/2\beta\hbar^2}, \tag{8.36}$$

and

$$\begin{aligned} &\int \mathrm{d}\mathbf{p}_j \; e^{-\beta p_j^2/2m} e^{-\mathbf{q}_{jk}\cdot\mathbf{p}_j/i\hbar} \mathbf{p}_j \cdot \nabla_j U \\ &\quad = (2\pi m k_\mathrm{B} T)^{3/2} e^{-m q_{kj}^2/2\beta\hbar^2} \frac{im}{\hbar\beta} \mathbf{q}_{jk} \cdot \nabla_j U. \end{aligned} \tag{8.37}$$

Also

$$\int \mathrm{d}\mathbf{p}_\ell \; e^{-\beta p_\ell^2/2m} p_\ell^2 = 3\, m k_\mathrm{B} T (2\pi m k_\mathrm{B} T)^{3/2}, \tag{8.38}$$

$$\begin{aligned} &\int \mathrm{d}\mathbf{p}_k \; e^{-\beta p_k^2/2m} e^{-\mathbf{q}_{kj}\cdot\mathbf{p}_k/i\hbar} p_k^2 \\ &\quad = (2\pi m k_\mathrm{B} T)^{3/2} e^{-m q_{kj}^2/2\beta\hbar^2} \left[\frac{-m^2 q_{kj}^2}{\hbar^2\beta^2} + 3\, m k_\mathrm{B} T \right], \end{aligned} \tag{8.39}$$

and

$$\begin{aligned} &\int \mathrm{d}\mathbf{p}_j \; e^{-\beta p_j^2/2m} e^{-\mathbf{q}_{jk}\cdot\mathbf{p}_j/i\hbar} \mathbf{p}_j \cdot \nabla_j U p_j^2 \\ &= e^{-m q_{kj}^2/2\beta\hbar^2} (2\pi m k_\mathrm{B} T)^{3/2} \left\{ \frac{5im^2}{\hbar\beta^2} \mathbf{q}_{jk} \cdot \nabla_j U - \frac{im^3}{\hbar^3\beta^3} q_{jk}^2 \mathbf{q}_{jk} \cdot \nabla_j U \right\}. \end{aligned} \tag{8.40}$$

8.2.4.2 Primary quantum correction

The primary correction comes from the single dimer transposition for the symmetrization function combined with the first quantum correction for the commutation function. This is

$$\begin{aligned} \langle \mathcal{K} \rangle_{\mu,V,T} &= \frac{\langle \mathcal{K} \rangle_\mathrm{cl} + \hbar \left\langle \mathcal{K} \omega_{\mathrm{p},1} \mathring{\eta}_\mathrm{q}^{\pm(2)} \right\rangle_\mathrm{cl} + \mathcal{O}(\hbar^2)}{1 + \hbar \left\langle \omega_{\mathrm{p},1} \mathring{\eta}_\mathrm{q}^{\pm(2)} \right\rangle_\mathrm{cl} + \mathcal{O}(\hbar^2)} \\ &= \langle \mathcal{K} \rangle_\mathrm{cl} + \hbar \left\langle \mathcal{K} \omega_{\mathrm{p},1} \mathring{\eta}_\mathrm{q}^{\pm(2)} \right\rangle_\mathrm{cl} - \hbar \left\langle \omega_{\mathrm{p},1} \mathring{\eta}_\mathrm{q}^{\pm(2)} \right\rangle_\mathrm{cl} \langle \mathcal{K} \rangle_\mathrm{cl} + \mathcal{O}(\hbar^2). \end{aligned} \tag{8.41}$$

The classical partition function is

$$Z_{\text{cl}} = \frac{1}{h^{3N}N!}\int \mathrm{d}\mathbf{q}\,\mathrm{d}\mathbf{p}\; e^{-\beta U(\mathbf{q})}e^{-\beta p^2/2m} = \frac{\Lambda^{-3N}}{N!}\int \mathrm{d}\mathbf{q}\; e^{-\beta U(\mathbf{q})}, \tag{8.42}$$

where the thermal wavelength is $\Lambda = \sqrt{2\pi\hbar^2/mk_{\text{B}}T}$. With this and the above integrals we have

$$\begin{aligned}
\langle \mathcal{K}\rangle_{\text{cl}} &= \frac{1}{Z_{\text{cl}}h^{3N}N!}\int \mathrm{d}\mathbf{q}\,\mathrm{d}\mathbf{p}\; e^{-\beta U(\mathbf{q})}e^{-\beta p^2/2m}\frac{1}{2m}\sum_{j=1}^{N} p_j^2 \\
&= \frac{3Nk_{\text{B}}T}{2}(2\pi m k_{\text{B}}T)^{3N/2}\frac{1}{Z_{\text{cl}}h^{3N}N!}\int \mathrm{d}\mathbf{q}\; e^{-\beta U(\mathbf{q})} \\
&= \frac{3Nk_{\text{B}}T}{2},
\end{aligned} \tag{8.43}$$

$$\begin{aligned}
&\left\langle \mathcal{K}\omega_{\text{p},1}\mathring{\eta}_{\text{q}}^{\pm(2)}\right\rangle_{\text{cl}} \\
&= \frac{1}{Z_{\text{cl}}h^{3N}N!}\int \mathrm{d}\mathbf{q}\,\mathrm{d}\mathbf{p}\; e^{-\beta U(\mathbf{q})}e^{-\beta p^2/2m} \\
&\times \frac{-\mathrm{i}\beta^2}{2m}\mathbf{p}\cdot\nabla U\left[\pm\sum_{j<k} e^{-\mathbf{q}_{jk}\cdot\mathbf{p}_j/\mathrm{i}\hbar}e^{-\mathbf{q}_{kj}\cdot\mathbf{p}_k/\mathrm{i}\hbar}\right]\frac{1}{2m}\sum_{\ell=1}^{N} p_\ell^2 \\
&= \frac{\mp\mathrm{i}\beta^2/4m^2}{Z_{\text{cl}}h^{3N}N!}\sum_{j<k}\int \mathrm{d}\mathbf{q}\,\mathrm{d}\mathbf{p}\; e^{-\beta U(\mathbf{q})}e^{-\beta p^2/2m} \\
&\times e^{-\mathbf{q}_{jk}\cdot\mathbf{p}_j/\mathrm{i}\hbar}e^{-\mathbf{q}_{kj}\cdot\mathbf{p}_k/\mathrm{i}\hbar}\left[\mathbf{p}_j\cdot\nabla_j U + \mathbf{p}_k\cdot\nabla_k U\right]\left[p_j^2 + p_k^2 + \sum_{\ell=1}^{N}{}^{(\ell\neq j,\,k)} p_\ell^2\right] \\
&= \frac{\mp\mathrm{i}\beta^2/4m^2}{Z_{\text{cl}}h^{3N}N!}\sum_{j<k}\int \mathrm{d}\mathbf{q}\; e^{-\beta U(\mathbf{q})}(2\pi m k_{\text{B}}T)^{3N}e^{-mq_{kj}^2/2\beta\hbar^2} \\
&\times\left\{(N-2)\frac{-m^2}{\mathrm{i}\hbar\beta^2}\mathbf{q}_{jk}\cdot[\nabla_j U - \nabla_k U] + \frac{-3m^2}{\mathrm{i}\hbar\beta^2}\mathbf{q}_{jk}\cdot[\nabla_j U - \nabla_k U]\right. \\
&\left. + \frac{5\mathrm{i}m^2}{\hbar\beta^2}\mathbf{q}_{jk}\cdot[\nabla_j U - \nabla_k U] - \frac{\mathrm{i}m^3}{\hbar^3\beta^3}q_{jk}^2\mathbf{q}_{jk}\cdot[\nabla_j U - \nabla_k U]\right\} \\
&= \frac{\pm\beta^2/4m^2}{Z_{\text{cl}}\Lambda^{3N}N!}\sum_{j<k}\int \mathrm{d}\mathbf{q}\; e^{-\beta U(\mathbf{q})}e^{-mq_{kj}^2/2\beta\hbar^2}\;\mathbf{q}_{jk}\cdot[\nabla_j U - \nabla_k U] \\
&\times\left\{(N-2)\frac{m^2}{\hbar\beta^2} + \frac{8m^2}{\hbar\beta^2} - \frac{m^3}{\hbar^3\beta^3}q_{jk}^2\right\}.
\end{aligned} \tag{8.44}$$

Also, from the result for $Z_{1,1}^{\pm(2)}$, equation (8.21),

$$\hbar\left\langle \omega_{\mathrm{p},1}\mathring{\eta}_{\mathrm{q}}^{\pm(2)}\right\rangle_{\mathrm{cl}} = \frac{\pm\beta/2}{Z_{\mathrm{cl}}\Lambda^{3N}N!}\sum_{j<k}\int \mathrm{d}\mathbf{q}\; e^{-\beta U(\mathbf{q})}e^{-mq_{kj}^2/2\beta\hbar^2}\mathbf{q}_{jk}\cdot\left[\nabla_j U - \nabla_k U\right]. \tag{8.45}$$

To leading order the quantum correction for the average kinetic energy is

$$\langle\mathcal{K}\rangle_{\omega_{\mathrm{p}}\eta_q^{\pm}} - \langle\mathcal{K}\rangle_{\mathrm{cl}} = \hbar\left\langle \mathcal{K}\omega_{\mathrm{p},1}\mathring{\eta}_{\mathrm{q}}^{\pm(2)}\right\rangle_{\mathrm{cl}} - \hbar\left\langle \omega_{\mathrm{p},1}\mathring{\eta}_{\mathrm{q}}^{\pm(2)}\right\rangle_{\mathrm{cl}}\langle\mathcal{K}\rangle_{\mathrm{cl}} + \mathcal{O}(\hbar^2). \tag{8.46}$$

Putting these together gives

$$\begin{aligned}
\langle\mathcal{K}\rangle_{\omega_{\mathrm{p}}\eta_q^{\pm}} - \langle\mathcal{K}\rangle_{\mathrm{cl}}
&= \frac{\pm 1}{Z_{\mathrm{cl}}\Lambda^{3N}N!}\sum_{j<k}\int \mathrm{d}\mathbf{q}\; e^{-mq_{kj}^2/2\beta\hbar^2}\mathbf{q}_{jk}\cdot\left[\nabla_j U - \nabla_k U\right]\\
&\quad\times\left[\frac{\hbar\beta^2}{4m^2}\left\{(N-2)\frac{m^2}{\hbar\beta^2} + \frac{8m^2}{\hbar\beta^2} - \frac{m^3}{\hbar^3\beta^3}q_{jk}^2\right\} - \frac{\beta 3Nk_{\mathrm{B}}T}{4}\right]\\
&= \frac{\pm 1}{Z_{\mathrm{cl}}\Lambda^{3N}N!}\sum_{j<k}\int \mathrm{d}\mathbf{q}\; e^{-\beta U(\mathbf{q})}e^{-mq_{kj}^2/2\beta\hbar^2}\\
&\quad\times\mathbf{q}_{jk}\cdot\left[\nabla_j U - \nabla_k U\right]\left\{\frac{3}{2} - \frac{m}{4\beta\hbar^2}q_{jk}^2\right\}\\
&= \pm\sum_{j<k}\left\langle e^{-mq_{kj}^2/2\beta\hbar^2}\mathbf{q}_{jk}\cdot\left[\nabla_j U - \nabla_k U\right]\left\{\frac{3}{2} - \frac{m}{4\beta\hbar^2}q_{jk}^2\right\}\right\rangle_{\mathrm{cl}}.
\end{aligned} \tag{8.47}$$

One can see by transforming $\mathbf{q}_{jk}/\hbar \Rightarrow \tilde{\mathbf{q}}_{jk}$ that this is $\mathcal{O}(\hbar)$. The integrand is only non-zero when $\mathbf{q}_j$ is close to $\mathbf{q}_k$, and so one loses one factor of volume in the average. However there are $N(N-1)/2$ identical contributions to the average, so that in net terms the quantum correction is extensive.

8.2.4.3 Second order correction

If one neglects symmetrization effects, $\eta_q^{\pm}(\mathbf{\Gamma}) = 1$, then the leading quantum correction comes from $\omega_{\mathrm{p},2}$, an expression for which was given above. We shall require the readily proven results

$$\begin{aligned}
\langle p^2\rangle_{\mathrm{cl}} &= 3Nmk_{\mathrm{B}}T,\\
\langle\mathbf{pp}\rangle_{\mathrm{cl}} &= mk_{\mathrm{B}}T\mathrm{I},\\
\langle p^2\mathbf{pp}\rangle_{\mathrm{cl}} &= [3N - 1 + 3](mk_{\mathrm{B}}T)^2\mathrm{I}.
\end{aligned} \tag{8.48}$$

With these one has

$$\begin{aligned}
&\left\langle \mathcal{K}\omega_{\mathrm{p},2}\right\rangle_{\mathrm{cl}} \\
&= \frac{1}{2m}\left\langle p^2\left\{\frac{-\beta^4}{8m^2}\mathbf{pp}\colon(\nabla U)(\nabla U) + \frac{\beta^3}{6m^2}(\mathbf{pp}\colon\nabla\nabla)U\right.\right. \\
&\quad \left.\left. - \frac{\beta^2}{4m}\nabla^2 U + \frac{\beta^3}{6m}\nabla U\cdot\nabla U\right\}\right\rangle_{\mathrm{cl}} \\
&= \frac{1}{2m}\left\langle \frac{\beta^3}{24m^2}[3N+2](mk_{\mathrm{B}}T)^2\nabla^2 U - \frac{2\beta^2}{24m}3N(mk_{\mathrm{B}}T)\nabla^2 U\right\rangle_{\mathrm{cl}} \\
&= \frac{\beta}{48m}[2-3N]\langle\nabla^2 U\rangle_{\mathrm{cl}}.
\end{aligned} \tag{8.49}$$

The penultimate equality uses the fact that $\langle\nabla U\cdot\nabla U\rangle_{\mathrm{cl}} = \beta^{-1}\langle\nabla^2 U\rangle_{\mathrm{cl}}$, which may be derived by an integration by parts, assuming that either $e^{-\beta U(\mathbf{q})}$ or $\nabla U(\mathbf{q})$ vanish on the boundary of the system. (Or else that boundary contributions are relatively negligible in the thermodynamic limit.) See section 8.4.1.1 below.

One also has

$$\begin{aligned}
&\left\langle \omega_{\mathrm{p},2}\right\rangle_{\mathrm{cl}} \\
&= \left\langle \frac{-\beta^4}{8m^2}\mathbf{pp}\colon(\nabla U)(\nabla U) + \frac{\beta^3}{6m^2}(\mathbf{pp}\colon\nabla\nabla)U - \frac{\beta^2}{4m}\nabla^2 U + \frac{\beta^3}{6m}\nabla U\cdot\nabla U\right\rangle_{\mathrm{cl}} \\
&= \left\langle \frac{-\beta^3}{8m}(\nabla U)\cdot(\nabla U) + \frac{\beta^2}{6m}\nabla^2 U - \frac{\beta^2}{4m}\nabla^2 U + \frac{\beta^3}{6m}\nabla U\cdot\nabla U\right\rangle_{\mathrm{cl}} \\
&= \frac{\beta^2}{24m}[-3+4-6+4]\langle\nabla^2 U\rangle_{\mathrm{cl}} \\
&= \frac{-\beta^2}{24m}\langle\nabla^2 U\rangle_{\mathrm{cl}}.
\end{aligned} \tag{8.50}$$

Hence the secondary quantum correction for the average kinetic energy that neglects symmetrization, $\eta_{\mathrm{q}}(\mathbf{q},\mathbf{p}) = 1$, is

$$\begin{aligned}
\langle\mathcal{K}\rangle_{\omega_{\mathrm{p},2}} - \langle\mathcal{K}\rangle_{\mathrm{cl}} &= \hbar^2\left\langle\mathcal{K}\omega_{\mathrm{p},2}\right\rangle_{\mathrm{cl}} - \hbar^2\left\langle\omega_{\mathrm{p},2}\right\rangle_{\mathrm{cl}}\langle\mathcal{K}\rangle_{\mathrm{cl}} + \mathcal{O}(\hbar^4) \\
&= \frac{\beta\hbar^2}{48m}[2-3N]\langle\nabla^2 U\rangle_{\mathrm{cl}} - \hbar^2\frac{-\beta^2}{24m}\langle\nabla^2 U\rangle_{\mathrm{cl}}\frac{3}{2}Nk_{\mathrm{B}}T \\
&= \frac{\beta\hbar^2}{24m}\langle\nabla^2 U\rangle_{\mathrm{cl}}.
\end{aligned} \tag{8.51}$$

This agrees with Wigner (1932, equation (30)).

8.3 Expansion 2

From the temperature derivative of the defining equation for the commutation function, equation (8.9), an expansion in powers of inverse temperature can be made

$$\omega_{\rm p} = \sum_{n=0}^{\infty} \tilde{\omega}_{{\rm p},n} \beta^n. \tag{8.52}$$

One has $\tilde{\omega}_{\rm p,0} = 1$, which is the classical limit, and $\tilde{\omega}_{\rm p,1} = 0$, since there are no terms of order β^0 on the right-hand side of the temperature derivative. Further,

$$\tilde{\omega}_{\rm p,2} = \frac{-\hbar^2}{4m}\nabla^2 U - \frac{{\rm i}\hbar}{2m}\mathbf{p}\cdot\nabla U, \tag{8.53}$$

and

$$\tilde{\omega}_{\rm p,3} = \frac{\hbar^2}{6m}\nabla U\cdot\nabla U - \frac{\hbar^4}{24m^2}\nabla^2\nabla^2 U - \frac{{\rm i}\hbar^3}{6m^2}\mathbf{p}\cdot\nabla\nabla^2 U + \frac{\hbar^2}{6m^2}\mathbf{pp}\colon \nabla\nabla U. \tag{8.54}$$

The recursion relation is

$$\begin{aligned} \tilde{\omega}_{{\rm p},n+1} =& \frac{-\hbar^2}{2(n+1)m}(\nabla^2 U)\tilde{\omega}_{{\rm p},n-1} - \frac{\hbar^2}{(n+1)m}\nabla U\cdot\nabla\tilde{\omega}_{{\rm p},n-1} \\ &+ \frac{\hbar^2}{2(n+1)m}(\nabla U\cdot\nabla U)\tilde{\omega}_{{\rm p},n-2} + \frac{\hbar^2}{2(n+1)m}\nabla^2\tilde{\omega}_{{\rm p},n} \\ &+ \frac{{\rm i}\hbar}{(n+1)m}\mathbf{p}\cdot\nabla\tilde{\omega}_{{\rm p},n} - \frac{{\rm i}\hbar}{(n+1)m}\mathbf{p}\cdot(\nabla U)\tilde{\omega}_{{\rm p},n-1}. \end{aligned} \tag{8.55}$$

8.4 Expansion 3

In the preceding sections expansions in powers of Planck's constant or in inverse temperature were given for the commutativity function $\omega_{\rm p}(\mathbf{\Gamma})$. The coefficients were determined from a recursion relation that arises from the temperature derivative of the defining equation. Alternatively, the commutation function can be re-defined as

$$\omega_{\rm p}(\mathbf{\Gamma}) \equiv e^{W_{\rm p}(\mathbf{\Gamma})}. \tag{8.56}$$

This was given as equation (8.6), where it was argued that in being extensive, W_p has an advantage over $\omega_{\rm p}$.

With this the temperature derivative of the defining equation (8.9) becomes

$$\begin{aligned} \frac{\partial W_{\rm p}}{\partial\beta} &= \frac{{\rm i}\hbar}{m}e^{\beta U - W_{\rm p}}\mathbf{p}\cdot\nabla\left\{e^{W_{\rm p}-\beta U}\right\} + \frac{\hbar^2}{2m}e^{\beta U - W_{\rm p}}\nabla^2\left\{e^{W_{\rm p}-\beta U}\right\} \\ &= \frac{{\rm i}\hbar}{m}\mathbf{p}\cdot\nabla(W_{\rm p} - \beta U) + \frac{\hbar^2}{2m}\{\nabla(W_{\rm p} - \beta U)\cdot\nabla(W_{\rm p} - \beta U) \\ &\quad + \nabla^2(W_{\rm p} - \beta U)\}. \end{aligned} \tag{8.57}$$

One can expand W_p in powers of Planck's constant,

$$W_{\mathrm{p}} \equiv \sum_{n=1}^{\infty} W_{\mathrm{p},n} \hbar^n. \tag{8.58}$$

This begins at $n = 1$ because the classical part must vanish, $\omega_{\mathrm{p}}(\hbar = 0) = 1 \Rightarrow W_{\mathrm{p}}(\hbar = 0) = 0$. Also $\omega_{\mathrm{p}}(\beta = 0) = 1 \Rightarrow W_{\mathrm{p},n}(\beta = 0) = 0$.

The first coefficient satisfies

$$\frac{\partial W_{\mathrm{p},1}}{\partial \beta} = \frac{-\mathrm{i}\beta}{m} \mathbf{p} \cdot \nabla U, \tag{8.59}$$

which gives

$$W_{\mathrm{p},1} = \frac{-\mathrm{i}\beta^2}{2m} \mathbf{p} \cdot \nabla U. \tag{8.60}$$

The second coefficient satisfies

$$\frac{\partial W_{\mathrm{p},2}}{\partial \beta} = \frac{\mathrm{i}}{m} \mathbf{p} \cdot \nabla W_{\mathrm{p},1} + \frac{1}{2m} \{ \beta^2 \nabla U \cdot \nabla U - \beta \nabla^2 U \}, \tag{8.61}$$

which gives

$$W_{\mathrm{p},2} = \frac{\beta^3}{6m^2} \mathbf{pp} \colon \nabla\nabla U + \frac{1}{2m} \left\{ \frac{\beta^3}{3} \nabla U \cdot \nabla U - \frac{\beta^2}{2} \nabla^2 U \right\}. \tag{8.62}$$

In general for $n > 2$

$$\begin{aligned} \frac{\partial W_{\mathrm{p},n}}{\partial \beta} = {} & \frac{\mathrm{i}}{m} \mathbf{p} \cdot \nabla W_{\mathrm{p},n-1} + \frac{1}{2m} \sum_{j=0}^{n-2} \nabla W_{\mathrm{p},n-2-j} \cdot \nabla W_{\mathrm{p},j} \\ & - \frac{\beta}{m} \nabla W_{\mathrm{p},n-2} \cdot \nabla U + \frac{1}{2m} \nabla^2 W_{\mathrm{p},n-2}. \end{aligned} \tag{8.63}$$

For $n = 3$ one can show that

$$W_{\mathrm{p},3} = \frac{\mathrm{i}\beta^4}{24m^3} \mathbf{ppp} \vdots \nabla\nabla\nabla U + \frac{5\mathrm{i}\beta^4}{24m^2} \mathbf{p}(\nabla U) \colon \nabla\nabla U - \frac{\mathrm{i}\beta^3}{6m^2} \mathbf{p} \cdot \nabla\nabla^2 U, \tag{8.64}$$

and for $n = 4$ one has

$$\begin{aligned} W_{\mathrm{p},4} = {} & \frac{-\beta^5}{5!m^4} (\mathbf{p} \cdot \nabla)^4 U - \frac{3\beta^5}{40m^3} (\nabla U) \mathbf{pp} \vdots \nabla\nabla\nabla U \\ & - \frac{\beta^5}{15m^2} (\nabla U)(\nabla U) \colon \nabla\nabla U + \frac{\beta^4}{16m^2} \nabla U \cdot \nabla\nabla^2 U \\ & + \frac{\beta^4}{16m^3} \mathbf{pp} \colon \nabla\nabla\nabla^2 U + \frac{\beta^4}{48m^2} \nabla^2 (\nabla U \cdot \nabla U) \\ & - \frac{\beta^3}{24m^2} \nabla^2\nabla^2 U - \frac{\beta^5}{15m^3} (\mathbf{p} \cdot \nabla\nabla U) \cdot (\mathbf{p} \cdot \nabla\nabla U). \end{aligned} \tag{8.65}$$

(Here $W_{\mathrm{p},4}$ is taken from Attard (2018a), which corrects equation (7.112) of Attard (2018b).)

Dealing with W has arguably two advantages over ω. First, experience shows that the expansion of an exponent converges more quickly than the expansion of an exponential. For example keeping only the terms W_1 and W_2 encompasses the terms ω_1 and ω_2, and an infinite number of higher order terms besides. Second, W is an extensive variable, as can be seen by inspection, whereas ω is a sum of terms of every power of the system size.

8.4.1 Second order analysis

In the first instance one can focus on the monomer grand potential, $\eta^{\pm}(\mathbf{\Gamma}) = 1$. (A completely analogous argument to that which follows can be applied to the loop grand potentials for $l \geqslant 2$.) The monomer grand potential is $\Omega_W = -k_{\mathrm{B}}T \ln \Xi_W$, with the monomer grand partition function being

$$\Xi_{e^W} = \sum_N \frac{z^N}{N!h^{3N}} \int \mathrm{d}\mathbf{\Gamma}\; e^{-\beta\mathcal{H}(\mathbf{\Gamma})} e^{W(\mathbf{\Gamma})}. \tag{8.66}$$

It makes no difference which of $W_{\mathrm{p}}(\mathbf{q}, \mathbf{p}) = W_{\mathrm{q}}(\mathbf{q}, \mathbf{p})^* = W_{\mathrm{p}}(\mathbf{q}, -\mathbf{p})^*$ is used here because the imaginary part integrates to zero.

The difference between the monomer grand potential and the classical grand potential is just the logarithm of the classical average of the commutation function,

$$-\beta[\Omega_{e^W} - \Omega_{\mathrm{cl}}] = \ln \left\langle e^W \right\rangle_{\mathrm{cl}}. \tag{8.67}$$

The left-hand side is extensive, which is to say that it scales with the volume of the system. Hence the grand potential density is independent of the size of the system. Since $W \propto V$, and since $W(\hbar = 0) = 0$, one can suppose that there exists a volume small enough that $W \ll 1$. For such a volume one can expand the exponential to obtain

$$\begin{aligned} -\beta[\Omega_{e^W} - \Omega_{\mathrm{cl}}] &= \ln \left\langle e^W \right\rangle_{\mathrm{cl}} = \ln \left\langle 1 + W + \frac{1}{2}W^2 + \cdots \right\rangle_{\mathrm{cl}} \\ &= \langle W \rangle_{\mathrm{cl}} + \frac{1}{2}\langle [W - \langle W \rangle_{\mathrm{cl}}]^2 \rangle_{\mathrm{cl}} + \cdots \end{aligned} \tag{8.68}$$

The first term on the right-hand side is extensive. The second term, as the fluctuation of an extensive variable, is also extensive. Having obtained the final result, one can now allow the volume to become as large as desired.

8.4.1.1 Explicit second order result

The first order in $\hbar$ correction vanishes, $-\beta\Delta\Omega_{e^W}^{(1)} = \hbar\langle W_1 \rangle_{\mathrm{cl}} = 0$, because W_1 is odd in $\mathbf{p}$. (Recall we have set the symmetrization function to unity.) The second order correction is given by

$$
\begin{aligned}
-\beta\Delta\Omega_{eW}^{(2)} &= \hbar^2\left\langle \frac{1}{2}W_1^2 + W_2\right\rangle_{\text{cl}}\\
&= \hbar^2\left\langle \frac{-1}{2}\frac{\beta^4}{4m^2}(\mathbf{p}\cdot\nabla U)^2 + \frac{\beta^3}{6m^2}\mathbf{pp}\colon\nabla\nabla U + \frac{\beta^3}{6m}\nabla U\cdot\nabla U - \frac{\beta^2}{4m}\nabla^2 U\right\rangle_{\text{cl}}\\
&= \hbar^2\left\langle \frac{-1}{2}\frac{\beta^4}{4m^2}mk_{\text{B}}T\nabla U\cdot\nabla U + \frac{\beta^3}{6m^2}mk_{\text{B}}T\nabla^2 U\right.\\
&\qquad \left. + \frac{\beta^3}{6m}\nabla U\cdot\nabla U - \frac{\beta^2}{4m}\nabla^2 U\right\rangle_{\text{cl}}
\end{aligned}
$$

$$
=\hbar^2\left\langle \frac{\beta^3}{24m}\nabla U\cdot\nabla U - \frac{\beta^2}{12m}\nabla^2 U\right\rangle_{\text{cl}} \tag{8.69a}
$$

$$
=\left\langle \frac{-\hbar^2\beta^2}{24m}\nabla^2 U\right\rangle_{\text{cl}}. \tag{8.69b}
$$

This is the same as the result given above, equation (8.26), which in turn agrees with that given by Wigner (1932) and by Kirkwood (1933). The present derivation is more satisfactory than the others because it explicitly takes into account extensivity. Also, the higher order terms lend themselves to practical implementation either as an expansion, or else as the original exponential combined with umbrella sampling.

The justification for the final equality is an integration by parts, assuming that surface integrals are negligible, as has been used above. This is tested explicitly by computer simulations in table 8.1, rows (a) and (b). The results labeled d correspond to averaging

Table 8.1. Quantum correction for monomers due to non-commutativity, $-\beta\Delta\Omega_{eW}^{(2)}\sigma^3/V$, for Lennard–Jones models of neon and helium in three dimensions at $k_{\text{B}}T/\epsilon = 1.5$ and $\rho\sigma^3 = 0.5$. For comparison, the classical pressure is $0.275\,75 \pm 0.000\,80$. The results are from Metropolis Monte Carlo simulations, with 50×100 configurations for averages, 20 trial steps per atom between averages, acceptance rate $\approx 50\%$, $R_{\text{cut}} = 3.5\sigma$, tail correction added, and no umbrella sampling, for a canonical system with $N =$ 250, 500, or 1000. The statistical error (68% confidence level) in the final digit is shown in parenthesis. Taken from Attard (2016).

	250	500	1000
	Neon		
a	−0.034 84(8)	−0.035 00(5)	−0.035 04(3)
b	−0.034 75(5)	−0.035 00(3)	−0.034 98(2)
d	−0.025 6(6)	−0.028 5(3)	−0.030 0(2)
	Helium		
a	−0.730(2)	−0.734(1)	−0.734 6(7)
b	−0.728(1)	−0.733 8(7)	−0.733 3(5)
d	−.38(1)	−0.518(7)	

the full exponential without umbrella sampling, $-\beta\Delta\Omega^{d}_{W_2} = \ln\langle \exp[\hbar W_1 + \hbar^2 W_2]\rangle_{\text{cl}}$. It can be seen that this approach has a significant system size dependence, and that in the case of helium numerical overflow occurs for the largest size studied.

8.4.2 Higher order analysis

8.4.2.1 Fourth order correction

Here the symmetrization function continues to be set to unity, $\eta^{\pm}(\mathbf{\Gamma}) = 1$. Again in this case it does not matters whether $W_{\text{p}}(\mathbf{\Gamma})$ or $W_{\text{q}}(\mathbf{\Gamma})$ is used.

The classical fluctuation of the commutation function exponent $W(\mathbf{\Gamma})$ is

$$\Delta_W(\mathbf{\Gamma}) \equiv W(\mathbf{\Gamma}) - \langle W\rangle_{\text{cl}}. \tag{8.70}$$

Define a phase space constant that is a series of averages of powers of the fluctuation

$$D[W] \equiv \langle W\rangle_{\text{cl}} + \frac{1}{2}\langle \Delta_W^2\rangle_{\text{cl}} + \frac{1}{3!}\langle \Delta_W^3\rangle_{\text{cl}} + \frac{1}{4!}\langle \Delta_W^4\rangle_{\text{cl}} - \frac{1}{8}\langle \Delta_W^2\rangle_{\text{cl}}^2. \tag{8.71}$$

This is constant in phase space, and so it can be taken in and out of averages. The average of the exponential of W less this particular constant is

$$\begin{aligned}
\langle e^{W-D[W]}\rangle_{\text{cl}} &= \left\langle \exp\left\{ \Delta_W - \frac{1}{2}\langle \Delta_W^2\rangle_{\text{cl}} - \frac{1}{3!}\langle \Delta_W^3\rangle_{\text{cl}} \right.\right.\\
&\qquad \left.\left. - \frac{1}{4!}\langle \Delta_W^4\rangle_{\text{cl}} + \frac{1}{8}\langle \Delta_W^2\rangle_{\text{cl}}^2 \right\}\right\rangle_{\text{cl}}\\
&= 1 + \langle \Delta_W\rangle_{\text{cl}} - \frac{1}{2}\langle \Delta_W^2\rangle_{\text{cl}} - \frac{1}{3!}\langle \Delta_W^3\rangle_{\text{cl}} - \frac{1}{4!}\langle \Delta_W^4\rangle_{\text{cl}}\\
&\quad + \frac{1}{8}\langle \Delta_W^2\rangle_{\text{cl}}^2 + \frac{1}{2}\left\langle \left[\Delta_W - \frac{1}{2}\langle \Delta_W^2\rangle_{\text{cl}}\right]^2\right\rangle_{\text{cl}}\\
&\quad + \frac{1}{3!}\left\langle \left[\Delta_W - \frac{1}{2}\langle \Delta_W^2\rangle_{\text{cl}}\right]^3\right\rangle_{\text{cl}} + \frac{1}{4!}\langle \Delta_W^4\rangle_{\text{cl}} + \cdots\\
&= 1 + \langle \mathcal{O}(\Delta_W^5)\rangle_{\text{cl}}.
\end{aligned} \tag{8.72}$$

This enables expansions to $\mathcal{O}(\hbar^4)$ to be obtained relatively painlessly. For example, the monomer potential is

$$\begin{aligned}
e^{-\beta[\Omega_{e^W} - \Omega_{\text{cl}}]} &= \frac{\Xi_{e^W}}{\Xi_{\text{cl}}} = \langle e^W\rangle_{\text{cl}} = e^{D[W]}\langle e^{W-D_W}\rangle_{\text{cl}}\\
&= e^{D[W]}\left[1 + \langle \mathcal{O}(\Delta_W^5)\rangle_{\text{cl}}\right].
\end{aligned} \tag{8.73}$$

Hence

$$-\beta\Omega_{e^W} = -\beta\Omega_{\rm cl} + D[W] + \mathcal{O}(h^5), \tag{8.74}$$

assuming that $W(\mathbf{\Gamma})$ itself has been obtained to at least fourth order.

8.4.2.2 Averages

The average of a 'pure' operator, $\hat{A} = A(\hat{\mathbf{q}}, \hat{\mathbf{p}}) = B(\hat{\mathbf{q}}) + C(\hat{\mathbf{p}})$ is

$$\begin{aligned}\langle \hat{A} \rangle_{\mu,V,T} &= \frac{1}{\Xi_{e^{W_{\rm p}}\eta_{\rm q}^{\pm}}} \sum_{N=0}^{\infty} \frac{1}{h^{3N} N!} \int \mathrm{d}\mathbf{\Gamma}\; e^{-\beta\mathcal{H}(\mathbf{\Gamma})} e^{W_{\rm p}(\mathbf{\Gamma})} \eta_{\rm q}^{\pm}(\mathbf{\Gamma}) A(\mathbf{q}, \mathbf{p}) \\ &= \frac{\left\langle e^{W_{\rm p}} \eta_{\rm q}^{\pm} A \right\rangle_{\rm cl}}{\left\langle e^{W_{\rm p}} \eta_{\rm q}^{\pm} \right\rangle_{\rm cl}}.\end{aligned} \tag{8.75}$$

Again neglect the symmetrization function, $\eta_{\rm q}^{\pm}(\mathbf{\Gamma}) = 1$, so it doesn't matter whether W_p or W_q is used. Define the fluctuation

$$\Delta_{AW} \equiv W - \langle W \rangle_A \equiv W - \frac{\langle AW \rangle_{\rm cl}}{\langle A \rangle_{\rm cl}}. \tag{8.76}$$

Define the phase space constant

$$D[AW] \equiv \langle W \rangle_A + \frac{1}{2}\langle \Delta_{AW}^2 \rangle_A + \frac{1}{3!}\langle \Delta_{AW}^3 \rangle_A + \frac{1}{4!}\langle \Delta_{AW}^4 \rangle_A - \frac{1}{8}\langle \Delta_{AW}^2 \rangle_A^2. \tag{8.77}$$

One has

$$\begin{aligned}\langle Ae^W \rangle_{\rm cl} &= \langle A \rangle_{\rm cl} \langle e^W \rangle_A \\ &= \langle A \rangle_{\rm cl} e^{D[AW]} \langle e^{W - D[AW]} \rangle_A \\ &= \langle A \rangle_{\rm cl} e^{D[AW]} [1 + \mathcal{O}(\hbar^5)].\end{aligned} \tag{8.78}$$

This result gives directly the monomer average of of a 'pure' operator.

For the loop potential one need only take $A \Rightarrow \mathring{\eta}_{\rm q}^{\pm(l)}$ (cf the discussion in section 7.2.5). With this the loop potential $l \geqslant 2$ can be expanded as

$$\begin{aligned}-\beta\Omega^{\pm(l)} &= \frac{\left\langle \mathring{\eta}_{\rm q}^{\pm(l)} e^{W_{\rm p}} \right\rangle_{\rm cl}}{\langle e^{W_{\rm p}} \rangle_{\rm cl}} = \frac{\left\langle \mathring{\eta}_{\rm q}^{\pm(l)} \right\rangle_{\rm cl} \langle e^{W_{\rm p}} \rangle_{\mathring{\eta}_{\rm q}^{\pm(l)}}}{\langle e^{W_{\rm p}} \rangle_{\rm cl}} \\ &= \left\langle \mathring{\eta}_{\rm q}^{\pm(l)} \right\rangle_{\rm cl} e^{D[\mathring{\eta}_{\rm q}^{\pm(l)} W_{\rm p}] - D[W_{\rm p}]} + \mathcal{O}(h^5).\end{aligned} \tag{8.79}$$

In equation (7.52), it was argued that the average of a pure operator could be factored into loop averages. Replacing ω by e^W the factorization can be written as

$$
\begin{aligned}
\langle \hat{A} \rangle_{\mu,V,T} &= \langle A \rangle_{e^{W_{\rm p}} \eta_{\rm q}^{\pm}} \\
&= \langle A \rangle_{e^{W_{\rm p}}} \prod_{\ell=2}^{\infty} \exp\left[\left\langle \overset{\circ}{\eta}{}_{\rm q}^{\pm(\ell)} \right\rangle_{Ae^{W_{\rm p}}} - \left\langle \overset{\circ}{\eta}{}_{\rm q}^{\pm(\ell)} \right\rangle_{e^{W_{\rm p}}} \right] \\
&= \frac{\langle Ae^{W_{\rm p}} \rangle_{\rm cl}}{\langle e^{W_{\rm p}} \rangle_{\rm cl}} \prod_{\ell=2}^{\infty} \exp\left[\frac{\left\langle \overset{\circ}{\eta}{}_{\rm q}^{\pm(\ell)} Ae^{W_{\rm p}} \right\rangle_{\rm cl}}{\langle Ae^{W_{\rm p}} \rangle_{\rm cl}} - \frac{\left\langle \overset{\circ}{\eta}{}_{\rm q}^{\pm(\ell)} e^{W_{\rm p}} \right\rangle_{\rm cl}}{\langle e^{W_{\rm p}} \rangle_{\rm cl}} \right] \\
&= \frac{\langle Ae^{W_{\rm p}} \rangle_{\rm cl}}{\langle e^{W_{\rm p}} \rangle_{\rm cl}} \prod_{\ell=2}^{\infty} \exp\left[\left\langle \overset{\circ}{\eta}{}_{\rm q}^{\pm(\ell)} \left[\frac{Ae^{W_{\rm p}}}{\langle Ae^{W_{\rm p}} \rangle_{\rm cl}} - \frac{e^{W_{\rm p}}}{\langle e^{W_{\rm p}} \rangle_{\rm cl}} \right] \right\rangle_{\rm cl} \right] \\
&\approx \frac{\langle Ae^{W_{\rm p}} \rangle_{\rm cl}}{\langle e^{W_{\rm p}} \rangle_{\rm cl}} \left[1 + \sum_{\ell=2}^{\infty} \left\langle \overset{\circ}{\eta}{}_{\rm q}^{\pm(\ell)} \left[\frac{Ae^{W_{\rm p}}}{\langle Ae^{W_{\rm p}} \rangle_{\rm cl}} - \frac{e^{W_{\rm p}}}{\langle e^{W_{\rm p}} \rangle_{\rm cl}} \right] \right\rangle_{\rm cl} \right].
\end{aligned}
\tag{8.80}
$$

The status of this as an exact or approximate result and its utility as a means to practical calculation are unclear. In the present context the classical averages that appear here involving $\overset{\circ}{\eta}{}_{\rm q}^{\pm(\ell)} Ae^{W_{\rm p}}$ can be evaluated using $D[\overset{\circ}{\eta}{}_{\rm q}^{\pm(l)} A\, W_{\rm p}]$. Whether there is any computational advantage in doing so is also unclear.

8.5 Fluctuation expansion

In this section an alternative derivation of the temperature expansion for the commutation function is given based on the fluctuations in the phase space expectation value of the energy operator. This lends itself rather directly to a recursion relation for the coefficients and to their physical interpretation. The fluctuation expansion may be rewritten as an exponentiated series, which becomes a temperature expansion for the commutation function exponent, the terms of which are extensive with system size and represent fluctuations of fluctuations.

8.5.1 Hamiltonian fluctuations as a phase function

Define

$$
\mathcal{H}^{(n)}(\mathbf{q}, \mathbf{p}) \equiv \frac{\langle \mathbf{q} | \hat{\mathcal{H}}^n | \mathbf{p} \rangle}{\langle \mathbf{q} | \mathbf{p} \rangle}. \tag{8.81}
$$

One has $\mathcal{H}^{(1)}(\mathbf{q}, \mathbf{p}) = \mathcal{H}(\mathbf{q}, \mathbf{p})$, the classical Hamiltonian. Also define the nth order fluctuation

$$
\Delta_{\mathcal{H}}^{(n)}(\mathbf{q}, \mathbf{p}) \equiv \frac{\langle \mathbf{q} | [\hat{\mathcal{H}} - \mathcal{H}(\mathbf{q}, \mathbf{p})]^n | \mathbf{p} \rangle}{\langle \mathbf{q} | \mathbf{p} \rangle}. \tag{8.82}
$$

One has $\Delta_{\mathcal{H}}^{(0)}(\mathbf{q}, \mathbf{p}) = 1$ and $\Delta_{\mathcal{H}}^{(1)}(\mathbf{q}, \mathbf{p}) = 0$.

The phase space density (sans symmetrization) may be expressed as a series of these fluctuations, namely

$$\begin{aligned} e^{-\beta\mathcal{H}(\mathbf{q},\,\mathbf{p})}\omega_{\mathrm{p}}(\mathbf{q},\,\mathbf{p}) &\equiv \frac{\langle \mathbf{q}|e^{-\beta\hat{\mathcal{H}}}|\mathbf{p}\rangle}{\langle \mathbf{q}|\mathbf{p}\rangle} \\ &= e^{-\beta\mathcal{H}(\mathbf{q},\,\mathbf{p})}\frac{\langle \mathbf{q}|e^{-\beta[\hat{\mathcal{H}}-\mathcal{H}(\mathbf{q},\,\mathbf{p})]}|\mathbf{p}\rangle}{\langle \mathbf{q}|\mathbf{p}\rangle} \\ &= e^{-\beta\mathcal{H}(\mathbf{q},\,\mathbf{p})}\sum_{n=0}^{\infty}\frac{(-\beta)^n}{n!}\Delta_{\mathcal{H}}^{(n)}(\mathbf{q},\,\mathbf{p}). \end{aligned} \tag{8.83}$$

The second equality follows because the Hamiltonian operator and Hamiltonian phase function commute, $[\hat{\mathcal{H}}(\mathbf{r}),\,\mathcal{H}(\mathbf{q},\,\mathbf{p})] = 0$. This gives the commutation function phase space density as a fluctuation expansion,

$$\omega_{\mathrm{p}}(\mathbf{q},\,\mathbf{p}) = 1 + \sum_{n=2}^{\infty}\frac{(-\beta)^n}{n!}\Delta_{\mathcal{H}}^{(n)}(\mathbf{q},\,\mathbf{p}). \tag{8.84}$$

Evidently this could also be called a high temperature expansion with a physical interpretation for the coefficients.

8.5.2 Exponentiated fluctuation series

One may modify this fluctuation series by writing it instead as

$$\frac{\langle \mathbf{q}|e^{-\beta\hat{\mathcal{H}}}|\mathbf{p}\rangle}{\langle \mathbf{q}|\mathbf{p}\rangle} = e^{-\beta\mathcal{H}(\mathbf{q},\,\mathbf{p})}\prod_{n=2}^{\infty} e^{((-\beta)^n/n!)\tilde{\Delta}_{\mathcal{H}}^{(n)}(\mathbf{q},\mathbf{p})}. \tag{8.85}$$

One has $\tilde{\Delta}_{\mathcal{H}}^{(2)} = \Delta_{\mathcal{H}}^{(2)}$, $\tilde{\Delta}_{\mathcal{H}}^{(3)} = \Delta_{\mathcal{H}}^{(3)}$, and

$$\tilde{\Delta}_{\mathcal{H}}^{(4)}(\mathbf{q},\,\mathbf{p}) = \Delta_{\mathcal{H}}^{(4)}(\mathbf{q},\,\mathbf{p}) - \frac{4!}{2^3}\Delta_{\mathcal{H}}^{(2)}(\mathbf{q},\,\mathbf{p})^2. \tag{8.86}$$

This is like the fluctuation of a fluctuation. As such it may be expected to be smaller than the fourth order fluctuation itself.

There are similar modifications for the higher order fluctuations. The series of modified fluctuations may be equated to the commutation function exponent, $W_{\mathrm{p}}(\mathbf{q},\,\mathbf{p}) = \sum_{n=2}^{\infty}((-\beta)^n/n!)\tilde{\Delta}_{\mathcal{H}}^{(n)}(\mathbf{q},\,\mathbf{p})$. Terminating the modified fluctuation series at order n can be expected to be yield more accurate results than terminating the original fluctuation series at the same order. This is because the first neglected term, $\tilde{\Delta}_{\mathcal{H}}^{(n+1)}$, is on average smaller in magnitude than $\Delta_{\mathcal{H}}^{(n+1)}$.

The fluctuation expansion coefficients $\Delta_{\mathcal{H}}^{(n)}(\mathbf{q},\,\mathbf{p})$ are not in general extensive with system size. (The $n = 2$ and $n = 3$ terms derived explicitly below are extensive with system size. The term $\Delta_{\mathcal{H}}^{(4)}$ is not extensive, but the term $\tilde{\Delta}_{\mathcal{H}}^{(4)}$ is.) The exponentiated fluctuation expansion is extensive with system size. The fluctuation expansion (8.84) is directly comparable to expansion 2 (section 8.3). The exponentiated expansion

(8.85) is indirectly comparable to expansion 3 (section 8.4). (The difference is that the former is in powers of inverse temperature, whereas the latter is in powers of Planck's constant.)

It should be mentioned that the author has been unsuccessful in the computational implementation of the fourth order exponentiated fluctuation series for a Lennard–Jones system. The problem appears to lie in the core region, where analysis indicates that the most divergent terms are the products of gradients, $\tilde{\Delta}_{\mathcal{H}}^{(n)} \sim \mathcal{O}((\nabla U)^{\lfloor 2n/3 \rfloor})$, and that the infinite series of these is divergent in the core region.

8.5.3 Explicit form for the fluctuations

One can write

$$\begin{aligned}
\mathcal{H}^{(n)}(\mathbf{q}, \mathbf{p}) &= \frac{\langle \mathbf{q} | \hat{\mathcal{H}}^n | \mathbf{p} \rangle}{\langle \mathbf{q} | \mathbf{p} \rangle} \\
&= \left[\mathcal{H}(\mathbf{q}, \mathbf{p}) - \frac{\hbar^2}{2m} \nabla^2 - \frac{\mathrm{i}\hbar}{m} \mathbf{p} \cdot \nabla \right] \mathcal{H}^{(n-1)}(\mathbf{q}, \mathbf{p}) \\
&= \left[\mathcal{H}(\mathbf{q}, \mathbf{p}) - \frac{\hbar^2}{2m} \nabla^2 - \frac{\mathrm{i}\hbar}{m} \mathbf{p} \cdot \nabla \right]^{(n-1)} \mathcal{H}(\mathbf{q}, \mathbf{p}).
\end{aligned} \tag{8.87}$$

Using this or directly, the phase space expectation value of the square of the energy operator is explicitly

$$\begin{aligned}
\mathcal{H}^{(2)}(\mathbf{q}, \mathbf{p}) &= \left[\mathcal{H}(\mathbf{q}, \mathbf{p}) - \frac{\hbar^2}{2m} \nabla^2 - \frac{\mathrm{i}\hbar}{m} \mathbf{p} \cdot \nabla \right] \mathcal{H}(\mathbf{q}, \mathbf{p}) \\
&= \mathcal{H}(\mathbf{q}, \mathbf{p})^2 - \frac{\hbar^2}{2m} \nabla^2 U(\mathbf{q}) - \frac{\mathrm{i}\hbar}{m} \mathbf{p} \cdot \nabla U(\mathbf{q}),
\end{aligned} \tag{8.88}$$

and that of the cube is

$$\begin{aligned}
&\mathcal{H}^{(3)}(\mathbf{q}, \mathbf{p}) \\
&= \mathcal{H}\mathcal{H}^{(2)} - \frac{\hbar^2}{m} \mathcal{H} \nabla^2 U - \frac{\hbar^2}{m} \nabla U \cdot \nabla U + \frac{\hbar^4}{4m^2} \nabla^2 \nabla^2 U + \frac{\mathrm{i}\hbar^3}{2m^2} \mathbf{p} \cdot \nabla \nabla^2 U \\
&\quad - \frac{2\mathrm{i}\hbar}{m} \mathcal{H} \mathbf{p} \cdot \nabla U + \frac{\mathrm{i}\hbar^3}{2m^2} \mathbf{p} \cdot \nabla \nabla^2 U - \frac{\hbar^2}{m^2} \mathbf{p}\mathbf{p} : \nabla \nabla U.
\end{aligned} \tag{8.89}$$

The second order fluctuation is explicitly

$$\begin{aligned}
\Delta_{\mathcal{H}}^{(2)}(\mathbf{q}, \mathbf{p}) &\equiv \frac{\langle \mathbf{q} | [\hat{\mathcal{H}} - \mathcal{H}(\mathbf{q}, \mathbf{p})]^2 | \mathbf{p} \rangle}{\langle \mathbf{q} | \mathbf{p} \rangle} \\
&= \mathcal{H}^{(2)}(\mathbf{q}, \mathbf{p}) - \mathcal{H}(\mathbf{q}, \mathbf{p})^2 \\
&= \frac{-\hbar^2}{2m} \nabla^2 U(\mathbf{q}) - \frac{\mathrm{i}\hbar}{m} \mathbf{p} \cdot \nabla U(\mathbf{q}).
\end{aligned} \tag{8.90}$$

This is clearly $\mathcal{O}(N)$, which proves that the (root mean square) fluctuations are relatively negligible in the thermodynamic limit. The third order fluctuation is explicitly

$$\begin{aligned}
\Delta_{\mathcal{H}}^{(3)}(\mathbf{q}, \mathbf{p}) \\
&= \frac{\langle \mathbf{q}|[\hat{\mathcal{H}} - \mathcal{H}(\mathbf{q}, \mathbf{p})]^3|\mathbf{p}\rangle}{\langle \mathbf{q}|\mathbf{p}\rangle} \\
&= \mathcal{H}^{(3)}(\mathbf{q}, \mathbf{p}) - 3\mathcal{H}^{(2)}(\mathbf{q}, \mathbf{p})\mathcal{H}(\mathbf{q}, \mathbf{p}) + 2\mathcal{H}(\mathbf{q}, \mathbf{p})^3 \\
&= \frac{-\hbar^2}{m}\nabla U \cdot \nabla U + \frac{\hbar^4}{4m^2}\nabla^2\nabla^2 U + \frac{i\hbar^3}{m^2}\mathbf{p} \cdot \nabla\nabla^2 U - \frac{\hbar^2}{m^2}\mathbf{pp}\colon \nabla\nabla U.
\end{aligned} \tag{8.91}$$

This is also clearly $\mathcal{O}(N)$.

8.5.4 Recursion relation

One can extend the definition of the fluctuation phase space function in a way that lends itself to a recursion relation, namely

$$\begin{aligned}
\Delta_{\mathcal{H}}^{(n)}(\mathbf{r}; \mathbf{q}, \mathbf{p}) \\
&\equiv e^{\mathbf{r}\cdot\mathbf{p}/i\hbar}[\hat{\mathcal{H}}(\mathbf{r}) - \mathcal{H}(\mathbf{q}, \mathbf{p})]^n e^{-\mathbf{r}\cdot\mathbf{p}/i\hbar} \\
&= e^{\mathbf{r}\cdot\mathbf{p}/i\hbar}[\hat{\mathcal{H}}(\mathbf{r}) - \mathcal{H}(\mathbf{q}, \mathbf{p})]\left(e^{-\mathbf{r}\cdot\mathbf{p}/i\hbar}\Delta_{\mathcal{H}}^{(n-1)}(\mathbf{r}; \mathbf{q}, \mathbf{p})\right) \\
&= -\mathcal{H}(\mathbf{q}, \mathbf{p})\Delta_{\mathcal{H}}^{(n-1)}(\mathbf{r}; \mathbf{q}, \mathbf{p}) + [U(\mathbf{r}) + \mathcal{K}(\mathbf{p})]\Delta_{\mathcal{H}}^{(n-1)}(\mathbf{r}; \mathbf{q}, \mathbf{p}) \\
&\quad - \frac{\hbar^2}{2m}\nabla_{\mathbf{r}}^2\Delta_{\mathcal{H}}^{(n-1)}(\mathbf{r}; \mathbf{q}, \mathbf{p}) - \frac{i\hbar}{m}\mathbf{p} \cdot \nabla_{\mathbf{r}}\Delta_{\mathcal{H}}^{(n-1)}(\mathbf{r}; \mathbf{q}, \mathbf{p}) \\
&= \left\{U(\mathbf{r}) - U(\mathbf{q}) - \frac{\hbar^2}{2m}\nabla_{\mathbf{r}}^2 - \frac{i\hbar}{m}\mathbf{p} \cdot \nabla_{\mathbf{r}}\right\}\Delta_{\mathcal{H}}^{(n-1)}(\mathbf{r}; \mathbf{q}, \mathbf{p}).
\end{aligned} \tag{8.92}$$

It is convenient to define $U_\Delta(\mathbf{r}; \mathbf{q}) \equiv U(\mathbf{r}) - U(\mathbf{q})$. The actual fluctuation is given by

$$\Delta_{\mathcal{H}}^{(n)}(\mathbf{q}, \mathbf{p}) = \Delta_{\mathcal{H}}^{(n)}(\mathbf{r}; \mathbf{q}, \mathbf{p})\Big|_{\mathbf{r}=\mathbf{q}}. \tag{8.93}$$

In section 8.3, the recursion relation derived from the temperature derivative and power series expression is given. It ought to be directly comparable to the present recursion relation, with $\beta^n\tilde{\omega}_{\mathrm{p},n} = ((-\beta)^n/n!)\Delta_{\mathcal{H}}^{(n)}(\mathbf{q}, \mathbf{p})$.

Starting with $\Delta_{\mathcal{H}}^{(0)}(\mathbf{r}; \mathbf{q}, \mathbf{p}) = 1$, the recursion relation yields the first order fluctuation as

$$\Delta_{\mathcal{H}}^{(1)}(\mathbf{r}; \mathbf{q}, \mathbf{p}) = \mathcal{H}(\mathbf{r}, \mathbf{p}) - \mathcal{H}(\mathbf{q}, \mathbf{p}) = U_\Delta(\mathbf{r}; \mathbf{q}), \tag{8.94}$$

which may be confirmed directly from the definition. The second order fluctuation function is

$$\begin{aligned}\Delta_{\mathcal{H}}^{(2)}(\mathbf{r};\,\mathbf{q},\,\mathbf{p}) &= \left\{ U_\Delta(\mathbf{r};\,\mathbf{q}) - \frac{\hbar^2}{2m}\nabla_\mathbf{r}^2 - \frac{i\hbar}{m}\mathbf{p}\cdot\nabla_\mathbf{r} \right\} U_\Delta(\mathbf{r};\,\mathbf{q}) \\ &= U_\Delta(\mathbf{r};\,\mathbf{q})^2 - \frac{\hbar^2}{2m}\nabla_\mathbf{r}^2 U(\mathbf{r}) - \frac{i\hbar}{m}\mathbf{p}\cdot\nabla_\mathbf{r} U(\mathbf{r}).\end{aligned} \tag{8.95}$$

At $\mathbf{r}=\mathbf{q}$ this agrees with the result obtained explicitly above. Note that $\nabla_\mathbf{r} U_\Delta(\mathbf{r};\,\mathbf{q}) = \nabla_\mathbf{r} U(\mathbf{r})$. The third order fluctuation function is

$$\begin{aligned}&\Delta_{\mathcal{H}}^{(3)}(\mathbf{r};\,\mathbf{q},\,\mathbf{p}) \\ &= \left\{ U_\Delta(\mathbf{r};\,\mathbf{q}) - \frac{\hbar^2}{2m}\nabla_\mathbf{r}^2 - \frac{i\hbar}{m}\mathbf{p}\cdot\nabla_\mathbf{r} \right\} \\ &\quad\times \left\{ U_\Delta(\mathbf{r};\,\mathbf{q})^2 - \frac{\hbar^2}{2m}\nabla_\mathbf{r}^2 U(\mathbf{r}) - \frac{i\hbar}{m}\mathbf{p}\cdot\nabla_\mathbf{r} U(\mathbf{r}) \right\} \\ &= U_\Delta(\mathbf{r};\,\mathbf{q}) \left\{ U_\Delta(\mathbf{r};\,\mathbf{q})^2 - \frac{\hbar^2}{2m}\nabla_\mathbf{r}^2 U(\mathbf{r}) - \frac{i\hbar}{m}\mathbf{p}\cdot\nabla_\mathbf{r} U(\mathbf{r}) \right\} \\ &\quad - \frac{\hbar^2}{2m}\nabla_\mathbf{r}^2 U_\Delta(\mathbf{r};\,\mathbf{q})^2 - \frac{i\hbar}{m}\mathbf{p}\cdot\nabla_\mathbf{r} U_\Delta(\mathbf{r};\,\mathbf{q})^2 \\ &\quad + \frac{\hbar^4}{4m^2}\nabla_\mathbf{r}^2\nabla_\mathbf{r}^2 U(\mathbf{r}) + \frac{i\hbar^3}{m^2}\mathbf{p}\cdot\nabla_\mathbf{r}\nabla_\mathbf{r}^2 U(\mathbf{r}) - \frac{\hbar^2}{m^2}\mathbf{pp}:\,\nabla_\mathbf{r}\nabla_\mathbf{r} U(\mathbf{r}).\end{aligned} \tag{8.96}$$

At $\mathbf{r}=\mathbf{q}$ this gives

$$\Delta_{\mathcal{H}}^{(3)}(\mathbf{q},\,\mathbf{p}) = \frac{-\hbar^2}{m}\nabla U\cdot\nabla U + \frac{\hbar^4}{4m^2}\nabla^2\nabla^2 U + \frac{i\hbar^3}{m^2}\mathbf{p}\cdot\nabla\nabla^2 U - \frac{\hbar^2}{m^2}\mathbf{pp}:\,\nabla\nabla U, \tag{8.97}$$

which agrees with the result obtained directly above.

The fourth order fluctuation function may be shown to be

$$\begin{aligned}\Delta_{\mathcal{H}}^{(4)}(\mathbf{q},\,\mathbf{p}) &= 3\Delta_{\mathcal{H}}^{(2)}(\mathbf{q},\,\mathbf{p})^2 + \frac{5\hbar^4}{2m^2}\nabla U\cdot\nabla\nabla^2 U + \frac{5i\hbar^3}{m^2}\mathbf{p}\nabla U:\,\nabla\nabla U \\ &\quad + \frac{\hbar^4}{m^2}\nabla\nabla U:\nabla\nabla U - \frac{\hbar^6}{8m^3}\nabla^2\nabla^2\nabla^2 U - \frac{3i\hbar^5}{4m^3}\mathbf{p}\cdot\nabla\nabla^2\nabla^2 U \\ &\quad + \frac{3\hbar^4}{2m^3}\mathbf{pp}:\,\nabla\nabla\nabla^2 U + \frac{i\hbar^3}{m^3}\mathbf{ppp}\,\vdots\,\nabla\nabla\nabla U.\end{aligned} \tag{8.98}$$

This can be shown to agree with the fourth order result independently obtained from the nested commutator expansion, equations (9.63) and (9.65). Apart from the first term, all terms here are $\mathcal{O}(N)$. One can see that the first term cancels in the modified fluctuation, which supports describing $\tilde{\Delta}_{\mathcal{H}}^{(4)}(\mathbf{q},\,\mathbf{p})$ as the fluctuation of a fluctuation. Expressions for the higher order gradients of singlet and pair power law potentials are given in sections 9.5.2 and 9.5.3.

The fourth order result was derived from the recursion in the form

$$\begin{aligned} \Delta_{\mathcal{H}}^{(n)}(\mathbf{q}, \mathbf{p}) &= \left\{ \frac{-\hbar^2}{2m} \nabla_{\mathbf{r}}^2 - \frac{\mathrm{i}\hbar}{m} \mathbf{p} \cdot \nabla_{\mathbf{r}} \right\} \Delta_{\mathcal{H}}^{(n-1)}(\mathbf{r}; \mathbf{q}, \mathbf{p}) \bigg|_{\mathbf{r}=\mathbf{q}} \\ &= \left\{ \frac{-\hbar^2}{2m} \nabla_{\mathbf{r}}^2 - \frac{\mathrm{i}\hbar}{m} \mathbf{p} \cdot \nabla_{\mathbf{r}} \right\} \\ &\quad \left[\left\{ U_{\Delta}(\mathbf{r}; \mathbf{q}) - \frac{\hbar^2}{2m} \nabla_{\mathbf{r}}^2 - \frac{\mathrm{i}\hbar}{m} \mathbf{p} \cdot \nabla_{\mathbf{r}} \right\} \Delta_{\mathcal{H}}^{(n-2)}(\mathbf{r}; \mathbf{q}, \mathbf{p}) \right] \bigg|_{\mathbf{r}=\mathbf{q}} \\ &= \left\{ \frac{-\hbar^2}{2m} \nabla^2 U - \frac{\mathrm{i}\hbar}{m} \mathbf{p} \cdot \nabla U \right\} \Delta_{\mathcal{H}}^{(n-2)}(\mathbf{q}, \mathbf{p}) - \frac{\hbar^2}{m} \nabla U \cdot \nabla \Delta_{\mathcal{H}}^{(n-2)}(\mathbf{q}, \mathbf{p}) \\ &\quad + \left[\left\{ \frac{\hbar^2}{2m} \nabla_{\mathbf{r}}^2 + \frac{\mathrm{i}\hbar}{m} \mathbf{p} \cdot \nabla_{\mathbf{r}} \right\}^2 \Delta_{\mathcal{H}}^{(n-2)}(\mathbf{r}; \mathbf{q}, \mathbf{p}) \right] \bigg|_{\mathbf{r}=\mathbf{q}} . \end{aligned} \tag{8.99}$$

Retaining only the first term of this yields

$$\Delta_{\mathcal{H}}^{(2n)}(\mathbf{q}, \mathbf{p}) \approx \left\{ \frac{-\hbar^2}{2m} \nabla^2 U - \frac{\mathrm{i}\hbar}{m} \mathbf{p} \cdot \nabla U \right\}^n , \tag{8.100}$$

and

$$\begin{aligned} \omega_{\mathrm{p}}(\mathbf{q}, \mathbf{p}) &\approx \sum_{n=0}^{\infty} \frac{(-\beta)^{2n}}{(2n)!} \left\{ \frac{-\hbar^2}{2m} \nabla^2 U - \frac{\mathrm{i}\hbar}{m} \mathbf{p} \cdot \nabla U \right\}^n \\ &= \cosh \left[\beta \left\{ \frac{-\hbar^2}{2m} \nabla^2 U - \frac{\mathrm{i}\hbar}{m} \mathbf{p} \cdot \nabla U \right\}^{1/2} \right] . \end{aligned} \tag{8.101}$$

This is of course complex, with an apparently immaterial branch cut for the square root. The utility of this approximation is not clear; it does not, for example, give the correct asymptote in the core region, section 11.4.2.

8.5.5 Advantages and disadvantages of the fluctuation expansion

The fluctuations are phase functions. Hence, for example, the commutation function can be evaluated during the course of a computer simulation. Methods for obtaining the commutation function by summing over energy states or by numerical temperature quadrature (see, for example, chapters 10 and 11) necessitate pre-computing the commutation function and storing it on a grid in phase space or few-body sub-space thereof.

The fluctuations are localized. In contrast to summing over energy states methods they don't need boundary conditions to be specified, and they don't require the calculation of a set of few-body energy eigenvalues and eigenfunctions.

The fluctuations involve the gradients of the potential. As such this form of the commutation function is more short-ranged than the potential, and is likely to be rapidly converging at large separations.

A possible disadvantage is that the required gradients of the potential diverge more rapidly than the potential itself in the core region. However, the fluctuation series is multiplied by the Maxwell–Boltzmann factor, which goes to zero exponentially in the core region. The increasingly divergent terms in the core region, $\tilde{\Delta}_{\mathcal{H}}^{(n)} \sim \mathcal{O}((\nabla U)^{\lfloor 2n/3 \rfloor})$, pose serious difficulties for the exponentiated form of the series. It is unclear whether or not the known analytic core asymptote (see section 11.4.2 for the core asymptote of the Lennard–Jones pair potential) can be achieved numerically.

The exponentiated fluctuation expansion is extensive with system size. This is likely advantageous both from a thermodynamic viewpoint and from a computational aspect. The exponentiation is equivalent to an infinite resummation, which means that it is likely more rapidly converging than a strict temperature expansion such as that given in sections 8.2 or 8.3. This idea is supported by the fact that the modified form (i.e., the expansion of the exponent) is in the form of a fluctuation of a fluctuation, which might be expected to be smaller at a given order than the fluctuation itself.

The fluctuation expansion (8.84) is directly comparable to expansion 2 (section 8.4). The exponentiated expansion (8.85) is indirectly comparable to expansion 3 (section 8.4). Therefore the present fluctuation series can be said to provide a physical interpretation and basis for the earlier expansions.

Possible disadvantages of the fluctuation expansion (and the related high temperature expansions) are that they may be slowly converging, and the higher order terms grow rapidly in complexity and number of contributions. For the case of a power law potential, evaluating the gradients analytically is not a problem. However for the case of a hard-sphere potential, some sort of resummation will have to be invoked for the potential gradients. Expressions for the higher order gradients of singlet and pair power law potentials are given in sections 9.5.2 and 9.5.3.

8.6 Numerical results

8.6.1 Interacting Lennard–Jones particles

The fluctuation form of the high temperature expansion has been used to perform computer simulations of interacting particles. The numerical result are for a one-dimensional system with the simple harmonic oscillator singlet potential and Lennard–Jones pair potential. The singlet potential is that of a simple harmonic oscillator,

$$u^{(1)}(q_j) = \frac{1}{2} m \omega^2 q_j^2, \tag{8.102}$$

and the pair potential is Lennard–Jones,

$$u^{(2)}(q_{jk}) = \varepsilon \left[\frac{r_{\mathrm{e}}^{12}}{q_{jk}^{12}} - 2 \frac{r_{\mathrm{e}}^{6}}{q_{jk}^{6}} \right]. \tag{8.103}$$

The system parameters are the same as in Hernando and Vaníček (2013). The so-called de Broglie wavelength is fixed at $\Lambda_{\text{dB}} \equiv 2^{1/6}\hbar/r_e\sqrt{m\varepsilon} = 0.16$, and the frequency is fixed by $\omega r_e\sqrt{m/\varepsilon} = 1/2$. These mean that $\varepsilon/\hbar\omega = 14.03$. The results scale the energy and temperature by the Lennard–Jones frequency, which equals $\omega_{\text{LJ}} \equiv \sqrt{u_{\text{LJ}}''(r_e)/m} = 16.97\omega$ for the present parameters.

The standard Metropolis algorithm in classical phase space was used, with the classical Maxwell–Boltzmann factor used for umbrella weight. The quantum contribution to the weight, both the commutation function and the symmetrization function, was explicitly included for the respective averages. Eighteen averages for each quantity were collected simultaneously: three averages for symmetrization (distinguishable particles, bosons, and fermions), and six high temperature averages (classical, linear $\mathcal{O}(\beta^2)$, $\mathcal{O}(\beta^3)$, $\mathcal{O}(\beta^4)$, and exponentiated $\mathcal{O}(\beta^3)$ and $\mathcal{O}(\beta^4)$).

It was found that umbrella sampling on the classical Maxwell–Boltzmann distribution was unsuitable for the exponentiated averages due to numerical overflow in the exponent of the quantum weight, and no results are reported for these. Using instead the real part of the fourth order fluctuation series exponent as the weight for the Metropolis sampling gave unphysical results as it favored overlap of the cores of the Lennard–Jones particles. These unpublished computational results of the author support the proposition that the finite order exponentiated fluctuation series is non-viable for the Lennard–Jones potential.

Table 8.2 shows the results of the Monte Carlo simulations of these Lennard–Jones particles. It can be seen that for the highest temperatures shown, $\beta\hbar\omega_{\text{LJ}} \lesssim 0.5$, the fluctuation series appears to converge, with successive quantum corrections decreasing in magnitude. For temperatures lower than this the corrections tend to be large and to change sign, and the final result is of doubtful reliability. One can conclude from the data in the table that the average energy for the quantum system is larger than the classical prediction, and that it approaches the classical limit from above as the temperature is increased.

At high temperatures, $\beta \to 0$, the Lennard–Jones potential has negligible effect, and so the particles behave like an ideal gas in a confining singlet simple harmonic

Table 8.2. Average energy, $\beta\langle\mathcal{H}\rangle$, of a canonical equilibrium one-dimensional Lennard–Jones system for $N = 4$ un-symmetrized particles. Monte Carlo results are given for the linear high temperature expansion, equation (8.84), to the given order. The statistical error (96% confidence level, twice the standard deviation of block sub-averages) is shown in parentheses for the final two digits. The exact limiting result is $\beta\langle\mathcal{H}\rangle \to N = 4$, $\beta \to 0$.

$\beta\hbar\omega_{\text{LJ}}$	classical	$\mathcal{O}(\beta^2)$	$\mathcal{O}(\beta^3)$	$\mathcal{O}(\beta^4)$
0.033 94	4.126 5(79)	4.125 7(79)	4.129 0(79)	4.128 1(79)
0.084 85	4.188 4(48)	4.185 7(48)	4.198 5(48)	4.200 5(48)
0.169 7	4.258 7(79)	4.252 6(80)	4.290 1(79)	4.320 6(78)
0.339 4	4.293 9(75)	4.285 6(78)	4.407 0(74)	4.597 5(70)
0.509 1	4.260 7(45)	4.270 3(48)	4.555 7(44)	4.879 7(43)
0.678 8	4.161 7(42)	4.237 6(48)	4.958 3(41)	5.016 7(44)
0.848 5	4.002(18)	4.296(24)	8.172(76)	4.991(19)
1.018 2	3.807 9(36)	5.417(11)	0.293 3(86)	4.863 9(41)

oscillator potential. One can't neglect the latter because the particles can always access sufficiently large displacements where it becomes comparable to or larger than k_BT. Hence in the high temperature limit the classical system has two quadratic degrees of freedom per particle (the kinetic energy, and the oscillator energy). According to the equipartition theorem (Attard 2012, section 3.7.4), these each contribute to the average energy an amount $k_BT/2$. Therefore the average energy of the system in the high temperature limit is Nk_BT, or $\beta\langle\mathcal{H}\rangle = 4$ in this case.

The data in table 8.2 indicate that with increasing temperature, the quantum results approach the respective classical result faster than the classical results approach the classical limit.

The preceding argument leads one to conclude that the spatial extent of this particular system increases with increasing temperature. This means that high temperatures pose numerical challenges for the calculation of the commutation function in the local state approximation, chapter 10, since it has to be stored on a grid that encompasses the likely configurations of the system. At lower temperatures than those in table 8.2 the calculations are much easier (see section 10.5.2).

8.6.2 Simple harmonic oscillator

Using $\hbar\omega$ as the unit of energy, and dimensionless momentum and position operators, $\hat{P} = \hat{p}/\sqrt{m\hbar\omega}$ and $\hat{Q} = \sqrt{m\omega/\hbar}\,\hat{q}$, the Hamiltonian operator for the simple harmonic oscillator may be written (Messiah 1961)

$$\hat{\mathcal{H}} = \frac{1}{2}\left\{\hat{P}^2 + \hat{Q}^2\right\}. \tag{8.104}$$

In these dimensionless units, $\hbar = m = 1$.

For one particle in d dimensions, $\nabla U = \mathbf{Q}$ and $\nabla^2 U = d$. Hence with $P^2 \equiv \mathbf{P}\cdot\mathbf{P}$, $Q^2 \equiv \mathbf{Q}\cdot\mathbf{Q}$, and $R \equiv \mathbf{P}\cdot\mathbf{Q}$, the recursion relation for the temperature expansion coefficients of the simple harmonic oscillator, $\omega_\mathrm{p} = \sum \omega_{\mathrm{p},n}\beta^n$, is

$$\begin{aligned}\omega_{\mathrm{p},n+1} = &\frac{-d}{2(n+1)}\omega_{\mathrm{p},n-1} - \frac{1}{n+1}\mathbf{Q}\cdot\nabla\omega_{\mathrm{p},n-1} + \frac{Q^2}{2(n+1)}\omega_{\mathrm{p},n-2}\\ &+ \frac{1}{2(n+1)}\nabla^2\omega_{\mathrm{p},n} + \frac{\mathrm{i}}{n+1}\mathbf{P}\cdot\nabla\omega_{\mathrm{p},n} - \frac{\mathrm{i}R}{n+1}\omega_{\mathrm{p},n-1}.\end{aligned} \tag{8.105}$$

(These coefficients for the temperature expansion were denoted with a tilde in section 8.3.) The first several expansion coefficients are explicitly $\omega_{\mathrm{p},0} = 1$, $\omega_{\mathrm{p},1} = 0$, and

$$\begin{aligned}\omega_{\mathrm{p},2} &= \frac{-d}{4} - \frac{\mathrm{i}}{2}R\\ \omega_{\mathrm{p},3} &= \frac{1}{6}P^2 + \frac{1}{6}Q^2\\ \omega_{\mathrm{p},4} &= \frac{3d^2+4d}{96} + \frac{3d+5}{24}\mathrm{i}R - \frac{1}{8}R^2\\ \omega_{\mathrm{p},5} &= -\frac{5d+8}{120}P^2 - \frac{5d+8}{120}Q^2 - \frac{1}{12}\mathrm{i}RQ^2 - \frac{1}{12}\mathrm{i}RP^2.\end{aligned} \tag{8.106}$$

Each of these is symmetric in **P** and **Q**, which is a useful check.

The first several coefficients in the exponential form of the commutation function expansion in powers of Planck's constant, $W_\mathrm{p} = \sum W_{\mathrm{p},n}\hbar^n$, for the simple harmonic oscillator for one particle in d dimensions are explicitly (with β dimensionless)

$$\begin{aligned} W_{\mathrm{p},1} &= \frac{-\mathrm{i}\beta^2}{2} R \\ W_{\mathrm{p},2} &= \frac{\beta^3}{6} P^2 + \frac{\beta^3}{6} Q^2 - \frac{d\beta^2}{4} \\ W_{\mathrm{p},3} &= \frac{5\mathrm{i}\beta^4}{24} R \\ W_{\mathrm{p},4} &= \frac{-\beta^5}{15} P^2 + \frac{d\beta^4}{24} - \frac{\beta^5}{15} Q^2. \end{aligned} \tag{8.107}$$

8.6.2.1 Series in energy states

Now the commutation function will be expressed in energy eigenfunctions, which will give a formally exact phase space representation.

The energy eigenfunctions and eigenvalues satisfy $\hat{\mathcal{H}}|\mathbf{n}\rangle = E_n|\mathbf{n}\rangle$. At this stage there is no need to be specific about the dimensionality or the number of particles. Formally the commutation function can be written as

$$\begin{aligned} e^{-\beta\mathcal{H}(\mathbf{p},\,\mathbf{q})}\omega_\mathrm{p}(\mathbf{p},\,\mathbf{q}) &= e^{\mathbf{p}\cdot\mathbf{q}/i\hbar} e^{-\beta\hat{\mathcal{H}}} e^{-\mathbf{p}\cdot\mathbf{q}/i\hbar} \\ &= e^{\mathbf{p}\cdot\mathbf{q}/i\hbar} L^{Nd/2} \sum_{\mathbf{n}} e^{-\beta\hat{\mathcal{H}}} |\mathbf{n}\rangle\,\langle\mathbf{n}|\mathbf{p}\rangle \\ &= e^{\mathbf{p}\cdot\mathbf{q}/i\hbar} L^{Nd/2} \sum_{\mathbf{n}} e^{-\beta E_\mathbf{n}}\,\langle\mathbf{n}|\mathbf{p}\rangle\,\phi_\mathbf{n}(\mathbf{q}). \end{aligned} \tag{8.108}$$

This expression is general and is not restricted to ideal systems or to the simple harmonic oscillator. In the summand appear in essence the energy eigenfunctions and their Fourier transform,

$$\begin{aligned} \langle\mathbf{n}|\mathbf{p}\rangle &= L^{-Nd/2} \int \mathrm{d}\mathbf{q}\; e^{-\mathbf{p}\cdot\mathbf{q}/i\hbar}\phi_\mathbf{n}(\mathbf{q}) \\ &\equiv L^{-Nd/2}\check{\phi}_\mathbf{n}(\mathbf{p}), \end{aligned} \tag{8.109}$$

With this the weighted commutation function may be written

$$e^{-\beta\mathcal{H}(\mathbf{p},\,\mathbf{q})}\omega_\mathrm{p}(\mathbf{p},\,\mathbf{q}) = e^{\mathbf{p}\cdot\mathbf{q}/i\hbar} \sum_{\mathbf{n}} e^{-\beta E_\mathbf{n}}\check{\phi}_\mathbf{n}(\mathbf{p})\phi_\mathbf{n}(\mathbf{q}). \tag{8.110}$$

The imaginary part of this is odd in **p**. This result is formally exact.

8.6.2.2 Simple harmonic oscillator commutation function

This form for the commutation function can be given explicitly for the simple harmonic oscillator. In dimensionless units, the energy eigenvalues are $E_{\mathbf{n}} = \sum_{j,\alpha}[n_{j\alpha} + 1/2]$, and the energy eigenfunctions are the Hermite functions,

$$\phi_{\mathbf{n}}(\mathbf{Q}) \equiv \prod_{j,\alpha} \frac{1}{\sqrt{2^{n_{j\alpha}} n_{j\alpha}! \sqrt{\pi}}} e^{-Q_{j\alpha}^2/2} \mathrm{H}_{n_{j\alpha}}(Q_{j\alpha}), \tag{8.111}$$

where $\mathrm{H}_n(Q)$ is the Hermite polynomial of degree n. The Hermite function is essentially its own Fourier transform,

$$\check{\phi}_{\mathbf{n}}(\mathbf{P}) = \prod_{j,\alpha} \frac{\mathrm{i}^{n_{j\alpha}} \sqrt{2\pi}}{\sqrt{2^{n_{j\alpha}} n_{j\alpha}! \sqrt{\pi}}} e^{-P_{j\alpha}^2/2} \mathrm{H}_{n_{j\alpha}}(P_{j\alpha}). \tag{8.112}$$

8.6.2.3 Numerical results

Figure 8.1 shows the simple harmonic oscillator commutation function for one particle in one dimension at $P = Q = 0$. One can see that the phase space weight is increasingly reduced from the classical Maxwell–Boltzmann weight as the temperature is decreased. At higher temperatures, $\beta \lesssim 1$, there is good agreement between the exact series form and the six (up to $\omega_{\mathrm{p},5}$) and five (up to $W_{\mathrm{p},4}$) term high temperature expansions for ω_{p} and W_{p}, respectively. The exact series used up to $n = 54$ terms for high temperatures, $\beta \to 0$. As the temperature is reduced the number of necessary terms declines: at $\beta = 1$ the tenth term was $\mathcal{O}(10^{-10})$, and at $\beta = 2$ the fifth term was $\mathcal{O}(10^{-10})$. The commutation function can differ significantly from unity at lower temperatures; at $\beta = 2$ and $P = Q = 0$, it reduces the phase space weight by about 50% from its classical value. It can be seen that the expansion

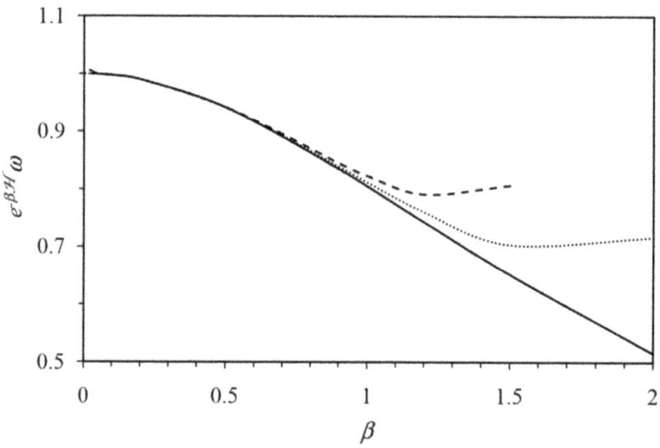

Figure 8.1. The commutation function and Maxwell–Boltzmann weight, $e^{-\beta\mathcal{H}(P,\,Q)}\omega_{\mathrm{p}}(P,\,Q)$, of the simple harmonic oscillator at $P = Q = 0$ as a function of inverse temperature β. The solid curve is the exact result, the dashed curve is the high temperature expansion using all coefficients up to $\omega_{\mathrm{p},5}$, and the dotted curve is the high temperature expansion using all coefficients up to $W_{\mathrm{p},4}$. Note that $\mathcal{H}(0,\,0) = 0$, and that $\omega_{\mathrm{p}}(0,\,0)$ is real.

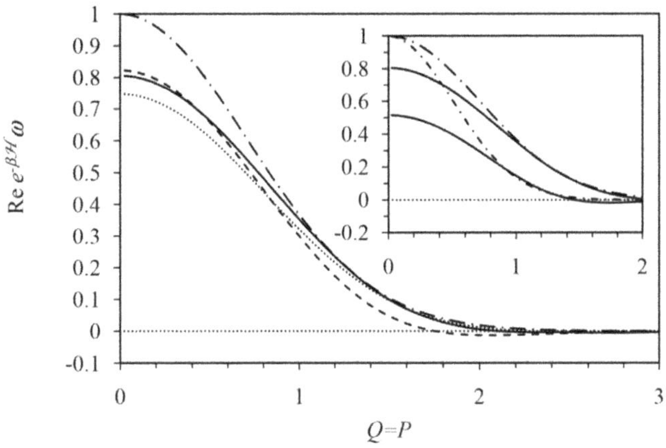

Figure 8.2. Real part of the commutation function and Maxwell–Boltzmann weight, Re $e^{-\beta\mathcal{H}(P,\,Q)}\omega_{\rm p}(P,\,Q)$, along the line $P = Q$ at $\beta = 1.0$. The dash–dotted curve is $e^{-\beta\mathcal{H}(P,\,Q)}$, the solid curve is the exact result using $n = 4$ terms, the dashed curve is the high temperature expansion using all coefficients up to $\omega_{{\rm p},5}$, the dotted curve is the high temperature expansion using all coefficients up to $W_{{\rm p},4}$. The horizontal dotted line is a guide to the eye. Inset: Results for $\beta = 1$ (above) and $\beta = 2$ (below).

in $W_{{\rm p},n}$ remains accurate to lower temperatures than the expansion in $\omega_{{\rm p},n}$ even though it uses fewer terms.

Figure 8.2 shows the phase space weight, Re $e^{-\beta\mathcal{H}(P,\,Q)}\omega_{\rm p}(P,\,Q)$, along the line $P = Q$ at a fixed temperature $\beta = 1$. The inset compares results for $\beta = 1$ with $\beta = 2$. The commutation function generally decreases the phase space weight from that given by the classical Maxwell–Boltzmann factor alone. The effect is most significant in the region of the potential minimum, in this case $P = Q \lesssim 1$, or, equivalently, $\mathcal{H} \lesssim 1$. It can be seen in the main figure that the high temperature expansions remain relatively accurate at $\beta = 1$. Fewer terms need be retained in the exact series as the temperature is decreased; at $\beta = 2$ results with $n = 4$ terms were indistinguishable from those with $n = 5$–10.

Some oscillatory behavior was observed at larger energies, which depended upon the number of terms retained in the series. One would like a probability density to be non-negative, but there is no fundamental requirement that the transformation of quantum statistical mechanics to classical phase space should yield an actual probability density. In any case, the numerical problems (i.e., the sensitivity to the number of retained terms) with the exact series for the commutation function at large energies and high temperatures are likely moot because the Maxwell–Boltzmann factor makes the total phase space weight negligible in this regime. Figure 8.3 compares the monomer grand potential obtained as a phase space integral of the weighted commutation function

$$\begin{aligned} -\beta\Omega_1 &= \ln \sum_{N=0}^{\infty} \frac{z^N}{h^{dN} N!} \int \mathrm{d}\mathbf{\Gamma}^N \, e^{-\beta\mathcal{H}^{(N)}(\mathbf{\Gamma}^N)} \, \omega_{\rm p}^{(N)}(\mathbf{\Gamma}^N) \\ &= \frac{z}{h^d} \int \mathrm{d}\mathbf{\Gamma}_1 \, e^{-\beta\mathcal{H}^{(1)}(\mathbf{\Gamma}_1)} \, \omega_{\rm p}^{(1)}(\mathbf{\Gamma}_1), \end{aligned} \tag{8.113}$$

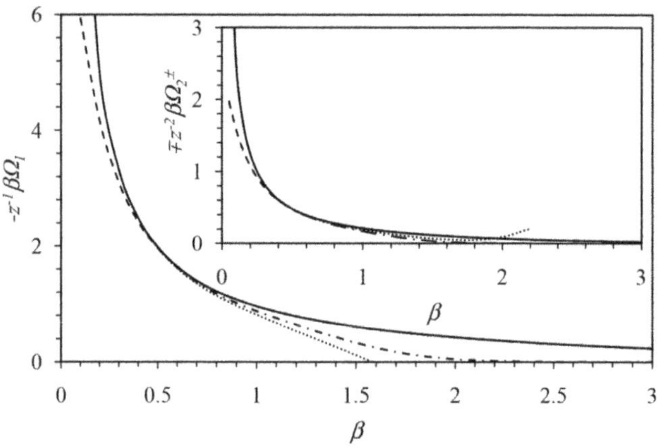

Figure 8.3. The monomer grand potential for ideal oscillators as a function of inverse temperature. The solid curve is the exact analytic result. The remaining curves result from numerical quadrature over phase space, using for the commutation function the series over energy states up to $n = 4$ (dashed curve, coincident with the analytic data at low temperatures), the high temperature expansions up to $\omega_{p,5}$ (dotted curve) and up to $W_{p,4}$ (dash–dotted curve), both coincident with the analytic data at high temperatures. Inset: Corresponding results for the dimer grand potential.

with the $\ell = 1$ term of the exact analytic result (derived in section 4.2),

$$-\beta\Omega^{\pm} = \sum_{l=1}^{\infty} \frac{(\pm 1)^{l-1} z^l}{l} \left[\frac{e^{-l\beta\hbar\omega/2}}{1 - e^{-l\beta\hbar\omega}} \right]^d. \tag{8.114}$$

The imaginary parts of the phase space expressions integrate to zero. As expected, the energy series expressions with a small number of terms works well at low temperatures, and the two high temperature expansions work well in the opposite regime. The purpose of the figure is to show that it is quite feasible to obtain computationally the quantum grand potential from an integral over classical phase space.

Because the grand potential diverges in the classical limit, $\beta \to 0$, it can be more challenging to obtain accurate numerical results in the high temperature regime. At $\beta = 0.2$, for the monomer grand potential $-z^{-1}\beta\Omega_1$, the exact analytic result is 4.99, using the energy series for the commutation function with $n = 4$ terms it is 3.16, and with $n = 8$ terms it is 4.17. Again at $\beta = 0.2$, for the dimer grand potential $\mp z^{-2}\beta\Omega_2^{\pm}$, the exact analytic result is 1.24, using the energy series with $n = 4$ terms it is 1.07 and with $n = 8$ terms it is 1.21. At high temperatures, the high temperature expansions are more sensitive to the limits used for the phase space integrals than is the energy series form.

The dimer grand potential is typically about 5–10 times smaller than the monomer grand potential at the same temperature. For the loop grand potential, the symmetrization factor induces an effective interaction between the adjacent particles around the loop, typically of the form $\cos(Q_{j,j+1}P_{j,j+1})$ or $\sin(Q_{j,j+1}P_{j,j+1})$. Since rapidly oscillating terms tend to cancel, it was found efficacious to introduce a cut-off $R_{\text{cut}}^{p/q}$

and to neglect configurations unless both $|Q_{j,j+1}| \leqslant R^q_{\rm cut}$ and $|P_{j,j+1}| \leqslant R^p_{\rm cut}$. For the present simple harmonic oscillator, a value of $R^{p/q}_{\rm cut} = 4$ was found to change the results by about 2% at $\beta = 0.1$ and by less than 0.1% at $\beta = 1$, while substantially reducing the computation time. (The results in the inset of figure 8.3 do not use a cut-off.)

References

Allen M P and Tildesley D J 1987 *Computer Simulations of Liquids* (Oxford: Oxford University Press)

Attard P 2002 *Thermodynamics and Statistical Mechanics: Equilibrium by Entropy Maximisation* (London: Academic)

Attard P 2012 *Non-Equilibrium Thermodynamics and Statistical Mechanics: Foundations and Applications* (Oxford: Oxford University Press)

Attard P 2016 Quantum statistical mechanics as an exact classical expansion with results for Lennard–Jones helium arXiv:1609.08178

Attard P 2018a Quantum statistical mechanics in classical phase space. Expressions for the multi-particle density, the average energy, and the virial pressure arXiv:1811.00730

Attard P 2018b Entropy Beyond the Second Law *Thermodynamics and Statistical Mechanics for Equilibrium, Non-Equilibrium, Classical, and Quantum Systems* (Bristol: IOP Publishing)

Hernando A 2013 Imaginary-time nonuniform mesh method for solving the multidimensional Schrödinger equation: fermionization and melting of quantum Lennard–Jones crystals arXiv:1304.8015v2

Kirkwood J G 1933 Quantum statistics of almost classical particles *Phys. Rev.* **44** 31

Messiah A 1961 *Quantum Mechanics* vols 1 and 2 (Amsterdam: North-Holland)

Onsager L 1931 Reciprocal relations in irreversible processes. I. *Phys. Rev.* **38** 2265

Pathria R K 1972 *Statistical Mechanics* (Oxford: Pergamon)

Uhlenbeck G E and Gropper L 1932 The equation of state of a non-ideal Einstein–Bose or Fermi–Dirac gas *Phys. Rev.* **41** 79

Wigner E 1932 On the quantum correction for thermodynamic equilibrium *Phys. Rev.* **40** 749

Chapter 9

Nested commutator expansion for the commutation function

The Maxwell–Boltzmann operator is factorized as exponentials of the potential energy operator, and the kinetic energy operator, betwixt a residual operator that is a series of nested commutators of the original two operators. This yields the classical Maxwell–Boltzmann weight as the leading order, high temperature limit. Substantial cancelation in the expansion of the commutation function exponent operator portends numerical efficiency. Explicit terms to fourth order are extensive. Numerical comparison is made with benchmark results for the one-dimensional harmonic crystal.

9.1 Introduction

One motivation for the phase space formulation of quantum statistical mechanics is that quantum effects are a small perturbation for most terrestrial condensed matter systems. However this is also the source of a significant computational challenge when it comes to obtaining the commutation function because it means that the dominant contribution in the defining equation is the classical Maxwell–Boltzmann factor, and this has to be subtracted or canceled to obtain the specific quantum effects of non-commutativity. Such cancelation demands high accuracy of the individual terms in any numerical calculation, which can be problematic. In the case of an analytic expansion, the contributions from the expansion of the Maxwell–Boltzmann factor can unduly complicate the individual terms and they have to be identified and removed to leave the specific quantum contribution that is really sought.

In the present chapter the defining equation for the commutation function is reformulated in such a way that the classical Maxwell–Boltzmann factor appears explicitly on both sides of the equation, which allows it to be canceled analytically. This leaves the commutation function expressed as the phase space expectation value solely of a commutation function operator, so that all quantum effects due to non-commutativity are explicit and no further numerical or analytic cancelation due to the Maxwell–Boltzmann factor is required.

doi:10.1088/978-0-7503-4055-7ch9

Compared to the high temperature expansion, this exponential commutator factorization method writes the terms in a very compact fashion. The first four terms are here obtained and written explicitly as nested commutators.

From these, the first three non-zero terms in the high temperature expansion of the commutation function exponent operator are obtained explicitly. A deal of cancelation occurs in this process, which suggests that the series is rapidly converging. The terms contain gradients of the potential, and their products and higher powers. These are more short-ranged than the potential itself, which is another good sign. Conversely, however, these are larger than the potential in the core region, which is precisely where one expects the effects of non-commutativity to be manifest. Whether or not this will pose a significant computational challenge is unclear.

9.2 Commutator factorization of exponentials

The exponential of the sum of two operators can be formally factorized as

$$\begin{aligned} e^{\hat{A}+\hat{B}} &= e^{\hat{A}}e^{C(\hat{A},\,\hat{B})}e^{\hat{B}} \\ &= e^{\hat{B}}e^{C(\hat{B},\,\hat{A})}e^{\hat{A}}. \end{aligned} \tag{9.1}$$

The product of the exponential and its inverse gives the identity operator,

$$\hat{\mathrm{I}} = e^{-\hat{A}-\hat{B}}e^{\hat{A}+\hat{B}} = e^{-\hat{B}}e^{C(-\hat{B},\,-\hat{A})}e^{-\hat{A}}e^{\hat{A}}e^{C(\hat{A},\,\hat{B})}e^{\hat{B}}, \tag{9.2}$$

or

$$e^{C(-\hat{B},\,-\hat{A})}e^{C(\hat{A},\,\hat{B})} = \hat{\mathrm{I}}. \tag{9.3}$$

It follows that the central exponent, or residual operator function, must satisfy

$$C(\hat{A},\,\hat{B}) = -C(-\hat{B},\,-\hat{A}). \tag{9.4}$$

9.2.1 Second order

Here $C(\hat{A},\,\hat{B})$ is obtained to quadratic order in the operators. Below the factorization is applied to the commutation function, in which case each of the two operators is linearly proportional to the inverse temperature. Hence quadratic order in the operators will be the same as quadratic order in the inverse temperature.

Expanding both sides of the factorization to quadratic order one has

$$e^{\hat{A}+\hat{B}} = 1 + \hat{A} + \hat{B} + \frac{1}{2}\hat{A}^2 + \frac{1}{2}\hat{B}^2 + \frac{1}{2}\hat{A}\hat{B} + \frac{1}{2}\hat{B}\hat{A}, \tag{9.5}$$

and

$$\begin{aligned} e^{\hat{A}}e^{C(\hat{A},\,\hat{B})}e^{\hat{B}} &= (1 + \hat{A} + \frac{1}{2}\hat{A}^2)(1 + C(\hat{A},\,\hat{B}))(1 + \hat{B} + \frac{1}{2}\hat{B}^2) \\ &= 1 + \hat{A} + \hat{B} + \frac{1}{2}\hat{A}^2 + \frac{1}{2}\hat{B}^2 + \hat{A}\hat{B} + C(\hat{A},\,\hat{B}). \end{aligned} \tag{9.6}$$

Equating these two gives the second order term for the residual operator function,

$$C_2(\hat{A},\,\hat{B}) = \frac{1}{2}[\hat{B},\,\hat{A}], \tag{9.7}$$

where the brackets indicate the usual quantum commutator, $[\hat{B},\,\hat{A}] = \hat{B}\hat{A} - \hat{A}\hat{B}$.

9.2.2 Third order

The cubic order terms in the exponentiated sum are

$$e^{\hat{A}+\hat{B}} = \frac{1}{6}\hat{A}^3 + \frac{1}{6}\hat{A}^2\hat{B} + \frac{1}{6}\hat{A}\hat{B}\hat{A} + \frac{1}{6}\hat{B}\hat{A}^2 + \frac{1}{6}\hat{A}\hat{B}^2 + \frac{1}{6}\hat{B}\hat{A}\hat{B} + \frac{1}{6}\hat{B}^2\hat{A} + \frac{1}{6}\hat{B}^3. \tag{9.8}$$

The factorized form with quadratic residual operator function expanded to cubic order is

$$\begin{aligned} &e^{\hat{A}}e^{C_2(\hat{A},\,\hat{B})}e^{\hat{B}} \\ &= \left\{1 + \hat{A} + \frac{1}{2}\hat{A}^2 + \frac{1}{6}\hat{A}^3\right\}\left\{1 + \frac{1}{2}[\hat{B},\,\hat{A}]\right\} \\ &\quad \left\{1 + \hat{B} + \frac{1}{2}\hat{B}^2 + \frac{1}{6}\hat{B}^3\right\} \\ &= 1 + \hat{A} + \frac{1}{2}\hat{A}^2 + \hat{B} + \frac{1}{2}\hat{B}^2 + \frac{1}{2}\hat{A}\hat{B} + \frac{1}{2}\hat{B}\hat{A} \\ &\quad + \frac{1}{6}\hat{A}^3 + \frac{1}{6}\hat{B}^3 + \frac{1}{2}\hat{A}\hat{B}\hat{A} + \frac{1}{2}\hat{B}\hat{A}\hat{B} + \cdots. \end{aligned} \tag{9.9}$$

The difference between the cubic parts of these two expressions is the third order residual operator function,

$$\begin{aligned} C_3(\hat{A},\,\hat{B}) &= \frac{-1}{3}\hat{A}\hat{B}\hat{A} - \frac{1}{3}\hat{B}\hat{A}\hat{B} + \frac{1}{6}\hat{A}^2\hat{B} + \frac{1}{6}\hat{B}\hat{A}^2 + \frac{1}{6}\hat{A}\hat{B}^2 + \frac{1}{6}\hat{B}^2\hat{A} \\ &= \frac{1}{6}[\hat{A},\,[\hat{A},\,\hat{B}]] + \frac{1}{6}[[\hat{A},\,\hat{B}],\,\hat{B}]. \end{aligned} \tag{9.10}$$

Hence to cubic order one has

$$e^{\hat{A}+\hat{B}} = e^{\hat{A}}e^{-[\hat{A},\,\hat{B}]/2+[\hat{A},[\hat{A},\hat{B}]]/6+[[\hat{A},\hat{B}],\hat{B}]/6}e^{\hat{B}}. \tag{9.11}$$

The neglected terms are $\mathcal{O}(\hat{A}^4)$, etc.

9.2.3 Fourth order

The expansion to fourth order from the third order commutator form are

$$
\begin{aligned}
&e^{\hat{A}}e^{C_2(\hat{A},\,\hat{B})+C_3(\hat{A},\,\hat{B})}e^{\hat{B}} \\
&\quad=\left\{1+\hat{A}+\frac{1}{2}\hat{A}^2+\frac{1}{6}\hat{A}^3+\frac{1}{24}\hat{A}^4\right\} \\
&\quad\times\left\{1+\frac{1}{2}[\hat{B},\,\hat{A}]+\frac{1}{6}[\hat{A},\,[\hat{A},\,\hat{B}]]+\frac{1}{6}[[\hat{A},\,\hat{B}],\,\hat{B}]+\frac{1}{8}[\hat{B},\,\hat{A}]^2\right\} \\
&\quad\times\left\{1+\hat{B}+\frac{1}{2}\hat{B}^2+\frac{1}{6}\hat{B}^3+\frac{1}{24}\hat{B}^4\right\}.
\end{aligned}
\tag{9.12}
$$

The fourth order terms here are

$$
\begin{aligned}
\hat{D}_4=&\frac{1}{24}\hat{A}^4+\frac{1}{24}\hat{B}^4+\frac{1}{6}\hat{A}^3\hat{B}+\frac{1}{6}\hat{A}\hat{B}^3+\frac{1}{4}\hat{A}^2\hat{B}^2 \\
&+\frac{1}{4}\hat{A}^2[\hat{B},\,\hat{A}]+\frac{1}{4}[\hat{B},\,\hat{A}]\hat{B}^2+\frac{1}{2}\hat{A}[\hat{B},\,\hat{A}]\hat{B} \\
&+\frac{1}{6}\hat{A}[\hat{A},\,[\hat{A},\,\hat{B}]]+\frac{1}{6}[\hat{A},\,[\hat{A},\,\hat{B}]]\hat{B}+\frac{1}{6}\hat{A}[[\hat{A},\,\hat{B}],\,\hat{B}] \\
&+\frac{1}{6}[[\hat{A},\,\hat{B}],\,\hat{B}]\hat{B}+\frac{1}{8}[\hat{B},\,\hat{A}]^2.
\end{aligned}
\tag{9.13}
$$

The expansion of the original $e^{\hat{A}+\hat{B}}$ form has fourth order terms

$$
\begin{aligned}
\hat{E}_4=&\frac{1}{24}(\hat{A}+\hat{B})^4 \\
=&\frac{1}{24}\hat{A}^4+\frac{1}{24}\hat{A}^3\hat{B}+\frac{1}{24}\hat{A}^2\hat{B}\hat{A}+\frac{1}{24}\hat{A}\hat{B}\hat{A}^2+\frac{1}{24}\hat{B}\hat{A}^3 \\
&+\frac{1}{24}\hat{A}^2\hat{B}^2+\frac{1}{24}\hat{A}\hat{B}\hat{A}\hat{B}+\frac{1}{24}\hat{A}\hat{B}^2\hat{A}+\frac{1}{24}\hat{B}\hat{A}^2\hat{B}+\frac{1}{24}\hat{B}\hat{A}\hat{B}\hat{A} \\
&+\frac{1}{24}\hat{B}^2\hat{A}^2+\frac{1}{24}\hat{B}^3\hat{A}+\frac{1}{24}\hat{B}^2\hat{A}\hat{B}+\frac{1}{24}\hat{B}\hat{A}\hat{B}^2+\frac{1}{24}\hat{A}\hat{B}^3+\frac{1}{24}\hat{B}^4.
\end{aligned}
\tag{9.14}
$$

The fourth order residual is $C_4(\hat{A},\,\hat{B})=\hat{E}_4-\hat{D}_4$. The term involving three $\hat{A}$ and one $\hat{B}$ is

$$
\begin{aligned}
C_{31}(\hat{A},\,\hat{B})=&\frac{1}{24}\hat{A}^3\hat{B}+\frac{1}{24}\hat{A}^2\hat{B}\hat{A}+\frac{1}{24}\hat{A}\hat{B}\hat{A}^2+\frac{1}{24}\hat{B}\hat{A}^3 \\
&-\frac{1}{6}\hat{A}^3\hat{B}-\frac{1}{4}\hat{A}^2[\hat{B},\,\hat{A}]-\frac{1}{6}\hat{A}[\hat{A},\,[\hat{A},\,\hat{B}]] \\
=&\frac{-1}{24}\hat{A}^3\hat{B}+\frac{3}{24}\hat{A}^2\hat{B}\hat{A}+\frac{-3}{24}\hat{A}\hat{B}\hat{A}^2+\frac{1}{24}\hat{B}\hat{A}^3 \\
=&\frac{-1}{24}[\hat{A},\,[\hat{A},\,[\hat{A},\,\hat{B}]]].
\end{aligned}
\tag{9.15}
$$

From equation (9.4), the term involving one $\hat{A}$ and three $\hat{B}$ is $C_{13}(\hat{A}, \hat{B}) = -C_{31}(-\hat{B}, -\hat{A}) = [\hat{B}, [\hat{B}, [\hat{B}, \hat{A}]]]/24$.

With this, the term involving two $\hat{A}$s and two $\hat{B}$ is

$$\begin{aligned} C_{22}(\hat{A}, \hat{B}) & \\ &= \frac{1}{24}\hat{A}^2\hat{B}^2 + \frac{1}{24}\hat{A}\hat{B}\hat{A}\hat{B} + \frac{1}{24}\hat{A}\hat{B}^2\hat{A} + \frac{1}{24}\hat{B}\hat{A}^2\hat{B} + \frac{1}{24}\hat{B}\hat{A}\hat{B}\hat{A} \\ &\quad + \frac{1}{24}\hat{B}^2\hat{A}^2 - \frac{1}{4}\hat{A}^2\hat{B}^2 - \frac{1}{2}\hat{A}[\hat{B}, \hat{A}]\hat{B} - \frac{1}{6}[\hat{A}, [\hat{A}, \hat{B}]]\hat{B} \\ &\quad - \frac{1}{6}\hat{A}[[\hat{A}, \hat{B}], \hat{B}] - \frac{1}{8}[\hat{B}, \hat{A}]^2 \\ &= \frac{-1}{24}\hat{A}^2\hat{B}^2 + \frac{2}{24}\hat{A}\hat{B}\hat{A}\hat{B} - \frac{2}{24}\hat{B}\hat{A}\hat{B}\hat{A} + \frac{1}{24}\hat{B}^2\hat{A}^2 \\ &= \frac{1}{24}[[[\hat{A}, \hat{B}], \hat{B}], \hat{A}]. \end{aligned} \tag{9.16}$$

How can one resist this?

Hence the factorization $e^{\hat{A}+\hat{B}} = e^{\hat{A}}e^{C(\hat{A}, \hat{B})}e^{\hat{B}}$, to fourth order is

$$\begin{aligned} C(\hat{A}, \hat{B}) & \\ &= C_2(\hat{A}, \hat{B}) + C_3(\hat{A}, \hat{B}) + C_4(\hat{A}, \hat{B}) \\ &= \frac{1}{2}[\hat{B}, \hat{A}] + \frac{1}{6}[[\hat{B}, \hat{A}], \hat{A}] + \frac{1}{6}[\hat{B}, [\hat{B}, \hat{A}]] \\ &\quad + \frac{1}{24}[[[\hat{B}, \hat{A}], \hat{A}], \hat{A}] + \frac{1}{24}[\hat{B}, [\hat{B}, [\hat{B}, \hat{A}]]] + \frac{1}{24}[[\hat{B}, [\hat{B}, \hat{A}]], \hat{A}]. \end{aligned} \tag{9.17}$$

9.3 Maxwell–Boltzmann operator factorized

9.3.1 Formal and algorithmic expressions

The energy operator, $\hat{\mathcal{H}} = U + \hat{\mathcal{K}}$, is the sum of the potential energy, $U(\mathbf{r})$, and the kinetic energy $\hat{\mathcal{K}}(\mathbf{r}) = (-\hbar^2/2m)\nabla_{\mathbf{r}}^2$, operators. Hence the Maxwell–Boltzmann operator can be written in factored form,

$$e^{-\beta\hat{\mathcal{H}}} = e^{-\beta U}e^{\hat{W}_p}e^{-\beta\hat{\mathcal{K}}}. \tag{9.18}$$

In the notation of the preceding section, the commutation function exponent operator is $\hat{W}_p = C(-\beta U, -\beta\hat{\mathcal{K}})$.

The motivation for choosing this factorization form is that setting $\hat{W}_p = 0$, gives the classical Maxwell–Boltzmann operator when it acts on the momentum eigenfunctions. This can be seen from the definition of the commutation function,

$$\begin{aligned} e^{-\beta \mathcal{H}(\mathbf{q},\,\mathbf{p})} e^{W_{\mathrm{p}}(\mathbf{q},\,\mathbf{p})} &= \frac{1}{\langle \mathbf{q}|\mathbf{p}\rangle}\langle \mathbf{q} \mid e^{-\beta\hat{\mathcal{H}}} \mid \mathbf{p}\rangle \\ &= \frac{1}{\langle \mathbf{q}|\mathbf{p}\rangle}\langle \mathbf{q} \mid e^{-\beta U} e^{\hat{W}_{\mathrm{p}}} e^{-\beta\hat{\mathcal{K}}} \mid \mathbf{p}\rangle \\ &= \frac{e^{-\beta\mathcal{H}(\mathbf{q},\,\mathbf{p})}}{\langle \mathbf{q}|\mathbf{p}\rangle}\langle \mathbf{q} \mid e^{\hat{W}_{\mathrm{p}}} \mid \mathbf{p}\rangle, \end{aligned} \tag{9.19}$$

or

$$e^{W_{\mathrm{p}}(\mathbf{q},\,\mathbf{p})} = \frac{1}{\langle \mathbf{q}|\mathbf{p}\rangle}\langle \mathbf{q} \mid e^{\hat{W}_{\mathrm{p}}} \mid \mathbf{p}\rangle. \tag{9.20}$$

It is likely to be of numerical benefit that the classical Maxwell–Boltzmann factor has been canceled in this formulation, and that the leading term in the residual exponent operator is of quadratic order in the inverse temperature.

The commutation function can be calculated from this in several ways. One way is to expand in eigenstates of the commutation function operator, $\hat{W}_{\mathrm{p}}(\mathbf{r})\phi_{\mathbf{n}}(\mathbf{r}) = W_{\mathrm{p},\mathbf{n}}\phi_{\mathbf{n}}(\mathbf{r})$. That is

$$e^{W_{\mathrm{p}}(\mathbf{q},\,\mathbf{p})} = \frac{1}{\langle \mathbf{q}|\mathbf{p}\rangle}\sum_{\mathbf{n}} e^{W_{\mathrm{p},\mathbf{n}}}\phi_{\mathbf{n}}(\mathbf{q})\check{\phi}_{\mathbf{n}}(\mathbf{p})^{*}, \tag{9.21}$$

where what is essentially the Fourier transform of the eigenfunction is $\check{\phi}_{\mathbf{n}}(\mathbf{p}) = \langle \mathbf{p}|\mathbf{n}\rangle$.

The challenge with this approach is that presumably obtaining the eigenstates of the commutation function operator is essentially a non-local computation that involves all the particles of the system and the specification of the conditions at the system boundaries. Possibly these hurdles could be ameliorated by an effective local field approach (chapter 10).

An alternative way to calculate the commutation function is to expand the exponential and to calculate the operation of the commutation function operator on the momentum eigenfunction directly. That is

$$\begin{aligned} e^{W_{\mathrm{p}}(\mathbf{q},\,\mathbf{p})} &= \frac{\langle \mathbf{q}|e^{\hat{W}_{\mathrm{p}}}|\mathbf{p}\rangle}{\langle \mathbf{q}|\mathbf{p}\rangle} \\ &= 1 + \frac{\langle \mathbf{q}|\hat{W}_{\mathrm{p}}|\mathbf{p}\rangle}{\langle \mathbf{q}|\mathbf{p}\rangle} + \frac{1}{2}\frac{\langle \mathbf{q}|\hat{W}_{\mathrm{p}}^{2}|\mathbf{p}\rangle}{\langle \mathbf{q}|\mathbf{p}\rangle} + \cdots \\ &= 1 + \frac{\langle \mathbf{q}|\beta^{2}\hat{W}_{\mathrm{p},2} + \beta^{3}\hat{W}_{\mathrm{p},3} + \beta^{4}\hat{W}_{\mathrm{p},4}|\mathbf{p}\rangle}{\langle \mathbf{q}|\mathbf{p}\rangle} \\ &\quad + \frac{1}{2}\frac{\langle \mathbf{q}|\beta^{4}\hat{W}_{\mathrm{p},\,2}^{2}|\mathbf{p}\rangle}{\langle \mathbf{q}|\mathbf{p}\rangle} + \mathcal{O}(\beta^{5}). \end{aligned} \tag{9.22}$$

Obviously this process can be continued to higher order. This gives a rather direct way of obtaining the commutation function and the operators whose expectation value is taken lend themselves to analytic formulae. This is one advantage that they have over the sum over eigenstates method. A second related advantage are that they are local since they involve no more than gradients and products of the potential energy.

Either of the above computational expressions for the commutation function can be combined with the expansion to a specified order in inverse temperature of the commutation function operator. The second, third, and fourth order expansions were obtained explicitly above for the general operators $\hat{A}$ and $\hat{B}$. Those formulae will now be used to obtain the respective expansion for the commutation function operator.

9.3.2 Commutation function operators

The commutation function exponent operator may be written as a series in inverse temperature,

$$\hat{W}_{\rm p} = \sum_{n=2}^{\infty} \beta^n \hat{W}_{{\rm p},n}. \tag{9.23}$$

We can obtain the coefficients of this from the above nested commutator expansion.

9.3.2.1 Second order

With $\hat{A} = -\beta U(\mathbf{r})$ and $\hat{B} = -\beta\hat{\mathcal{K}}(\mathbf{r}) = (\beta\hbar^2/2m)\nabla_{\mathbf{r}}^2$, equation (9.7) gives the second order term for the commutation function operator

$$\begin{aligned} \hat{W}_{{\rm p},2}(\mathbf{r}) &= \frac{-1}{2}[U, \hat{\mathcal{K}}] \\ &= \frac{-\hbar^2}{4m}\left\{(\nabla_{\mathbf{r}}^2 U(\mathbf{r})) + 2(\nabla_{\mathbf{r}} U(\mathbf{r})) \cdot \nabla_{\mathbf{r}}\right\}. \end{aligned} \tag{9.24}$$

9.3.2.2 Third order

Equation (9.10) gives the third order term for the commutation function operator

$$\begin{aligned} \hat{W}_{{\rm p},3}(\mathbf{r}) &= \frac{-1}{6}[U, [U, \hat{\mathcal{K}}]] - \frac{1}{6}[[U, \hat{\mathcal{K}}], \hat{\mathcal{K}}] \\ &= \frac{\hbar^2}{6m}(\nabla U) \cdot (\nabla U) \\ &\quad - \frac{\hbar^4}{24m^2}\left\{(\nabla^2\nabla^2 U) + 4(\nabla^2\nabla U) \cdot \nabla + 4(\nabla\nabla U) : \nabla\nabla\right\}. \end{aligned} \tag{9.25}$$

All of these terms are extensive.

9.3.2.3 Fourth order

The fourth order coefficient for the residual operator, equations (9.15) and (9.16), is

$$C_4(\hat{A}, \hat{B}) = \frac{-1}{24}[\hat{A}, [\hat{A}, [\hat{A}, \hat{B}]]] - \frac{1}{24}[[[\hat{A}, \hat{B}], \hat{B}], \hat{B}] + \frac{1}{24}[[[\hat{A}, \hat{B}], \hat{B}], \hat{A}]. \tag{9.26}$$

Writing the fourth order commutation function exponent operator as

$$\hat{W}_{\rm p,4} = \hat{W}_{\rm p,31} + \hat{W}_{\rm p,13} + \hat{W}_{\rm p,22}, \tag{9.27}$$

the first of these is

$$\begin{aligned}\hat{W}_{\rm p,31} &= \frac{-1}{24}\frac{-\hbar^2}{2m}[U, [U, [U, \nabla^2]]] \\ &= \frac{-\hbar^2}{48m}[U, [U, ((\nabla^2 U) + 2(\nabla U)\cdot\nabla)]] \\ &= 0.\end{aligned} \tag{9.28}$$

This vanishes because $(\nabla^2 U)$ and $[U, (2(\nabla U)\cdot\nabla)] = -2(\nabla U)\cdot(\nabla U)$ both commute with U.

This is in fact a generally useful result, since nested terms of this form occur at each order. One such term is

$$\hat{W}_{\rm p;n-1,\,1} \equiv \frac{-1}{n!}[U, [U, \ldots, [U, \hat{\mathcal{K}}]\ldots]] = \frac{1}{n}[U, \hat{W}_{\rm p;n-2,\,1}], \tag{9.29}$$

and one has

$$\hat{W}_{\rm p,11} = \frac{-1}{2}[U, \hat{\mathcal{K}}] = \frac{-\hbar^2}{4m}(\nabla^2 U) - (\nabla U)\cdot\nabla, \tag{9.30}$$

$$\hat{W}_{\rm p,21} = \frac{-1}{3!}[U, [U, \hat{\mathcal{K}}]] = \frac{\hbar^2}{6m}(\nabla U)\cdot(\nabla U), \tag{9.31}$$

and

$$\hat{W}_{\rm p,n1} = 0, \quad n \geqslant 3. \tag{9.32}$$

This also means that any nest of commutators with an inner kernel of three U and one $\hat{\mathcal{K}}$ that form $\hat{W}_{\rm p,31}(U, \hat{\mathcal{K}})$ vanishes.

The second term in the fourth order commutation function exponent operator is

$$\begin{aligned}\hat{W}_{\rm p,13} &= \frac{-1}{24}\frac{-\hbar^6}{8m^3}[[[U, \nabla^2], \nabla^2], \nabla^2] \\ &= \frac{-\hbar^6}{192m^3}((\nabla^2\nabla^2\nabla^2 U) + 6(\nabla^2\nabla^2\nabla U)\cdot\nabla \\ &\quad + 12(\nabla^2\nabla\nabla U):\nabla\nabla + 8(\nabla\nabla\nabla U)\vdots\nabla\nabla\nabla).\end{aligned} \tag{9.33}$$

The third term is

$$\begin{aligned}\hat{W}_{\rm p,22} &= \frac{-1}{24}\frac{\hbar^4}{4m^2}[U, [[U, \nabla^2], \nabla^2]] \\ &= \frac{\hbar^4}{96m^2}(4(\nabla^2\nabla U)\cdot(\nabla U) + 4(\nabla\nabla U):(\nabla\nabla U) \\ &\quad + 8(\nabla\nabla U):(\nabla U)\nabla).\end{aligned} \tag{9.34}$$

9.4 Temperature derivative of the commutation function operator

9.4.1 Nested commutator form

The nested commutator form for the commutation function exponent operator can alternatively be obtained from the temperature derivative of the defining equation. In view of the factorization (9.18), the commutation function exponent operator can be written as

$$e^{\hat{W}_{\rm p}} = e^{\beta U} e^{-\beta\hat{\mathcal{H}}} e^{\beta\hat{\mathcal{K}}}. \tag{9.35}$$

Differentiation with respect to inverse temperature yields

$$\begin{aligned}\frac{\partial e^{\hat{W}_{\rm p}}}{\partial\beta} &= e^{\beta U} U e^{-\beta\hat{\mathcal{H}}} e^{\beta\hat{\mathcal{K}}} - e^{\beta U}\hat{\mathcal{H}} e^{-\beta\hat{\mathcal{H}}} e^{\beta\hat{\mathcal{K}}} + e^{\beta U} e^{-\beta\hat{\mathcal{H}}}\hat{\mathcal{K}} e^{\beta\hat{\mathcal{K}}} \\ &= e^{\beta U}[e^{-\beta\hat{\mathcal{H}}}, \hat{\mathcal{K}}] e^{\beta\hat{\mathcal{K}}}.\end{aligned} \tag{9.36}$$

The right-hand side can also be written $e^{\beta U}[U, e^{-\beta\hat{\mathcal{H}}}]e^{\beta\hat{\mathcal{K}}}$, but this is less convenient for our purposes.

Using the temperature series for the commutation function exponent operator, equation (9.23), to fifth order one has

$$\begin{aligned}e^{\hat{W}_{\rm p}} &= 1 + \beta^2\hat{W}_{\rm p,2} + \beta^3\hat{W}_{\rm p,3} + \beta^4\hat{W}_{\rm p,4} + \beta^5\hat{W}_{\rm p,5} \\ &\quad + \frac{1}{2}\beta^4\hat{W}_{\rm p,2}^2 + \frac{1}{2}\beta^5\hat{W}_{\rm p,2}\hat{W}_{\rm p,3} + \frac{1}{2}\beta^5\hat{W}_{\rm p,3}\hat{W}_{\rm p,2} + \mathcal{O}(\beta^6).\end{aligned} \tag{9.37}$$

The derivative is

$$\begin{aligned}\frac{\partial e^{\hat{W}_{\rm p}}}{\partial\beta} &= 2\beta\hat{W}_{\rm p,2} + 3\beta^2\hat{W}_{\rm p,3} + 4\beta^3\hat{W}_{\rm p,4} + 5\beta^4\hat{W}_{\rm p,5} \\ &\quad + \frac{4}{2}\beta^3\hat{W}_{\rm p,2}^2 + \frac{5}{2}\beta^4\hat{W}_{\rm p,2}\hat{W}_{\rm p,3} + \frac{5}{2}\beta^4\hat{W}_{\rm p,3}\hat{W}_{\rm p,2} + \mathcal{O}(\beta^5).\end{aligned} \tag{9.38}$$

The right-hand side to fourth order is

$$\begin{aligned}\frac{\partial e^{\hat{W}_{\rm p}}}{\partial\beta} &= e^{\beta U}[e^{-\beta\hat{\mathcal{H}}}, \hat{\mathcal{K}}] e^{\beta\hat{\mathcal{K}}} \\ &= \left\{1 + \beta U + \frac{1}{2}\beta^2 U^2 + \frac{1}{3!}\beta^3 U^3\right\} \\ &\quad \times\left[\left(-\beta\hat{\mathcal{H}} + \frac{1}{2}\beta^2\hat{\mathcal{H}}^2 - \frac{1}{3!}\beta^3\hat{\mathcal{H}}^3 + \frac{1}{4!}\beta^4\hat{\mathcal{H}}^4\right), \hat{\mathcal{K}}\right] \\ &\quad \times\left\{1 + \beta\hat{\mathcal{K}} + \frac{1}{2}\beta^2\hat{\mathcal{K}}^2 + \frac{1}{3!}\beta^3\hat{\mathcal{K}}^3\right\} + \mathcal{O}(\beta^5).\end{aligned} \tag{9.39}$$

Equating the coefficients of β gives

$$\hat{W}_{p,2} = \frac{-1}{2}[\hat{\mathcal{H}}, \hat{\mathcal{K}}] = \frac{-1}{2}[U, \hat{\mathcal{K}}]. \tag{9.40}$$

Equating the coefficients of β^2 gives

$$\begin{aligned} \hat{W}_{p,3} &= \frac{1}{3}\left\{-U[\hat{\mathcal{H}}, \hat{\mathcal{K}}] - [\hat{\mathcal{H}}, \hat{\mathcal{K}}]\hat{\mathcal{K}} + \frac{1}{2}[\hat{\mathcal{H}}^2, \hat{\mathcal{K}}]\right\} \\ &= \frac{-1}{6}[U, [U, \hat{\mathcal{K}}]] - \frac{1}{6}[[U, \hat{\mathcal{K}}], \hat{\mathcal{K}}]. \end{aligned} \tag{9.41}$$

Equating the coefficients of β^3 gives

$$\begin{aligned} \hat{W}_{p,4} = \frac{-1}{8}[U, \hat{\mathcal{K}}]^2 &+ \frac{1}{4}\Big\{-U^2\hat{\mathcal{K}}^2 + U\hat{\mathcal{K}}U\hat{\mathcal{K}} - \frac{1}{2}U^2[U, \hat{\mathcal{K}}] - \frac{1}{2}[U, \hat{\mathcal{K}}]\hat{\mathcal{K}}^2 \\ &+ \frac{1}{2}U[(U^2 + U\hat{\mathcal{K}} + \hat{\mathcal{K}}U), \hat{\mathcal{K}}] + \frac{1}{2}[(U^2 + U\hat{\mathcal{K}} + \hat{\mathcal{K}}U), \hat{\mathcal{K}}]\hat{\mathcal{K}} \\ &- \frac{1}{3!}\left[\left(U^3 + U^2\hat{\mathcal{K}} + U\hat{\mathcal{K}}U + \hat{\mathcal{K}}U^2 + \hat{\mathcal{K}}^2U + \hat{\mathcal{K}}U\hat{\mathcal{K}} + U\hat{\mathcal{K}}^2\right), \hat{\mathcal{K}}\right]\Big\}. \end{aligned} \tag{9.42}$$

The term involving three U and one $\hat{\mathcal{K}}$ is $\hat{W}_{p,31} \equiv W_{31}(U, \hat{\mathcal{K}})$, with

$$\begin{aligned} W_{31}(U, \hat{\mathcal{K}}) &= \frac{-1}{8}U^3\hat{\mathcal{K}} + \frac{1}{8}U^2\hat{\mathcal{K}}U + \frac{1}{8}U^3\hat{\mathcal{K}} - \frac{1}{8}U\hat{\mathcal{K}}U^2 - \frac{1}{24}U^3\hat{\mathcal{K}} + \frac{1}{24}\hat{\mathcal{K}}U^3 \\ &= \frac{-1}{4!}[U, [U, [U, \hat{\mathcal{K}}]]]. \end{aligned} \tag{9.43}$$

As shown in equation (9.32), this vanishes. The term involving one U and three $\hat{\mathcal{K}}$ may be confirmed explicitly as obeying the general rule, equation (9.4), $W_{13}(U, \hat{\mathcal{K}}) = -W_{31}(-\hat{\mathcal{K}}, -U)$. The term involving two U and two $\hat{\mathcal{K}}$ is

$$\begin{aligned} W_{22}(U, \hat{\mathcal{K}}) &= \frac{-1}{4}U^2\hat{\mathcal{K}}^2 + \frac{1}{4}U\hat{\mathcal{K}}U\hat{\mathcal{K}} - \frac{1}{8}U\hat{\mathcal{K}}U\hat{\mathcal{K}} + \frac{1}{8}U\hat{\mathcal{K}}^2U + \frac{1}{8}\hat{\mathcal{K}}U^2\hat{\mathcal{K}} \\ &\quad - \frac{1}{8}\hat{\mathcal{K}}U\hat{\mathcal{K}}U + \frac{1}{8}U^2\hat{\mathcal{K}}^2 - \frac{1}{8}U\hat{\mathcal{K}}U\hat{\mathcal{K}} + \frac{1}{8}U\hat{\mathcal{K}}U\hat{\mathcal{K}} - \frac{1}{8}U\hat{\mathcal{K}}^2U \\ &\quad + \frac{1}{8}U^2\hat{\mathcal{K}}^2 - \frac{1}{8}\hat{\mathcal{K}}U^2\hat{\mathcal{K}} - \frac{1}{24}U^2\hat{\mathcal{K}}^2 + \frac{1}{24}\hat{\mathcal{K}}U^2\hat{\mathcal{K}} \\ &\quad - \frac{1}{24}U\hat{\mathcal{K}}U\hat{\mathcal{K}} + \frac{1}{24}\hat{\mathcal{K}}U\hat{\mathcal{K}}U - \frac{1}{24}\hat{\mathcal{K}}U^2\hat{\mathcal{K}} + \frac{1}{24}\hat{\mathcal{K}}^2U^2 \\ &= \frac{2}{4!}U\hat{\mathcal{K}}U\hat{\mathcal{K}} + \frac{-2}{4!}\hat{\mathcal{K}}U\hat{\mathcal{K}}U + \frac{-1}{4!}U^2\hat{\mathcal{K}}^2 + \frac{1}{4!}\hat{\mathcal{K}}^2U^2 \\ &= \frac{1}{4!}[\hat{\mathcal{K}}, [[\hat{\mathcal{K}}, U], U]]. \end{aligned} \tag{9.44}$$

Equating the coefficients of β^4 gives

$$\begin{aligned}\hat{W}_{p,5} = & \frac{-1}{2}\hat{W}_{p,2}\hat{W}_{p,3} - \frac{1}{2}\hat{W}_{p,3}\hat{W}_{p,2} \\ & + \frac{1}{5}\left\{\frac{-1}{6}[U, \hat{\mathcal{K}}]\hat{\mathcal{K}}^3 - \frac{1}{2}U[U, \hat{\mathcal{K}}]\hat{\mathcal{K}}^2 - \frac{1}{2}U^2[U, \hat{\mathcal{K}}]\hat{\mathcal{K}} - \frac{1}{6}U^3[U, \hat{\mathcal{K}}]\right. \\ & + \frac{1}{4}[(U^2 + \hat{\mathcal{K}}U + U\hat{\mathcal{K}}), \hat{\mathcal{K}}]\hat{\mathcal{K}}^2 + \frac{1}{2}U[(U^2 + \hat{\mathcal{K}}U + U\hat{\mathcal{K}}), \hat{\mathcal{K}}]\hat{\mathcal{K}} \\ & + \frac{1}{4}U^2[(U^2 + \hat{\mathcal{K}}U + U\hat{\mathcal{K}}), \hat{\mathcal{K}}] \\ & - \frac{1}{6}[(U^3 + U^2\hat{\mathcal{K}} + U\hat{\mathcal{K}}U + \hat{\mathcal{K}}U^2 + U\hat{\mathcal{K}}^2 + \hat{\mathcal{K}}U\hat{\mathcal{K}} + \hat{\mathcal{K}}^2U), \hat{\mathcal{K}}]\hat{\mathcal{K}} \\ & - \frac{1}{6}U[(U^3 + U^2\hat{\mathcal{K}} + U\hat{\mathcal{K}}U + \hat{\mathcal{K}}U^2 + U\hat{\mathcal{K}}^2 + \hat{\mathcal{K}}U\hat{\mathcal{K}} + \hat{\mathcal{K}}^2U), \hat{\mathcal{K}}] \\ & + \frac{1}{24}[(U^4 + U^3\hat{\mathcal{K}} + U^2\hat{\mathcal{K}}U + U\hat{\mathcal{K}}U^2 + \hat{\mathcal{K}}U^3 + U^2\hat{\mathcal{K}}^2 \\ & + U\hat{\mathcal{K}}U\hat{\mathcal{K}} + \hat{\mathcal{K}}U\hat{\mathcal{K}}U + \hat{\mathcal{K}}U^2\hat{\mathcal{K}} + U\hat{\mathcal{K}}^2U + \hat{\mathcal{K}}^2U^2 \\ & \left. + U\hat{\mathcal{K}}^3 + \hat{\mathcal{K}}U\hat{\mathcal{K}}^2 + \hat{\mathcal{K}}^2U\hat{\mathcal{K}} + \hat{\mathcal{K}}^3U), \hat{\mathcal{K}}]\right\}.\end{aligned} \tag{9.45}$$

Now

$$\hat{W}_{p,2}\hat{W}_{p,3} = \frac{1}{12}[U, \hat{\mathcal{K}}]\left\{[\hat{\mathcal{K}}, [\hat{\mathcal{K}}, U]] + [U, [U, \hat{\mathcal{K}}]]\right\}. \tag{9.46}$$

The term involving four U and one $\hat{\mathcal{K}}$ is

$$\begin{aligned}W_{41}(U, \hat{\mathcal{K}}) &= \frac{-1}{5!}U^4\hat{\mathcal{K}} + \frac{4}{5!}U^3\hat{\mathcal{K}}U - \frac{6}{5!}U^2\hat{\mathcal{K}}U^2 + \frac{4}{5!}U\hat{\mathcal{K}}U^3 - \frac{1}{5!}\hat{\mathcal{K}}U^4 \\ &= \frac{-1}{5!}[U, [U, [U, [U, \hat{\mathcal{K}}]]]].\end{aligned} \tag{9.47}$$

As shown in equation (9.32), this vanishes. Obviously

$$W_{14}(U, \hat{\mathcal{K}}) = -W_{41}(-\hat{\mathcal{K}}, -U) = \frac{-1}{5!}[\hat{\mathcal{K}}, [\hat{\mathcal{K}}, [\hat{\mathcal{K}}, [\hat{\mathcal{K}}, U]]]]. \tag{9.48}$$

It is tedious but straightforward to derive the term involving three U and two $\hat{\mathcal{K}}$,

$$\begin{aligned}W_{32}(U, \hat{\mathcal{K}}) &= \frac{1}{5!}\left\{-U\hat{\mathcal{K}}^2U^2 - 6U\hat{\mathcal{K}}U^2\hat{\mathcal{K}} - 6\hat{\mathcal{K}}U^2\hat{\mathcal{K}}U + 4\hat{\mathcal{K}}U\hat{\mathcal{K}}U^2\right. \\ &\quad + 4U^2\hat{\mathcal{K}}U\hat{\mathcal{K}} \\ &\quad \left. - U^2\hat{\mathcal{K}}^2U - U^3\hat{\mathcal{K}}^2 + 4\hat{\mathcal{K}}U^3\hat{\mathcal{K}} - \hat{\mathcal{K}}^2U^3 + 4U\hat{\mathcal{K}}U\hat{\mathcal{K}}U\right\} \\ &= \frac{-1}{5!}[U, [U, [[U, \hat{\mathcal{K}}], \hat{\mathcal{K}}]]] - \frac{2}{5!}[[[U, \hat{\mathcal{K}}], U], [U, \hat{\mathcal{K}}]].\end{aligned} \tag{9.49}$$

Note that the first term on the right-hand side of the final equality does *not* contain a nest of three U and one $\hat{\mathcal{K}}$ in the sense of equation (9.32) (i.e., $W_{31}(U, \hat{\mathcal{K}})$), and so it cannot be assumed to vanish. Obviously $W_{23}(U, \hat{\mathcal{K}}) = -W_{32}(-\hat{\mathcal{K}}, -U) = W_{32}(\hat{\mathcal{K}}, U)$.

9.4.2 Expanded operator form

The above procedures are a little tedious in having to equate the same orders of the two expansions, and then having to reduce the expanded result into nested commutator form. Undoubtedly there is some virtue in the compact nature of the latter, but manually obtaining the higher order terms seems unrealistic. Here a much quicker way of obtaining the expanded higher order form is given.

We have

$$e^{-\beta(U+\hat{\mathcal{K}})} = e^{-\beta U} e^{C(-\beta U,\, -\beta\hat{\mathcal{K}})}\, e^{-\beta\hat{\mathcal{K}}}. \tag{9.50}$$

An abbreviated notation writes this in more compact form as

$$\hat{E} = e^{-\beta U}\mathcal{E}(-\beta U, -\beta\hat{\mathcal{K}})e^{-\beta\hat{\mathcal{K}}}. \tag{9.51}$$

The commutation function exponent operator is of the form

$$C(-\beta U, -\beta\hat{\mathcal{K}}) = \sum_{n=2}^{\infty}(-\beta)^n C_n(U, \hat{\mathcal{K}}), \tag{9.52}$$

and its exponential has expansion

$$\mathcal{E}(-\beta U, -\beta\hat{\mathcal{K}}) = 1 + \sum_{n=2}^{\infty}(-\beta)^n \mathcal{E}_n(U, \hat{\mathcal{K}}). \tag{9.53}$$

The two sets of coefficients are related as

$$\begin{aligned}\mathcal{E}_n = C_n &+ \frac{1}{2}\sum_{j=2}^{n-2} C_j C_{n-j} + \frac{1}{3!}\sum_{j=2}^{n-4}\sum_{k=2}^{n-2-j} C_j C_k C_{n-j-k} \\ &+ \frac{1}{4!}\sum_{j=2}^{n-6}\sum_{k=2}^{n-4-j}\sum_{\ell=2}^{n-2-j-k} C_j C_k C_\ell C_{n-j-k-\ell} + \cdots\end{aligned} \tag{9.54}$$

This allows C_n to be obtained from $\mathcal{E}_n$ and the lower order commutation function exponent coefficient operators. (In practice it seems more convenient to compute directly with the $\mathcal{E}_n$.)

Now the nth derivative of the commutation function operator is

$$\frac{\partial^n \mathcal{E}(-\beta U, -\beta\hat{\mathcal{K}})}{\partial(-\beta)^n} = e^{\beta U} f_n(\hat{E}) e^{\beta\hat{\mathcal{K}}}. \tag{9.55}$$

Here $f_n(\hat{E})$ is a linear function of $\hat{E}(-\beta)$, which is its sole dependence on β. (It also depends on U and $\hat{\mathcal{K}}$.) With this the nth coefficient of the expansion of the commutation function operator is

$$\mathcal{E}_n = \frac{1}{n!}\frac{\partial^n \mathcal{E}(-\beta U, -\beta\hat{\mathcal{K}})}{\partial(-\beta)^n}\bigg|_{\beta=0} = \frac{1}{n!}f_n(1). \tag{9.56}$$

One has

$$\begin{aligned}&\frac{\partial^{n+1}\mathcal{E}(-\beta U, -\beta\hat{\mathcal{K}})}{\partial(-\beta)^{n+1}}\\ &= -e^{\beta U}Uf_n(\hat{E})e^{\beta\hat{\mathcal{K}}} - e^{\beta U}f_n(\hat{E})\hat{\mathcal{K}}e^{\beta\hat{\mathcal{K}}} + e^{\beta U}f_n(\hat{\mathcal{H}}\hat{E})e^{\beta\hat{\mathcal{K}}}.\end{aligned} \tag{9.57}$$

This gives the recursion relation

$$f_{n+1}(\hat{E}) = f_n(\hat{\mathcal{H}}\hat{E}) - Uf_n(\hat{E}) - f_n(\hat{E})\hat{\mathcal{K}}. \tag{9.58}$$

The initial functions are

$$f_1(\hat{E}) = [\hat{\mathcal{K}}, \hat{E}], \tag{9.59}$$

and

$$f_2(\hat{E}) = [\hat{\mathcal{K}}, \hat{\mathcal{H}}\hat{E}] - U[\hat{\mathcal{K}}, \hat{E}] - [\hat{\mathcal{K}}, \hat{E}]\hat{\mathcal{K}}. \tag{9.60}$$

At $\beta = 0$, $\hat{E} = 1$, these give $f_1(1) = 0$ and $f_2(1) = [\hat{\mathcal{K}}, U]$. The latter gives $C_2(U, \hat{\mathcal{K}}) = -[U, \hat{\mathcal{K}}]/2$, which is the same result as was obtained several times above.

The recursion relation allows higher order expanded expressions for the commutator exponent operator to be generated in a mechanical fashion. Of course the number of terms grows exponentially with the order.

9.5 Evaluation of the commutation function

9.5.1 Expectation values

Using the fact that $\nabla|\mathbf{p}\rangle = (\mathrm{i}\mathbf{p}/\hbar)|\mathbf{p}\rangle$, one has for the second order expectation value

$$\frac{\langle\mathbf{q}|\hat{W}_{\mathrm{p},2}|\mathbf{p}\rangle}{\langle\mathbf{q}|\mathbf{p}\rangle} = \frac{-\hbar^2}{4m}\left\{(\nabla_{\mathbf{q}}^2 U(\mathbf{q})) + \frac{2\mathrm{i}}{\hbar}\mathbf{p}\cdot(\nabla_{\mathbf{q}} U(\mathbf{q}))\right\}. \tag{9.61}$$

The parenthesis around the gradient operators are now redundant. Here and below the potential $U(\mathbf{q})$ and gradient $\nabla_{\mathbf{q}}$ are functions of the phase space position $\mathbf{q}$.

The third order expectation value is

$$\begin{aligned}\frac{\langle\mathbf{q}|\hat{W}_{\mathrm{p},3}|\mathbf{p}\rangle}{\langle\mathbf{q}|\mathbf{p}\rangle} = \frac{\hbar^2}{6m}(\nabla U)\cdot(\nabla U) - \frac{\hbar^4}{24m^2}\Big\{&(\nabla^2\nabla^2 U) + \frac{4\mathrm{i}}{\hbar}\mathbf{p}\cdot(\nabla^2\nabla U)\\ &- \frac{4}{\hbar^2}\mathbf{pp}:(\nabla\nabla U)\Big\}.\end{aligned} \tag{9.62}$$

The fourth order expectation value is

$$\begin{aligned}\frac{\langle\mathbf{q}|\hat{W}_{\mathrm{p},4}|\mathbf{p}\rangle}{\langle\mathbf{q}|\mathbf{p}\rangle} &= \frac{\langle\mathbf{q}|\hat{W}_{\mathrm{p},31}+\hat{W}_{\mathrm{p},13}+\hat{W}_{\mathrm{p},22}|\mathbf{p}\rangle}{\langle\mathbf{q}|\mathbf{p}\rangle}\\ &= \frac{-\hbar^6}{192m^3}\bigg((\nabla^2\nabla^2\nabla^2 U)+\frac{6\mathrm{i}}{\hbar}\mathbf{p}\cdot(\nabla^2\nabla^2\nabla U)\\ &\quad -\frac{12}{\hbar^2}\mathbf{pp}\colon(\nabla^2\nabla\nabla U)-\frac{8\mathrm{i}}{\hbar^3}\mathbf{ppp}\,\vdots\,(\nabla\nabla\nabla U)\bigg)\\ &\quad +\frac{\hbar^4}{96m^2}\bigg(4(\nabla^2\nabla U)\cdot(\nabla U)+4(\nabla\nabla U)\colon(\nabla\nabla U)\\ &\quad +\frac{8\mathrm{i}}{\hbar}(\nabla\nabla U)\colon(\mathbf{p}\nabla U)\bigg).\end{aligned} \tag{9.63}$$

These three expectation values are extensive.

The square of the second order commutation function operator is given by

$$\begin{aligned}&\frac{16m^2}{\hbar^4}\hat{W}_{\mathrm{p},2}^2\\ &=\left\{(\nabla_{\mathbf{r}}^2 U(\mathbf{r}))+2(\nabla_{\mathbf{r}}U(\mathbf{r}))\cdot\nabla_{\mathbf{r}}\right\}^2\\ &=(\nabla_{\mathbf{r}}^2 U(\mathbf{r}))^2+2(\nabla_{\mathbf{r}}^2 U(\mathbf{r}))(\nabla_{\mathbf{r}}U(\mathbf{r}))\cdot\nabla_{\mathbf{r}}\\ &\quad+2(\nabla_{\mathbf{r}}U(\mathbf{r}))\cdot(\nabla_{\mathbf{r}}^2\nabla_{\mathbf{r}}U(\mathbf{r}))+2(\nabla_{\mathbf{r}}U(\mathbf{r}))\cdot(\nabla_{\mathbf{r}}^2 U(\mathbf{r}))\nabla_{\mathbf{r}}\\ &\quad+4(\nabla_{\mathbf{r}}U(\mathbf{r}))\cdot(\nabla_{\mathbf{r}}\nabla_{\mathbf{r}}U(\mathbf{r}))\cdot\nabla_{\mathbf{r}}+4(\nabla_{\mathbf{r}}U(\mathbf{r}))(\nabla_{\mathbf{r}}U(\mathbf{r}))\colon\nabla_{\mathbf{r}}\nabla_{\mathbf{r}}.\end{aligned} \tag{9.64}$$

The expectation value is

$$\begin{aligned}\frac{\langle\mathbf{q}|\hat{W}_{\mathrm{p},2}^2|\mathbf{p}\rangle}{\langle\mathbf{q}|\mathbf{p}\rangle} &= \frac{\hbar^4}{16m^2}\bigg\{(\nabla^2 U)^2+\frac{4\mathrm{i}}{\hbar}(\nabla^2 U)\mathbf{p}\cdot(\nabla U)+2(\nabla U)\cdot(\nabla^2\nabla U)\\ &\quad+\frac{4\mathrm{i}}{\hbar}\mathbf{p}(\nabla U)\colon(\nabla\nabla U)-\frac{4}{\hbar^2}\mathbf{pp}\colon(\nabla U)(\nabla U)\bigg\}.\end{aligned} \tag{9.65}$$

9.5.2 Gradients of the singlet and pair potential

Consider a system of N particles with potential energy comprising singlet and pair potentials,

$$U(\mathbf{q}) = U^{(1)}(\mathbf{q}) + U^{(2)}(\mathbf{q}) = \sum_{j=1}^{N} u^{(1)}(\mathbf{q}_j) + \frac{1}{2}\sum_{j=1}^{N}\sum_{k=1}^{N}{}^{(k\neq j)}u^{(2)}(\mathbf{q}_j, \mathbf{q}_k). \tag{9.66}$$

The pair potential is necessarily symmetric, $u^{(2)}(\mathbf{q}_j, \mathbf{q}_k) = u^{(2)}(\mathbf{q}_k, \mathbf{q}_j)$, and it is often central, $u^{(2)}(\mathbf{q}_j, \mathbf{q}_k) = u^{(2)}(q_{jk})$, with $q_{jk} = |\mathbf{q}_{jk}|$ and $\mathbf{q}_{jk} = \mathbf{q}_j - \mathbf{q}_k$. For brevity we may write $u_j^{(1)} \equiv u^{(1)}(\mathbf{q}_j)$ and $u_{jk}^{(2)} \equiv u^{(2)}(\mathbf{q}_j, \mathbf{q}_k)$. The gradient operator is $\nabla = \sum_{j=1}^{N}\hat{\mathbf{j}}\,\nabla_j$.

For the second order expectation value one requires

$$\nabla U(\mathbf{q}) = \sum_{j=1}^{N} \hat{\mathbf{j}}\, \nabla_j u^{(1)}(\mathbf{q}_j) + \sum_{j=1}^{N} \sum_{k=1}^{N}{}^{(k \neq j)} \hat{\mathbf{j}}\, \nabla_j u^{(2)}(\mathbf{q}_j, \mathbf{q}_k), \tag{9.67}$$

and

$$\nabla^2 U(\mathbf{q}) = \sum_{j=1}^{N} \nabla_j^2 u^{(1)}(\mathbf{q}_j) + \sum_{j=1}^{N} \sum_{k=1}^{N}{}^{(k \neq j)} \nabla_j^2 u^{(2)}(\mathbf{q}_j, \mathbf{q}_k). \tag{9.68}$$

Here and below these are functions, not operators, which is to say that the gradients do not operate beyond the function to their right.

For the third order expectation value one requires

$$\begin{aligned}
\nabla U \cdot \nabla U
&= \sum_{j=1}^{N} \left\{ \nabla_j u_j^{(1)} + \sum_{k=1}^{N}{}^{(k \neq j)} \nabla_j u_{jk}^{(2)} \right\} \cdot \left\{ \nabla_j u_j^{(1)} + \sum_{k'=1}^{N}{}^{(k' \neq j)} \nabla_j u_{jk'}^{(2)} \right\} \\
&= \sum_{j=1}^{N} \nabla_j u_j^{(1)} \cdot \nabla_j u_j^{(1)} + 2 \sum_{j=1}^{N} \sum_{k=1}^{N}{}^{(k \neq j)} \nabla_j u_j^{(1)} \cdot \nabla_j u_{jk}^{(2)} \\
&\quad + \sum_{j=1}^{N} \sum_{k=1}^{N}{}^{(k \neq j)} \sum_{k'=1}^{N}{}^{(k' \neq j)} \nabla_j u_{jk}^{(2)} \cdot \nabla_j u_{jk'}^{(2)},
\end{aligned} \tag{9.69}$$

$$\begin{aligned}
\nabla^2 \nabla^2 U &= \sum_{\ell=1}^{N} \nabla_\ell^2 \left\{ \sum_{j=1}^{N} \nabla_j^2 u_j^{(1)} + \sum_{j=1}^{N} \sum_{k=1}^{N}{}^{(k \neq j)} \nabla_j^2 u_{jk}^{(2)} \right\} \\
&= \sum_{j=1}^{N} \nabla_j^2 \nabla_j^2 u_j^{(1)} + \sum_{j=1}^{N} \sum_{k=1}^{N}{}^{(k \neq j)} \left\{ \nabla_j^2 \nabla_j^2 u_{jk}^{(2)} + \nabla_j^2 \nabla_k^2 u_{jk}^{(2)} \right\},
\end{aligned} \tag{9.70}$$

$$\begin{aligned}
\nabla^2 \nabla U &= \sum_{\ell=1}^{N} \hat{\ell} \nabla_\ell \left\{ \sum_{j=1}^{N} \nabla_j^2 u_j^{(1)} + \sum_{j=1}^{N} \sum_{k=1}^{N}{}^{(k \neq j)} \nabla_j^2 u_{jk}^{(2)} \right\} \\
&= \sum_{j=1}^{N} \hat{\mathbf{j}} \nabla_j \nabla_j^2 u_j^{(1)} + \sum_{j=1}^{N} \sum_{k=1}^{N}{}^{(k \neq j)} \left\{ \hat{\mathbf{j}} \nabla_j \nabla_j^2 u_{jk}^{(2)} + \hat{\mathbf{j}} \nabla_j \nabla_k^2 u_{jk}^{(2)} \right\},
\end{aligned} \tag{9.71}$$

and

$$\nabla \nabla U = \sum_{j=1}^{N} \hat{\mathbf{j}}\hat{\mathbf{j}} \nabla_j \nabla_j u_j^{(1)} + \sum_{j=1}^{N} \sum_{k=1}^{N}{}^{(k \neq j)} \left\{ \hat{\mathbf{j}}\hat{\mathbf{j}} \nabla_j \nabla_j u_{jk}^{(2)} + \hat{\mathbf{j}}\hat{\mathbf{k}} \nabla_j \nabla_k u_{jk}^{(2)} \right\}. \tag{9.72}$$

For the fourth order expectation value one requires

$$\begin{aligned}\nabla^2\nabla^2\nabla^2 U = &\sum_{j=1}^{N} \nabla_j^2\nabla_j^2\nabla_j^2 u_j^{(1)}\\ &+\sum_{j=1}^{N}\sum_{k=1}^{N}{}^{(k\neq j)}\left\{\nabla_j^2\nabla_j^2\nabla_j^2 u_{jk}^{(2)} + 3\nabla_k^2\nabla_j^2\nabla_j^2 u_{jk}^{(2)}\right\},\end{aligned} \tag{9.73}$$

$$\begin{aligned}\nabla^2\nabla^2\nabla U = &\sum_{j=1}^{N}\hat{\mathbf{j}}\ \nabla_j\nabla_j^2\nabla_j^2 u_j^{(1)} + \sum_{j=1}^{N}\sum_{k=1}^{N}{}^{(k\neq j)}\left\{\hat{\mathbf{j}}\ \nabla_j\nabla_j^2\nabla_j^2 u_{jk}^{(2)}\right.\\ &\left.+\hat{\mathbf{k}}\ \nabla_k\nabla_j^2\nabla_j^2 u_{jk}^{(2)} + 2\hat{\mathbf{j}}\ \nabla_j\nabla_j^2\nabla_k^2 u_{jk}^{(2)}\right\},\end{aligned} \tag{9.74}$$

$$\begin{aligned}\nabla^2\nabla\nabla U = &\sum_{j=1}^{N}\hat{\mathbf{j}}\hat{\mathbf{j}}\nabla_j^2\nabla_j\nabla_j u_j^{(1)} + \sum_{j=1}^{N}\sum_{k=1}^{N}{}^{(k\neq j)}\left\{\hat{\mathbf{j}}\hat{\mathbf{j}}\nabla_j^2\nabla_j\nabla_j u_{jk}^{(2)}\right.\\ &\left.+\hat{\mathbf{j}}\hat{\mathbf{j}}\nabla_k^2\nabla_j\nabla_j u_{jk}^{(2)} + 2\hat{\mathbf{j}}\hat{\mathbf{k}}\nabla_j^2\nabla_j\nabla_k u_{jk}^{(2)}\right\},\end{aligned} \tag{9.75}$$

and

$$\begin{aligned}\nabla\nabla\nabla U = &\sum_{j=1}^{N}\hat{\mathbf{j}}\hat{\mathbf{j}}\hat{\mathbf{j}}\nabla_j\nabla_j\nabla_j u_j^{(1)} + \sum_{j=1}^{N}\sum_{k=1}^{N}{}^{(k\neq j)}\left\{\hat{\mathbf{j}}\hat{\mathbf{j}}\hat{\mathbf{j}}\nabla_j\nabla_j\nabla_j u_{jk}^{(2)}\right.\\ &\left.+\hat{\mathbf{j}}\hat{\mathbf{j}}\hat{\mathbf{k}}\nabla_j\nabla_j\nabla_k u_{jk}^{(2)} + \hat{\mathbf{j}}\hat{\mathbf{k}}\hat{\mathbf{j}}\nabla_j\nabla_j\nabla_k u_{jk}^{(2)} + \hat{\mathbf{k}}\hat{\mathbf{j}}\hat{\mathbf{j}}\nabla_k\nabla_j\nabla_j u_{jk}^{(2)}\right\}.\end{aligned} \tag{9.76}$$

9.5.3 Central pair potential

In d dimensions, $\mathbf{q}_j = \sum_{\alpha=1}^{d} q_{j\alpha}\hat{\mathbf{x}}_\alpha$, the separation squared is $q_{jk}^2 = \mathbf{q}_{jk}\cdot\mathbf{q}_{jk} = \sum_{\alpha=1}^{d} q_{jk\alpha}^2$, with $\mathbf{q}_{jk} = \mathbf{q}_j - \mathbf{q}_k$. A central pair potential is a function of the separation, $u^{(2)}(\mathbf{q}_j, \mathbf{q}_k) = u(q_{jk})$. Note the superscript (2) on the central pair potential has been dropped because in this section we shall only deal with the pair potential. Note that because the pair potential is a function of separation, $q_{jk} = |\mathbf{q}_j - \mathbf{q}_k|$, one can always make the replacement $\nabla_k \Rightarrow -\nabla_j$. Everywhere below, $j \neq k$.

For the second order expectation value one has

$$\nabla_j u(q_{jk}) = u'(q_{jk})\nabla_j q_{jk} = \frac{\mathbf{q}_{jk}}{q_{jk}}u'(q_{jk}), \tag{9.77}$$

and

$$\nabla_j^2 u(q_{jk}) = \nabla_j\cdot\left(u'(q_{jk})\frac{\mathbf{q}_{jk}}{q_{jk}}\right) = u''(q_{jk}) + \frac{d-1}{q_{jk}}u'(q_{jk}). \tag{9.78}$$

For the third order expectation value one has

$$
\begin{aligned}
\nabla_j^2\nabla_j^2 u_{jk}^{(2)} &= \nabla_j^2\nabla_k^2 u_{jk}^{(2)} \\
&= u^{\mathrm{iv}}(q_{jk}) + \frac{2(d-1)}{q_{jk}}u'''(q_{jk}) + \frac{d^2-4d+3}{q_{jk}^2}u''(q_{jk}) \\
&\quad - \frac{d^2-4d+3}{q_{jk}^3}u'(q_{jk}),
\end{aligned}
\tag{9.79}
$$

$$
\begin{aligned}
\nabla_j\nabla_j^2 u_{jk}^{(2)} &= \nabla_j\nabla_k^2 u_{jk}^{(2)} \\
&= \frac{\mathbf{q}_{jk}}{q_{jk}}u'''(q_{jk}) + \frac{(d-1)\mathbf{q}_{jk}}{q_{jk}^2}u''(q_{jk}) - \frac{(d-1)\mathbf{q}_{jk}}{q_{jk}^3}u'(q_{jk}),
\end{aligned}
\tag{9.80}
$$

and

$$
\begin{aligned}
\nabla_j\nabla_j u_{jk}^{(2)} &= -\nabla_k\nabla_j u_{jk}^{(2)} \\
&= \frac{\mathbf{q}_{jk}\mathbf{q}_{jk}}{q_{jk}^2}u''(q_{jk}) - \frac{\mathbf{q}_{jk}\mathbf{q}_{jk}}{q_{jk}^3}u'(q_{jk}) + \frac{1}{q_{jk}}u'(q_{jk})\hat{\mathrm{I}}.
\end{aligned}
\tag{9.81}
$$

For the fourth order expectation value one has

$$
\begin{aligned}
&\nabla_j\nabla_j^2\nabla_j^2 u_{jk}^{(2)} \\
&= \left\{\frac{\mathbf{q}_{jk}}{q_{jk}}u^{\mathrm{v}}(q_{jk}) + \frac{2(d-1)\mathbf{q}_{jk}}{q_{jk}^2}u^{\mathrm{iv}}(q_{jk}) + \frac{(d^2-6d+5)\mathbf{q}_{jk}}{q_{jk}^3}u'''(q_{jk})\right. \\
&\quad \left. - \frac{(3d^2-12d+9)\mathbf{q}_{jk}}{q_{jk}^4}u''(q_{jk}) + \frac{(3d^2-12d+9)\mathbf{q}_{jk}}{q_{jk}^5}u'(q_{jk})\right\},
\end{aligned}
\tag{9.82}
$$

$$
\begin{aligned}
&\nabla_j^2\nabla_j^2\nabla_j^2 u_{jk}^{(2)} \\
&= u^{\mathrm{vi}}(q_{jk}) + \frac{3d-3}{q_{jk}}u^{\mathrm{v}}(q_{jk}) + \frac{(3d^2-12d+9)}{q_{jk}^2}u^{\mathrm{iv}}(q_{jk}) \\
&\quad + \frac{(d^3-12d^2+35d-24)}{q_{jk}^3}u'''(q_{jk}) - \frac{3d^3-27d^2+69d-45}{q_{jk}^4}u''(q_{jk}) \\
&\quad + \frac{(3d^2-12d+9)(d-5)}{q_{jk}^5}u'(q_{jk}),
\end{aligned}
\tag{9.83}
$$

$$
\begin{aligned}
&\nabla_j \nabla_j \nabla_j^2 u_{jk}^{(2)} \\
&= \frac{\mathbf{q}_{jk}\mathbf{q}_{jk}}{q_{jk}^2} u^{\mathrm{iv}}(q_{jk}) + \frac{q_{jk}^2 \hat{\mathrm{I}} - \mathbf{q}_{jk}\mathbf{q}_{jk}}{q_{jk}^3} u'''(q_{jk}) \\
&+ \frac{(d-1)\mathbf{q}_{jk}\mathbf{q}_{jk}}{q_{jk}^3} u'''(q_{jk}) + \frac{(d-1)[q_{jk}^2 \hat{\mathrm{I}} - 2\mathbf{q}_{jk}\mathbf{q}_{jk}]}{q_{jk}^4} u''(q_{jk}) \\
&- \frac{(d-1)\mathbf{q}_{jk}\mathbf{q}_{jk}}{q_{jk}^4} u''(q_{jk}) - \frac{(d-1)[q_{jk}^2 \hat{\mathrm{I}} - 3\mathbf{q}_{jk}\mathbf{q}_{jk}]}{q_{jk}^5} u'(q_{jk}),
\end{aligned} \tag{9.84}
$$

and

$$
\begin{aligned}
&\nabla_{j,\alpha} \nabla_{j,\beta} \nabla_{j,\gamma} u_{jk}^{(2)} \\
&= \frac{q_{jk,\alpha} q_{jk,\beta} q_{jk,\gamma}}{q_{jk}^3} u'''(q_{jk}) \\
&+ \frac{[\delta_{\alpha\beta} q_{jk,\gamma} + \delta_{\alpha\gamma} q_{jk,\beta} + \delta_{\beta\gamma} q_{jk,\alpha}] q_{jk}^2 - 3 q_{jk,\beta} q_{jk,\gamma} q_{jk,\alpha}}{q_{jk}^4} u''(q_{jk}) \\
&- \frac{[\delta_{\alpha\beta} q_{jk,\gamma} + \delta_{\alpha\gamma} q_{jk,\beta} + \delta_{\beta\gamma} q_{jk,\alpha}] q_{jk}^2 - 3 q_{jk,\beta} q_{jk,\gamma} q_{jk,\alpha}}{q_{jk}^5} u'(q_{jk}).
\end{aligned} \tag{9.85}
$$

The Greek indices belong to $\{x, y, \ldots, d\}$.

9.6 Results for the one-dimensional harmonic crystal

9.6.1 Expectation value of high temperature expansion operators

The expectation values of the nested commutator operators in the high temperature expansion of the commutation function were given in section 9.5.1.

The expectation value of the second order operator is

$$
\frac{\langle \mathbf{q} | \hat{W}_{\mathrm{p},2} | \mathbf{p} \rangle}{\langle \mathbf{q} | \mathbf{p} \rangle} = \frac{-\hbar^2}{4m} \left\{ (\nabla_{\mathbf{q}}^2 U(\mathbf{q})) + \frac{2\mathrm{i}}{\hbar} \mathbf{p} \cdot (\nabla_{\mathbf{q}} U(\mathbf{q})) \right\}. \tag{9.86}
$$

The expectation value of the third order operator is

$$
\begin{aligned}
\frac{\langle \mathbf{q} | \hat{W}_{\mathrm{p},3} | \mathbf{p} \rangle}{\langle \mathbf{q} | \mathbf{p} \rangle} = &\frac{\hbar^2}{6m} (\nabla U) \cdot (\nabla U) - \frac{\hbar^4}{24m^2} \left\{ (\nabla^2 \nabla^2 U) + \frac{4\mathrm{i}}{\hbar} \mathbf{p} \cdot (\nabla^2 \nabla U) \right. \\
&\left. - \frac{4}{\hbar^2} \mathbf{p}\mathbf{p} : (\nabla \nabla U) \right\}.
\end{aligned} \tag{9.87}
$$

The expectation value of the fourth order operator is

$$\begin{aligned}\frac{\langle \mathbf{q}|\hat{W}_{\mathrm{p},4}|\mathbf{p}\rangle}{\langle \mathbf{q}|\mathbf{p}\rangle} &= \frac{\langle \mathbf{q}|\hat{W}_{\mathrm{p},4;31}+\hat{W}_{\mathrm{p},4;13}+\hat{W}_{\mathrm{p},4;22}|\mathbf{p}\rangle}{\langle \mathbf{q}|\mathbf{p}\rangle} \\ &= \frac{-\hbar^6}{192m^3}\Bigg((\nabla^2\nabla^2\nabla^2 U) + \frac{6\mathrm{i}}{\hbar}\mathbf{p}\cdot(\nabla^2\nabla^2\nabla U) \\ &\quad - \frac{12}{\hbar^2}\mathbf{pp}\colon(\nabla^2\nabla\nabla U) - \frac{8\mathrm{i}}{\hbar^3}\mathbf{ppp}\,\vdots\,(\nabla\nabla\nabla U)\Bigg) \\ &\quad + \frac{\hbar^4}{96m^2}\Bigg(4(\nabla^2\nabla U)\cdot(\nabla U) + 4(\nabla\nabla U)\colon(\nabla\nabla U) \\ &\quad + \frac{8\mathrm{i}}{\hbar}(\nabla\nabla U)\colon(\mathbf{p}\nabla U)\Bigg).\end{aligned} \tag{9.88}$$

These three expectation values are extensive.

The expectation value of the square of the second order operator is

$$\begin{aligned}\frac{\langle \mathbf{q}|\hat{W}^2_{\mathrm{p},2}|\mathbf{p}\rangle}{\langle \mathbf{q}|\mathbf{p}\rangle} = \frac{\hbar^4}{16m^2}\Bigg\{&(\nabla^2 U)^2 + \frac{4\mathrm{i}}{\hbar}(\nabla^2 U)\mathbf{p}\cdot(\nabla U) + 2(\nabla U)\cdot(\nabla^2\nabla U) \\ &+ \frac{4\mathrm{i}}{\hbar}\mathbf{p}(\nabla U)\colon(\nabla\nabla U) - \frac{4}{\hbar^2}\mathbf{pp}\colon(\nabla U)(\nabla U)\Bigg\}.\end{aligned} \tag{9.89}$$

9.6.2 Harmonic crystal potential and gradients

Analytic expressions for the energy eigenvalues and eigenfunctions of a one-dimensional harmonic crystal are derived in section 6.2, following earlier work (Attard 2002, 2019a). The model has particles attached by linear springs to each neighbor and to lattice sites. Let the coordinate of the jth particle be q_j. Let the jth lattice position, the center of the external potential acting on particle j, be $\bar{q}_j = j\Delta_q$, which are evidently ordered along the line. The lattice spacing is also the relaxed inter-particle spring length. There are fixed 'wall' particles at $q_0 = 0$ and $q_{N+1} = (N+1)\Delta_q$. Let $d_j \equiv q_j - \bar{q}_j$ be the displacement from the lattice position; for the wall particles, $d_0 = d_{N+1} = 0$. The system has number density $\rho = \Delta_q^{-1}$.

In this model, there is an external harmonic potential of spring constant κ acting on each particle centered at its lattice site. The inter-particle spring has strength λ and relaxed length Δ_q. With these the potential energy is

$$\begin{aligned}U(\mathbf{q}) &= \frac{\kappa}{2}\sum_{j=1}^{N}\left[q_j - \bar{q}_j\right]^2 + \frac{\lambda}{2}\sum_{j=0}^{N}\left[q_{j+1} - q_j - \Delta_q\right]^2 \\ &= \frac{\kappa}{2}\sum_{j=1}^{N} d_j^2 + \frac{\lambda}{2}\sum_{j=0}^{N}\left[d_{j+1} - d_j\right]^2,\end{aligned} \tag{9.90}$$

Obviously any derivative beyond the second is zero. Also, the gradients only involve the particles, $j \in \{1, 2, \ldots, N\}$. For the pair potential, $k = j \pm 1$.

As discussed in section 6.2, the potential is non-symmetric with respect to particle permutations. Therefore the particles are not identical and the wave function for this system does not obey any particular symmetrization rules; the particles cannot be regarded as fermions or bosons.

One has for $j \in \{1, 2, \dots, N\}$,

$$\nabla_j U = \kappa d_j + \lambda(2d_j - d_{j+1} - d_{j-1}), \tag{9.91}$$

and for $k \in \{1, 2, \dots, N\}$

$$\nabla_j \nabla_k U = \kappa \delta_{jk} + 2\lambda\delta_{jk} - \lambda\delta_{k,j+1} - \lambda\delta_{k,j-1}. \tag{9.92}$$

Here for the last atom, $k = N$ for example, one of the neighbor derivatives is zero, $\delta_{k,j+1} = 0$. Higher derivatives vanish. Obviously $\nabla^2 U = N\kappa + 2N\lambda$, $\nabla U \cdot \nabla U = \sum_{j=1}^{N}\{\kappa d_j + \lambda[2d_j - d_{j+1} - d_{j-1}]\}^2$, and

$$\begin{aligned} \mathbf{pp}: (\nabla\nabla U) &= \sum_{j=1}^{N}\sum_{k=1}^{N} p_j p_k \left\{\kappa\delta_{jk} + 2\lambda\delta_{jk} - \lambda\delta_{k,j+1} - \lambda\delta_{k,j-1}\right\} \\ &= \sum_{j=1}^{N}\left\{[\kappa + 2\lambda]p_j^2 + 2\lambda\delta_{jk} - \lambda p_j p_{j+1} - \lambda p_j p_{j-1}\right\}. \end{aligned} \tag{9.93}$$

Also

$$\begin{aligned} (\nabla\nabla U): (\nabla\nabla U) &= \sum_{j,k=1}^{N} \left\{\kappa\delta_{jk} + 2\lambda\delta_{jk} - \lambda\delta_{k,j+1} - \lambda\delta_{k,j-1}\right\}^2 \\ &= N(\kappa + 2\lambda)^2 + 2(N-1)\lambda^2, \end{aligned} \tag{9.94}$$

and

$$\begin{aligned} (\nabla\nabla U): (\mathbf{p}\nabla U) &= \sum_{j,k=1}^{N} \left\{\kappa\delta_{jk} + 2\lambda\delta_{jk} - \lambda\delta_{k,j+1} - \lambda\delta_{k,j-1}\right\} \\ &\quad \times p_k \left\{\kappa d_j + \lambda(2d_j - d_{j+1} - d_{j-1})\right\} \\ &= \sum_{j=1}^{N}\left\{(\kappa + 2\lambda)p_j - \lambda p_{j+1} - \lambda p_{j-1}\right\} \\ &\quad \times \left[(\kappa + 2\lambda)d_j - \lambda(d_{j+1} + d_{j-1})\right]. \end{aligned} \tag{9.95}$$

Note that in these, $p_{N+1} = p_0 = 0$, and $d_{N+1} = d_0 = 0$.

9.6.3 Simulation results

A simulation of the one-dimensional quantum harmonic crystal in classical phase space using the commutation function given by the above high temperature expansion was performed with a simulation algorithm similar to that described in detail elsewhere (Attard 2018, 2019b). Briefly the Metropolis algorithm in position and momentum space was used with the usual classical Maxwell–Boltzmann weight. (The Gaussian integrals over the momenta were not performed analytically.) Averages were evaluated by using each specific commutation function as weight for each of the four averages that were collected (unity (i.e., classical), $\mathcal{O}(\beta^2)$, $\mathcal{O}(\beta^3)$, $\mathcal{O}(\beta^4)$). Since the particles that comprise the harmonic crystal are not identical, the symmetrization function was set to unity.

Typically enough configurations were generated to make the statistical error less than 1%, sometimes much less. Interestingly enough, for an accuracy of about 1%, the Monte Carlo algorithm was a factor of about 2000 times more efficient (in terms of total computer time) than the quasi-analytic exact phonon method described in section 6.2 (Attard 2019a). The main bottleneck in the latter was the crude numerical quadrature method that was used to evaluate the density profile, and this was exacerbated by the large number of energy levels that were required for accurate results at higher temperatures. The phonon method is more efficient in one respect, namely that it requires negligible computer time for each additional temperature point; the simulations give results for only one temperature at a time.

The results that follow have been made dimensionless using parameters for a realistic Lennard–Jones potential model of neon (van Sciver 2012). The mass is $m = 3.35 \times 10^{-26}$ kg, the well-depth is $\varepsilon = 4.93 \times 10^{-22}$ J, and the equilibrium separation is $r_e = 3.13 \times 10^{-10}$ m. These give the Lennard–Jones frequency as $\omega_{LJ} = 3.28 \times 10^{12}$ Hz.

In table 9.1, results for the average energy given by the high temperature expansion are compared with the exact results. These results invoke a symmetrization function equal to unity (i.e., unsymmetrized wave function; no difference between bosons and fermions). Because of the limited number of energy levels used,

Table 9.1. Canonical equilibrium average energy, $\beta\langle\mathcal{H}\rangle$, of the harmonic crystal for $N = 4$ unsymmetrized particles, and $\lambda = \kappa = m\omega_{LJ}^2$. Columns 2, 3 and 4 are the exact results using either the number of energy levels at the top of each column, or else the analytic expression, equation (6.32). The final three columns are the high temperature expansion taken to the order given at the top of the column. The statistical error (96% confidence level, twice the standard deviation of block sub-averages) is shown in parentheses for the final two digits. The classical result is $\beta\langle\mathcal{H}\rangle = N = 4$.

$\beta\hbar\omega_{LJ}$	5000	10 000	analytic	$\mathcal{O}(\beta^2)$	$\mathcal{O}(\beta^3)$	$\mathcal{O}(\beta^4)$
0.1	2.146	2.459	4.010 0	3.999 2(19)	4.038 8(19)	4.009 3(19)
0.2	3.440	3.695	4.040 0	3.999 8(15)	4.153 6(17)	4.040 4(15)
0.3	3.961	4.046	4.090 0	4.001 3(14)	4.331 8(16)	4.091 0(14)
0.4	4.140	4.155	4.159 0	3.999 8(14)	4.551 4(17)	4.147 9(15)
0.5	4.244	4.246	4.246 4	4.000 0(14)	4.799 8(18)	4.197 2(16)
0.6	4.353	4.353	4.352 7	4.000 0(13)	5.058 3(19)	4.221 3(16)
0.7	4.477	4.477	4.476 6	4.000 0(13)	5.316 1(19)	4.216 8(16)
0.8	4.618	4.617	4.617 4	4.000 8(12)	5.560 9(20)	4.198 8(15)
0.9	4.774	4.774	4.774 4	3.999 9(13)	5.789 3(21)	4.172 6(15)
1.0	4.947	4.947	4.946 5	3.999 3(13)	5.999 6(21)	4.147 8(14)

those exact results become inaccurate as the temperature is increased. As mentioned in the analysis of the normal modes of the harmonic crystal, section 6.2, the average energy can be obtained most simply from the quantum harmonic oscillator analytic expression using the mode frequencies, equation (6.32); this is the fourth column in the table. Comparing the three columns, it can be seen that exact results obtained with 10 000 energy eigenvalues are reliable for $\beta\hbar\omega_{\rm LJ} \gtrsim 0.4$. Conversely, the high temperature expansions are reliable at high temperatures. One can see that there is some overlap between the regimes of validity of the two approaches; the fourth order results and the exact results for 10 000 levels agree to within about 1% for $0.4 \lesssim \beta\hbar\omega_{\rm LJ} \lesssim 0.5$.

Note that for this harmonic crystal, Re $W_2(\mathbf{q}, \mathbf{p})$ is a constant in phase space. This means that it does not contribute to any average, and so the $\mathcal{O}(\beta^2)$ result should be identical to the classical result, $\beta\langle\mathcal{H}\rangle = N = 4$. The variation in this column in the table is statistical error.

It is worth emphasizing the large number of energy eigenvalues that are required to obtain reliable exact results for even moderately high temperatures $\beta\hbar\omega_{\rm LJ} \approx 0.5$. This is just for $N = 4$ particles. By using Lennard–Jones parameters to make the results dimensionless, the present calculations can be more or less directly compared to calculations for Lennard–Jones systems with similar number of particles at similar densities. Hernando and Vaníček (2013) for just such a similar system obtained, heroically, the first 50 energy eigenvalues. (It should be noted that the statistical averages that these authors presented for the density profiles were effectively at temperatures an order of magnitude lower than this.)

The data from table 9.1 is presented in visual form in figure 9.1. One can see that the average energy of the quantum harmonic crystal is higher than the classical prediction. The exact calculations with 10 000 energy levels appears reliable for $\beta\hbar\omega_{\rm LJ} \gtrsim 0.2$, and the calculations with the $\mathcal{O}(\beta^4)$ commutation function appear

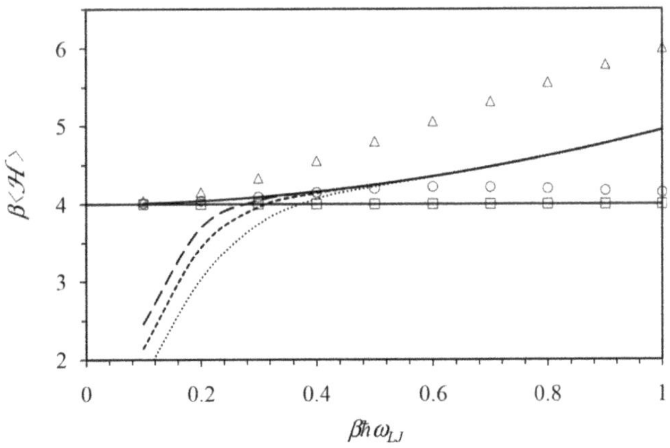

Figure 9.1. Same data as table 9.1. The broken curves are the exact results for 2000 (dotted), 5000 (short dashed), and 10 000 (long dashed) energy levels. The solid curve is the exact analytic result, equation (6.32). The symbols are the high temperature expansions to $\mathcal{O}(\beta^2)$ (squares), $\mathcal{O}(\beta^3)$ (triangles), and $\mathcal{O}(\beta^4)$ (circles). The straight line is the classical result.

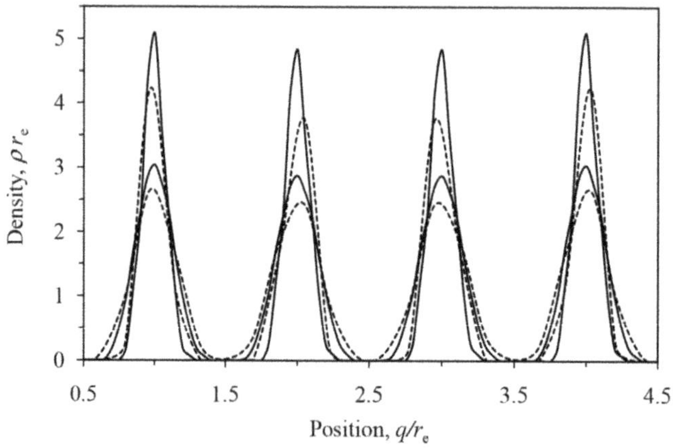

Figure 9.2. Density profiles for $\beta\hbar\omega_{LJ}$= 0.2 (lower) and 0.8 (upper) ($N = 4$, $\Delta_q = r_e$, $\lambda = \kappa = m\omega_{LJ}^2$). The solid curves are exact results with $l_{max} = 10\,000$, and the dashed curves result from the fourth order high temperature expansion.

reliable for $\beta\hbar\omega_{LJ} \lesssim 0.6$. (This is quantitatively similar to the regime of applicability of the high temperature expansion established for a Lennard–Jones system based on data presented in table 8.2.) Evidently the departure from the classical prediction is due to the non-commutativity of the momentum and position operators.

Figure 9.2 shows the density profiles for the harmonic crystal with equal inter-particle and lattice spring constants, $\lambda = \kappa = m\omega_{LJ}^2$. This is a relatively stiff system, with the particles largely confined to their lattice sites, as signified by the sharp peaks in the density profiles. At the higher temperature, $\beta\hbar\omega_{LJ} = 0.2$, the peaks are broader and the density is not quite zero between the lattice sites. As the temperature decreases the particles are increasingly localized to the lattice positions. The high temperature expansion tends to give broader peaks than the exact calculations at each of the two temperatures shown. The classical results would be almost indistinguishable from the high temperature expansions on the scale of the figure. The reason for the lack of quantum effect on the density profile in the high temperature expansion is unclear. It should be mentioned that results for the density profile at temperatures intermediate between those shown in figure 9.2 show no better agreement between the fourth order high temperature expansion and the exact modal calculation, nor any great change in the former from the classical prediction.

References

Attard P 2002 *Thermodynamics and Statistical Mechanics: Equilibrium by Entropy Maximisation* (London: Academic)

Attard P 2018 Quantum statistical mechanics in classical phase space. III. Mean field approximation benchmarked for interacting Lennard–Jones particles arXiv:1812.03635

Attard P 2019a Fermionic phonons: exact analytic results and quantum statistical mechanics for a one dimensional harmonic crystal arXiv:1903.06866

Attard P 2019b Quantum Monte Carlo in classical phase space. Mean field and exact results for a one dimensional harmonic crystal arXiv:1904.10650

Hernando A 2013 Imaginary-time nonuniform mesh method for solving the multidimensional Schrödinger equation: fermionization and melting of quantum Lennard–Jones crystals arXiv:1304.8015v2
van Sciver S W 2012 *Helium Cryogenics* 2nd edn (New York: Springer)

Chapter 10

Local state expansion for the commutation function

The commutation function for the full system is approximated as the sum of localized single-particle commutation functions composed from the singlet energy eigenvalues and eigenfunctions. These are obtained for an effective singlet Hamiltonian operator based on the effective local field for a fixed configuration of neighbor particles. The singlet effective local field is similar to the mean field used in the Gross–Pitaevskii non-linear Schrödinger equation. Temperature-dependent corrections may be systematically added that extend the singlet Maxwell–Boltzmann operator to higher order in inverse temperature. Pair, triplet, etc approximations can be similarly developed. An harmonic approximation to the effective local field enables the simple harmonic oscillator commutation function, which is analytic, to be used. The approach is tested numerically against benchmarks in the one-dimensional cases of a harmonic crystal and of Lennard–Jones particles.

10.1 Effective local field and operator

The aim of this chapter is to derive a practical algorithm for computing the commutation function at all temperatures. The algorithm is localized as it writes the commutation function as the sum over energy eigenfunctions for one variable particle in the effective local field from fixed neighbor particles. This is a little like a Born–Oppenheimer approach, except that the effective rather than the actual local field is used. In the first instance the energy eigenfunctions of the total system are the product of singlet energy eigenfunctions. These give a realistic singlet commutation function, the sum of which gives the phase space weight of the full system. Systematic improvements within the singlet ansatz are derived. Although the commutation function is referred to as singlet, the presence of the fixed neighbors really make it a many-body effective potential (Attard 2020).

This approach turns out to be almost identical to an earlier harmonic local field (then called mean field) approach, with one improvement. The singlet commutation

doi:10.1088/978-0-7503-4055-7ch10

function is here calculated numerically for the exact effective local field. The earlier approach used a harmonic approximation to the effective local field (Attard 2018a, 2018b). What the two approaches have in common is the way the effective local field is defined (see section 10.1.2). Tests are made against the analytic expressions for a one-dimensional harmonic crystal derived in section 6.2 following earlier work (Attard 2002, 2019a).

10.1.1 Recapitulation of phase space formulation

The quantum probability density in classical phase space for the canonical equilibrium system, as shown in chapter 7, is

$$\wp^{\pm}(\mathbf{q}, \mathbf{p}) = \frac{e^{-\beta\mathcal{H}(\mathbf{q}, \mathbf{p})}}{N!h^{3N}Z^{\pm}(T)}\,\omega(\mathbf{q}, \mathbf{p})\,\eta^{\pm}(\mathbf{p}, \mathbf{q}). \tag{10.1}$$

Here N is the number of particles, $\mathbf{p} = \{\mathbf{p}_1, \mathbf{p}_2, \dots, \mathbf{p}_N\}$ are their momenta, with $\mathbf{p}_j = \{p_{jx}, p_{jy}, p_{jz}\}$, $\mathbf{q} = \{\mathbf{q}_1, \mathbf{q}_2, \dots, \mathbf{q}_N\}$ are the positions, $\beta = 1/k_BT$ is the inverse temperature, with T the temperature and k_B Boltzmann's constant, and h is Planck's constant. The classical Hamiltonian is $\mathcal{H}(\mathbf{q}, \mathbf{p}) = \mathcal{K}(\mathbf{p}) + U(\mathbf{q})$, with $\mathcal{K} = p^2/2m$ being the kinetic energy, and U being the potential energy. The partition function $Z^{\pm}(T)$ normalizes the phase space probability density. The quantum aspects are embodied in the symmetrization function $\eta^{\pm}$ and the commutation function density ω.

The symmetrization and commutation functions are defined in terms of the unsymmetrized position and momentum eigenfunctions, which in the position representation $\mathbf{r}$ are respectively (Messiah 1961)

$$|\mathbf{q}\rangle = \delta(\mathbf{r} - \mathbf{q}), \quad \text{and} \quad |\mathbf{p}\rangle = \frac{1}{V^{N/2}}e^{-\mathbf{p}\cdot\mathbf{r}/i\hbar}. \tag{10.2}$$

where $\hbar = h/2\pi$, and V is the volume of the system. (The normalization of the momentum eigenfunction is unimportant as only their ratio appears below.) The distinction between the position representation coordinate $\mathbf{r}$ and the position eigenvalue $\mathbf{q}$ is crucial for what follows.

The symmetrization function is formally (see chapter 7)

$$\eta_q^{\pm}(\mathbf{p}, \mathbf{q}) \equiv \frac{1}{\langle\mathbf{p}|\mathbf{q}\rangle}\sum_{\hat{\mathrm{P}}}(\pm 1)^p\,\langle\hat{\mathrm{P}}\mathbf{p}|\mathbf{q}\rangle, \tag{10.3}$$

with $\hat{\mathrm{P}}$ the permutation operator and p its parity. The upper sign is for bosons and the lower sign is for fermions.

The commutation function density ω, which is related to similar functions introduced by Wigner (1932) and analyzed by Kirkwood (1933), is defined by (see chapter 7)

$$e^{-\beta\mathcal{H}(\mathbf{q}, \mathbf{p})}\omega_p(\mathbf{q}, \mathbf{p}) = \frac{\langle\mathbf{q}|e^{-\beta\hat{\mathcal{H}}}|\mathbf{p}\rangle}{\langle\mathbf{q}|\mathbf{p}\rangle}. \tag{10.4}$$

Using the completeness of the energy eigenfunctions, $\hat{\mathcal{H}}|\mathbf{n}\rangle = E_{\mathbf{n}}|\mathbf{n}\rangle$, equation (7.9) writes this as

$$e^{-\beta\mathcal{H}(\mathbf{q},\mathbf{p})}\omega_{\mathrm{p}}(\mathbf{q},\mathbf{p}) = \frac{1}{\langle\mathbf{q}|\mathbf{p}\rangle}\sum_{\mathbf{n}} e^{-\beta E_{\mathbf{n}}}\langle\mathbf{q}|\mathbf{n}\rangle\,\langle\mathbf{n}|\mathbf{p}\rangle. \tag{10.5}$$

The importance of the non-commutativity of the position and momentum operators in a particular configuration is reflected in the extent that the commutation function departs from unity. Since the system is classical in the high temperature limit, one must have that $\omega(\mathbf{q},\mathbf{p}) \to 1$, $\beta \to 0$.

The commutation function is usefully cast as a temperature-dependent effective potential, equation (8.6),

$$\omega_{\mathrm{p}}(\mathbf{q},\mathbf{p}) \equiv e^{W_{\mathrm{p}}(\mathbf{q},\mathbf{p})}. \tag{10.6}$$

The rationale for this is that W is extensive with system size, which is a useful property that is widely exploited in thermodynamics. Obviously, $W_{\mathrm{p}}(\mathbf{q},\mathbf{p}) \to 0$, $\beta \to 0$.

10.1.2 Effective local field

In this section is derived the first order, singlet, effective local field. This will be used as the first term in an expansion for the Maxwell–Boltzmann operator and subsequently for the commutation function. The meaning, justification, and motivation for the choice of local field will be clarified in the derivation of the expansion for the Maxwell–Boltzmann operator, subsection 10.1.3. Higher order extensions to second and third order singlet approximations, and also to the first order pair commutation function will be given in section 10.2.

10.1.2.1 Definition

Consider a system of N particles subject to singlet and pair potentials,

$$U(\mathbf{q}) = \sum_{j=1}^{N} u^{(1)}(\mathbf{q}_j) + \sum_{j<k}^{N} u^{(2)}(q_{jk}). \tag{10.7}$$

A central pair potential is assumed, $q_{jk} = |\mathbf{q}_j - \mathbf{q}_k|$. The addition of many-body potentials is discussed below.

In view of the position eigenfunctions, equation (10.2), one can draw a distinction between the eigenvalue $\mathbf{q}$ and the representation coordinate $\mathbf{r}$. The Dirac δ-function means that ultimately these are close to each other. Accordingly, the local potential felt by particle j in configuration $\mathbf{q}$ can be written as

$$\begin{aligned} u_j(\mathbf{r}_j|\mathbf{q}) &= u^{(1)}(\mathbf{r}_j) + \frac{1}{2}\sum_{k=1}^{N}{}^{(k\neq j)} u^{(2)}(|\mathbf{r}_j - \mathbf{r}_k|) \\ &\approx u^{(1)}(\mathbf{r}_j) + \frac{1}{2}\sum_{k=1}^{N}{}^{(k\neq j)} u^{(2)}(|\mathbf{r}_j - \mathbf{q}_k|). \end{aligned} \tag{10.8}$$

The factor of one half is essential, and it is what distinguishes this effective local field from the actual local field. It accounts for the fact that summing over the effective local fields counts each pair interaction twice in the total potential,

$$U(\mathbf{r}|\mathbf{q}) = \sum_{j=1}^{N} u_j(\mathbf{r}_j|\mathbf{q}). \tag{10.9}$$

This is a sum of singlet potentials that are functions of the position representation coordinates. The factor of one half in the definition of the effective local field means that $U(\mathbf{q}|\mathbf{q}) = U(\mathbf{q})$, which is crucial for understanding the nature of the expansion to be derived below and for its convergence properties.

The approximation in the second line of equation (10.8) makes particle j effectively independent of all the other particles in the system since the representation coordinates of particles other than j do not appear. In turn this makes the total system potential energy the sum of independent single-particle potentials. The mathematical consequences of this, and the method for systematizing the approximation as an expansion, are discussed in section 10.1.3.

10.1.2.2 Neighborhood and tail correction

In practice instead of all particles in the system, the effective local potential would be limited to contributions from near neighbors of particle j, $u(\mathbf{r}_j|\mathbf{q}_{/j})$. The notation $\mathbf{q}_{/j} = \{\mathbf{q}_{j_1}, \mathbf{q}_{j_2}, \ldots\}$ signifies the set of positions of neighbors of particle j. For example, $\mathbf{q}_k \in \mathbf{q}_{/j} \Leftrightarrow q_{jk} \leqslant q_{\text{cut}}$. Typically for a given configuration $\{\mathbf{q}, \mathbf{p}\}$, the position eigenvalue $\mathbf{q}_j$ is known, and the representation coordinates $\mathbf{r}_j$ are required over a certain encompassing domain in order to calculate the commutation function $W(\mathbf{q}, \mathbf{p})$. The reason for limiting the effective local potential to near neighbors is that it reduces the amount of calculation and memory required to obtain and to store the subsequent eigenvalues and eigenfunctions. The justification for doing so is that in general the pair potential decays with increasing separation, and so the total potential is dominated by the singlet potentials and the nearest neighbor pair potentials. It is this that makes the theory local. The remaining contributions can be neglected or else perhaps included in a mean field fashion, for example,

$$u(\mathbf{r}_j|\mathbf{q}_{/j}) = u^{(1)}(\mathbf{r}_j) + \frac{1}{2}\sum_{k}^{(\mathbf{q}_k \in \mathbf{q}_{/j})} u^{(2)}(|\mathbf{r}_j - \mathbf{q}_k|) + u_{\text{tail}}, \tag{10.10}$$

where the tail contribution is

$$u_{\text{tail}} = \frac{4\pi\rho}{2}\int_{q_{\text{cut}}}^{\infty} \mathrm{d}q\; q^2 u^{(2)}(q). \tag{10.11}$$

Here ρ is the number density and three-dimensional space is assumed. One could go further and include the radial distribution function, either classical or self-consistent quantum, in the integrand here. In fact since u_{tail} is a constant independent of the configuration $\mathbf{q}$ or of the representation coordinate $\mathbf{r}_j$, it has no non-trivial effect on the commutation function, and so it can be neglected.

10.1.2.3 Effective local field for many-body potentials

In general the effective local field is determined by two principles:

- for a given configuration $\mathbf{q}$, the total effective energy equals the actual energy, $U(\mathbf{q}|\mathbf{q}) = U(\mathbf{q})$
- identical particles share identically.

The first principle means that the fractional weights of the interaction potential in each local field to which it contributes must add up to one. The second principle means that the fractional weight must be the same for each identical particle that shares an interaction potential. These two principles can only be satisfied with weight one half for the pair potential contribution to the local field for identical particles. Using instead the actual local field for each particle would violate the first principle.

These principles can be applied as well for many-body potentials. For example, the contribution from a three-body potential to the effective singlet local field for particle j is

$$\frac{1}{6}\sum_{k=1}^{N}{}^{(k\neq j)}\sum_{\ell=1}^{N}{}^{(\ell\neq j,\,k)}u^{(3)}(\mathbf{r}_j,\,\mathbf{q}_k,\,\mathbf{q}_\ell). \tag{10.12}$$

For particles with different masses, the share of the effective local field felt by a particle is inversely proportional to its mass, (i.e., the field acting on particle j from the pair potential with particle k has fraction $m_k/(m_j + m_k)$).

10.1.3 Expansion of the Maxwell–Boltzmann operator

The effective local field provides the basis for approximating the Maxwell–Boltzmann operator and subsequently the commutation function. Whereas in conventional direct quantum methods it is common to approximate the eigenvalue and eigenfunctions of an operator, here we instead develop an approximation for the operator that lends itself to systematic correction. We find the exact eigenvalues and eigenfunctions, which turns out to be a local rather than a global process.

The singlet effective local field allows an effective Hamiltonian operator to be defined that is the sum of independent singlet Hamiltonian operators,

$$\hat{\mathcal{H}}^{(1,\,1)}(\mathbf{r}|\mathbf{q}) = \sum_j \hat{\mathcal{H}}_j^{(1,\,1)}(\mathbf{r}_j),\quad \hat{\mathcal{H}}_j^{(1,\,1)}(\mathbf{r}_j) \equiv \frac{-\hbar^2}{2m}\nabla_{\mathbf{r}_j}^2 + u_j(\mathbf{r}_j|\mathbf{q}). \tag{10.13}$$

The notation for the operator $\hat{\mathcal{H}}^{(n,\,m)}$ signifies the n-particle operator (i.e., n variable particles in the fixed effective field), and the m signifies that it is the mth term in the mth order approximation (i.e., the mth order approximation to the n-particle Hamiltonian operator is $\hat{\mathcal{H}}^{(n,\,1)} + \hat{\mathcal{H}}^{(n,\,2)} + \cdots + \hat{\mathcal{H}}^{(n,\,m)}$); this will be detailed shortly. Here $\hat{\mathcal{H}}^{(1,\,1)}$ is the singlet operator in first order approximation. Typically, $u_j(\mathbf{r}_j|\mathbf{q}) = u(\mathbf{r}_j|\mathbf{q}_{/j})$. The singlet Hamiltonian operators are independent of each other because each only depends upon the representation coordinates of one particle.

Let the eigenvalues and eigenfunctions of the effective Hamiltonian be

$$\hat{\mathcal{H}}^{(1,1)}(\mathbf{r}|\mathbf{q})\Phi_{\mathbf{n}}^{(1,1)}(\mathbf{r}|\mathbf{q}) = E_{\mathbf{n}}^{(1,1)}(\mathbf{q})\Phi_{\mathbf{n}}^{(1,1)}(\mathbf{r}|\mathbf{q}), \tag{10.14}$$

or

$$\hat{\mathcal{H}}_j^{(1,1)}(\mathbf{r}_j)\phi_{j,\mathbf{n}_j}^{(1,1)}(\mathbf{r}_j|\mathbf{q}) = E_{j,\mathbf{n}_j}^{(1,1)}(\mathbf{q})\phi_{j,\mathbf{n}_j}^{(1,1)}(\mathbf{r}_j|\mathbf{q}). \tag{10.15}$$

Obviously $\Phi_{\mathbf{n}}^{(1,1)}(\mathbf{r}|\mathbf{q}) = \prod_{j=1}^{N} \phi_{j,\mathbf{n}_j}^{(1,1)}(\mathbf{r}_j|\mathbf{q})$ and $E_{\mathbf{n}}^{(1,1)}(\mathbf{q}) = \sum_{j=1}^{N} E_{j,\mathbf{n}_j}^{(1,1)}(\mathbf{q})$.

Recall the essential requirement of the singlet effective local field, $U(\mathbf{q}|\mathbf{q}) = U(\mathbf{q})$. Define $U_\Delta(\mathbf{r}|\mathbf{q}) \equiv U(\mathbf{r}) - U(\mathbf{r}|\mathbf{q})$, so that the exact energy operator is $\hat{\mathcal{H}}(\mathbf{r}) = \hat{\mathcal{H}}^{(1,1)}(\mathbf{r}|\mathbf{q}) + U_\Delta(\mathbf{r}|\mathbf{q})$. The actual energy operator acting on the singlet, first order eigenfunction yields

$$\begin{aligned}\hat{\mathcal{H}}(\mathbf{r})\Phi_{\mathbf{n}}^{(1,1)}(\mathbf{r}|\mathbf{q}) &= [E_{\mathbf{n}}^{(1,1)}(\mathbf{q}) + U_\Delta(\mathbf{r}|\mathbf{q})]\Phi_{\mathbf{n}}^{(1,1)}(\mathbf{r}|\mathbf{q}) \\ &= E_{\mathbf{n}}^{(1,1)}(\mathbf{q})\Phi_{\mathbf{n}}^{(1,1)}(\mathbf{q}|\mathbf{q}), \quad \mathbf{r} = \mathbf{q}.\end{aligned} \tag{10.16}$$

In so far as $U_\Delta(\mathbf{r}|\mathbf{q}) = 0$ at $\mathbf{r} = \mathbf{q}$, one can see that the exact eigenfunction of the singlet, first order energy operator is also an approximate eigenfunction of the actual energy operator in the vicinity of $\mathbf{r} = \mathbf{q}$.

This conclusion is only valid at the linear level. At the quadratic level one has

$$\begin{aligned}&\hat{\mathcal{H}}^{(1,1)}(\mathbf{r}|\mathbf{q})^2 \\ &= [\hat{\mathcal{H}}(\mathbf{r}) - U_\Delta(\mathbf{r}|\mathbf{q})]^2 \\ &= \hat{\mathcal{H}}(\mathbf{r})^2 - \hat{\mathcal{H}}(\mathbf{r})U_\Delta(\mathbf{r}|\mathbf{q}) - U_\Delta(\mathbf{r}|\mathbf{q})\hat{\mathcal{H}}(\mathbf{r}) + U_\Delta(\mathbf{r}|\mathbf{q})^2 \\ &= \hat{\mathcal{H}}(\mathbf{r})^2 + \frac{\hbar^2}{2m}(\nabla_{\mathbf{r}}^2 U_\Delta(\mathbf{r}|\mathbf{q})) + 2\frac{\hbar^2}{2m}(\nabla_{\mathbf{r}} U_\Delta(\mathbf{r}|\mathbf{q})) \cdot \nabla_{\mathbf{r}} \\ &\quad - 2U_\Delta(\mathbf{r}|\mathbf{q})\hat{\mathcal{H}}(\mathbf{r}) + U_\Delta(\mathbf{r}|\mathbf{q})^2 \\ &\xrightarrow{\mathbf{r}\to\mathbf{q}} \hat{\mathcal{H}}(\mathbf{r})^2 + \frac{\hbar^2}{2m}(\nabla_{\mathbf{r}}^2 U_\Delta(\mathbf{r}|\mathbf{q})) + 2\frac{\hbar^2}{2m}(\nabla_{\mathbf{r}} U_\Delta(\mathbf{r}|\mathbf{q})) \cdot \nabla_{\mathbf{r}}.\end{aligned} \tag{10.17}$$

These excess terms do not vanish at $\mathbf{r} = \mathbf{q}$. Hence the singlet, first order eigenfunction is not an eigenfunction of the square of singlet, first order energy operator, $\hat{\mathcal{H}}^{(1,1)}(\mathbf{r}|\mathbf{q})^2\Phi_{\mathbf{n}}^{(1,1)}(\mathbf{r}|\mathbf{q}) \neq E_{\mathbf{n}}^{(1,1)}(\mathbf{q})^2\Phi_{\mathbf{n}}^{(1,1)}(\mathbf{r}|\mathbf{q})$. In view of this, we may describe the singlet, first order energy operator $\hat{\mathcal{H}}^{(1,1)}(\mathbf{r}|\mathbf{q})$ as a linear approximation to the actual energy operator of the system. We may similarly call its exact eigenfunctions $\Phi_{\mathbf{n}}^{(1,1)}(\mathbf{r}|\mathbf{q})$ and eigenvalues $E_{\mathbf{n}}^{(1,1)}(\mathbf{q})$ as linear approximations to the energy eigenstates of the actual system.

This linear interpretation of the meaning of the present effective local field opens the door to a systematic expansion, the next two terms of which are given explicitly in section 10.2. It is directly relevant to the derivation of the local field expansion of the commutation function. Since the latter invokes the Maxwell–Boltzmann operator, we may call on the linear interpretation of the local field to write

$$\begin{aligned} e^{-\beta\hat{\mathcal{H}}(\mathbf{r})} &\approx 1 - \beta\hat{\mathcal{H}}(\mathbf{r}) \\ &= 1 - \beta\hat{\mathcal{H}}^{(1,\,1)}(\mathbf{r}|\mathbf{q}) \\ &\approx e^{-\beta\hat{\mathcal{H}}^{(1,\,1)}(\mathbf{r}|\mathbf{q})}. \end{aligned} \tag{10.18}$$

As has just been shown, the passage from the first to the second equality is exact at $\mathbf{r}_{lj} = \mathbf{q}_{lj}$. The subsequent re-exponentiation gives a singlet approximation for the Maxwell–Boltzmann operator.

The reason for showing the linearization step is that it explains using the present explicit formula for the approximation to the singlet Hamiltonian operator that ultimately gives the singlet Maxwell–Boltzmann operator. In section 10.2, we shall proceed directly from the singlet ansatz that is the final equality to develop a series expansion for the singlet Hamiltonian operator of which the present approximation is the first order term.

In evaluating the merits of the effective local field approach developed in this chapter, the question isn't whether the system energy eigenvalues and eigenfunctions have been accurately obtained in general, but rather whether $\hat{\mathcal{H}}(\mathbf{r})$, is well-approximated by $\hat{\mathcal{H}}^{(1,\,1)}(\mathbf{r}|\mathbf{q})$ so that the subsequent $E_{\mathbf{n}}^{(1,\,1)}(\mathbf{q})$ and $\Phi_{\mathbf{n}}^{(1,\,1)}(\mathbf{r}|\mathbf{q})$ are accurate enough for the purposes of calculating the commutation function.

Because the above derivation involves a linearization of the exponential, it is strictly exact at high temperatures, $\beta \to 0$. The subsequent re-exponentiation has the virtue of making the procedure applicable at lower temperatures, because it effectively creates an infinite series in powers of inverse temperature, albeit with approximate coefficients. One should not underestimate the worth of this re-exponentiation in going beyond a strictly linear approach and in extending its applicability to lower temperatures. For example, amongst other contributions it gives the exponential of the kinetic and potential energies to form the classical Maxwell–Boltzmann factor, which is essential in the exact phase space formulation at all temperatures. It is shown in the derivation of the cubic term in section 10.2 that exponentiating the linear term contributes substantially to the exact higher order low temperature terms.

The efficiency of the linear formulation and resummation can be seen quantitatively as follows. As above, the difference between the exact and the linear Hamiltonian operator is

$$\Delta\hat{\mathcal{H}} \equiv \hat{\mathcal{H}}(\mathbf{r}) - \hat{\mathcal{H}}^{(1,\,1)}(\mathbf{r}|\mathbf{q}) = U(\mathbf{r}) - U(\mathbf{r}|\mathbf{q}) \equiv U_\Delta. \tag{10.19}$$

Obviously $U_\Delta = 0$ at $\mathbf{r} = \mathbf{q}$. One can write the exact Maxwell–Boltzmann operator as

$$\begin{aligned} e^{-\beta\hat{\mathcal{H}}(\mathbf{r})} &= e^{-\beta\left[\hat{\mathcal{H}}^{(1,\,1)}(\mathbf{r}|\mathbf{q}) + U_\Delta\right]} \\ &= \sum_{n=0}^{\infty} \frac{(-\beta)^n}{n!}\left[\hat{\mathcal{H}}^{(1,\,1)}(\mathbf{r}|\mathbf{q}) + U_\Delta\right]^n. \end{aligned} \tag{10.20}$$

A binomial expansion of the nth term in the series has first term in each case $\hat{\mathcal{H}}^{(1,1)}(\mathbf{r}|\mathbf{q})^n$, which will resum to $e^{-\beta\hat{\mathcal{H}}^{(1,1)}(\mathbf{r}|\mathbf{q})}$, as if $\Delta\hat{\mathcal{H}} = U_\Delta$ had been neglected entirely. The remaining terms in the binomial expansion are of the form $U(\mathbf{r}|\mathbf{q})^j U_\Delta^k \nabla^l U_\Delta^m$, and more complicated. But because $U_\Delta = 0$ at $\mathbf{r} = \mathbf{q}$, one sees that this is zero unless there are no factors of U_Δ remaining after the various gradients have been evaluated. So the only non-zero correction terms consist solely of factors of powers of $(\nabla U_\Delta) \cdot \nabla$, $(\nabla^2 U_\Delta)$, $(\nabla\nabla^2 U_\Delta) \cdot \nabla$, etc (and also U), and no factor of the form U_Δ^k. Hence significantly many terms in the excess sum vanish; for example only 14% of the possible corrections for the cubic term are non-zero. This, as well as the fact that higher gradients of the potential vanish rapidly at large separations, explains why the non-linear resummation of the linear formulation, $e^{-\beta\hat{\mathcal{H}}(\mathbf{r})} \approx e^{-\beta\hat{\mathcal{H}}^{(1,1)}(\mathbf{r}|\mathbf{q})}$, can be expected to be accurate even at low temperatures.

10.1.3.1 Harmonic local field

In section 10.3 an harmonic approximation to the effective local field will be derived (Attard 2018a, 2018b, 2019b). Whereas for the actual effective local field discussed here the energy eigenvalues, eigenfunctions and commutation function are obtained numerically, in the harmonic approximation, which was originally called a mean field approach (Attard 2018a, 2018b, 2019b), a quadratic expansion of the instantaneous effective local field is made, and the corresponding analytic simple harmonic oscillator energy eigenvalues, eigenfunctions, and commutation function are used. Both approaches are linear resummed in the above sense.

Numerical results for this harmonic effective local field theory are presented below for a one-dimensional harmonic crystal, Section 10.5.1 (Attard 2019b), and for Lennard–Jones particles also in one-dimension, section 10.5.2 (Attard 2018a). For the harmonic crystal the anharmonic contributions are limited, and so it mainly tests the effective local field and linear level Hamiltonian themselves. (Actually, whilst the singlet potential is purely harmonic, the pair interaction is technically of the form of a double oscillator, W-shaped, potential, and so the results also test the efficacy of the harmonic approximation for it.) Those results for the harmonic crystal show that the singlet harmonic theory is accurate at both high and low temperatures, even for temperatures where the ground state is dominant, and that the pair theory is an improvement. For the Lennard–Jones system, the harmonic effective local field is less successful at low temperatures.

10.1.3.2 Boundary conditions

Penultimately, one has to obtain the singlet energy eigenvalues and eigenfunctions for the particle in the effective local field. This means that boundary conditions have to be specified. In one-dimension it is relatively easy: because of the repulsive potential between particles at close separations, one can simply specify the nearest neighbors at $q_{j\pm1}$ as the fixed particles, and demand that $\psi(r_j|q_{j+1}, q_{j-1}) \to 0$, as $r_j \to q_{j\pm1}$.

In two- or three-dimensions it is more difficult. One might imagine selecting fixed neighbors that surround the nominated particle and creating a closed polyhedron with these at the vertices. It might be sufficient to demand that the wave function and perhaps its derivatives vanish in the vicinity of each vertex, but this is by no means clear. Or one might instead demand that the wave function or its derivative vanish on the faces of the polyhedron, but again it is unclear whether this is sufficient to provide reliable results. In view of the localization property of the commutation function, perhaps it is only necessary to impose the condition that $\psi(\mathbf{r}_j|\mathbf{q}_{/j}) \to 0$, $|\mathbf{r}_j - \mathbf{q}_{0j}| \to \infty$, where $\mathbf{q}_{0j}$ is, for example, the centroid of the neighbors located at $\mathbf{q}_{/j}$. In any case one has to establish the sensitivity of the resultant commutation function to the type and location of the boundary conditions that are applied. One argument in favor of these various suggestions is that the commutation function must be a local property of the fluid, and therefore obtaining it as a temperature-weighted sum over localized energy eigenstates cannot be overly sensitive to the conditions imposed on the energy eigenfunctions at the boundaries of the local region.

10.1.3.3 Relationship to the Born–Oppenheimer method

The present effective local field approach is a little like the Born–Oppenheimer method of quantum chemistry, in that certain coordinates are fixed in the calculation of the wave function for the remainder. One difference is that the Born–Oppenheimer method uses the actual local field for the electrons, whereas the present method uses an effective local field. A second difference is that in the Born–Oppenheimer method the electron energy eigenvalues are fed into the nuclear Schrödinger equation as an effective potential, whereas here the singlet wave functions and singlet states for any one particle are fully independent of the states of the other particles. But perhaps the major difference between the two is that the Born–Oppenheimer method seeks the energy eigenfunction of the total system, whereas the present method at the singlet level only aspires to the single particle energy eigenfunction. This is used in a one-particle quantum average to obtain the commutation function that will subsequently be used for classical many-body averages; the combined singlet energy eigenfunctions don't approximate the total system energy eigenfunction (not least because they have a parametric dependence on $\mathbf{q}$ that the actual eigenfunctions don't have).

10.1.4 Singlet, first order commutation function

The linear approximation using the effective local field just explained can be called the first order approximation, because the corresponding exponent of the Maxwell–Boltzmann operator is linear in inverse temperature. Also it gives the primary quantum correction for non-commutativity. To recapitulate, the Hamiltonian operator of the system for a particular configuration is the sum of first order single-particle Hamiltonian operators,

$$\hat{\mathcal{H}}_j^{(1,\,1)}(\mathbf{r}_j) = \frac{-\hbar^2}{2m}\partial^2_{\mathbf{r}_j} + u(\mathbf{r}_j|\mathbf{q}_{/j}), \tag{10.21}$$

the energy eigenfunctions of the full system are simply products of single particle energy eigenfunctions,

$$\Phi_{\mathbf{n}}^{(1,1)}(\mathbf{r}|\mathbf{q}) = \prod_{j=1}^{N} \phi_{j,\mathbf{n}_j}^{(1,1)}(\mathbf{r}_j), \quad \phi_{j,\mathbf{n}_j}^{(1,1)}(\mathbf{r}_j) \equiv \phi_{\mathbf{n}_j}^{(1,1)}(\mathbf{r}_j|\mathbf{q}_{/j}), \tag{10.22}$$

and the eigenvalue equation is

$$\hat{\mathcal{H}}_j^{(1,1)}(\mathbf{r}_j)\, \phi_{j,\mathbf{n}_j}^{(1,1)}(\mathbf{r}_j) = E_{j,\mathbf{n}_j}^{(1,1)}\, \phi_{j,\mathbf{n}_j}^{(1,1)}(\mathbf{r}_j). \tag{10.23}$$

The subscript j on the energy operator, eigenfunction, and eigenvalue signifies the dependence on the neighbor configuration, e.g., $E_{j,\mathbf{n}_j}^{(1,1)} \equiv E_{\mathbf{n}_j}^{(1,1)}(\mathbf{q}_{/j})$. It is straightforward to solve this single particle eigenvalue equation numerically, at least in one-dimension (Attard 2020).

This definition of the one-particle energy eigenfunctions means that the commutation function of the total system can be written as the sum of singlet commutation functions,

$$W_{\mathrm{p}}^{(1,1)}(\mathbf{q}, \mathbf{p}) = \sum_{j=1}^{N} w_{\mathrm{p}}^{(1,1)}(\mathbf{q}_j, \mathbf{p}_j|\mathbf{q}_{/j}). \tag{10.24}$$

This would be exact if the singlet potential were the only potential in the system; it is an approximation for the present case of singlet and pair potentials.

From the definition, the singlet commutation function for particle j is given by

$$\begin{aligned} e^{-\beta\mathcal{H}^{(1,1)}(\mathbf{q}_j, \mathbf{p}_j|\mathbf{q}_{/j})} e^{w_{\mathrm{p}}^{(1,1)}(\mathbf{q}_j, \mathbf{p}_j|\mathbf{q}_{/j})} &= \frac{1}{\langle \mathbf{q}_j|\mathbf{p}_j\rangle} \left\langle \mathbf{q}_j \,\middle|\, e^{-\beta\hat{\mathcal{H}}_j^{(1,1)}(r_j)} \,\middle|\, \mathbf{p}_j \right\rangle \\ &= \frac{1}{\langle \mathbf{q}_j|\mathbf{p}_j\rangle} \sum_{\mathbf{n}_j} e^{-\beta E_{\mathbf{n}_j}^{(1,1)}(\mathbf{q}_{/j})} \phi_{\mathbf{n}_j}^{(1,1)}(\mathbf{q}_j|\mathbf{q}_{/j})\, \check{\phi}_{\mathbf{n}_j}^{(1,1)}(\mathbf{p}_j|\mathbf{q}_{/j}). \end{aligned} \tag{10.25}$$

Here $\phi_{\mathbf{n}_j}^{(1,1)}(\mathbf{r}_j|\mathbf{q}_{/j}) = \langle \mathbf{r}_j|\mathbf{n}_j, \mathbf{q}_{/j}\rangle$, is the $\mathbf{n}_j$ th energy eigenfunction for the effective singlet local Hamiltonian operator with the potential $u(\mathbf{r}_j|\mathbf{q}_{/j})$, and $\check{\phi}_{\mathbf{n}_j}^{(1,1)}(\mathbf{p}_j|\mathbf{q}_{/j}) \equiv \langle \mathbf{n}_j, \mathbf{q}_{/j}|\mathbf{p}_j\rangle = V^{-1/2} \int \mathrm{d}\mathbf{r}_j\, e^{-\mathbf{r}_j\cdot\mathbf{p}_j/i\hbar} \phi_{\mathbf{n}_j}^{(1,1)}(\mathbf{r}_j|\mathbf{q}_{/j})^*$ is essentially its Fourier transform. The eigenfunctions of the effective singlet local Hamiltonian operator are also eigenfunctions of the effective singlet local Maxwell–Boltzmann operator. This expression for the commutation function as a sum over a complete set of singlet energy states shows the necessity of the re-exponentiation step above, equation (10.18).

The structure of equation (10.25) makes it convenient to define the combined singlet commutation function as $\tilde{w}_{\mathrm{p}}^{(1,1)}(\mathbf{q}_j, \mathbf{p}_j|\mathbf{q}_{/j}) \equiv w_{\mathrm{p}}^{(1,1)}(\mathbf{q}_j, \mathbf{p}_j|\mathbf{q}_{/j}) - \beta\mathcal{H}^{(1,1)}(\mathbf{q}_j, \mathbf{p}_j|\mathbf{q}_{/j})$. The classical singlet Hamiltonian that appears here and in the preceding equation is

$$\mathcal{H}^{(1,1)}(\mathbf{q}_j, \mathbf{p}_j|\mathbf{q}_{/j}) = \frac{p_j^2}{2m} + u(\mathbf{q}_j|\mathbf{q}_{/j}). \tag{10.26}$$

The weighting of one half for the pair potentials ensures that the classical Hamiltonian of the full system is exactly

$$\mathcal{H}(\mathbf{q}, \mathbf{p}) = \sum_{j=1}^{N} \mathcal{H}^{(1,1)}(\mathbf{q}_j, \mathbf{p}_j|\mathbf{q}_{/j}). \tag{10.27}$$

The total commutation function exponent is similarly given by

$$W_\mathrm{p}^{(1,1)}(\mathbf{q}, \mathbf{p}) = \sum_{j=1}^{N} w_\mathrm{p}^{(1,1)}(\mathbf{q}_j, \mathbf{p}_j|\mathbf{q}_{/j}). \tag{10.28}$$

With these results and suppressing for the moment the symmetrization function, the phase space weight is proportional to

$$e^{-\beta\mathcal{H}(\mathbf{q},\mathbf{p})}e^{W_\mathrm{p}^{(1,1)}(\mathbf{q},\mathbf{p})} = \prod_{j=1}^{N} e^{-\beta\mathcal{H}^{(1,1)}(\mathbf{q}_j, \mathbf{p}_j|\mathbf{q}_{/j})}e^{w_\mathrm{p}^{(1,1)}(\mathbf{q}_j, \mathbf{p}_j|\mathbf{q}_{/j})}. \tag{10.29}$$

10.2 Higher order local fields

10.2.1 Singlet, second order

In the preceding section it was shown that the effective local field gave a singlet Hamiltonian operator that coincided with the actual Hamiltonian operator exactly to linear order. In this subsection the next correction in inverse temperature to this singlet operator that makes it exact to quadratic order in the expression for the commutation function. Here and throughout the gradient operator means a derivative with respect to $\mathbf{r}$, and these are all performed *before* setting $\mathbf{r}_{/j}$ equal to $\mathbf{q}_{/j}$.

One expands the singlet energy operator in powers of inverse temperature expansion

$$\hat{\mathcal{H}}^{(1)}(\mathbf{r}_j|\mathbf{q}_{/j}) = \sum_{\alpha=1}^{\infty} \beta^{\alpha-1}\hat{\mathcal{H}}^{(1,\alpha)}(\mathbf{r}_j|\mathbf{q}_{/j}). \tag{10.30}$$

The leading term is the first order term derived above,

$$\hat{\mathcal{H}}^{(1,1)}(\mathbf{r}_j|\mathbf{q}_{/j}) = \frac{-\hbar^2}{2m}\nabla_{\mathbf{r}_j}^2 + u(\mathbf{r}_j|\mathbf{q}_{/j}). \tag{10.31}$$

The first order total energy operator, $\hat{\mathcal{H}}^{(1,1)}(\mathbf{r}|\mathbf{q}) = \sum_{j=1}^{N} \hat{\mathcal{H}}^{(1,1)}(\mathbf{r}_j|\mathbf{q}_{/j})$, is the leading approximation to the total energy operator, $\hat{\mathcal{H}}(\mathbf{r})$.

The corresponding eigenfunctions, $\Phi_\mathbf{n}^{(1,1)}(\mathbf{r}|\mathbf{q}) = \prod_{j=1}^{N} \phi_\mathbf{n}^{(1,1)}(\mathbf{r}_j|\mathbf{q}_{/j})$, satisfy

$$\hat{\mathcal{H}}^{(1,1)}(\mathbf{r}|\mathbf{q})\, \Phi_\mathbf{n}^{(1,1)}(\mathbf{r}|\mathbf{q}) = E_\mathbf{n}^{(1,1)}(\mathbf{q})\, \Phi_\mathbf{n}^{(1,1)}(\mathbf{r}|\mathbf{q}), \tag{10.32}$$

or

$$\hat{\mathcal{H}}^{(1,1)}(\mathbf{r}_j|\mathbf{q}_{/j})\,\phi_{\mathbf{n}}^{(1,1)}(\mathbf{r}_j|\mathbf{q}_{/j}) = E_{\mathbf{n}}^{(1,1)}(\mathbf{q}_{/j})\,\phi_{\mathbf{n}}^{(1,1)}(\mathbf{r}_j|\mathbf{q}_{/j}). \tag{10.33}$$

These first order energy eigenvalues and eigenfunctions give the zeroth order commutation function, $W^{(1,1)}(\mathbf{q},\mathbf{p}) = \sum_{j=1}^{N} w^{(1,1)}(\mathbf{q}_j, \mathbf{p}_j|\mathbf{q}_{/j})$ or more precisely its exponent. The aim in this subsection is to make corrections to the eigenvalues and eigenfunctions so that they give a better approximation to the commutation function within the singlet ansatz.

To this end we now obtain a correction to the singlet, first order eigenfunctions that reduces the error in them as eigenfunctions of the exact Maxwell–Boltzmann operator. (In section 10.2.2, an alternative approach reduces the second order error in the singlet energy operator itself, which can be combined with the present correction.) We require that they be accurate at $\mathbf{r} = \mathbf{q}$, *after* the gradient operator has been applied.

One has to quadratic order

$$\begin{aligned} & e^{-\beta\hat{\mathcal{H}}(\mathbf{r})}\Phi_{\mathbf{n}}^{(1,1)}(\mathbf{r}|\mathbf{q}) \\ &= e^{-\beta\left[\hat{\mathcal{H}}^{(1,1)}(\mathbf{r}|\mathbf{q}) + U(\mathbf{r}) - U(\mathbf{r}|\mathbf{q})\right]}\Phi_{\mathbf{n}}^{(1,1)}(\mathbf{r}|\mathbf{q}) \\ &= \left\{1 - \beta\left[\hat{\mathcal{H}}^{(1,1)} + U_\Delta\right] + \frac{\beta^2}{2}\left[\hat{\mathcal{H}}^{(1,1)} + U_\Delta\right]^2\right\}\Phi_{\mathbf{n}}^{(1,1)}(\mathbf{r}|\mathbf{q}) \\ &= \left\{1 - \beta\left[E_{\mathbf{n}} + U_\Delta\right] + \frac{\beta^2}{2}\left[E_{\mathbf{n}}^2 + \hat{\mathcal{H}}^{(1,1)}U_\Delta + U_\Delta E_{\mathbf{n}} + U_\Delta^2\right]\right\} \\ & \quad \times \Phi_{\mathbf{n}}^{(1,1)}(\mathbf{r}|\mathbf{q}). \end{aligned} \tag{10.34}$$

All instances of $U_\Delta = U(\mathbf{r}) - U(\mathbf{r}|\mathbf{q})$ vanish at $\mathbf{r} = \mathbf{q}$ except where its gradient is taken. For example

$$\begin{aligned} &= \left[\frac{-\hbar^2}{2m}\nabla_{\mathbf{r}}^2 + U(\mathbf{r}|\mathbf{q})\right]U_\Delta(\mathbf{r}|\mathbf{q})\,\Phi_{\mathbf{n}}^{(1,1)}(\mathbf{r}|\mathbf{q}) \\ &= U_\Delta\left[\frac{-\hbar^2}{2m}\nabla_{\mathbf{r}}^2 + U\right]\Phi_{\mathbf{n}}^{(1,1)} - \frac{\hbar^2}{m}(\nabla U_\Delta)\cdot(\nabla\Phi_{\mathbf{n}}^{(1,1)}) - \frac{\hbar^2}{2m}\Phi_{\mathbf{n}}^{(1,1)}\nabla^2 U_\Delta. \end{aligned} \tag{10.35}$$

The second and third terms are non-zero at $\mathbf{r} = \mathbf{q}$.

The potential difference $U_\Delta(\mathbf{r}|\mathbf{q}) = U(\mathbf{r}) - U(\mathbf{r}|\mathbf{q}) = U^{(2)}(\mathbf{r}) - U^{(2)}(\mathbf{r}|\mathbf{q})$ only depends upon the pair potential. For the pair potential one has

$$\begin{aligned} \nabla^2 U^{(2)}(\mathbf{r}) &= \sum_j \nabla_j^2 \sum_{k<l} u^{(2)}(\mathbf{r}_k, \mathbf{r}_l) \\ &= \sum_{j,k}{}^{(j\neq k)}\nabla_j^2 u^{(2)}(\mathbf{r}_j, \mathbf{r}_k). \end{aligned} \tag{10.36}$$

Similarly, $\nabla_j U^{(2)}(\mathbf{r}) = \sum_k^{(k\neq j)} \nabla_j u(\mathbf{r}_j, \mathbf{r}_k)$. For the pair part of the effective local potential one has

$$
\begin{aligned}
\nabla^2 U^{(2)}(\mathbf{r}|\mathbf{q}) &= \sum_j \nabla_j^2 \sum_{k=1}^{N} \frac{1}{2} \sum_l^{(l\neq k)} u^{(2)}(\mathbf{r}_k, \mathbf{q}_l) \\
&= \frac{1}{2} \sum_j \sum_k^{(k\neq j)} \nabla_j^2 u^{(2)}(\mathbf{r}_j, \mathbf{q}_k).
\end{aligned}
\tag{10.37}
$$

One similarly has $\nabla_j U^{(2)}(\mathbf{r}, \mathbf{q}) = \sum_l^{(l\neq j)} \nabla_j u^{(2)}(\mathbf{r}_j, \mathbf{q}_l)/2$. For each particle j, at $\mathbf{r}_k = \mathbf{q}_k$, $k \neq j$, these are half the respective previous expressions. Therefore at $\mathbf{r}_{l/j} = \mathbf{q}_{l/j}$ they take the form of singlet results

$$
\nabla^2 U_\Delta(\mathbf{r}|\mathbf{q}) = \sum_{j=1}^{N} \nabla_{\mathbf{r}_j}^2 u^{(2)}(\mathbf{r}_j|\mathbf{q}_{l/j}) \tag{10.38}
$$

and

$$
(\nabla U_\Delta(\mathbf{r}|\mathbf{q})) \cdot \nabla = \sum_{j=1}^{N} (\nabla_{\mathbf{r}_j} u^{(2)}(\mathbf{r}_j|\mathbf{q}_{l/j})) \cdot \nabla_{\mathbf{r}_j}. \tag{10.39}
$$

Neglecting U_Δ, which vanishes at $\mathbf{r} = \mathbf{q}$, but keeping its gradients as functions of $\mathbf{r}$ as just derived, to quadratic order one has

$$
\begin{aligned}
e^{-\beta\hat{\mathcal{H}}(\mathbf{r})}\Phi_{\mathbf{n}}^{(1,1)}(\mathbf{r}|\mathbf{q}) &= \left\{1 - \beta E_{\mathbf{n}} + \frac{\beta^2}{2} E_{\mathbf{n}}^2\right\} \Phi_{\mathbf{n}}^{(1,1)}(\mathbf{r}|\mathbf{q}) \\
&\quad - \frac{\beta^2\hbar^2}{4m}\{(\nabla^2 U_\Delta(\mathbf{r}|\mathbf{q})) + 2(\nabla U_\Delta(\mathbf{r}|\mathbf{q})) \cdot \nabla\} \Phi_{\mathbf{n}}^{(1,1)}(\mathbf{r}|\mathbf{q}).
\end{aligned}
\tag{10.40}
$$

From this one sees that the second order term (or correction) to the singlet Hamiltonian operator is

$$
\hat{\mathcal{H}}^{(1,2)}(\mathbf{r}|\mathbf{q}) = \frac{\hbar^2}{4m}\left\{\left(\nabla_{\mathbf{r}}^2 U_\Delta(\mathbf{r}|\mathbf{q})\right) + 2\left(\nabla_{\mathbf{r}} U_\Delta(\mathbf{r}|\mathbf{q})\right) \cdot \nabla_{\mathbf{r}}\right\}. \tag{10.41}
$$

The corresponding eigenvalue equation is

$$
\left[\hat{\mathcal{H}}^{(1,1)}(\mathbf{r}|\mathbf{q}) + \beta\hat{\mathcal{H}}^{(1,2)}(\mathbf{r}|\mathbf{q})\right]\Phi_{\mathbf{n}}^{(1,2)}(\mathbf{r}|\mathbf{q}) = E_{\mathbf{n}}^{(1,2)}(\mathbf{q})\Phi_{\mathbf{n}}^{(1,2)}(\mathbf{r}|\mathbf{q}). \tag{10.42}
$$

With this solved one can obtain the singlet, second order commutation function $W^{(1,2)}(\mathbf{q}, \mathbf{p})$.

The notation used here is not entirely consistent. For the n-particle operator $\hat{\mathcal{H}}^{(n,m)}$, the m signifies the mth term in the mth order approximation. However for the n-let eigenfunctions and eigenvalues, $\Phi_\ell^{(n,m)}$ and $E_\ell^{(n,m)}$, the m signifies the mth order approximation itself.

10.2.2 Singlet, modified second order

Above the effective local field for the pair part of the potential is

$$U^{(2)}(\mathbf{r}|\mathbf{q}) = \frac{1}{2}\sum_{j,k}^{(k\neq j)} u^{(2)}(\mathbf{r}_j, \mathbf{q}_k) = \sum_{j=1}^{N} u^{(2)}(\mathbf{r}_j|\mathbf{q}_{/j}). \tag{10.43}$$

This has the properties

$$\begin{aligned} U^{(2)}(\mathbf{q}|\mathbf{q}) &= U^{(2)}(\mathbf{q}), \\ \nabla_{\mathbf{r}} U^{(2)}(\mathbf{r}|\mathbf{q})\big|_{\mathbf{r}=\mathbf{q}} &= \frac{1}{2}\nabla_{\mathbf{q}} U^{(2)}(\mathbf{q}) \quad \text{and} \\ \nabla_{\mathbf{r}}^2 U^{(2)}(\mathbf{r}|\mathbf{q})\big|_{\mathbf{r}=\mathbf{q}} &= \frac{1}{2}\nabla_{\mathbf{q}}^2 U^{(2)}(\mathbf{q}). \end{aligned} \tag{10.44}$$

The aim of this section is to modify the pair part of the singlet effective potential so that its gradients equal those of the original pair part of the potential. (There is no need to modify the one-body contribution to the effective potential because it is identical to the actual one-body potential.)

To this end define the modified effective local pair potential for particle j,

$$\begin{aligned} \tilde{u}_j^{(2)}(\mathbf{r}_j|\mathbf{q}_{/j}) \equiv{}& u_j^{(2)}(\mathbf{r}_j|\mathbf{q}_{/j}) + \frac{1}{2}(\mathbf{r}_j - \mathbf{q}_j)\cdot \nabla_{\mathbf{q}_j} u_j^{(2)}(\mathbf{q}_j|\mathbf{q}_{/j}) \\ &+ \frac{1}{4}\sum_{\alpha=x}^{d} (r_{j\alpha} - q_{j\alpha})^2 \nabla_{q_{j\alpha}}^2 u_j^{(2)}(\mathbf{q}_j|\mathbf{q}_{/j}). \end{aligned} \tag{10.45}$$

This gives

$$\begin{aligned} \tilde{U}^{(2)}(\mathbf{q}|\mathbf{q}) &= U^{(2)}(\mathbf{q}), \\ \nabla_{\mathbf{r}} \tilde{U}^{(2)}(\mathbf{r}|\mathbf{q})\big|_{\mathbf{r}=\mathbf{q}} &= \nabla_{\mathbf{q}} U^{(2)}(\mathbf{q}) \\ \nabla_{\mathbf{r}}^2 \tilde{U}^{(2)}(\mathbf{r}|\mathbf{q})\big|_{\mathbf{r}=\mathbf{q}} &= \nabla_{\mathbf{q}}^2 U^{(2)}(\mathbf{q}). \end{aligned} \tag{10.46}$$

One could continue this to higher order.

The modified effective local field that this gives, $\tilde{U}(\mathbf{r}|\mathbf{q})$, can be used to defined a modified singlet first order energy operator $\hat{\tilde{\mathcal{H}}}^{(1,1)}(\mathbf{r}|\mathbf{q})$. As in equation (10.17), the square of this is

$$\begin{aligned} \hat{\tilde{\mathcal{H}}}^{(1,1)}(\mathbf{r}|\mathbf{q})^2 ={}& \hat{\mathcal{H}}(\mathbf{r})^2 + \frac{\hbar^2}{2m}(\nabla_{\mathbf{r}}^2 \tilde{U}_\Delta(\mathbf{r}|\mathbf{q})) + 2\frac{\hbar^2}{2m}(\nabla_{\mathbf{r}} \tilde{U}_\Delta(\mathbf{r}|\mathbf{q}))\cdot\nabla_{\mathbf{r}} \\ &- 2\tilde{U}_\Delta(\mathbf{r}|\mathbf{q})\hat{\mathcal{H}}(\mathbf{r}) + \tilde{U}_\Delta(\mathbf{r}|\mathbf{q})^2 \\ \xrightarrow{\mathbf{r}\to\mathbf{q}}{}& \hat{\mathcal{H}}(\mathbf{r})^2 + \frac{\hbar^2}{2m}(\nabla_{\mathbf{r}}^2 \tilde{U}_\Delta(\mathbf{r}|\mathbf{q})) + 2\frac{\hbar^2}{2m}(\nabla_{\mathbf{r}} \tilde{U}_\Delta(\mathbf{r}|\mathbf{q}))\cdot\nabla_{\mathbf{r}}. \end{aligned} \tag{10.47}$$

One could argue that the two final terms here can be neglected because by design they vanish at $\mathbf{r} = \mathbf{q}$, and therefore they must be small in the vicinity $\mathbf{r} \approx \mathbf{q}$. In this case one can simply use the singlet first order modified eigenvalues, $\tilde{E}_{\mathbf{n}}^{(1,\,1)}(\mathbf{q})$, and eigenfunctions, $\tilde{\Phi}_{\mathbf{n}}^{(1,\,1)}(\mathbf{r}|\mathbf{q})$, to obtain the commutation function. Alternatively, one can keep these two terms and define the singlet second order energy operator as in equation (10.41), with tildes everywhere.

10.2.3 Singlet, third order

The exact Maxwell–Boltzmann operator to cubic order in inverse temperature is

$$\begin{aligned} e^{-\beta\hat{\mathcal{H}}(\mathbf{r})} &= 1 - \beta\left[\frac{-\hbar^2}{2m}\nabla_{\mathbf{r}}^2 + U\right] + \frac{\beta^2}{2}\left[\frac{-\hbar^2}{2m}\nabla_{\mathbf{r}}^2 + U\right]^2 - \frac{\beta^3}{3!}\left[\frac{-\hbar^2}{2m}\nabla_{\mathbf{r}}^2 + U\right]^3 \\ &= 1 - \beta\left[\frac{-\hbar^2}{2m}\nabla_{\mathbf{r}}^2 + U\right] + \frac{\beta^2}{2}\left[\frac{\hbar^4}{4m^2}\nabla_{\mathbf{r}}^2\nabla_{\mathbf{r}}^2 - \frac{\hbar^2}{2m}\nabla_{\mathbf{r}}^2 U - \frac{\hbar^2}{2m}U\nabla_{\mathbf{r}}^2 \right. \\ &\quad \left. + U^2\right] - \frac{\beta^3}{3!}\left[\frac{-\hbar^6}{8m^3}\nabla_{\mathbf{r}}^6 + \frac{\hbar^4}{4m^2}\nabla_{\mathbf{r}}^2\nabla_{\mathbf{r}}^2 U + \frac{\hbar^4}{4m^2}\nabla_{\mathbf{r}}^2 U\nabla_{\mathbf{r}}^2 \right. \\ &\quad \left. + \frac{\hbar^4}{4m^2}U\nabla_{\mathbf{r}}^2\nabla_{\mathbf{r}}^2 - \frac{\hbar^2}{2m}\nabla_{\mathbf{r}}^2 U^2 - \frac{\hbar^2}{2m}U\nabla_{\mathbf{r}}^2 U - \frac{\hbar^2}{2m}U^2\nabla_{\mathbf{r}}^2 + U^3\right]. \end{aligned} \tag{10.48}$$

Here $U \equiv U(\mathbf{r})$. The gradient operator is 'open', which is to say that it acts on everything to its right, including any wave function subsequently inserted. For example, $\nabla_{\mathbf{r}}^2 U \equiv (\nabla_{\mathbf{r}}^2 U) + 2(\nabla_{\mathbf{r}} U) \cdot \nabla_{\mathbf{r}} + U\nabla_{\mathbf{r}}^2$.

Everything here has to be (eventually) written in singlet effective potential form, such as $U(\mathbf{r}) \Rightarrow U(\mathbf{r}|\mathbf{q})$,

$$\nabla^2 U(\mathbf{r}) \Rightarrow \sum_{j=1}^{N} \nabla_{\mathbf{r}_j}^2\left[u^{(1)}(\mathbf{r}_j) + 2u^{(2)}(\mathbf{r}_j|\mathbf{q}_{/j})\right], \tag{10.49}$$

and

$$(\nabla U(\mathbf{r})) \cdot \nabla \Rightarrow \sum_{j=1}^{N}\left(\nabla_{\mathbf{r}_j}\left[u^{(1)}(\mathbf{r}_j) + 2u^{(2)}(\mathbf{r}_j|\mathbf{q}_{/j})\right]\right) \cdot \nabla_{\mathbf{r}_j}. \tag{10.50}$$

As above, the singlet Hamiltonian operator has inverse temperature expansion of the form

$$\hat{\mathcal{H}}^{(1)}(\mathbf{r}_j|\mathbf{q}_{/j}) = \sum_{\alpha=1}^{\infty} \beta^{\alpha-1}\hat{\mathcal{H}}^{(1,\,\alpha)}(\mathbf{r}_j|\mathbf{q}_{/j}). \tag{10.51}$$

The inverse temperature expansion of the singlet Maxwell–Boltzmann operator to cubic order is

$$\begin{aligned}
e^{-\beta\hat{\mathcal{H}}^{(1)}(\mathbf{r}|\mathbf{q})} &= 1 - \beta\left[\hat{\mathcal{H}}^{(1,1)} + \beta\hat{\mathcal{H}}^{(1,2)} + \beta^2\hat{\mathcal{H}}^{(1,3)}\right] + \frac{\beta^2}{2}\left[\hat{\mathcal{H}}^{(1,1)} + \beta\hat{\mathcal{H}}^{(1,2)}\right]^2 \\
&\quad - \frac{\beta^3}{3!}\hat{\mathcal{H}}^{(1,1)3} = 1 - \beta\hat{\mathcal{H}}^{(1,1)} - \beta^2\left[\hat{\mathcal{H}}^{(1,2)} - \frac{1}{2}\hat{\mathcal{H}}^{(1,1)2}\right] \\
&\quad - \beta^3\left[\hat{\mathcal{H}}^{(1,3)} - \frac{1}{2}\hat{\mathcal{H}}^{(1,1)}\hat{\mathcal{H}}^{(1,2)} - \frac{1}{2}\hat{\mathcal{H}}^{(1,2)}\hat{\mathcal{H}}^{(1,1)} + \frac{1}{3!}\hat{\mathcal{H}}^{(1,1)3}\right].
\end{aligned} \tag{10.52}$$

Equating each order of inverse temperature, one obtains for β,

$$\hat{\mathcal{H}}^{(1,1)} = \frac{-\hbar^2}{2m}\nabla_{\mathbf{r}}^2 + U(\mathbf{r}|\mathbf{q}), \tag{10.53}$$

which agrees with the above first order result. For β^2 one has

$$\begin{aligned}
\hat{\mathcal{H}}^{(1,2)} &= \frac{1}{2}\left[\frac{\hbar^4}{4m^2}\nabla_{\mathbf{r}}^2\nabla_{\mathbf{r}}^2 - \frac{\hbar^2}{2m}\nabla_{\mathbf{r}}^2 U(\mathbf{r}|\mathbf{q}) - \frac{\hbar^2}{2m}U(\mathbf{r}|\mathbf{q})\nabla_{\mathbf{r}}^2 + U(\mathbf{r}|\mathbf{q})^2\right] \\
&\quad - \frac{1}{2}\left[\frac{\hbar^4}{4m^2}\nabla_{\mathbf{r}}^2\nabla_{\mathbf{r}}^2 - \frac{\hbar^2}{2m}\nabla_{\mathbf{r}}^2 U(\mathbf{r}) - \frac{\hbar^2}{2m}U(\mathbf{r})\nabla_{\mathbf{r}}^2 + U(\mathbf{r})^2\right] \\
&= \frac{\hbar^2}{4m}\nabla_{\mathbf{r}}^2[U(\mathbf{r}) - U(\mathbf{r}|\mathbf{q})] \\
&= \frac{\hbar^2}{4m}(\nabla_{\mathbf{r}}^2 U_\Delta) + \frac{\hbar^2}{2m}(\nabla_{\mathbf{r}} U_\Delta)\cdot\nabla_{\mathbf{r}}.
\end{aligned} \tag{10.54}$$

The final equality follows because $U(\mathbf{r}) = U(\mathbf{r}|\mathbf{q})$ at $\mathbf{r}_{lj} = \mathbf{q}_{lj}$. This agrees with the second order result obtained in the preceding subsection.

For β^3 one has

$$\hat{\mathcal{H}}^{(1,3)} = \frac{1}{2}\hat{\mathcal{H}}^{(1,1)}\hat{\mathcal{H}}^{(1,2)} + \frac{1}{2}\hat{\mathcal{H}}^{(1,2)}\hat{\mathcal{H}}^{(1,1)} - \frac{1}{3!}\hat{\mathcal{H}}^{(1,1)3} + \frac{1}{3!}\hat{\mathcal{H}}^3. \tag{10.55}$$

With $U_{\mathbf{r}} \equiv U(\mathbf{r})$, $U_{\mathbf{q}} \equiv U(\mathbf{r}|\mathbf{q})$, and $U_\Delta \equiv U(\mathbf{r}) - U(\mathbf{r}|\mathbf{q})$, one has

$$\begin{aligned}
\hat{\mathcal{H}}^3 = \Bigg[&\frac{-\hbar^6}{8m^3}\nabla_{\mathbf{r}}^6 + \frac{\hbar^4}{4m^2}\nabla_{\mathbf{r}}^2\nabla_{\mathbf{r}}^2 U_{\mathbf{r}} + \frac{\hbar^4}{4m^2}\nabla_{\mathbf{r}}^2 U_{\mathbf{r}}\nabla_{\mathbf{r}}^2 + \frac{\hbar^4}{4m^2}U_{\mathbf{r}}\nabla_{\mathbf{r}}^2\nabla_{\mathbf{r}}^2 \\
&- \frac{\hbar^2}{2m}\nabla_{\mathbf{r}}^2 U_{\mathbf{r}}^2 - \frac{\hbar^2}{2m}U_{\mathbf{r}}\nabla_{\mathbf{r}}^2 U_{\mathbf{r}} - \frac{\hbar^2}{2m}U_{\mathbf{r}}^2\nabla_{\mathbf{r}}^2 + U_{\mathbf{r}}^3\Bigg].
\end{aligned} \tag{10.56}$$

The cube of the first order operator, $\left(\hat{\mathcal{H}}^{(1,1)}\right)^3$, is the same as this with $U_{\mathbf{r}} \Rightarrow U_{\mathbf{q}}$. Subtracting these two and canceling any term that does not contain a gradient of the potential one obtains

$$\hat{\mathcal{H}}^3 - \left(\hat{\mathcal{H}}^{(1,1)}\right)^3 = \frac{\hbar^4}{4m^2}\nabla_{\mathbf{r}}^2\nabla_{\mathbf{r}}^2[U_{\mathbf{r}} - U_{\mathbf{q}}] + \frac{\hbar^4}{4m^2}\nabla_{\mathbf{r}}^2[U_{\mathbf{r}} - U_{\mathbf{q}}]\nabla_{\mathbf{r}}^2$$
$$- \frac{\hbar^2}{2m}\nabla_{\mathbf{r}}^2[U_{\mathbf{r}}^2 - U_{\mathbf{q}}^2] - \frac{\hbar^2}{2m}U_{\mathbf{r}}\nabla_{\mathbf{r}}^2 U_{\mathbf{r}} + \frac{\hbar^2}{2m}U_{\mathbf{q}}\nabla_{\mathbf{r}}^2 U_{\mathbf{q}}. \tag{10.57}$$

Notice the substantial cancelation that occurs here. For the remaining terms one has

$$\begin{aligned}
&\hat{\mathcal{H}}^{(1,1)}\hat{\mathcal{H}}^{(1,2)} + \hat{\mathcal{H}}^{(1,2)}\hat{\mathcal{H}}^{(1,1)} \\
&= \left[\frac{-\hbar^2}{2m}\nabla_{\mathbf{r}}^2 + U_{\mathbf{q}}\right]\left[\frac{\hbar^2}{4m}\nabla_{\mathbf{r}}^2 U_\Delta\right] + \left[\frac{\hbar^2}{4m}\nabla_{\mathbf{r}}^2 U_\Delta\right]\left[\frac{-\hbar^2}{2m}\nabla_{\mathbf{r}}^2 + U_{\mathbf{q}}\right] \\
&= \frac{-\hbar^4}{8m^2}\nabla_{\mathbf{r}}^2\nabla_{\mathbf{r}}^2 U_\Delta + \frac{\hbar^2}{4m}U_{\mathbf{q}}\nabla_{\mathbf{r}}^2 U_\Delta - \frac{\hbar^4}{8m^2}\nabla_{\mathbf{r}}^2 U_\Delta\nabla_{\mathbf{r}}^2 + \frac{\hbar^2}{4m}\nabla_{\mathbf{r}}^2 U_{\mathbf{q}} U_\Delta.
\end{aligned} \tag{10.58}$$

With these for the third order singlet Hamiltonian operator one has

$$\begin{aligned}
\hat{\mathcal{H}}^{(1,3)} &= \frac{1}{2}\hat{\mathcal{H}}^{(1,1)}\hat{\mathcal{H}}^{(1,2)} + \frac{1}{2}\hat{\mathcal{H}}^{(1,2)}\hat{\mathcal{H}}^{(1,1)} - \frac{1}{3!}(\hat{\mathcal{H}}^{(1,1)})^3 + \frac{1}{3!}\hat{\mathcal{H}}^3 \\
&= \frac{1}{2}\left\{\frac{-\hbar^4}{8m^2}\nabla_{\mathbf{r}}^2\nabla_{\mathbf{r}}^2 U_\Delta + \frac{\hbar^2}{4m}U_{\mathbf{q}}\nabla_{\mathbf{r}}^2 U_\Delta - \frac{\hbar^4}{8m^2}\nabla_{\mathbf{r}}^2 U_\Delta\nabla_{\mathbf{r}}^2 + \frac{\hbar^2}{4m}\nabla_{\mathbf{r}}^2 U_{\mathbf{q}} U_\Delta\right\} \\
&\quad + \frac{1}{3!}\left[\frac{\hbar^4}{4m^2}\nabla_{\mathbf{r}}^2\nabla_{\mathbf{r}}^2 U_\Delta + \frac{\hbar^4}{4m^2}\nabla_{\mathbf{r}}^2 U_\Delta\nabla_{\mathbf{r}}^2\right. \\
&\quad \left. - \frac{\hbar^2}{2m}\nabla_{\mathbf{r}}^2\left[U_{\mathbf{r}}^2 - U_{\mathbf{q}}^2\right] - \frac{\hbar^2}{2m}U_{\mathbf{q}}\nabla_{\mathbf{r}}^2 U_\Delta\right] \\
&= \frac{-\hbar^4}{48m^2}\nabla_{\mathbf{r}}^2\nabla_{\mathbf{r}}^2 U_\Delta + \frac{\hbar^2}{24m}U_{\mathbf{q}}\nabla_{\mathbf{r}}^2 U_\Delta - \frac{\hbar^4}{48m^2}\nabla_{\mathbf{r}}^2 U_\Delta\nabla_{\mathbf{r}}^2 + \frac{\hbar^2}{8m}\nabla_{\mathbf{r}}^2 U_{\mathbf{q}} U_\Delta \\
&\quad - \frac{\hbar^2}{12m}\nabla_{\mathbf{r}}^2\left[U_{\mathbf{r}}^2 - U_{\mathbf{q}}^2\right].
\end{aligned} \tag{10.59}$$

Recall that $U_{\mathbf{r}} \equiv U(\mathbf{r})$, $U_{\mathbf{q}} \equiv U(\mathbf{r}|\mathbf{q})$, and $U_\Delta \equiv U(\mathbf{r}) - U(\mathbf{r}|\mathbf{q})$. The gradient operators act on everything to the right, including any wave functions later added. After the gradients are evaluated one has to turn this into a singlet operator by making the replacement $\mathbf{r}_{lj} \Rightarrow \mathbf{q}_{lj}$ for each particle j. For example

$$\begin{aligned}
(\nabla_{\mathbf{r}}^2\nabla_{\mathbf{r}}^2 U_\Delta) &= \frac{1}{2}\sum_{j,k}\nabla_j^2\nabla_k^2\sum_{m,n}^{(m\neq n)}\left[u^{(2)}(\mathbf{r}_m, \mathbf{r}_n) - u^{(2)}(\mathbf{r}_m, \mathbf{q}_n)\right] \\
&= \frac{1}{2}\sum_j\sum_n^{(n\neq j)}\nabla_j^2\nabla_j^2\left[u^{(2)}(\mathbf{r}_j, \mathbf{r}_n) + u^{(2)}(\mathbf{r}_n, \mathbf{r}_j) - u^{(2)}(\mathbf{r}_j, \mathbf{q}_n)\right] \\
&\quad + \frac{1}{2}\sum_{j,k}^{(j\neq k)}\nabla_j^2\nabla_k^2\left[u^{(2)}(\mathbf{r}_j, \mathbf{r}_k) + u^{(2)}(\mathbf{r}_k, \mathbf{r}_j)\right] \\
&= \frac{1}{2}\sum_{j,k}^{(k\neq j)}\nabla_{\mathbf{r}_j}^2\nabla_{\mathbf{r}_j}^2 u^{(2)}(\mathbf{r}_j, \mathbf{q}_k) + \sum_{j,k}^{(j\neq k)}\nabla_{\mathbf{r}_j}^2\nabla_{\mathbf{q}_k}^2 u^{(2)}(\mathbf{r}_j, \mathbf{q}_k).
\end{aligned} \tag{10.60}$$

The final equality is in singlet form.

There are some terms in the expression for $\hat{\mathcal{H}}^{(1,3)}$ that are difficult to cast in singlet form. In view of this, perhaps it is better to pursue a modified approach, as in section 10.2.2, where the gradients of the difference between the actual potential and the modified effective potential vanish at $\mathbf{r} = \mathbf{q}$.

10.2.4 Pair, first order

Consider a system with singlet and pair potentials

$$U(\mathbf{r}) = \sum_{j=1}^{N} u^{(1)}(\mathbf{r}_j) + \frac{1}{2} \sum_{j,k}^{(j \neq k)} u^{(2)}(\mathbf{r}_j - \mathbf{r}_k). \tag{10.61}$$

Assume three-dimensional space.

For any configuration $\mathbf{q}$, one can relabel the particles and group them into disjoint pairs, $\{2j, 2j-1\}$. The particles assigned to a pair may vary from configuration to configuration. Define the six-dimensional pair configuration as $\mathbf{Q}_j \equiv \{\mathbf{q}_{2j}, \mathbf{q}_{2j-1}\}$, and similarly for the representation coordinate, $\mathbf{R}_j \equiv \{\mathbf{r}_{2j}, \mathbf{r}_{2j-1}\}$.

Define the pair-pair interaction potential as

$$\begin{aligned} u^{(4)}(\mathbf{R}_j, \mathbf{R}_k) \equiv\ & u^{(2)}(\mathbf{r}_{2j}, \mathbf{r}_{2k}) + u^{(2)}(\mathbf{r}_{2j-1}, \mathbf{r}_{2k}) \\ & + u^{(2)}(\mathbf{r}_{2j}, \mathbf{r}_{2k-1}) + u^{(2)}(\mathbf{r}_{2j-1}, \mathbf{r}_{2k-1}). \end{aligned} \tag{10.62}$$

The singlet energies and the internal pair energies can be included in the total energy by defining

$$u^{(2)}(\mathbf{R}_j) \equiv u^{(1)}(\mathbf{r}_{2j}) + u^{(1)}(\mathbf{r}_{2j-1}) + u^{(2)}(\mathbf{r}_{2j}, \mathbf{r}_{2j-1}). \tag{10.63}$$

The total energy may now be written as

$$U(\mathbf{R}) = \sum_{j=1}^{N/2} u^{(2)}(\mathbf{R}_j) + \sum_{j<k}^{N/2} u^{(4)}(\mathbf{R}_j, \mathbf{R}_k). \tag{10.64}$$

This has exactly the same functional form as the singlet theory, with $u^{(2)}$ playing the role of the singlet potential and $u^{(4)}$ that of the pair potential, now in six- rather than three-dimensional space.

By analogy with the singlet formulation, define the effective pair local potential as

$$\begin{aligned} u_j^{(2)}(\mathbf{R}_j) &\equiv u^{(2)}(\mathbf{R}_j | \mathbf{Q}_{/j}) \\ &= u^{(2)}(\mathbf{R}_j) + \frac{1}{2} \sum_{k=1}^{N/2}{}^{(k \neq j)} u^{(4)}(\mathbf{R}_j, \mathbf{Q}_k). \end{aligned} \tag{10.65}$$

With this the total effective local pair energy is

$$\begin{aligned} U^{(2)}(\mathbf{R}|\mathbf{Q}) &\equiv \sum_{j=1}^{N/2} u_j^{(2)}(\mathbf{R}_j) \\ &= \sum_{j=1}^{N/2} u^{(2)}(\mathbf{R}_j) + \frac{1}{2}\sum_{j=1}^{N/2}\sum_{k=1}^{N/2}{}^{(k\neq j)} u^{(4)}(\mathbf{R}_j, \mathbf{Q}_k). \end{aligned} \tag{10.66}$$

Clearly, $U^{(2)}(\mathbf{Q}|\mathbf{Q}) = U(\mathbf{Q})$.

The effective pair Hamiltonian operator for pair j is

$$\hat{\mathcal{H}}_j^{(2)}(\mathbf{R}_j) = \frac{-\hbar^2}{2m}\nabla_j^2 + u^{(2)}(\mathbf{R}_j|\mathbf{Q}_{/j}). \tag{10.67}$$

The gradient operator is of course six-dimensional. With this Schrödinger's equation can be solved to obtain the eigenvalues $E_{n_j}^{(2)}(\mathbf{Q}_{/j})$ and eigenfunctions $\phi_{n_j}^{(2)}(\mathbf{R}_j|\mathbf{Q}_{/j})$.

The total effective Hamiltonian operator is

$$\hat{\mathcal{H}}^{(2)}(\mathbf{R}|\mathbf{Q}) = \sum_{j=1}^{N/2} \hat{\mathcal{H}}_j^{(2)}(\mathbf{R}_j), \tag{10.68}$$

with eigenfunction $\Phi^{(2)}(\mathbf{R}|\mathbf{Q}) = \prod_{j=1}^{N/2} \phi_{n_j}^{(2)}(\mathbf{R}_j|\mathbf{Q}_{/j})$ and eigenvalue $E_{\mathbf{n}}^{(2)}(\mathbf{Q}) = \sum_{j=1}^{N/2} E_{n_j}^{(2)}(\mathbf{Q}_{/j})$.

The effective pair eigenvalues, $E_{n_j}^{(2)}(\mathbf{Q}_{/j})$, and eigenfunctions, $\phi_{n_j}^{(2)}(\mathbf{Q}_j|\mathbf{Q}_{/j})$, allow the pair commutation function to be obtained, $w^{(2)}(\mathbf{Q}_j, \mathbf{P}_j|\mathbf{Q}_{/j})$. These give the total commutation function as a sum over two-body commutation functions.

One suspects that the pair theory will be more accurate than the singlet theory. This indeed turned out to be the case for the first order, pair numerical results already obtained for the harmonic crystal (Attard 2019b).

10.3 Harmonic local field

10.3.1 Singlet harmonic field approximation

The singlet local field approximation described above requires the singlet energy eigenvalues and eigenfunctions to be obtained, from which the singlet commutation function can be calculated. One can simplify this procedure somewhat by making a quadratic expansion of the effective local field, which effectively models the system in any configuration as a set of independent harmonic oscillators. Since the energy eigenvalues and eigenfunctions of the simple harmonic oscillator is known analytically, it is a relatively trivial matter to obtain their commutation function (Attard 2018a), and hence the commutation function of the full system in this singlet, harmonic approximation.

This method has previously been tested for the simulation of a Lennard–Jones system. (Attard 2018b) and of a harmonic crystal (Attard 2019b).

In general, the particles of the sub-system interact via the potential energy, which is the sum of one-body, two-body, three-body terms, etc,

$$U(\mathbf{q}) = \sum_{j=1}^{N} u^{(1)}(\mathbf{q}_j) + \sum_{j<k}^{N} u^{(2)}(\mathbf{q}_j, \mathbf{q}_k) + \sum_{j<k<l}^{N} u^{(3)}(\mathbf{q}_j, \mathbf{q}_k, \mathbf{q}_l) + \cdots \tag{10.69}$$

Here we assume N particles in d-dimensional space. Distributing the energy equally as above, the effective local field of particle j in configuration $\mathbf{q}$ is defined as

$$\begin{aligned} U_j(\mathbf{q}_j|\mathbf{q}) = u^{(1)}(\mathbf{q}_j) + \frac{1}{2}\sum_{k=1}^{N}{}^{(k\neq j)}\, u^{(2)}(\mathbf{q}_j, \mathbf{q}_k) \\ + \frac{1}{3}\sum_{k<l}^{N}{}^{(k,\, l\neq j)}\, u^{(3)}(\mathbf{q}_j, \mathbf{q}_k, \mathbf{q}_l) + \cdots \end{aligned} \tag{10.70}$$

with $U(\mathbf{q}) = \sum_{j=1}^{N} U_j(\mathbf{q}_j|\mathbf{q})$. We do not need to replace $\mathbf{q}_j$ by $\mathbf{r}_j$ in the harmonic field approach because we do not need to solve numerically for the energy eigenfunctions.

The potential energy of particle j in configuration $\mathbf{q}$ may be expanded to second order about its local minimum at $\bar{\mathbf{q}}_j(\mathbf{q})$,

$$U_j(\mathbf{q}_j|\mathbf{q}) = \bar{U}_j(\mathbf{q}) + \frac{1}{2}[\mathbf{q}_j - \bar{\mathbf{q}}_j][\mathbf{q}_j - \bar{\mathbf{q}}_j]\colon \underline{\underline{U}}_j'', \tag{10.71}$$

where the minimum value of the potential is $\bar{U}_j(\mathbf{q}) \equiv U_j(\bar{\mathbf{q}}_j|\mathbf{q})$. The $d \times d$ second derivative matrix for particle j at the minimum, $\underline{\underline{U}}_j'' = \nabla_{\mathbf{r}_j}\nabla_{\mathbf{r}_j} U_j(\mathbf{r}_j|\mathbf{q})\big|_{\mathbf{r}_j=\bar{\mathbf{q}}_j}$, is assumed positive definite.

For configurations $\mathbf{q}$ that have no local minimum in the potential, or that have too large a displacement $|\mathbf{q}_j - \bar{\mathbf{q}}_j|$, the corresponding single particle commutation function can be set to unity, $W_j(\mathbf{\Gamma}) = 1$, (or, in the multi-dimensional case, the commutation function of the corresponding mode). This is justified by analytic results for the simple harmonic oscillator (Attard 2018a).

The positive definite second derivative matrix has d eigenvalues $\lambda_{j\alpha}(\mathbf{q}) > 0$, and orthonormal eigenvectors, $\underline{\underline{U}}_j''\mathbf{X}_{j\alpha} = \lambda_{j\alpha}\mathbf{X}_{j\alpha}$, $\alpha = x, y, \ldots, d$. For particle j in configuration $\mathbf{q}$ the eigenvalues define the frequencies

$$\omega_{j\alpha}(\mathbf{q}) = \sqrt{\lambda_{j\alpha}(\mathbf{q})/m}, \quad \alpha = x, y, \ldots, d. \tag{10.72}$$

With this the potential energy is

$$\begin{aligned} U(\mathbf{r}|\mathbf{q}) &= \sum_{j=1}^{N} \bar{U}_j + \frac{1}{2}\sum_{j=1}^{N}(\mathbf{q}_j - \bar{\mathbf{q}}_j)(\mathbf{q}_j - \bar{\mathbf{q}}_j)\colon \underline{\underline{U}}_j'' \\ &= \sum_{j=1}^{N} \bar{U}_j + \frac{1}{2}\sum_{j,\alpha} \hbar\omega_{j\alpha} Q_{j\alpha}^2. \end{aligned} \tag{10.73}$$

Here $Q_{j\alpha} \equiv \sqrt{m\omega_{j\alpha}/\hbar}\, Q'_{j\alpha}$, and $\mathbf{Q}'_j = \underline{\underline{X}}_j^{\mathrm{T}}[\mathbf{q}_j - \bar{\mathbf{q}}_j]$.

The harmonic field approximation combined with the second order expansion about the local minima maps each configuration $\mathbf{\Gamma}$ to a system of dN independent harmonic oscillators with frequencies $\omega_{j\alpha}$ displacements $Q_{j\alpha}$, and momenta $P_{j\alpha} = \left\{\underline{\underline{X}}_j^{\mathrm{T}}\mathbf{p}_j\right\}_\alpha \Big/ \sqrt{m\hbar\omega_{j\alpha}}$.

With this harmonic approximation for the potential energy, the effective Hamiltonian in a particular configuration can be written

$$\mathcal{H}^{\mathrm{SHO}}(\mathbf{p}, \mathbf{q} - \bar{\mathbf{q}}) = \sum_{j=1}^{N} \bar{U}_j + \frac{1}{2}\sum_{j,\alpha} \hbar\omega_{j\alpha}\left[P_{j\alpha}^2 + Q_{j\alpha}^2\right]. \tag{10.74}$$

The commutation function for the interacting system for a particular configuration can be approximated as the product of commutation functions for effective non-interacting harmonic oscillators which have the local displacement as their argument. With this the harmonic approximation to the commutation function is

$$\begin{aligned} W_p(\mathbf{\Gamma}) &\approx W_p^{\mathrm{SHO}}(\mathbf{p}, \mathbf{q} - \bar{\mathbf{q}}) \\ &= e^{\beta\mathcal{H}^{\mathrm{SHO}}(\mathbf{p},\, \mathbf{r}-\bar{\mathbf{q}})} \frac{\left\langle \mathbf{q} - \bar{\mathbf{q}} \middle| e^{-\beta\hat{\mathcal{H}}^{\mathrm{SHO}}} \middle| \mathbf{p} \right\rangle}{\langle \mathbf{q} - \bar{\mathbf{q}} | \mathbf{p}\rangle} \\ &= \prod_{j,\alpha} W_{p,j\alpha}^{\mathrm{SHO}}(P_{j\alpha}, Q_{j\alpha}). \end{aligned} \tag{10.75}$$

The harmonic oscillator commutation function for a single mode is (Attard 2018b)

$$\begin{aligned} W_{p,j\alpha}^{\mathrm{SHO}}(P_{j\alpha}, Q_{j\alpha}) &= \sqrt{2}\, e^{-\mathrm{i}P_{j\alpha}Q_{j\alpha}} e^{\beta\hbar\omega_{j\alpha}\left[P_{j\alpha}^2+Q_{j\alpha}^2\right]/2} e^{-\left[P_{j\alpha}^2+Q_{j\alpha}^2\right]/2} \\ &\times \sum_{n_{j\alpha}=0}^{\infty} \frac{\mathrm{i}^{n_{j\alpha}} e^{-\beta\hbar\omega_{j\alpha}(n_{j\alpha}+1/2)}}{2^{n_{j\alpha}} n_{j\alpha}!} \mathrm{H}_{n_{j\alpha}}(P_{j\alpha})\mathrm{H}_{n_{j\alpha}}(Q_{j\alpha}). \end{aligned} \tag{10.76}$$

Here $\mathrm{H}_n(z)$ is the Hermite polynomial of degree n. The imaginary terms here are odd in momentum. As justified by analytic results for the simple harmonic oscillator (Attard 2018b) for configurations such that $\underline{\underline{U}}_j''(\mathbf{q})$ is not positive definite (i.e., a particular eigenvalue is not positive, $\lambda_{j\alpha} \leqslant 0$), or that the displacement $Q_{j\alpha}$ exceeds a predetermined cut-off, the corresponding commutation function can be set to unity, $W_{p,j\alpha}^{\mathrm{SHO}} = 1$.

For the averages, the momentum integrals can be performed analytically, both here and in combination with the symmetrization function. This considerably reduces computer time and substantially increases accuracy.

10.3.2 Cluster harmonic field approximation

Any configuration $\mathbf{q}$ can be decomposed into disjoint clusters labeled $\alpha = 1, 2, \ldots$. Of the different criteria that can be used to define a cluster, perhaps the simplest is that two particles belong to the same cluster if, and only if, they are connected by at least one path of bonds. Two particles are bonded if their separation is less than a nominated length. Some clusters, perhaps the great majority, will consist of a single particle.

An even simpler definition can be made in one dimension. In this case define the pair cluster α, $\alpha = 1, 2, \ldots, N/2$, as the nearest neighbors $\{2\alpha - 1, 2\alpha\}$, irrespective of their actual separation. This criteria is used in the results presented below.

Using a separation-based criterion for the definition of a cluster is useful not only for the harmonic field approximation to the commutation function, but also for the calculation of the symmetrization function. Depending on the chosen bond length, only permutations of particles within the same cluster need to be considered. (This idea is not used in the results presented below.)

The cluster energy is the internal energy plus the relevant proportion of the interaction energy with other clusters: half for pair interactions, one third for triplet interactions, etc. The total potential energy is

$$U(\mathbf{q}) = \sum_{j=1}^{N} u^{(1)}(\mathbf{q}_j) + \sum_{j<k} u^{(2)}(\mathbf{q}_j, \mathbf{q}_k) + \cdots = \sum_{\alpha} U_\alpha(\mathbf{q}_\alpha; \mathbf{q}), \tag{10.77}$$

where the energy of cluster α is

$$U_\alpha(\mathbf{q}_\alpha|\mathbf{q}) = \sum_{j\in\alpha} u^{(1)}(\mathbf{q}_j) + \sum_{j<k}^{(j,\, k\in\alpha)} u^{(2)}(\mathbf{q}_j, \mathbf{q}_k) + \frac{1}{2}\sum_{j\in\alpha}\sum_{\beta}^{(\beta\neq\alpha)}\sum_{k\in\beta} u^{(2)}(\mathbf{q}_j, \mathbf{q}_k) + \cdots. \tag{10.78}$$

There are N_α particles in cluster α, with positions $\mathbf{q}_\alpha = \{\mathbf{q}_{\alpha,1}, \mathbf{q}_{\alpha,2}, \ldots, \mathbf{q}_{\alpha,N_\alpha}\}$, where $\mathbf{q}_{\alpha,j} = \mathbf{q}_k$ for one of the $k \in \alpha$. This is a (dN_α)-dimensional vector.

The second order expansion about the minimum energy cluster configuration, $\bar{\mathbf{q}}_\alpha$, is

$$U_\alpha(\mathbf{q}_\alpha|\mathbf{q}) = \bar{U}_\alpha(\mathbf{q}) + \frac{1}{2}\underline{\underline{U}}_\alpha'' : [\mathbf{q}_\alpha - \bar{\mathbf{q}}_\alpha][\mathbf{q}_\alpha - \bar{\mathbf{q}}_\alpha]. \tag{10.79}$$

The second derivative matrix is $\underline{\underline{U}}_\alpha'' \equiv \nabla_{\mathbf{r}_\alpha}\nabla_{\mathbf{r}_\alpha} U_\alpha(\mathbf{r}_\alpha|\mathbf{q})\big|_{\mathbf{r}_\alpha=\bar{\mathbf{q}}_\alpha}$, which is $(dN_\alpha) \times (dN_\alpha)$. One has to find the eigenvalues, assumed positive, and eigenvectors of this. These give the cluster phonon mode frequencies $\omega_{\alpha,b}$, $b = 1, 2, \ldots, dN_\alpha$, and mode amplitudes, $\mathbf{Q}_\alpha = \underline{\underline{\omega}}_\alpha \underline{\underline{X}}_\alpha^{\mathrm{T}}[\mathbf{q}_\alpha - \bar{\mathbf{q}}_\alpha]$. The frequency matrix is diagonal with elements $\omega_{\alpha;bc} = \sqrt{m\omega_{\alpha,b}/\hbar}\,\delta_{bc}$. As before, the momentum of mode b in cluster α is $P_{\alpha,b} = \left\{\underline{\underline{X}}_\alpha^{\mathrm{T}}\mathbf{p}_\alpha\right\}_b / \sqrt{m\hbar\omega_{\alpha,b}}$.

This formulation is essentially the same as the singlet harmonic field theory, and one may similarly define the cluster harmonic field commutation function as the product of simple harmonic oscillator commutation functions, one for each phonon mode of each cluster. The cluster harmonic field commutation function of the configuration is

$$W(\mathbf{\Gamma}) = \prod_{\alpha} W_{\alpha}^{\mathrm{SHO}}(\mathbf{P}_{\alpha}, \mathbf{Q}_{\alpha}) = \prod_{\alpha,b} W_{\alpha,b}^{\mathrm{SHO}}(P_{\alpha,b}, Q_{\alpha,b}). \tag{10.80}$$

From the computational point of view, a felicitous aspect of the cluster harmonic field approximation is that there are exactly as many modes as in the singlet harmonic field approximation. This means that all of the sub-routines called to obtain the various statistical averages in the singlet approximation can be called without change in the cluster harmonic field approximation.

10.4 Gross–Pitaevskii mean field Schrödinger equation

In this section is developed a mean field singlet non-linear Schrödinger equation. This is closely related to the Gross–Pitaevskii equation (Gross 1961, Pitaevskii 1961). The way that the singlet mean field is formulated from the pair potential in the most general case is similar to the singlet effective local field derived earlier in this chapter.

Consider an isolated system containing N particles. Let $\psi(\mathbf{r})$ be the wave function, with the position vector being $\mathbf{r} = \{\mathbf{r}_1, \mathbf{r}_2, \ldots, \mathbf{r}_N\}$. This is taken to be normalized, $\langle\psi|\psi\rangle = 1$, but unsymmetrized. The probability density for particles being at those positions is just the square of the amplitude, $\wp(\mathbf{r}) = \psi(\mathbf{r})^* \psi(\mathbf{r})$. As usual integrating over the remaining positions gives the pair probability, $\wp^{(2)}(\mathbf{r}_1, \mathbf{r}_2) = \int \mathrm{d}\mathbf{r}_3\, \mathrm{d}\mathbf{r}_4 \ldots \mathrm{d}\mathbf{r}_N\, \psi(\mathbf{r})^* \psi(\mathbf{r})$.

We now invoke the singlet approximation for the wave function,

$$\psi_{\mathbf{n}}(\mathbf{r}) = \prod_{j=1}^{N} \psi_{\mathbf{n}_j}^{(1)}(\mathbf{r}_j). \tag{10.81}$$

Here particle j is in the single particle state $\mathbf{n}_j$. This is obviously not symmetrized. The pair probability is $\wp^{(2)}(\mathbf{r}_1, \mathbf{r}_2) = |\psi_{\mathbf{n}_1}^{(1)}(\mathbf{r}_1)|^2\, |\psi_{\mathbf{n}_2}^{(1)}(\mathbf{r}_2)|^2$.

We suppose that the potential energy is the sum of a singlet potential and a pair potential. The latter is subsumed into an effective singlet potential,

$$\tilde{u}_j^{(1)}\left(\mathbf{r}_j; \left[\underline{\psi}^{(1)}\right]\right) = \frac{1}{2} \sum_{k=1}^{N}{}^{(k\neq j)} \int \mathrm{d}\mathbf{r}_k\, \psi_{\mathbf{n}_k}^{(1)}(\mathbf{r}_k)^*\, \psi_{\mathbf{n}_k}^{(1)}(\mathbf{r}_k)\, u^{(2)}(\mathbf{r}_j, \mathbf{r}_k). \tag{10.82}$$

Apart from the factor of one half, this is the mean field experienced by particle j. It is usually the case that the pair potential depends only upon the separation, $u(\mathbf{r}_j, \mathbf{r}_k) = u(r_{jk})$. For this to be meaningful it is essential that the singlet wave function be normalized to unity, $\langle\psi_{\mathbf{n}_k}^{(1)}|\psi_{\mathbf{n}_k}^{(1)}\rangle = 1$. Because we are dealing with unsymmetrized wave functions (i.e., the $\mathbf{n}_k$ may be different for different k), this singlet potential depends upon j, which may pose problems for the Hamiltonian operator, which must be symmetric with respect to the particle labels.

Gross (1961) and Pitaevskii (1961) replace the full pair potential by a contact form (Pathria 1972, chapter 10),

$$u^{(2)}(\mathbf{r}_j, \mathbf{r}_k) = g\delta(\mathbf{r}_j - \mathbf{r}_k), \quad g \equiv \frac{4\pi\hbar^2 a}{m}. \tag{10.83}$$

Here a is the scattering length, which is a gross measure of the strength and range of the pair potential. For a hard-sphere potential it is equal to the diameter. It can be negative for attractive potentials. The contact form is valid at low densities, $a\rho^{1/3} \ll 1$, and low temperatures when the thermal wavelength is large, $\Lambda a \gg 1$, in which regime the pair potential is subordinate to wave function symmetrization effects. With this the effective singlet potential becomes

$$\tilde{u}_j^{(1)}\left(\mathbf{r}_j; \left[\underline{\psi}^{(1)}\right]\right) = \frac{g}{2} \sum_{k=1}^{N}{}^{(k\neq j)} \psi_{\mathbf{n}_k}^{(1)}(\mathbf{r}_j)^* \, \psi_{\mathbf{n}_k}^{(1)}(\mathbf{r}_j). \tag{10.84}$$

With this effective singlet potential, the total Hamiltonian operator is

$$\begin{aligned} \hat{\mathcal{H}}\left(\mathbf{r}; \left[\underline{\psi}^{(1)}\right]\right) &= \sum_{j=1}^{N} \left\{ \frac{-\hbar^2}{2m} \nabla_j^2 + u^{(1)}(\mathbf{r}_j) + \tilde{u}_j^{(1)}\left(\mathbf{r}_j; \left[\underline{\psi}^{(1)}\right]\right) \right\} \\ &= \sum_{j=1}^{N} \left\{ \frac{-\hbar^2}{2m} \nabla_j^2 + u^{(1)}(\mathbf{r}_j) + \frac{g}{2} \sum_{k=1}^{N}{}^{(k\neq j)} \psi_{\mathbf{n}_k}^{(1)}(\mathbf{r}_j)^* \, \psi_{\mathbf{n}_k}^{(1)}(\mathbf{r}_j) \right\}. \end{aligned} \tag{10.85}$$

The second equality uses the Gross–Pitaevskii contact form of the pair potential. The factor of one half used in the definition of the effective mean field potential ensures that the total contribution from this term equals the average total pair contribution. Unless all the single-particle states are the same, $\mathbf{n}_k = \mathbf{n}_1$, this is not symmetric with respect to particle permutations.

Using the jth term, the time-dependent Schrödinger equation for the singlet wave function is

$$\begin{aligned} \mathrm{i}\hbar \frac{\partial \psi_{\mathbf{n}_j}^{(1)}(\mathbf{r}_j)}{\partial t} &= \left[\frac{-\hbar^2}{2m} \nabla_j^2 + u^{(1)}(\mathbf{r}_j) + \tilde{u}_j^{(1)}(\mathbf{r}_j; [\underline{\psi}^{(1)}]) \right] \psi_{\mathbf{n}_j}^{(1)}(\mathbf{r}_j) \\ &= \left[\frac{-\hbar^2}{2m} \nabla_j^2 + u^{(1)}(\mathbf{r}_j) + \frac{g}{2} \sum_{k=1}^{N}{}^{(k\neq j)} \psi_{\mathbf{n}_k}^{(1)}(\mathbf{r}_j)^* \, \psi_{\mathbf{n}_k}^{(1)}(\mathbf{r}_j) \right] \psi_{\mathbf{n}_j}^{(1)}(\mathbf{r}_j). \end{aligned} \tag{10.86}$$

This is a coupled set of non-linear partial differential equations for the time evolution of the effective singlet wave function. It depends upon the cube of the wave function.

10.5 Numerical results in one-dimension

10.5.1 Harmonic crystal

Analytic expressions for the energy eigenvalues and eigenfunctions of a one-dimensional harmonic crystal are derived in section 6.2, following earlier work (Attard 2002, 2019a). The exact results for the average energy were compared with Monte Carlo simulations in classical phase space using the high temperature expansion in section 5.4. The numerical results in this section for the harmonic effective local field are taken from Attard (2019b).

10.5.1.1 Model

In summary, the model has particles attached by linear springs to each neighbor and to lattice sites. Let the coordinate of the *j*th particle be q_j. Here for simplicity the particles are ordered along the line, $q_j < q_{j+1}$. Let the *j*th lattice position, where the particle would sit in the lowest energy state, be $\bar{q}_j = j\Delta_q$. The lattice spacing is also the relaxed inter-particle spring length. There are fixed 'wall' particles at $q_0 = 0$ and $q_{N+1} = (N + 1)\Delta_q$. Let $d_j \equiv q_j - \bar{q}_j$ be the displacement from the lattice position; for the wall particles, $d_0 = d_{N+1} = 0$. The system has number density $\rho = \Delta_q^{-1}$.

In this model, there is an external harmonic potential of spring constant κ acting on each particle centered at its lattice site. The inter-particle spring has strength λ and relaxed length Δ_q. With these the potential energy is

$$\begin{aligned} U(\mathbf{q}) &= \frac{\kappa}{2}\sum_{j=1}^{N}\left[q_j - \bar{q}_j\right]^2 + \frac{\lambda}{2}\sum_{j=0}^{N}\left[q_{j+1} - q_j - \Delta_q\right]^2 \\ &= \frac{\kappa}{2}\sum_{j=1}^{N} d_j^2 + \frac{\lambda}{2}\sum_{j=0}^{N}\left[d_{j+1} - d_j\right]^2. \end{aligned} \tag{10.87}$$

10.5.1.2 Simulation algorithm

The simulation algorithm was as described by Attard (2018b). The Metropolis algorithm in position space was used with the usual classical Maxwell–Boltzmann weight. The various momentum integrals were performed analytically. Averages were evaluated by umbrella sampling using the commutation function as weight. Three versions of the commutation function (unity (i.e., classical), singlet mean field, pair mean field) were tested. The particles in the model harmonic crystal are non-identical and so the symmetrization function was set to unity.

Typically enough configurations were generated to make the statistical error less than 1%, sometimes much less. The total simulation time depends on how many Hermite polynomials are used for the commutation function ($M = 8$ in the results reported below; tests with $M = 6$ and $M = 12$ showed no great effect), and the cut-off for the mode amplitude beyond which the commutation function was set to unity ($Q^{\text{cut}} = 1$ in the results below; tests with $Q^{\text{cut}} = 2$ showed no great effect). The number of polynomials used equates to the number of energy levels per particle. However the number of energy levels effectively used to characterize the system is much greater than N^M because this is the number per phase space point, and effectively a continuum of such points is accessible in the simulations.

Interestingly enough, for an accuracy of about 1%, the Monte Carlo algorithm was a factor of about 2000 times more efficient (in terms of total computer time) than the quasi-analytic exact phonon method (Attard 2019a). The main bottleneck in the latter was the crude numerical quadrature method that was used to evaluate the density profile, and this was exacerbated by the large number of energy levels that were required for accurate results at higher temperatures. For this particular comparison at $\beta\hbar\omega_{LJ} = 2$, $l^{max} = 20\,000$ energy levels were necessary; grid parameters $\Delta_Q = 0.15$ and $Q^{max} = 6$ gave a quadrature error of about 0.8%. The phonon method is more efficient in one respect, namely that it requires negligible computer time for each additional temperature point; the simulations give results for only one temperature at a time.

The Lennard–Jones frequency used to scale the results below is $\omega_{LJ} = 3.28 \times 10^{12}$Hz, the mass is $m = 3.35 \times 10^{-26}$kg, the well-depth is $\varepsilon = 4.93 \times 10^{-22}$J, and the equilibrium separation is $r_e = 3.13 \times 10^{-10}$m. These parameters are appropriate for neon (van Sciver 2012).

10.5.1.3 Results

Figure 10.1 shows the average energy for a canonical equilibrium system as a function of inverse temperature using exact calculations and classical phase space simulations. At low temperatures the phonon modes are in the ground state, $E_1 = \sum_{n=1}^{N} \hbar\omega_n/2$, and at high temperatures the system approaches the classical result, $\langle \mathcal{H} \rangle_{cl} = N/\beta$. It can be seen that both the singlet and pair harmonic effective local field approximations are in relatively good agreement with these limiting results and with the exact calculations over the whole temperature regime shown. The commutation function provides a significant correction to the classical results at low temperatures.

The exact calculations in figure 10.1 use a finite number of energy levels, namely 10 000. (See table 9.1 for the effects of the number of energy levels on the average energy at various temperatures.) A numerical quadrature is required for the density profiles, and this is affected by the chosen grid spacing and number of points.

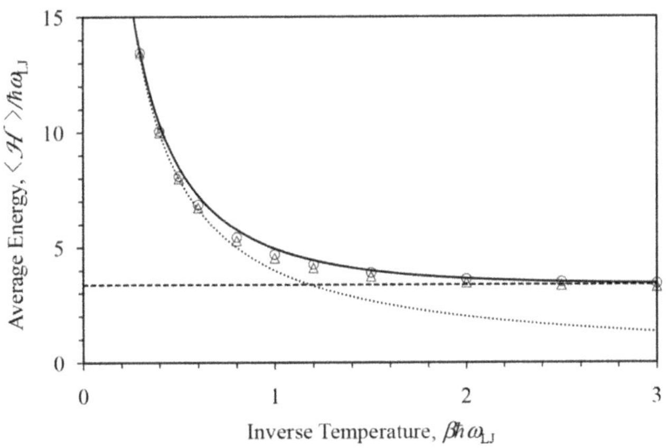

Figure 10.1. Average energy versus inverse temperature of the harmonic crystal ($N = 4$, $\Delta_q = r_e$, $\lambda = \kappa = m\omega_{LJ}^2$, and no symmetrization). The solid curve is the exact result using $l_{max} = 10\,000$ energy levels. The triangles are the singlet and the circles are the pair harmonic effective local field Monte Carlo simulation results. The statistical error is less than the size of the symbols. The dotted curve is the classical result, $\langle E \rangle_{cl} = N/\beta$. The dashed line is the ground state, $E_1 = \sum_{n=1}^{N} \hbar\omega_n/2$.

Despite these limitations, it is appropriate to designate the phonon mode calculations as exact because they use explicit analytic expressions for the energy eigenvalues and eigenfunctions.

According to the data in table 9.1, the exact calculations of the average energy, using 10 000 energy levels, as in figure 10.1, are reliable for $\beta\hbar\omega_{\rm LJ} \gtrsim 0.4$. The exact data begins to underestimate the classical result for higher temperatures than this. One might speculate that the exact quantum result for the average energy of the harmonic crystal should approach the classical limit from above. Using fewer energy levels reduces the domain of inverse temperatures in which the exact calculations are reliable.

Both the singlet and pair harmonic effective local field simulations are practically exact at higher temperatures, $\beta\hbar\omega_{\rm LJ} \lesssim 0.5$ in this case. As the temperature is decreased, the singlet harmonic effective local field energy lies between the exact energy and the classical energy. The pair harmonic effective local field result lies between the singlet energy and the exact energy. It can be seen that at low temperatures the classical energy is substantially less than the exact ground state energy, but the pair harmonic effective local field energy is only slightly less than the exact ground state energy. It may be concluded that the harmonic effective local field approach is better than a high temperature expansion in that it yields the dominant quantum correction to the classical result over the entire range of temperatures.

In figure 10.2, the average energy is scaled by the inverse temperature, which focuses attention on the low temperature regime. For inverse temperatures $\beta\hbar\omega_{\rm LJ} \gtrsim 3$, the exact results with 10 000 energy levels are practically indistinguishable from the ground state energy. At low temperatures, both harmonic effective local field classical phase space approximations give a lower energy than the ground state. For example, in the case treated in the figure, the ground state energy is $E_1/\hbar\omega_{\rm LJ} = 3.385$. At $\beta\hbar\omega_{\rm LJ} = 10$, the singlet harmonic effective local field theory gives $\langle\hat{\mathcal{H}}\rangle/\hbar\omega_{\rm LJ} = 3.116 \pm .002$, and the pair harmonic effective local field theory gives $3.305 \pm .001$. At this temperature the classical result given by the simulation

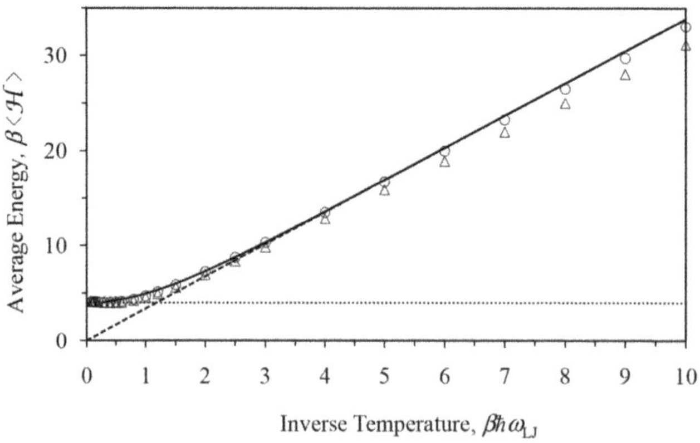

Figure 10.2. Same as the preceding figure, but with the average energy normalized by the inverse temperature and extending to lower temperatures.

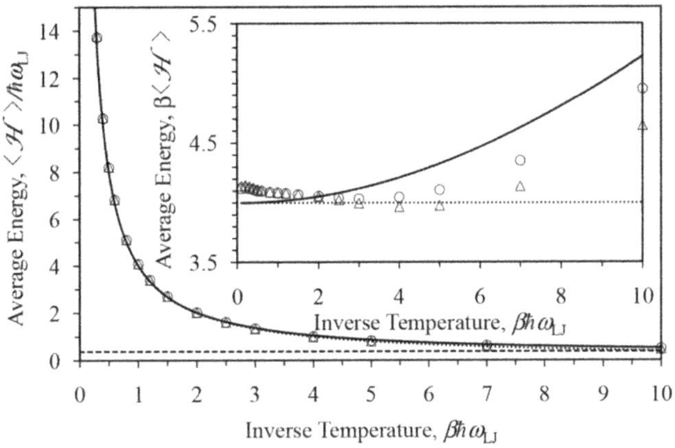

Figure 10.3. Average energy versus inverse temperature ($N = 4$, $\Delta_q = r_e$, $\kappa = 0$, $\lambda/m\omega_{LJ}^2 = 0.02$). The solid curve is the exact analytic result, equation (6.32). The triangles are the singlet and the circles are the pair harmonic effective local field Monte Carlo simulations, respectively. The dotted curve is the classical result, $\langle E\rangle_{cl} = N/\beta$, and the dashed line is the ground state energy. Inset: Average energy scaled by the inverse temperature.

was 0.400 2 ± .0002, which is rather close to the exact classical result of 0.4. Here and throughout, the statistical error quoted for the simulations is twice the standard error on the mean, which is the 96% confidence level.

One can conclude from the data that at low temperatures such that the system is close to the ground state, the harmonic effective local field approximations remain viable. The pair harmonic effective local field approach substantially reduces the error in the singlet harmonic effective local field approach. In the absence of exact data, the difference between the pair and singlet harmonic effective local field results would give a guide to the quantitative accuracy of the former.

It is worth mentioning that at each temperature the classical phase space simulations took about five minutes on a desktop personal computer to obtain the quoted accuracy. In comparison, it took about two days to obtain the exact results with these energy levels and quadrature grid, the latter being the time limiting part of the computations. (Once the exact results are obtained and saved, they can be applied at any temperature with negligible additional computing time.)

Figure 10.3 shows the average energy for a weakly coupled crystal. There is no singlet potential, $\kappa = 0$, and the nearest neighbor spring constant has been decreased by a factor of 50, $\lambda/m\omega_{LJ}^2 = 0.02$. The lattice spacing and spring length is unchanged. The ground state energy is $E_1 = 0.375\,7$, which is approximately one tenth of that for the parameters of figure 10.1.

In the main body of figure 10.3, the results of both harmonic effective local field approximations appear indistinguishable from the exact results. At $\beta\hbar\omega_{LJ} = 3$, the exact energy is 1.373 0 (1.370 7 for $l_{max} = 20\,000$; 1.362 7 for 10 000), the singlet harmonic effective local field gives 1.332 2 ± 0.0009 and the pair harmonic effective local field gives 1.344 3 ± 0.0008. The exact classical result here is 4/3, compared to

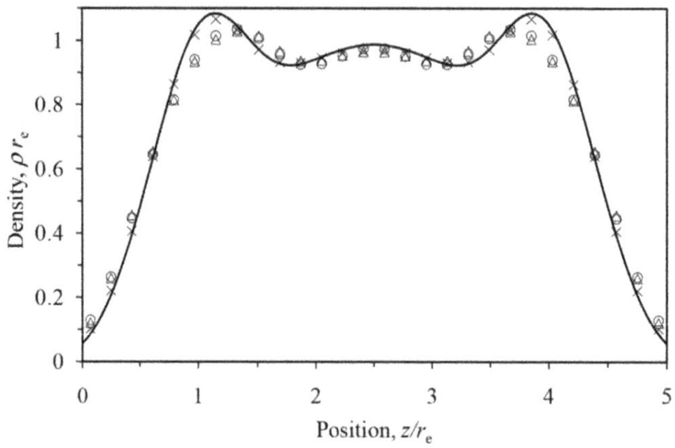

Figure 10.4. Density profile at $\beta\hbar\omega_{LJ} = 2$ ($N = 4$, $\Delta_q = r_e$, $\kappa = 0$, $\lambda/m\omega_{LJ}^2 = 0.02$, no symmetrization). The solid curve is the exact result using $l_{max} = 20\,000$ energy levels, and the symbols are the classical phase space Monte Carlo simulations, with the crosses using $W = 1$ (i.e., classical, no commutation function), the triangles using the singlet and the circles using the pair harmonic effective local field commutation function.

$1.332\,9 \pm 0.0007$ given by the classical simulations (i.e., commutation function set to unity). For inverse temperatures $\beta\hbar\omega_{LJ} \lesssim 1.8$, the harmonic effective local field algorithms are more reliable than the exact results with $l_{max} = 20\,000$ (not shown).

The inset of figure 10.3 shows the average energy scaled by the inverse temperature. This emphasizes the departure from the classical result, which is constant with this scaling. The exact calculations with 20 000 energy levels would approach the classical results from the low temperature side, but begin to fail for temperature higher than $\beta\hbar\omega_{LJ} \lesssim 3$ (not shown). Interestingly enough the harmonic effective local field results show a small departure from the classical result at the highest temperatures. It is apparent from the data in the inset that the pair harmonic approximation about halves the error of the singlet harmonic approximation at the lowest temperatures shown.

Figure 10.4 shows the density profile for this weak coupling case at $\beta\hbar\omega_{LJ} = 2$. The density peaks are rather broad, with the central two particles merging into a single peak. Interestingly enough, the density profile spills over beyond the wall particles at $q_0 = 0$ and $q_5 = 5r_e$. There is good agreement between all four methods, although the harmonic effective local field methods tend to underestimate the density at the shoulders of the density profile.

10.5.2 Lennard–Jones fluid

The results in this subsection are taken from Attard (2020). They are for a one-dimensional Lennard–Jones fluid. Singlet effective local field results are given, both directly and in harmonic approximation (Attard 2018a). The algorithm for the former is now briefly outlined.

10.5.2.1 Energy eigenstates

The particles of the one-dimensional system experience one-body and two-body potentials. The effective local singlet potential is taken to be

$$U(r|q', q'') = u^{(1)}(r) + \frac{1}{2}[u^{(2)}(r - q') + u^{(2)}(q'' - r)]. \tag{10.88}$$

Here only nearest neighbor pair interactions are included; r is the representation position of the particle of interest, and q' and q'' are the position coordinates (eigenvalues) of the nearest neighbors on either side. (If the particle of interest is a terminal one, then one of the pair potentials is absent.) The singlet potential is that of a simple harmonic oscillator,

$$u^{(1)}(r) = \frac{1}{2}m\omega^2 r^2, \tag{10.89}$$

and the pair potential is Lennard–Jones,

$$u^{(2)}(r - q') = \varepsilon\left[\frac{r_e^{12}}{(r - q')^{12}} - 2\frac{r_e^{6}}{(r - q')^{6}}\right]. \tag{10.90}$$

The singlet wave function is obtained on a three-dimensional grid for q, q', and q'', with a fourth dimension for the energy level n. One has $q' + r_{\rm min} \leqslant r \leqslant q'' \quad r_{\rm min}$, with the fixed particles in the range $-L/2 \leqslant q' \leqslant L/2 - r_e$, and $-L/2 + r_e \leqslant q'' \leqslant L/2$. Typically, $r_{\rm min} = 0.75 r_e$ and $L = 10 r_e$.

The nth energy eigenvalue $E_n(q', q'')$ and eigenfunction $\psi_n(r|q', q'')$ are obtained by minimizing the energy expectation value $\langle n|\hat{\mathcal{H}}|n\rangle$ using a Gram–Schmidt procedure to ensure that $\langle n|k\rangle = 0$, $k = 1, 2, \ldots, n - 1$. An iterative procedure is used where the change in wave function is proportional to the force at each grid point. A three-point finite difference formula is used for the Laplacian for the kinetic energy, with the gradient of the wave function at the terminal grid points being used as part of the minimization procedures. The first fifty singlet energy states are obtained in each case. Roughly speaking, this equates to on the order of N^{50} energy levels for the full system for each configuration in phase space; there is a continuum of such configurations.

The interior wave function is saved as a matrix $\phi(j, n, j', j'')$ on a $100 \times 50 \times 50 \times 50$ grid. To obtain the entire set of energy eigenfunctions takes about 18 hours on an elderly personal computer. No effort was made to optimize the process because this is a once only computation as the wave functions are independent of temperature and of the size of the ultimate system.

10.5.2.2 Commutation functions

The singlet commutation function is constructed according to equation (10.25). The Fourier integral is evaluated directly rather than by fast techniques. Typically 100 momentum points are used on a uniform grid up to a maximum kinetic energy $\beta\mathcal{K} = 50$. The singlet commutation function has to be calculated once for each temperature explored. This computation is straightforward, taking less time than the several minutes spent writing the 412 MB output file to disc.

10.5.2.3 Monte Carlo simulations

The standard Metropolis algorithm is used, with the umbrella weight being the real part of the total combined commutation function, including the beyond nearest neighbor potential contributions. The imaginary contribution is added for taking quantum averages, and the commutation part is subtracted for taking classical averages. This differs from the algorithm described above for the harmonic crystal, where the Metropolis algorithm was used in conjunction with the Maxwell–Boltzmann umbrella weight, and the entire commutation function was added for the quantum averages.

It is found that at lower temperatures it is more statistically efficient to use umbrella sampling with the real part of the commutation function and the Maxwell–Boltzmann weight, than umbrella sampling using the classical Maxwell–Boltzmann weight alone, or broader variant thereof.

The commutation functions for a configuration are obtained by linear interpolation of the combined singlet commutation function exponent that are stored on the 3- or 4-dimensional grids. Recall that the combined commutation function exponent is $\tilde{W} = W - \beta\mathcal{H}$. This is more accurate and reliable than interpolation of the commutation functions directly, as a delicate cancelation has to occur between some rather large numbers in the core region

Each simulation for a given temperature and system size produces 6 averages of any given quantity: either classical or quantum, combined with either distinguishable particles, or bosons, or fermions. A typical simulation for $N = 4$ particles takes about 35 minutes on the aforementioned personal computer, for about a 1% statistical error at the 96% confidence level of the average total energy for distinguishable particles. The same set of energy eigenfunctions are used for simulations for a dozen different temperatures and for $N = 4$ and $N = 5$, with several repeated with different simulation parameters.

10.5.2.4 Results

The numerical result reported here are for a one dimensional system with the simple harmonic oscillator singlet potential and Lennard–Jones pair potential given above. Quantitative comparison is made with the benchmark results obtained by Hernando and Vaníček (2013), and following them the so-called de Broglie wavelength is fixed at $\Lambda_{\rm dB} \equiv 2^{1/6}\hbar/r_{\rm e}\sqrt{m\varepsilon} = 0.16$, and the frequency is fixed by $\omega r_{\rm e}\sqrt{m/\varepsilon} = 1/2$. These mean that $\varepsilon/\hbar\omega = 14.03$. The energy and temperature are normalized by the Lennard–Jones frequency, $\omega_{\rm LJ} = \sqrt{u_{\rm LJ}''(r_{\rm e})/m} = 16.97\omega$ for the present parameters.

Figure 10.5 compares the present prediction for the average energy for a system of $N = 4$ distinguishable particles with the benchmark results derived from Hernando and Vaníček (2013). Unfortunately there is an unknown shift in the energy eigenvalues published by Hernando and Vaníček (2013). To make the comparison in the figure, the average energy derived from the published energy eigenvalues has

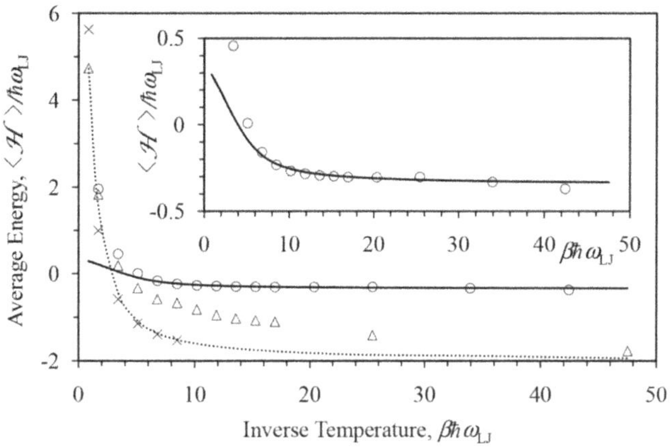

Figure 10.5. Average energy for the Lennard–Jones system (N=4, no symmetrization). The circles result from classical phase space Monte Carlo simulations with the effective local field singlet commutation function. The triangles use the harmonic effective local field singlet commutation function (Attard 2018a), the crosses use a high temperature expansion (Attard 2019b), the dotted curve is the classical result, and the solid curve are benchmark results derived from the 50 energy eigenvalues given by Hernando and Vaníček (2013), shifted vertically to coincide with the singlet results at intermediate temperatures. The statistical error is less than the symbol size.

been shifted vertically to coincide with the effective local field theory at intermediate temperatures, where it is believed to be the most reliable. This of course limits the utility of the benchmarks in absolute terms. Nevertheless the slope and curvature of the effective local field results agree with those of the benchmark results at intermediate temperatures, $5 \lesssim \beta\hbar\omega_{\mathrm{LJ}} \lesssim 40$ (inset), which tends to confirm their quantitative validity.

Following a relatively flat regime at intermediate temperatures, the data in figure 10.5 show that the effective local field results start to decrease below the benchmark ground state energy. This is possibly an artifact of the singlet, linear approximation. In any case one can see that including quantum effects makes a very great difference to the average energy at low temperatures; the data in the figure show that the ground state energy is significantly greater than the classical prediction at absolute zero.

At high temperatures the effective local field method is exact, as can be seen by the agreement with the classical results in figure 10.5. The failure of the data derived from Hernando and Vaníček (2013) in this regime is due to the fact that they are based on only 50 energy eigenvalues for the full system. The present calculations used 50 energy eigenvalues for each particle for each configuration, and they were used to construct the singlet commutation function. Since the total commutation function is the sum of the latter, (equivalently, the total wave function is the product of singlet wave functions) the present method effectively characterizes the full system with 4^{50} energy levels *for each configuration in phase space.*

It can be seen in figure 10.5 that the local field theory in harmonic approximation, correctly gives the high temperature classical limit, but at intermediate

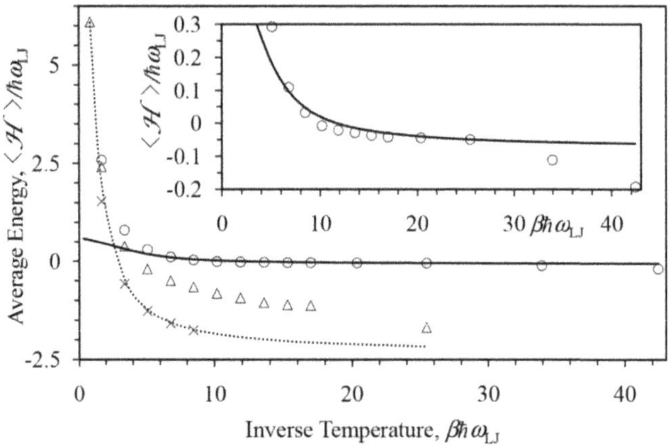

Figure 10.6. Same as the preceding figure but for $N = 5$.

and low temperatures it disagrees with the local field approach based on the actual potential. This, presumably, indicates that the anharmonic contributions to the local potential field experienced by a Lennard–Jones particle becomes increasingly significant as the temperature is decreased. In this case it is better to obtain the exact energy eigenfunctions for the actual local field rather than to use the simple harmonic oscillator energy eigenfunctions for the quadratic expansion of the local field. Figure 10.5 also includes the result of a high temperature expansion for the commutation function (Attard 2019b). Although the terms in this expansion are formally exact, it can be seen that the result is practically indistinguishable from the classical result on the scale of the figure. The negligible size of the high temperature quantum corrections to the classical result for the energy observed here is reminiscent of a similar negligibility for the density profile of the harmonic crystal obtained in figure 9.2. Given the complexity of the terms and the brute force nature of the expansion, this does not appear to be a promising approach for the quantitative description of quantum effects in terrestrial condensed matter.

Figure 10.6 shows the average energy for $N = 5$ particles (no symmetrization). Again it can be seen (inset) that the slope and curvature of the energy curve given by the present theory agree with that derived from the eigenvalue data of Hernando and Vaníček (2013) in the intermediate temperature regime, $6 \lesssim \beta\hbar\omega_{\rm LJ} \lesssim 30$. The singlet effective local field approach (between two fixed neighbors), the harmonic local field approach, and the high temperature expansion, all go over to the classical limit at high temperatures. The quadratic approximation to the local field again appears inadequate at lower temperatures. For the ground state energy, the singlet effective local field result again appears to underestimate the shifted benchmark result.

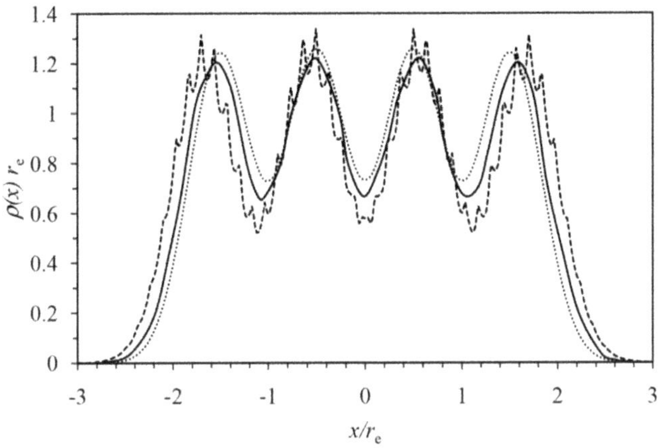

Figure 10.7. Density profile at $\beta\hbar\omega_{\mathrm{LJ}} = 11.9$ for $N = 4$ (no symmetrization). The dashed curve is the singlet effective local field, the dotted curve is the classical prediction, and the full curve are benchmark results given by Hernando and Vaníček (2013).

Density profiles at $\beta\hbar\omega_{\mathrm{LJ}} = 11.9$ are shown in figure 10.7. All approaches predict molecular-sized oscillations in the density profile. The structure is much more well-defined at this low temperature than at a temperature twice as high (Hernando and Vaníček 2013, Attard 2020), and the departure from classical close-packing is relatively small. It is remarkable just how much of the density structure is due to purely classical considerations even at this low temperature.

Symmetrization effects due to nearest neighbor transpositions have been calculated but are not included in figure 10.7. On the scale of the figure no difference would be seen between this density profile and that of bosons or fermions.

The high frequency oscillatory fine structure evident in the profile given by the singlet effective local field theory is statistically significant. It is not an artifact of the grids used to collect the density profile or to store the singlet commutation function.

References

Attard P 2002 *Thermodynamics and Statistical Mechanics: Equilibrium by Entropy Maximisation* (London: Academic)

Attard P 2018a Quantum statistical mechanics in classical phase space. Test results for quantum harmonic oscillators arXiv:1811.02032

Attard P 2018b Quantum statistical mechanics in classical phase space. III. Mean field approximation benchmarked for interacting Lennard–Jones particles arXiv:1812.03635

Attard P 2019a Fermionic phonons: exact analytic results and quantum statistical mechanics for a one dimensional harmonic crystal arXiv:1903.06866

Attard P 2019b Quantum Monte Carlo in classical phase space. Mean field and exact results for a one dimensional harmonic crystal arXiv:1904.10650

Attard P 2020 Quantum statistical mechanics in classical phase space. V. Quantum local, average global arXiv:2005.06165

Gross E P 1961 Structure of a quantized vortex in boson systems *Il Nuovo Cimento* **20** 454
Hernando A and Vaníček J 2013 Imaginary-time nonuniform mesh method for solving the multidimensional Schrödinger equation: fermionization and melting of quantum Lennard–Jones crystals *Phys. Rev.* A **88** 062107
Kirkwood J G 1933 Quantum statistics of almost classical particles *Phys. Rev.* **44** 31
Messiah A 1961 *Quantum Mechanics* vols 1 and 2 (Amsterdam: North-Holland)
Pathria R K 1972 *Statistical Mechanics* (Oxford: Pergamon)
Pitaevskii L P 1961 Vortex lines in an imperfect Bose gas *Sov. Phys. JETP* **13** 451
Wigner E 1932 On the quantum correction for thermodynamic equilibrium *Phys. Rev.* **40** 749
van Sciver S W 2012 *Helium Cryogenics* 2nd edn (New York: Springer)

IOP Publishing

Phil Attard

Chapter 11

Many-body expansion for the commutation function

The commutation function exponent is formally written as a sum of one-body, two-body, three-body, etc temperature-dependent effective potentials, which allow the phase space weight to be expressed in terms of generalized Mayer f-functions. There are stability issues with the temperature quadrature of the pair commutation function for the Lennard–Jones potential, whereas a sum over the energy eigenstates is feasible. Numerical results suggest that termination of the series at the pair term is unreliable. An exponential series expansion for the phase space weight of a quantum system can be cast in terms of the Lee–Yang binary kernel and Ursell connected clusters.

11.1 Commutation function

This chapter presents a many-body expansion of the commutation function exponent. The original rationale for expanding the commutation function as a series of singlet, pair, triplet, etc terms is that these take the form of temperature- and momentum-dependent effective potentials with which phase space has the appearance of a Maxwell–Boltzmann classical weight. As an effective potential the commutation function is by definition localized. Depending on the reliability of the many-body expansion with the effective potentials, one can take over all of the theorems, expansions, integral equations, asymptotes, algorithms, etc that have been invented for classical statistical mechanics and apply them directly to quantum systems. Although perhaps not as quickly converging as originally hoped, to a limited extent the many-body expansion appears feasible.

In what follows the formal analysis is spelled out, along with the details by which the singlet and pair commutation functions are calculated. In section 11.4, these are applied to the one-dimensional Lennard–Jones and compared with the singlet commutation function given by the local state expansion from chapter 10. In the final section 11.5 of the chapter, the Ursell cluster expansion and the Lee–Yang binary kernel are applied to classical phase space.

11.1.1 Background

The commutation function ω, which is related to similar functions introduced by Wigner (1932) and analyzed by Kirkwood (1933), is defined in equation (7.7),

$$e^{-\beta\mathcal{H}(\mathbf{q},\,\mathbf{p})}\omega_{\mathrm{p}}(\mathbf{q},\,\mathbf{p}) = \frac{\langle\mathbf{q}|e^{-\beta\hat{\mathcal{H}}}|\mathbf{p}\rangle}{\langle\mathbf{q}|\mathbf{p}\rangle}. \tag{11.1}$$

In this chapter this is the only form of the commutation function that will be considered, and henceforth the subscript p will be dropped.

Quantum effects, specifically the non-commutativity of the position and momentum operators, are reflected in the extent to which ω departs from unity. To see this simply note that *if* position and momentum commuted, *then* the Maxwell–Boltzmann operator would factorize, $e^{-\beta\hat{\mathcal{H}}(\mathbf{r})} = e^{-\beta U(\mathbf{r})}e^{-\beta\hat{\mathcal{K}}(\mathbf{r})}$. Then since $e^{-\beta\hat{\mathcal{K}}}|\mathbf{p}\rangle = e^{-\beta\mathcal{K}(\mathbf{p})}|\mathbf{p}\rangle$, the right-hand side would reduce to the classical Maxwell–Boltzmann factor $e^{-\beta\mathcal{H}(\mathbf{p},\,\mathbf{q})}$, in which case $\omega(\mathbf{q},\,\mathbf{p}) = 1$.

Classical behavior dominates in the high temperature limit, $\omega(\mathbf{q},\,\mathbf{p}) \to 1$, $\beta \to 0$. This limit and the quantum corrections to it were discussed in chapters 8 and 9.

Further, the potential energy decays to zero at large separation (at least the inter-particle part does), and the gradients of the potential are negligible compared to the potential itself. This means that at large separations the potential energy and the kinetic energy operator effectively commute, and so one similarly must have $\omega(\mathbf{q},\,\mathbf{p}) \to 1$, all $q_{jk} \to \infty$.

Following Kirkwood (1933), differentiation of the defining equation with respect to inverse temperature gives equation (8.9),

$$\begin{aligned}\frac{\partial\omega}{\partial\beta} = \frac{-\beta\hbar^2}{2m}(\nabla^2 U)\omega - \frac{\beta\hbar^2}{m}(\nabla U)\cdot(\nabla\omega) + \frac{\beta^2\hbar^2}{2m}(\nabla U)\cdot(\nabla U)\omega \\ + \frac{\hbar^2}{2m}\nabla^2\omega + \frac{\mathrm{i}\hbar}{m}\mathbf{p}\cdot(\nabla\omega) - \frac{\mathrm{i}\hbar\beta}{m}\mathbf{p}\cdot(\nabla U)\omega.\end{aligned} \tag{11.2}$$

One can see from this that since $\nabla U \to 0$ at large separations, the putative asymptote, $\omega(\mathbf{q},\,\mathbf{p}) \to 1$, makes the right-hand side vanish, which is consistent with the asymptote being a temperature-independent constant. This partial differential equation provides the basis for the high temperature expansions that were derived in chapter 8 (Wigner 1932, Kirkwood 1933, Attard 2018a, 2018b). The multiplicity of gradients on the right-hand side is the origin for the rapid increase of number and complexity of the contributions to each order of the expansions.

In section 8.1.4, it was argued that it is more useful to cast the commutation function as an effective potential (Attard 2018a),

$$W(\mathbf{q},\,\mathbf{p}) \equiv \ln\omega(\mathbf{p},\,\mathbf{q}), \quad \text{or} \quad \omega(\mathbf{q},\,\mathbf{p}) \equiv e^{W(\mathbf{q},\,\mathbf{p})}. \tag{11.3}$$

In this form W is a temperature- and momentum-dependent effective potential. The original rationale for this was that W is extensive with system size, which is an exceedingly useful property in thermodynamics. It was also hoped that the expansion of W in powers of Planck's constant or inverse temperature might have better convergence properties than those of ω itself, a hope more happy in prospect than in actual experience.

11.1.2 Many-body expansion

Defining the combined commutation function exponent as $\tilde{W} \equiv W - \beta\mathcal{H}$, the temperature derivative of the definition leads to equation (8.7),

$$\frac{\partial \tilde{W}}{\partial \beta} = -\mathcal{H} + \frac{i\hbar}{m}\mathbf{p} \cdot \nabla \tilde{W} + \frac{\hbar^2}{2m}\nabla^2 \tilde{W} + \frac{\hbar^2}{2m}\nabla \tilde{W} \cdot \nabla \tilde{W}, \tag{11.4}$$

with $\tilde{W} \sim -\beta\mathcal{H} + \mathcal{O}(\beta^2)$, $\beta \to 0$. The right-hand side contains two terms linear in $\tilde{W}$ and one non-linear, quadratic term.

In general the potential energy is the sum of one-body, two-body, three-body, etc potentials,

$$U(\mathbf{q}) = \sum_{j=1}^{N} u(\mathbf{q}_j) + \sum_{j<k}^{N} u(\mathbf{q}_j, \mathbf{q}_k) + \sum_{j<k<\ell}^{N} u(\mathbf{q}_j, \mathbf{q}_k, \mathbf{q}_\ell) + \cdots \tag{11.5}$$

The different potentials are distinguished by the number of their arguments. In this chapter three-body and higher potentials will be neglected. Initially it will also be assumed that there is no one-body potential, that the system is homogeneous, and that the pair potential is a function only of separation,

$$U(\mathbf{q}) = \sum_{j<k}^{N} u(q_{jk}) = \frac{1}{2}\sum_{j,k}^{(k \neq j)} u(q_{jk}), \tag{11.6}$$

where the particle separation is $q_{jk} = |\mathbf{q}_j - \mathbf{q}_k|$.

For the effective potential form of the commutation function, a similar many-body decomposition can be formally made, with momentum now being included

$$\begin{aligned} W(\mathbf{q}, \mathbf{p}) = &\sum_{j=1}^{N} w(\mathbf{q}_j, \mathbf{p}_j) + \sum_{j<k}^{N} w(\mathbf{q}_j, \mathbf{p}_j; \mathbf{q}_k, \mathbf{p}_k) \\ &+ \sum_{j<k<\ell}^{N} w(\mathbf{q}_j, \mathbf{p}_j; \mathbf{q}_k, \mathbf{p}_k; \mathbf{q}_\ell, \mathbf{p}_\ell) + \cdots \end{aligned}$$

Again the different many-body commutation functions are distinguished by the number of their arguments.

11.1.3 Pair term

For the homogeneous case (i.e., no singlet potential), the singlet contribution vanishes ($W^{(1)} = 0$, $\tilde{W}^{(1)} = -\beta\mathcal{K}(\mathbf{p})$, and $\nabla \tilde{W}^{(1)} = 0$). At this stage the three-body and higher contributions, will be neglected. Assuming homogeneity, the pair commutation function for particles j and k is a function of the magnitudes of the relative momentum and separation, and the angle between them, $w^{(2)}(q_{jk}, p_{jk}, \theta_{jk})$. Alternatively, the relative momentum can be aligned with the z axis, $\mathbf{p}_{jk} = p_{jk}\hat{\mathbf{z}}$, and the separation can be rotated to the xz-plane, $w^{(2)}(q_{jk,x}, q_{jk,z}, p_{jk})$, or $w^{(2)}(q_\perp, q_\parallel, p)$. In any case the commutation function that will be considered here is

$$W(\mathbf{q}, \mathbf{p}) = \sum_{j<k}^{N} w^{(2)}(\mathbf{q}_{jk}, \mathbf{p}_{jk}) = \frac{1}{2}\sum_{j,k}^{(k\neq j)} w_{jk}. \tag{11.7}$$

Inserting this pair ansatz into the temperature derivative equation, one sees that the left-hand side, and the constant and linear terms on the right-hand side, are all the sum of pair functions. However the non-linear quadratic term on the right-hand side is

$$\begin{aligned}
\nabla\tilde{W}\cdot\nabla\tilde{W} &= \sum_{\ell} \nabla_\ell\tilde{W}\cdot\nabla_\ell\tilde{W} \\
&= \frac{1}{4}\sum_{\ell}\sum_{j,k}^{(k\neq j)}\sum_{j',k'}^{(k'\neq j')} \nabla_\ell\tilde{w}_{jk}\cdot\nabla_\ell\tilde{w}_{j'k'} \\
&= \frac{1}{4}\sum_{\ell,k,k'}^{(k,\,k'\neq\ell)} \nabla_\ell\tilde{w}_{\ell k}\cdot\nabla_\ell\tilde{w}_{\ell k'} + \frac{1}{4}\sum_{j,\ell,k'}^{(j,\,k'\neq\ell)} \nabla_\ell\tilde{w}_{j\ell}\cdot\nabla_\ell\tilde{w}_{\ell k'} \\
&\quad + \frac{1}{4}\sum_{\ell,k,j'}^{(k,\,j'\neq\ell)} \nabla_\ell\tilde{w}_{\ell k}\cdot\nabla_\ell\tilde{w}_{j'\ell} + \frac{1}{4}\sum_{\ell,j,j'}^{(j,\,j'\neq\ell)} \nabla_\ell\tilde{w}_{j\ell}\cdot\nabla_\ell\tilde{w}_{j'\ell} \\
&= \sum_{j,k,k'}^{(k,\,k'\neq j)} \nabla_j\tilde{w}_{jk}\cdot\nabla_j\tilde{w}_{jk'}.
\end{aligned} \tag{11.8}$$

This is in essence a three-body term. However, to the extent that the 'force' on particle j due to particle k is uncorrelated with that due to k', $k' \neq k$, the sum of forces over k' may be argued to be small, perhaps averaging to zero. In this case the $k' = k$ terms dominate, and this becomes

$$\begin{aligned}
\nabla\tilde{W}\cdot\nabla\tilde{W} &\approx \sum_{j,k}^{(k\neq j)} \nabla_j\tilde{w}_{jk}\cdot\nabla_j\tilde{w}_{jk} \\
&= 2\sum_{j<k} \nabla_j\tilde{w}_{jk}\cdot\nabla_j\tilde{w}_{jk}.
\end{aligned} \tag{11.9}$$

This is now a two-body term, which is expected to be accurate at low densities and high temperatures where correlations are reduced. The two-body solution can test the magnitude of the neglected three-body terms, and successively improve the pairwise additive approximation.

Applying the pair ansatz to the temperature derivative and equating both sides term by term one obtains

$$\frac{\partial\tilde{w}_{jk}}{\partial\beta} = -u_{jk} + \frac{i\hbar}{m}\mathbf{p}_{jk}\cdot\nabla_j\tilde{w}_{jk} + \frac{\hbar^2}{m}\nabla_j^2\tilde{w}_{jk} + \frac{\hbar^2}{m}\nabla_j\tilde{w}_{jk}\cdot\nabla_j\tilde{w}_{jk}. \tag{11.10}$$

Here and above the symmetry $\tilde{w}_{jk} = \tilde{w}_{kj}$ has been exploited. Note that $\tilde{w}_{jk} \sim -\beta u_{jk}$, $\beta \to 0$.

Write $\mathbf{p}_{jk} = p_z\hat{\mathbf{z}}$, and $\mathbf{q}_{jk} = q_x\hat{\mathbf{x}} + q_z\hat{\mathbf{z}}$, and $q_{jk} = \sqrt{q_x^2 + q_z^2}$. This then becomes

$$\frac{\partial \tilde{w}(q_x, q_z, p_z)}{\partial \beta} = -u(q) + \frac{\mathrm{i}\hbar}{m} p_z \tilde{w}_z + \frac{\hbar^2}{m}\{\tilde{w}_{xx} + \tilde{w}_{zz}\} + \frac{\hbar^2}{m}\left\{\tilde{w}_x^2 + \tilde{w}_z^2\right\}. \tag{11.11}$$

The subscripts on $\tilde{w}$ signify spatial derivatives.

Define the two-dimensional Fourier transform pair

$$\begin{aligned} \hat{w}(k_x, k_y, p_z) &= \int \mathrm{d}\mathbf{q}\; e^{-\mathrm{i}\mathbf{k}\cdot\mathbf{q}} \tilde{w}(q_x, q_z, p_z) \\ \tilde{w}(q_x, q_z, p_z) &= \frac{1}{(2\pi)^2} \int \mathrm{d}\mathbf{k}\; e^{\mathrm{i}\mathbf{k}\cdot\mathbf{q}} \hat{w}(k_x, k_y, p_z). \end{aligned} \tag{11.12}$$

The Fourier transform of a gradient such as $\tilde{w}_x(\mathbf{q})$ is just $\mathrm{i}k_x\hat{w}(\mathbf{q})$. Note that here $\mathbf{q}$ is a two-dimensional vector.

The Fourier transform of the temperature derivative is

$$\begin{aligned} \frac{\partial \hat{w}(k_x, k_y, p_z)}{\partial \beta} &= -\,\hat{u}(k) - \left[\frac{\hbar}{m} p_z k_z + \frac{\hbar^2}{m} k^2\right] \hat{w}(\mathbf{k}, \mathbf{p}) \\ &\quad - \frac{\hbar^2}{(2\pi)^2 m} \int \mathrm{d}\mathbf{k}'\; \mathbf{k}' \cdot (\mathbf{k} - \mathbf{k}')\, \hat{w}(\mathbf{k}', \mathbf{p})\, \hat{w}(\mathbf{k} - \mathbf{k}', \mathbf{p}). \end{aligned} \tag{11.13}$$

Here $k^2 = k_x^2 + k_z^2$. Write $b(\mathbf{k}) \equiv -\hbar p_z k_z/m - \hbar^2 k^2/m$, which is negative for large k.

Of course these expressions implicitly assume that the Fourier transform exists. For the case of a Lennard–Jones pair potential, which at short range goes like $u(r) \sim r^{-12}$, $r \to 0$, this assumption is problematic.

11.1.4 Linear solution

In the case that w is small, as occurs at high temperatures or at large separations, one can simply neglect the quadratic term, so that the differential equation is linear, $\partial\hat{w}_{\mathrm{lin}}/\partial\beta = -\hat{u} + b\hat{w}_{\mathrm{lin}}$. This has the solution

$$\hat{w}_{\mathrm{lin}}^{(2)}(\mathbf{k}) = \frac{-\hat{u}(k)}{b(\mathbf{k})}[e^{\beta b(\mathbf{k})} - 1]. \tag{11.14}$$

The linear approximation is valid at high temperatures, or at large separations. The pair ansatz is exact in the linear regime.

In the limit of large k, this gives $\hat{w}_{\mathrm{lin}} \sim \hat{u}(k)/b(k) \sim k^{-2}\hat{u}(k)$, $k \to \infty$. This says that at small separations $w^{(2)}(q)$ to leading order cancels the classical $-\beta u^{(2)}(q)$, which is essential because the quantum effects of non-commutativity have to dominate on small length scales. (For example, for an atom quantum effects have to dominate at short range the electron–nucleus interaction; the classical Maxwell–Boltzmann weight alone would lead to catastrophe.) The asymptotic behavior in the core region is derived in section 11.4.2.

For small k, $b(k) \to 0$, and $\hat{\tilde{w}}_{\text{lin}}(k) \sim -\beta\hat{u}(k)$. This confirms that $w(q)$ decays more quickly at large separations than the pair potential itself. (This can be shown to be true in the non-linear case as well.)

The linear solution can be inserted into the convolution integral and the temperature integration performed explicitly. This gives the first non-linear correction and tells how reliable the linear solution is.

For any value of $p_{jk} = p_{jk,z}$, using the two-dimensional fast Fourier transform, one can numerically invert $\hat{\tilde{w}}^{(2)}_{\text{lin}}(k_x, k_z, p_{jk})$ giving $\tilde{w}^{(2)}_{\text{lin}}(q_{jk,x}, q_{jk,z}, p_{jk,z})$. Summing over pairs gives the full commutation function for any phase space point, $W_{\text{lin}}(\mathbf{q}, \mathbf{p})$. Obviously if many phase space points are required, it would be most efficient to evaluate $\tilde{w}^{(2)}_{\text{lin}}(q_\perp, q_\parallel, p)$ once only, storing this on a three-dimensional grid. The full commutation function can be evaluated for any phase space point by summing over all pairs each interpolated value of the stored function.

11.1.5 Non-linear solution

The non-linear partial differential equation can be attempted to be solved by stepping forward in inverse temperature, starting from a high temperature with the linear solution,

$$\hat{\tilde{w}}^{(2)}(\mathbf{k}; \beta_{\alpha+1}) = \hat{\tilde{w}}^{(2)}(\mathbf{k}; \beta_\alpha) + \Delta\beta \left. \frac{\partial \hat{\tilde{w}}^{(2)}(\mathbf{k}; \beta)}{\partial \beta} \right|_{\beta=\beta_\alpha}, \tag{11.15}$$

Runge–Kutta procedures might be used to accelerate and to stabilize the temperature integration. Either evaluate the convolution integral with the two-dimensional fast Fourier transform, or else use finite differences in real space. As in the linear case, precalculate the non-linear $w^{(2)}(q_\perp, q_\parallel, p)$ and store it on a three-dimensional grid.

It must be mentioned that the author has spent a great deal of time exploring the numerical temperature integration of the two-body term for the case of a Lennard–Jones potential. Although the procedures converge at large and intermediate separations, numerical instabilities always occurred approaching the core region at lower temperatures that precluded a reliable solution. The author has never found an algorithm that resolved these problems for the direct temperature quadrature for the two-body commutation function. In the numerical results, section 11.4, the two-body commutation function is instead obtained by summing over the energy states of two particles interacting with the Lennard–Jones pair potential.

11.1.6 Singlet plus pair potential

When the potential contains singlet and pair terms, the differential equations for the singlet and pair commutation functions are

$$\frac{\partial \tilde{w}_j^{(1)}}{\partial \beta} = \frac{-p_j^2}{2m} - u_j^{(1)} + \frac{\mathrm{i}\hbar}{m}\mathbf{p}_j \cdot \nabla_j \tilde{w}_j^{(1)} + \frac{\hbar^2}{2m}\nabla_j^2 \tilde{w}_j^{(1)} + \frac{\hbar^2}{2m}\nabla_j \tilde{w}_j^{(1)} \cdot \nabla_j \tilde{w}_j^{(1)}, \tag{11.16}$$

and

$$\begin{aligned}\frac{\partial \tilde{w}_{jk}^{(2)}}{\partial \beta} = & - u_{jk}^{(2)} + \frac{\mathrm{i}\hbar}{2m}\left\{\mathbf{p}_j \cdot \nabla_j \tilde{w}_{jk}^{(2)} + \mathbf{p}_k \cdot \nabla_k \tilde{w}_{jk}^{(2)}\right\} \\ & + \frac{\hbar^2}{2m}\left\{\nabla_j^2 \tilde{w}_{jk}^{(2)} + \nabla_k^2 \tilde{w}_{jk}^{(2)}\right\} \\ & + \frac{\hbar^2}{2m}\left\{\nabla_j \tilde{w}_{jk}^{(2)} \cdot \nabla_j \tilde{w}_{jk}^{(2)} + \nabla_k \tilde{w}_{jk}^{(2)} \cdot \nabla_k \tilde{w}_{jk}^{(2)}\right\} \\ & + \frac{\hbar^2}{m}\left\{\nabla_j \tilde{w}_j^{(1)} \cdot \nabla_j \tilde{w}_{jk}^{(2)} + \nabla_k \tilde{w}_k^{(1)} \cdot \nabla_k \tilde{w}_{jk}^{(2)}\right\},\end{aligned} \tag{11.17}$$

where $w_{jk}^{(2)} = w^{(2)}(\mathbf{q}_j, \mathbf{p}_j; \mathbf{q}_k, \mathbf{p}_k) = w_{kj}^{(2)}$. This neglects the three-body contribution, as in equation (11.9).

In section 11.4.1, the singlet and pair commutation functions are expressed as a sum over energy states, and the resultant functions are tested for a Lennard–Jones system.

11.1.7 Three-body term

For the preceding case (i.e., only singlet and pair potentials), the neglected term contributes to the temperature derivative of the three-body commutation function. The three-body contributions to the latter are

$$\begin{aligned}\frac{\partial \tilde{w}_{jk\ell}^{(3)}}{\partial \beta} \approx & \frac{\mathrm{i}\hbar}{2m}\left\{\mathbf{p}_j \cdot \nabla_j \tilde{w}_{jk\ell}^{(3)} + \mathbf{p}_k \cdot \nabla_k \tilde{w}_{jk\ell}^{(3)} + \mathbf{p}_\ell \cdot \nabla_\ell \tilde{w}_{jk\ell}^{(3)}\right\} \\ & + \frac{\hbar^2}{2m}\left\{\nabla_j^2 \tilde{w}_{jk\ell}^{(3)} + \nabla_k^2 \tilde{w}_{jk\ell}^{(3)} + \nabla_\ell^2 \tilde{w}_{jk\ell}^{(3)}\right\} \\ & + \frac{\hbar^2}{2m}\left\{\nabla_j \tilde{w}_{jk\ell}^{(3)} \cdot \nabla_j \tilde{w}_{jk\ell}^{(3)} + \nabla_k \tilde{w}_{jk\ell}^{(3)} \cdot \nabla_k \tilde{w}_{jk\ell}^{(3)} + \nabla_\ell \tilde{w}_{jk\ell}^{(3)} \cdot \nabla_\ell \tilde{w}_{jk\ell}^{(3)}\right\} \\ & + \frac{\hbar^2}{m}\left\{\nabla_j \tilde{w}_j^{(1)} \cdot \nabla_j \tilde{w}_{jk\ell}^{(3)} + \nabla_k \tilde{w}_k^{(1)} \cdot \nabla_k \tilde{w}_{jk\ell}^{(3)} + \nabla_\ell \tilde{w}_\ell^{(1)} \cdot \nabla_\ell \tilde{w}_{jk\ell}^{(3)}\right\} \\ & + \frac{\hbar^2}{m}\left\{\nabla_j \tilde{w}_{jk}^{(2)} \cdot \nabla_j \tilde{w}_{jk\ell}^{(3)} + \nabla_j \tilde{w}_{j\ell}^{(2)} \cdot \nabla_j \tilde{w}_{jk\ell}^{(3)} + \nabla_k \tilde{w}_{jk}^{(2)} \cdot \nabla_k \tilde{w}_{jk\ell}^{(3)}\right. \\ & \left. + \nabla_k \tilde{w}_{k\ell}^{(2)} \cdot \nabla_k \tilde{w}_{jk\ell}^{(3)} + \nabla_\ell \tilde{w}_{j\ell}^{(2)} \cdot \nabla_\ell \tilde{w}_{jk\ell}^{(3)} + \nabla_\ell \tilde{w}_{k\ell}^{(2)} \cdot \nabla_\ell \tilde{w}_{jk\ell}^{(3)}\right\} \\ & + \frac{\hbar^2}{m}\left\{\nabla_j \tilde{w}_{jk}^{(2)} \cdot \nabla_j \tilde{w}_{j\ell}^{(2)} + \nabla_k \tilde{w}_{jk}^{(2)} \cdot \nabla_k \tilde{w}_{k\ell}^{(2)} + \nabla_\ell \tilde{w}_{j\ell}^{(2)} \cdot \nabla_\ell \tilde{w}_{k\ell}^{(2)}\right\}.\end{aligned} \tag{11.18}$$

The four-body and higher terms on the right-hand side have obviously been neglected here.

Alternatively, and likely better, for $N = 3$ the effective triplet potential is given by

$$e^{\tilde{w}^{(3)}_{123}} = \frac{e^{\tilde{W}^{(3)}_{123}}}{e^{\tilde{w}^{(1)}_1} e^{\tilde{w}^{(1)}_2} e^{\tilde{w}^{(1)}_3} e^{\tilde{w}^{(2)}_{12}} e^{\tilde{w}^{(2)}_{23}} e^{\tilde{w}^{(2)}_{31}}}, \tag{11.19}$$

where $\{123\} = \{\mathbf{q}_1, \mathbf{p}_1; \mathbf{q}_2, \mathbf{p}_2; \mathbf{q}_3, \mathbf{p}_3\}$. The total phase space weight for three particles can be obtained from the sum over their energy states,

$$\begin{aligned} e^{\tilde{W}^{(3)}_{123}} &= \frac{1}{\langle \mathbf{q}_{1,2,3} | \mathbf{p}_{1,2,3} \rangle} e^{-\beta \hat{\mathcal{H}}^{(3)}(q_{1,2,3})} \langle \mathbf{q}_{1,2,3} | \mathbf{p}_{1,2,3} \rangle \\ &= \frac{1}{\langle \mathbf{q}_{1,2,3} | \mathbf{p}_{1,2,3} \rangle} \sum_{\ell} e^{-\beta E^{(3)}_{\ell}} \langle \mathbf{q}_{1,2,3} | \ell \rangle \, \langle \ell | \mathbf{p}_{1,2,3} \rangle. \end{aligned} \tag{11.20}$$

See section 11.4.3 for a discussion of a numerical approach to evaluating this.

11.1.8 Effective Mayer *f*-functions

The above many-body expansion casts the commutation function as a series of n-body temperature- and momentum-dependent effective potentials. The latter goes to zero if any of the n particles is separated far from the rest. This makes the formulation suitable for a many-body Mayer f-function cluster expansion, originally introduced for classical statistical mechanics by Mayer *et al* (1937a, 1937b, 1938, Harrison and Mayer 1938). The Mayer f-function expansions have provided the basis for many powerful exact and approximate approaches to classical statistical mechanics (Morita and Hiroike 1961, Stell 1964, Hansen and McDonald 1986).

For a quantum system, the present many-body effective potential expansion of the commutation function directly leads to a many-body effective Mayer f-function cluster expansion. For N particles the total (unsymmetrized) phase space weight is formally

$$\begin{aligned} \Omega^{(N)}(\mathbf{q}, \mathbf{p}) &= \prod_{j=1}^{N} e^{\tilde{w}_j} \prod_{j<k}^{N} e^{\tilde{w}_{jk}} \prod_{j<k<\ell}^{N} e^{\tilde{w}_{jk\ell}} \ldots \\ &= \prod_{j=1}^{N} e^{\tilde{w}_j} \prod_{j<k}^{N} \left[1 + \tilde{f}_{jk}\right] \prod_{j<k<\ell}^{N} \left[1 + \tilde{f}_{jk\ell}\right] \ldots \end{aligned} \tag{11.21}$$

Hopefully use of the symbol Ω for the phase space weight, as here, and for the grand potential, as in chapters 3 and 7, will not cause confusion. Here and below the superscript denotes the total number of particles in the system and a subscript denotes the position and momentum of a particle: $j \equiv \{\mathbf{q}_j, \mathbf{p}_j\}$. Here the pair effective Mayer f-function is $\tilde{f}_{jk} \equiv e^{\tilde{w}_{jk}} - 1$, the triplet is $\tilde{f}_{jk\ell} \equiv e^{\tilde{w}_{jk\ell}} - 1$, etc. Recall that $\tilde{w}_j \equiv w_j - \beta \mathcal{H}_j$, $\tilde{w}_{jk} \equiv w_{jk} - \beta u_{jk}$, etc. The effective many-body potentials go to zero if any member of the cluster is far separated from the rest (in general at and beyond the pair level they go to zero faster than the pair potential does), and therefore the many-body effective Mayer f-functions go to zero at the same rate.

In order to see the similarities and differences with the Ursell cluster expansion discussed in the following subsection, one can write out explicitly the effective Mayer f-function for the first several many-body effective potentials. For the phase space total weight, for $N = 1$,

$$\Omega_1^{(1)} = e^{\tilde{w}_1}. \tag{11.22}$$

Here and below the superscript denotes the total number of particles in the system and a subscript denotes the position and momentum of a particle: $j \equiv \{\mathbf{q}_j, \mathbf{p}_j\}$. For $N = 2$ the (unsymmetrized) phase space total weight is

$$\Omega_{12}^{(2)} = e^{\tilde{w}_1}e^{\tilde{w}_2}e^{\tilde{w}_{12}} = \Omega_1^{(1)}\Omega_2^{(1)}[1 + \tilde{f}_{12}]. \tag{11.23}$$

Therefore one can identify a connected pair weight as

$$\Omega_{c,\,12}^{(2)} = \Omega_{12}^{(2)} - \Omega_1^{(1)}\Omega_2^{(1)} = \Omega_1^{(1)}\Omega_2^{(1)}\tilde{f}_{12}. \tag{11.24}$$

This turns out to be the same as the Ursell pair cluster (see section 11.5).

For $N = 3$ the (unsymmetrized) phase space total weight is

$$\begin{aligned}\Omega_{123}^{(3)} &= e^{\tilde{w}_1}e^{\tilde{w}_2}e^{\tilde{w}_3}e^{\tilde{w}_{12}}e^{\tilde{w}_{23}}e^{\tilde{w}_{31}}e^{\tilde{w}_{123}} \\ &= \Omega_1^{(1)}\Omega_2^{(1)}\Omega_3^{(1)}[1 + \tilde{f}_{12}][1 + \tilde{f}_{23}][1 + \tilde{f}_{31}][1 + \tilde{f}_{123}].\end{aligned} \tag{11.25}$$

The triplet effective Mayer f-function, $\tilde{f}_{123} = e^{\tilde{w}_{123}} - 1$, has the same form as that given for a triplet potential (Attard 2002, equation (8.66)). This goes to zero if any particle is separated from the other two, and so one can identify a connected triplet weight as

$$\begin{aligned}\Omega_{c,\,123}^{(3)} &= \Omega_1^{(1)}\Omega_2^{(1)}\Omega_3^{(1)}\tilde{f}_{123} \\ &= \Omega_1^{(1)}\Omega_2^{(1)}\Omega_3^{(1)}\left[\frac{\Omega_{123}^{(3)}}{\Omega_1^{(1)}\Omega_2^{(1)}\Omega_3^{(1)}}\frac{(\Omega_1^{(1)}\Omega_2^{(1)}\Omega_3^{(1)})^2}{\Omega_{12}^{(2)}\Omega_{23}^{(2)}\Omega_{31}^{(2)}} - 1\right].\end{aligned} \tag{11.26}$$

In contrast, the Ursell cluster for $N = 3$ is

$$\begin{aligned}\Omega_{U,\,123}^{(3)} &= \Omega_{123}^{(3)} - \Omega_{12}^{(2)}\Omega_3^{(1)} - \Omega_{23}^{(2)}\Omega_1^{(1)} - \Omega_{31}^{(2)}\Omega_2^{(1)} + 2\Omega_1^{(1)}\Omega_2^{(1)}\Omega_3^{(1)} \\ &= \Omega_1^{(1)}\Omega_2^{(1)}\Omega_3^{(1)}\big\{[1 + \tilde{f}_{12}][1 + \tilde{f}_{23}][1 + \tilde{f}_{31}][1 + \tilde{f}_{123}] \\ &\quad - \tilde{f}_{12} - \tilde{f}_{23} - \tilde{f}_{31} - 1\big\} \\ &= \Omega_1^{(1)}\Omega_2^{(1)}\Omega_3^{(1)}\big\{\big(1 + \tilde{f}_{12} + \tilde{f}_{23} + \tilde{f}_{31}\big)\tilde{f}_{123} \\ &\quad + \big(\tilde{f}_{12}\tilde{f}_{23} + \tilde{f}_{12}\tilde{f}_{31} + \tilde{f}_{23}\tilde{f}_{31} + \tilde{f}_{12}\tilde{f}_{23}\tilde{f}_{31}\big)[1 + \tilde{f}_{123}]\big\}.\end{aligned} \tag{11.27}$$

Although both the Ursell triplet cluster and the triplet effective Mayer f-function go to zero if any one particle is separated far from the rest, it can be seen that the respective definitions of the connected part of the total phase space weight are different.

11.1.9 Ursell clusters and Mayer *f*-functions

Historically, the Mayer *f*-functions of classical statistical mechanics did not prove suitable for quantum statistical mechanics. Instead, Kahn and Uhlenbeck (1938) introduced the classical cluster functions of Ursell (1927) into quantum statistical mechanics. Loosely speaking, the Mayer *f*-functions provide a route to the statistical properties of a system at the molecular level, whereas the Ursell cluster functions are useful for the macroscopic, thermodynamic properties of a system. Nevertheless, as Pathria (1972 chapter 9) explains, there is a direct connection between the coefficients of the density expansion of the pressure of a classical system in terms of integrals over Mayer *f*-functions, and the coefficients in the fugacity expansion, which are proportional to the Ursell cluster integrals. Kahn and Uhlenbeck (1938) based their approach to quantum systems on the latter. Lee and Yang (1959a, 1959b, 1960a, 1960b, 1960c) developed the diagrammatic form, including the binary kernel resummation. A brief mathematical presentation of the Lee–Yang approach together with an application to the present classical phase space formulation of quantum statistical mechanics is given in section 11.5.

The expansions of Lee and Yang require nested temperature integrations over the binary kernel (section 11.5.3). Each integration is reminiscent of the quadrature required for the temperature derivative of the formal definition of the commutation function, equation (11.2). The present author's experience is that such temperature quadratures are inefficient, unwieldy, and numerically unstable. It is possible that resummation of the pair potential as the binary kernel, which has the advantage of going to zero at both large and small separations, may ameliorate the numerical challenges. This possibility notwithstanding, the author's conclusion is that obtaining the n-body effective potential via the sum over n-body energy states (e.g., section 11.4.1) rather than by temperature integration is the best way to proceed.

However the n-body energy states require the specification of boundary conditions. This is problematic for reasons connected with the non-local nature of quantum mechanics. In the preface, the point is made that quantum mechanics is inherently global in nature, as exemplified by the instantaneous collapse of an entangled wave function over macroscopic distances, or the quantization of an N-particle system due to the boundary conditions applied to the wave function. The contrast with classical systems is stark, since it is an empirical axiom born out of historical experience and measurement that individual particle dynamics are not directly influenced by particles beyond the range of the interaction potential. In this sense classical systems are localized.

The treatment of classical systems is often facilitated by breaking them up into subsystems, with interactions across the boundaries treated in a gross or mean-field or thermodynamic fashion. Such an approach relies upon the inherently local nature of classical systems. One can in fact develop formally exact expansions for classical systems based on such dominant local interactions, an example being the cluster expansion just mentioned.

The application of such local concepts to quantum systems is highly problematic. On the one hand there is no formal basis, or attempt to provide a formal basis, for the break up of a quantum system to one or more independent subsystems that at the same time addresses the problem of the non-local nature of quantum mechanics. On the other hand such a break up, as detailed in the Lee–Yang cluster expansions just mentioned, has been used with success, albeit limited, in the analysis of quantum many-particle systems.

One cannot but conclude that what motivates the factorization of quantum systems into quasi-independent subsystems is the success of such factorizations for classical systems, together with the realization that at high enough temperatures or low enough densities quantum systems must be dominated by their classical aspects. The answer to the more fundamental question—how precisely do quantum many-body systems become localized—seems likely to be found in the reasons for the formulation of quantum statistical mechanics itself: the total wave function is decoherent due to its projection onto a sub-space (see chapter 12).

11.2 Symmetrization function

The symmetrization function was treated in chapters 3 and 7. Here a brief summary of the permutation loop expansion is given because it is directly relevant to the application of the many-body expansion of the commutation function. Note that what follows is the symmetrization function $\eta_{\rm q}^{\pm}(\mathbf{\Gamma})$, which is to be combined with $\omega_{\rm p}(\mathbf{\Gamma})$.

Decomposing the permutations into loops, the symmetrization function is

$$\begin{aligned}\eta_{\rm q}^{\pm}(\mathbf{p},\mathbf{q}) &\equiv \frac{1}{\langle\mathbf{p}|\mathbf{q}\rangle}\sum_{\hat{\rm P}}(\pm 1)^{p}\,\langle\hat{\rm P}\mathbf{p}|\mathbf{q}\rangle \\ &= \frac{1}{\langle\mathbf{p}|\mathbf{q}\rangle}\left\{\langle\mathbf{p}|\mathbf{q}\rangle \pm \sum_{j,k}{}'\langle\hat{\rm P}_{jk}\mathbf{p}|\mathbf{q}\rangle + \sum_{j,k,\ell}{}'\langle\hat{\rm P}_{jk}\hat{\rm P}_{k\ell}\mathbf{p}|\mathbf{q}\rangle \right.\\ &\qquad \left. + \sum_{j,k,\ell,m}{}'\langle\hat{\rm P}_{jk}\hat{\rm P}_{\ell m}\mathbf{p}|\mathbf{q}\rangle \pm \ldots\right\} \\ &= 1 + \sum_{j<k}\eta_{{\rm q},jk}^{\pm(2)} + \sum_{j<k<\ell}\eta_{{\rm q},jk\ell}^{\pm(3)} + \sum_{j,k,\ell,m}{}'\eta_{{\rm q},jk}^{\pm(2)}\eta_{{\rm q},\ell m}^{\pm(2)} \pm \ldots\end{aligned} \tag{11.28}$$

The prime on the summations indicates that all the labels are different *and* that each permutation is counted once only. Here $\hat{\rm P}_{jk}$ is the transposition of particles j and k, which has odd parity. (Actually, it is the transposition of the respective elements in the vectors of the particles' characteristics, which makes up with precision what it lacks in concision.) The loops are successive transpositions. For example, a three-loop or trimer is $\hat{\rm P}_{jk}\hat{\rm P}_{k\ell}\{\ldots x_j\ldots x_k\ldots x_\ell\ldots\} = \{\ldots x_\ell\ldots x_j\ldots x_k\ldots\}$. For N objects, the number of distinct permutations consisting of m_ℓ ℓ-loops (i.e., $\sum_{\ell=1}^{N}\ell m_\ell = N$) is $N!/\prod_\ell \ell^{m_\ell}m_\ell!$.

With $\langle\mathbf{q}|\mathbf{p}\rangle = V^{-N/2}\prod_{j=1}^{N} e^{-\mathbf{p}_j\cdot\mathbf{q}_j/i\hbar}$, the dimer symmetrization loop for two specific particles is

$$
\begin{aligned}
\eta_{\mathrm{q},\,jk}^{\pm(2)} &\equiv \frac{\pm\langle \hat{\mathrm{P}}_{jk}\mathbf{p}|\mathbf{q}\rangle}{\langle \mathbf{p}|\mathbf{q}\rangle} \\
&= \frac{\pm e^{\mathbf{p}_k\cdot\mathbf{q}_j/i\hbar} e^{\mathbf{p}_j\cdot\mathbf{q}_k/i\hbar}}{e^{\mathbf{p}_j\cdot\mathbf{q}_j/i\hbar} e^{\mathbf{p}_k\cdot\mathbf{q}_k/i\hbar}} \\
&= \pm\, e^{[\mathbf{p}_k-\mathbf{p}_j]\cdot\mathbf{q}_j/i\hbar} e^{[\mathbf{p}_j-\mathbf{p}_k]\cdot\mathbf{q}_k/i\hbar} \\
&= \pm\, e^{-\mathbf{p}_{jk}\cdot\mathbf{q}_{jk}/i\hbar}.
\end{aligned} \tag{11.29}
$$

Similarly the trimer loop for three specific particles is

$$
\begin{aligned}
\eta_{\mathrm{q},\,jk\ell}^{\pm(3)} &\equiv \frac{\langle \hat{\mathrm{P}}_{jk}\hat{\mathrm{P}}_{k\ell}\mathbf{p}|\mathbf{q}\rangle}{\langle \mathbf{p}|\mathbf{q}\rangle} \\
&= \frac{e^{\mathbf{p}_\ell\cdot\mathbf{q}_j/i\hbar} e^{\mathbf{p}_j\cdot\mathbf{q}_k/i\hbar} e^{\mathbf{p}_k\cdot\mathbf{q}_\ell/i\hbar}}{e^{\mathbf{p}_j\cdot\mathbf{q}_j/i\hbar} e^{\mathbf{p}_k\cdot\mathbf{q}_k/i\hbar} e^{\mathbf{p}_\ell\cdot\mathbf{q}_\ell/i\hbar}} \\
&= e^{\mathbf{p}_{\ell j}\cdot\mathbf{q}_j/i\hbar} e^{\mathbf{p}_{jk}\cdot\mathbf{q}_k/i\hbar} e^{\mathbf{p}_{k\ell}\cdot\mathbf{q}_\ell/i\hbar} \\
&= e^{\mathbf{q}_{\ell j}\cdot\mathbf{p}_j/i\hbar} e^{\mathbf{q}_{jk}\cdot\mathbf{p}_k/i\hbar} e^{\mathbf{q}_{k\ell}\cdot\mathbf{p}_\ell/i\hbar}.
\end{aligned} \tag{11.30}
$$

In general a specific l-loop is

$$
\eta_{\mathrm{q},\,j_1,\,j_2,\,\ldots,\,j_\ell}^{\pm(\ell)} = e^{\mathbf{p}_{j_1 j_\ell}\cdot\mathbf{q}_{j_\ell}/i\hbar} \prod_{k=1}^{\ell-1} e^{\mathbf{p}_{j_{k+1} j_k}\cdot\mathbf{q}_{j_k}/i\hbar}. \tag{11.31}
$$

The sums of these specific loops are phase functions, such as $\eta_{\mathrm{q}}^{\pm(2)}(\mathbf{p},\mathbf{q}) \equiv \sum_{j<k}\eta_{\mathrm{q},\,jk}^{\pm(2)}$, $\eta_{\mathrm{q}}^{\pm(3)}(\mathbf{p},\mathbf{q}) \equiv \sum_{j<k<\ell}\eta_{\mathrm{q},\,jk\ell}^{\pm(3)}$, etc.

Because the complex exponentials are highly oscillatory, they average to zero unless the exponents are small. Hence the loops are compact in phase space, and the phase functions are extensive, $\eta_{\mathrm{q}}^{\pm(\ell)}(\mathbf{p},\mathbf{q}) = \mathcal{O}(N)$.

The fourth term in the expansion written out explicitly above contains the products of two dimer loops. They are not actually independent because all labels must be different. Since each permutation appears once only, it can be written as half the product of independent dimer functions plus a lower order correction,

$$
\begin{aligned}
&\sum_{j,k,\ell,m}{}' \eta_{\mathrm{q},\,jk}^{\pm(2)}\eta_{\mathrm{q},\,\ell m}^{\pm(2)} \\
&= \frac{1}{8}\sum_{j,k,\ell,m} \bar{\eta}_{\mathrm{q},\,jk}^{\pm(2)}\bar{\eta}_{\mathrm{q},\,\ell m}^{\pm(2)}\bar{\delta}_{j\ell}\bar{\delta}_{jm}\bar{\delta}_{k\ell}\bar{\delta}_{km} \\
&= \frac{1}{8}\sum_{j,k,\ell,m} \bar{\eta}_{\mathrm{q},\,jk}^{\pm(2)}\bar{\eta}_{\mathrm{q},\,\ell m}^{\pm(2)} - \frac{4}{8}\sum_{j,k,\ell} \bar{\eta}_{\mathrm{q},\,jk}^{\pm(2)}\bar{\eta}_{\mathrm{q},\,j\ell}^{\pm(2)} + \frac{2}{8}\sum_{j,k}(\bar{\eta}_{\mathrm{q},\,jk}^{\pm(2)})^2 \\
&= \frac{1}{2}\eta_{\mathrm{q}}^{\pm(2)}(\mathbf{p},\mathbf{q})^2 + \mathcal{O}(N),
\end{aligned} \tag{11.32}
$$

where $\bar{\delta}_{jk} \equiv 1 - \delta_{jk}$ and $\bar{\eta}_{\mathrm{q},\,jk}^{\pm(2)} \equiv \bar{\delta}_{jk}\eta_{\mathrm{q},\,jk}^{\pm(2)}$. Analogous results hold for all the products of loops that appear in the loop permutation expansion. Keeping only the leading order of independent products in each case, the symmetrization function can be written

$$\begin{aligned}
&\eta_{\rm q}^{\pm}(\mathbf{p}, \mathbf{q}) \\
&\approx 1 + \eta_{\rm q}^{\pm(2)}(\mathbf{p}, \mathbf{q}) + \eta_{\rm q}^{\pm(3)}(\mathbf{p}, \mathbf{q}) + \frac{1}{2}\eta_{\rm q}^{\pm(2)}(\mathbf{p}, \mathbf{q})^2 + \cdots \\
&= \exp\left\{\eta_{\rm q}^{\pm(2)}(\mathbf{p}, \mathbf{q}) + \eta_{\rm q}^{\pm(3)}(\mathbf{p}, \mathbf{q}) + \eta_{\rm q}^{\pm(4)}(\mathbf{p}, \mathbf{q}) + \cdots\right\} \\
&= \prod_{\ell=2}^{\infty} e^{\eta_{\rm q}^{\pm(\ell)}(\mathbf{p}, \mathbf{q})}.
\end{aligned} \tag{11.33}$$

The symmetrization loop functions that are retained in the exponent are extensive, $\mathcal{O}(N)$.

This loop expansion and exponential form for the symmetrization function has the same character as in chapters 3 and 7, in which the quantum grand potential was expressed as a series of loop potentials. The main difference is that here the exponentiation of the series of extensive single loops is done for the phase function itself rather than after statistical averaging.

11.3 Generalized Mayer *f*-function

The symmetrization loop phase functions are related to the many-body formulation of the commutation function. Retaining only the pair term for the latter is valid at high temperatures or low densities. Retaining only the dimer term in the symmetrization function is valid at low densities. The quantitative meaning of 'high' and 'low' obviously depends on the particular system, but it seems likely to encompass most of condensed matter at terrestrial densities and temperatures.

Assuming a pairwise additive potential (and no singlet potential), and keeping only the pair terms in both the commutation and symmetrization functions, the grand canonical partition function is (see chapter 7)

$$\begin{aligned}
\Xi &= \sum_{N=0}^{\infty} \frac{e^{\beta\mu N}}{N! h^{3N}} \int \mathrm{d}\boldsymbol{\Gamma}\; e^{-\beta\mathcal{H}(\boldsymbol{\Gamma})} e^{W_{\rm p}(\boldsymbol{\Gamma})} \prod_{\ell=2}^{\infty} e^{\eta_{\rm q}^{\pm(\ell)}(\boldsymbol{\Gamma})} \\
&\approx \sum_{N=0}^{\infty} \frac{1}{N! h^{3N}} \int \mathrm{d}\boldsymbol{\Gamma} \prod_{j=1}^{N} [e^{\beta\mu} e^{-\beta p_j^2/2m}] \prod_{j<k}^{N} [e^{-\beta u_{jk}^{(2)}} e^{w_{{\rm p},jk}^{(2)}} e^{\eta_{{\rm q},jk}^{\pm(2)}}] \\
&\equiv \sum_{N=0}^{\infty} \frac{1}{N! h^{3N}} \int \mathrm{d}\boldsymbol{\Gamma} \prod_{j=1}^{N} z_j \prod_{j<k}^{N} \left[1 + f_{jk}^{(2)}\right].
\end{aligned} \tag{11.34}$$

Here $\boldsymbol{\Gamma} = \{\mathbf{p}, \mathbf{q}\}$ is a point in classical phase space, and $z_j \equiv e^{\beta\mu} e^{-\beta p_j^2/2m}$ is a generalized fugacity. The quantity $f_{jk}^{(2)} \equiv f^{(2)}(\mathbf{p}_{jk}, \mathbf{q}_{jk}) = e^{-\beta u_{jk}^{(2)}} e^{w_{{\rm p},jk}^{(2)}} e^{\eta_{{\rm q},jk}^{\pm(2)}} - 1$ is a generalized pair Mayer *f*-function that depends upon the relative position and momentum of the two particles (see also subsection 11.1.7). It has the desirable property that $f^{(2)}(\mathbf{p}_{jk}, \mathbf{q}_{jk}) \to 0$, $q_{jk} \to \infty$. (If there is a singlet potential, then its Maxwell–Boltzmann factor would be included in the fugacity, along with the singlet part of the commutation function. The pair part of the commutation function would need to be modified.)

This generalized Mayer f-function allows quantum systems to be treated with the powerful techniques that have advanced the field of classical statistical mechanics. Examples include cluster diagrams, density functional theory, integral equation methods, and asymptotic analysis (Pathria 1972, Hansen and McDonald 1986, Attard 2002). The Mayer f-function and cluster diagrams are not restricted to pairwise additive potentials (Morita and Hiroike 1961, Stell 1964, Attard 1992), although this is certainly the most common case.

For example, the quantum pair Ornstein–Zernike equation can just be written down,

$$h(\Gamma_1, \Gamma_2) = c(\Gamma_1, \Gamma_2) + \int \mathrm{d}\Gamma_3\, \rho(\Gamma_3)c(\Gamma_1, \Gamma_3)h(\Gamma_3, \Gamma_2), \tag{11.35}$$

where $\Gamma_j = \{\mathbf{p}_j, \mathbf{q}_j\}$, the singlet density is $\rho_j^{(1)} = Ne^{-\beta p_j^2/2m}/(2\pi m/\beta)^{3/2}V$, $h_{12} = g_{12} - 1$ is the total correlation function, c_{12} is the direct correlation function, and the pair distribution function is related to the pair density as $\rho_{12}^{(2)} = \rho_1^{(1)}\rho_2^{(1)}g_{12}$. This may be combined with, for example, the hypernetted chain approximation,

$$h_{12} = -1 + e^{h_{12} - c_{12} - \beta u_{12}^{(2)} + w_{\mathrm{p},12}^{(2)} + \eta_{\mathrm{q},12}^{\pm(2)}}, \tag{11.36}$$

to give a closed system of equations. The asymptote is

$$c_{12} \sim -\beta u_{12}^{(2)} + w_{\mathrm{p},\,12}^{(2)} + \eta_{\mathrm{q},\,12}^{\pm(2)}, \quad q_{12} \to \infty. \tag{11.37}$$

Of course the various quantities are complex. (Recall that these are quantum probabilities, albeit in classical phase space.) Because of the oscillations at large separations, it may be worth adding numerically a damping or cut-off factor, and checking that the final results are independent of its precise value or form. One should also be aware that the requisite commutation function can vary with the quantity being averaged (chapter 7).

11.4 Numerical analysis

11.4.1 Two-body commutation function

Integrating the temperature derivative of the commutation function from the high temperature limit can be unstable. This is the method suggested in subsection 11.1.4 for extending the linear solution. (The instability in the quadrature is not limited to the many-body expansion.) Although it works for the simple harmonic oscillator, it does not, for example, give reliable results for the Lennard–Jones pair potential. Experience in a variety of cases has shown that obtaining the energy eigenstates and summing over them is by far the most efficient and reliable way to obtain the commutation function. Here this alternative method to obtain the singlet and pair commutation functions is outlined.

The singlet combined commutation function, given above as the partial differential equation for the temperature derivative, equation (11.16), can be obtained for a one-particle system, $N = 1$, from the fundamental definition

$$\begin{aligned} e^{\tilde{w}^{(1)}(q_1, p_1)} &= \frac{1}{\langle q_1|p_1\rangle} e^{-\beta\hat{\mathcal{H}}^{(1)}(q_1)}\langle q_1|p_1\rangle \\ &= \frac{1}{\langle q_1|p_1\rangle} \sum_\ell e^{-\beta E_\ell^{(1)}}\langle q_1|\ell\rangle \, \langle \ell|p_1\rangle. \end{aligned} \tag{11.38}$$

Here the singlet Hamiltonian operator is $\hat{\mathcal{H}}^{(1)}(q) = (-\hbar^2/2m)\nabla^2 + u^{(1)}(q)$, and the singlet energy eigenfunctions satisfy $\hat{\mathcal{H}}^{(1)}|\ell\rangle = E_\ell^{(1)}|\ell\rangle$. For the simple harmonic oscillator potential these are known analytically.

For the case of two particles, the total combined commutation function is $\tilde{W}^{(2)}(q_1, p_1; q_2, p_2) = \tilde{w}^{(1)}(q_1, p_1) + \tilde{w}^{(1)}(q_2, p_2) + \tilde{w}^{(2)}(q_1, p_1; q_2, p_2)$. Hence the pair combined commutation function given above as the partial differential equation for the temperature derivative, equation (11.17), can instead be obtained for a two-particle system, $N = 2$, from the fundamental definition

$$\begin{aligned} & e^{\tilde{w}^{(2)}(q_1, p_1; q_2, p_2)} \\ &= e^{\tilde{W}^{(2)}(q_1, p_1; q_2, p_2) - \tilde{w}^{(1)}(q_1, p_1) - \tilde{w}^{(1)}(q_2, p_2)} \\ &= \frac{e^{-\tilde{w}^{(1)}(q_1, p_1)} e^{-\tilde{w}^{(1)}(q_2, p_2)}}{\langle q_1, q_2|p_1, p_2\rangle} e^{-\beta\hat{\mathcal{H}}^{(2)}(q_1, q_2)}\langle q_1, q_2|p_1, p_2\rangle \\ &= \frac{e^{-\tilde{w}^{(1)}(q_1, p_1)} e^{-\tilde{w}^{(1)}(q_2, p_2)}}{\langle q_1, q_2|p_1, p_2\rangle} \sum_\ell e^{-\beta E_\ell^{(2)}}\langle q_1, q_2|\ell\rangle\langle \ell|p_1, p_2\rangle. \end{aligned} \tag{11.39}$$

Here the pair Hamiltonian operator is $\hat{\mathcal{H}}^{(2)}(q_1, q_2) = (-\hbar^2/2m)[\nabla_1^2 + \nabla_2^2] + u^{(1)}(q_1) + u^{(1)}(q_2) + u^{(2)}(q_1, q_2)$. Again trial and much error has shown that it is most efficient to obtain the pair combined commutation function from the sum over energy states. The latter were obtained by standard minimization and orthogonalization techniques. The Fourier transform methods proposed above do not work for the Lennard–Jones potential.

Two points: the system these were applied to consists of a singlet simple harmonic oscillator potential and a Lennard–Jones pair potential in one dimension. For the two-particle eigenfunctions in this system the factorization into center of mass and interaction coordinates is exact (Messiah 1961, Merzbacher 1970, Pathria 1972). This means that one has to solve two one-dimensional systems (one of which is the already solved simple harmonic oscillator case), rather than a two-dimensional system. Second, for the interaction coordinate, the Lennard–Jones core repulsion makes the energy eigenfunction vanish at $q = 0$, and the simple harmonic oscillator potential makes it vanish at large q. Hence one can choose a system size $q_{\rm max}$, with $q \in [0, q_{\rm max}]$ and set $\phi_{{\rm int},\ell}(q_{\rm max}) = \phi'_{{\rm int},\ell}(q_{\rm max}) = 0$. (This is done by setting the values beyond the boundary to zero in the central difference formulae for the Laplacian.) Then one can impose periodic boundary conditions, and extend the eigenfunctions found on $q \in [0, q_{max}]$ to negative values, taking them to have the same energy for the even and odd extensions.

There is a further conceptual point that can be made that is connected with the non-local nature of quantum mechanics that was discussed in the preface. The energy states that are referred to above are for a two-particle system. The quantization of the states is dependent upon the boundary conditions that are applied. On the one hand it seems natural to apply the same boundary conditions as for the full N-particle system that the singlet and pair commutation functions will be used for. (As just mentioned, in the present specific case of a simple harmonic oscillator singlet potential, the wave function was taken to be periodic, and it and its derivative were taken to vanish at some large L.) On the other hand it seems unnatural that the pair commutation function, which is local in nature, should depend upon boundaries that in the general case may be at macroscopic distances. Possibly in general the sum over energy states is insensitive to the actual spectrum of states. (The sensitivity of the results below to the boundary conditions was not checked.) Alternatively, it might prove best practice to implement mean-field or effective local field (chapter 10) boundary conditions.

11.4.2 Core asymptote

In this subsection the behavior of the energy eigenfunction in the Lennard–Jones core region is derived. The full potential is $u(r) = \varepsilon[(r_e/r)^{12} - 2(r_e/r)^6]$, $r > 0$, and the particle mass is m. (For the effective local field, chapter 10, replace ε by $\varepsilon/2$, and for the interaction coordinate, replace m by $m/2$.)

Take the energy eigenfunction function to be of the form

$$\psi(r) = g(r)\exp\left\{a(r_e/r)^5\right\}, \tag{11.40}$$

with $a = -\sqrt{2m\varepsilon r_e^2/25\hbar^2}$. The energy eigenvalue equation is then

$$\begin{aligned} E\psi(r) = \varepsilon[(r_e/r)^{12} - 2(r_e/r)^6]\psi(r) - \frac{\hbar^2}{2m}[25a^2 r_e^{10} r^{-12} g(r) \\ + 30 a r_e^5 r^{-7} g(r) - 10 a r_e^5 r^{-6} g'(r) + g''(r)]e^{a(r_e/r)^5}. \end{aligned} \tag{11.41}$$

This gives

$$\frac{-2mE}{\hbar^2}g(r) = \frac{4m\varepsilon}{\hbar^2}r_e^6 r^{-6} g(r) + 30 a r_e^5 r^{-7} g(r) - 10 a r_e^5 r^{-6} g'(r) + g''(r). \tag{11.42}$$

Insert $g(r) = \sum_{n=3}^{\infty} g_n (r/r_e)^n$, equate coefficients of $(r/r_e)^n$, and rearrange to obtain for $n \geqslant -3$,

$$g_{n+7} = \frac{1}{10a(n+4)}\left\{\frac{4m\varepsilon r_e^2}{\hbar^2}g_{n+6} + (n+2)(n+1)g_{n+2} + \frac{2mEr_e^2}{\hbar^2}g_n\right\}, \tag{11.43}$$

with $g_n = 0$, $n < 3$. This is homogeneous in g_3, which can be used as a scale factor to match the asymptote to the actual eigenfunction at a boundary.

For the case of the interaction commutation function, one has

$$\omega_{\text{int}}(q, p) \sim f(q) e^{\beta U_{\text{int}}(q) + a(r_e/q)^5}, \quad f(q) = \sum_{n=3}^{\infty} f_n r_e^{-n} q^n, \quad q \to 0. \tag{11.44}$$

The coefficients for $j \geqslant -3$ satisfy

$$\begin{aligned}
&\left\{ \frac{5(j+7)a\hbar^2}{\mu r_e^{j+2}} - \frac{15a\hbar^2}{\mu r_e^{j+2}} \right\} f_{j+7} \\
&= \left\{ \frac{2\varepsilon}{r_e^j} - \frac{5ia\hbar p}{\mu r_e^{j+1}} \right\} f_{j+6} + \frac{(j+2)(j+1)\hbar^2}{2\mu r_e^{j+2}} f_{j+2} \\
&\quad + \frac{(j+1)i\hbar p}{\mu r_e^{j+1}} f_{j+1} - r_e^{-j} \dot{f}_j - \frac{\kappa}{2 r_e^{j-2}} f_{j-2},
\end{aligned} \tag{11.45}$$

with $f_n = 0$, $n < 3$, the over-dot signifying the derivative with respect to β, the interaction mass being $\mu = m_1 m_2/(m_1 + m_2)$, the interaction momentum being $p = (m_2 p_1 - m_1 p_2)/(m_1 + m_2)$, and the separation being $q = q_1 - q_2$. The interaction potential used here is $U_{\text{int}}(q) = \varepsilon[(r_e/q)^{12} - 2(r_e/q)^6] + \kappa q^2/2$, with $\kappa = \mu\omega^2$. (The corresponding center of mass potential is $U_{\text{cm}}(Q) = M\omega^2 Q^2/2$, with the center of mass being $Q = (m_1 q_1 + m_2 q_2)/(m_1 + m_2)$, and the total mass being $M = m_1 + m_2$.) The coefficients f_n, $4 \leqslant n \leqslant 9$, are proportional to f_3, and, apart from that, independent of temperature.

This gives a phase space weight that vanishes in the core much more slowly than that due to the Lennard–Jones potential itself. This is qualitatively in agreement with the results now shown.

11.4.3 Three-body commutation function

The total three-body commutation function, $\tilde{W}^{(3)}_{\text{p},123} \equiv -\beta\mathcal{H}^{(3)}_{123} + W^{(3)}_{\text{p},123}$, is expressed in equation (11.20) as a sum over energy states of the energy eigenfunctions weighted by the Maxwell–Boltzmann factor. In this section is briefly discussed an efficient way to organize the computation and storage of the three-body eigenfunctions and commutation function by invoking center of mass and interaction coordinates.

Messiah (1961, chapter 9, section 13) gives a procedure for constructing the center of mass and relative coordinate transformation in general for an $(N + 1)$-particle system. The center of mass and interaction or relative coordinates for two particles of arbitrary mass are well known (Messiah 1961, Merzbacher 1970, Pathria 1972). Messiah (1961) successively applies this two-particle result, creating the jth relative particle from the center of mass variables of the first j particles relative to the actual variables of the $j + 1$ st particle.

For $N + 1$ particles define the center of mass position $\mathbf{Q}$, momentum $\mathbf{P}$, and mass M by

$$\mathbf{Q} = \frac{1}{M} \sum_{j=1}^{N+1} m_j \mathbf{q}_j, \quad \mathbf{P} = \sum_{j=1}^{N+1} \mathbf{p}_j, \quad \text{and} \quad M = \sum_{j=1}^{N+1} m_j. \tag{11.46}$$

Let $\boldsymbol{\rho}_j$, $\boldsymbol{\pi}_j$, and μ_j be the corresponding quantities for the jth relative particle. These satisfy

$$\begin{aligned}\prod_{j=1}^{N+1} m_j &= M \prod_{j=1}^{N} \mu_j,\\ \sum_{j=1}^{N+1} \frac{1}{2m_j} p_j^2 &= \frac{1}{2M} P^2 + \sum_{j=1}^{N} \frac{1}{2\mu_j} \pi_j^2\\ \sum_{j=1}^{N+1} \mathbf{q}_j \cdot \mathbf{p}_j &= \mathbf{Q} \cdot \mathbf{P} + \sum_{j=1}^{N} \boldsymbol{\rho}_j \cdot \boldsymbol{\pi}_j\\ \sum_{j=1}^{N+1} \mathbf{q}_j \times \mathbf{p}_j &= \mathbf{Q} \times \mathbf{P} + \sum_{j=1}^{N} \boldsymbol{\rho}_j \times \boldsymbol{\pi}_j.\end{aligned} \tag{11.47}$$

These invariants are readily shown to hold for two particles. Messiah (1961) asserts that they must also hold for $N + 1$ particles because of the method by which the relative coordinates are successively constructed in a pairwise fashion.

Applying this method to three particles, one first selects particles 1 and 2 and creates their center of mass variables

$$M' = m_1 + m_2, \quad \mathbf{Q}' = \frac{m_1 \mathbf{q}_1 + m_2 \mathbf{q}_2}{m_1 + m_2}, \quad \text{and} \quad \mathbf{P}' = \mathbf{p}_1 + \mathbf{p}_2, \tag{11.48}$$

and their interaction variables

$$\mu_1 = \frac{m_1 m_2}{m_1 + m_2}, \quad \boldsymbol{\rho}_1 = \mathbf{q}_1 - \mathbf{q}_2, \quad \text{and} \quad \boldsymbol{\pi}_1 = \frac{m_2 \mathbf{p}_1 - m_1 \mathbf{p}_2}{m_1 + m_2}. \tag{11.49}$$

This is the first relative particle.

The second relative particle is created from the first center of mass particle and the third particle. It has center of mass variables

$$\begin{aligned}M'' &= m_1 + m_2 + m_3,\\ \mathbf{Q}'' &= \frac{M'\mathbf{Q}' + m_3 \mathbf{q}_3}{M' + m_3} = \frac{m_1 \mathbf{q}_1 + m_2 \mathbf{q}_2 + m_3 \mathbf{q}_3}{m_1 + m_2 + m_3}, \quad \text{and}\\ \mathbf{P}'' &= \mathbf{P}' + \mathbf{p}_3 = \mathbf{p}_1 + \mathbf{p}_2 + \mathbf{p}_3,\end{aligned} \tag{11.50}$$

and interaction variables

$$\begin{aligned}\mu_2 &= \frac{M' m_3}{M' + m_3} = \frac{m_1 m_3 + m_2 m_3}{m_1 + m_2 + m_3},\\ \boldsymbol{\rho}_2 &= \mathbf{Q}' - \mathbf{q}_3 = \frac{m_1[\mathbf{q}_1 - \mathbf{q}_3] + m_2[\mathbf{q}_2 - \mathbf{q}_3]}{m_1 + m_2}, \quad \text{and}\\ \boldsymbol{\pi}_2 &= \frac{m_3 \mathbf{P}' - M' \mathbf{p}_3}{M' + m_3} = \frac{m_3 \mathbf{p}_1 + m_3 \mathbf{p}_2 - m_1 \mathbf{p}_3 - m_2 \mathbf{p}_3}{m_1 + m_2 + m_3}.\end{aligned} \tag{11.51}$$

This is the second relative particle. Notice that there is an asymmetry between the relative particles. The double prime on the center of mass variables may be dropped.

We now apply the transformation to a system of three identical particles with a singlet potential that is harmonic (i.e., the third and higher derivatives vanish) and with a pair potential that is a function of separation.

For three particles with the same mass, $\mathbf{Q}' = [\mathbf{q}_1 + \mathbf{q}_2]/2$ and $\mathbf{Q}'' = [\mathbf{q}_1 + \mathbf{q}_2 + \mathbf{q}_3]/3$. In this case the total singlet potential to quadratic order is

$$\begin{aligned} u(\mathbf{q}_1) + u(\mathbf{q}_2) + u(\mathbf{q}_3) \\ &= 2u(\mathbf{Q}') + \frac{1}{2}[\mathbf{q}_1 - \mathbf{Q}'][\mathbf{q}_1 - \mathbf{Q}']: u''(\mathbf{Q}') + \frac{1}{2}[\mathbf{q}_2 - \mathbf{Q}'][\mathbf{q}_2 - \mathbf{Q}']: u''(\mathbf{Q}') \\ &\quad + u(\mathbf{Q}'') + [\mathbf{q}_3 - \mathbf{Q}'']: u'(\mathbf{Q}'') + \frac{1}{2}[\mathbf{q}_3 - \mathbf{Q}''][\mathbf{q}_3 - \mathbf{Q}'']: u''(\mathbf{Q}'') \\ &= 3u(\mathbf{Q}'') + \frac{1}{4}\rho_1\rho_1 : u''(\mathbf{Q}'') \\ &\quad + 2[\mathbf{Q}' - \mathbf{Q}''] \cdot u'(\mathbf{Q}'') + \frac{1}{2}2[\mathbf{Q}' - \mathbf{Q}''][\mathbf{Q}' - \mathbf{Q}'']: u''(\mathbf{Q}'') \\ &\quad + [\mathbf{q}_3 - \mathbf{Q}'']: u'(\mathbf{Q}'') + \frac{1}{2}[\mathbf{q}_3 - \mathbf{Q}''][\mathbf{q}_3 - \mathbf{Q}'']: u''(\mathbf{Q}''). \end{aligned} \tag{11.52}$$

Now $\mathbf{Q}' - \mathbf{Q}'' = \rho_2/3$, and $\mathbf{q}_3 - \mathbf{Q}'' = -2\rho_2/3$. Also, for a harmonic potential the second derivative is constant, $u''(\mathbf{Q}'') = u''$. Hence

$$u(\mathbf{q}_1) + u(\mathbf{q}_2) + u(\mathbf{q}_3) = 3u(\mathbf{Q}'') + \frac{1}{4}\rho_1\rho_1 : u'' + \frac{1}{3}\rho_2\rho_2 : u''. \tag{11.53}$$

Hence for a harmonic singlet potential, the transformation to center of mass and relative coordinates is exact.

The total separation-dependent pair potential for three identical particles is

$$\begin{aligned} u^{(2)}(\mathbf{q}_1 - \mathbf{q}_2) + u^{(2)}(\mathbf{q}_2 - \mathbf{q}_3) + u^{(2)}(\mathbf{q}_3 - \mathbf{q}_1) \\ = u^{(2)}(\rho_1) + u^{(2)}(\rho_2 - \rho_1) + u^{(2)}(\rho_1 - \rho_2). \end{aligned} \tag{11.54}$$

This is a function solely of the interaction coordinates, as required.

For three particles in one dimension, the interaction energy eigenfunctions are $\phi^{\text{int}}_{n_1, n_2}(\rho_1, \rho_2)$, which have to be stored on a four-dimensional grid. The symmetry $U^{\text{int}}(\rho_1, \rho_2) = U^{\text{int}}(-\rho_1, -\rho_2)$ can be used to reduce the computational burden. The interaction part of the three-particle commutation function is $\omega^{\text{int}}(\rho_1, \pi_1; \rho_2, \pi_2)$, which is also four-dimensional. The center of mass parts are the usual simple harmonic oscillator functions, which are two-dimensional.

11.4.4 Results

Figure 11.1 shows the real part of the exponent for the phase space weight for three particles, Re $\tilde{W}^{(3)}(q_1, p_1; q_2, p_2; q_3, p_3)$, with the two exterior particles fixed at $\pm r_e$

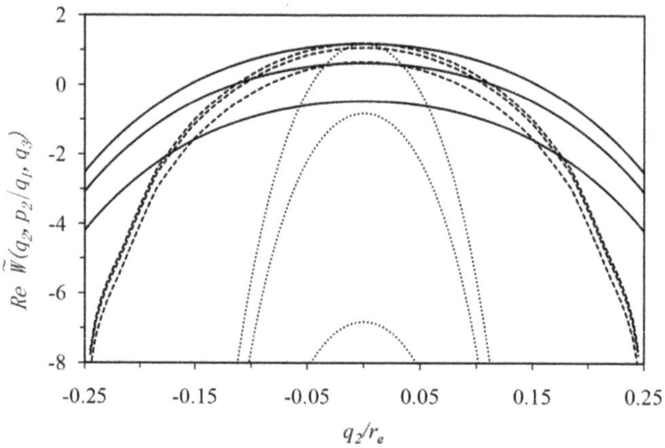

Figure 11.1. Real part of the combined total commutation function, $\tilde{W} = W - \beta\mathcal{H}$, for three particles, with $q_1 = -q_3 = -r_e$ and $p_1 = p_3 = 0$, at $\beta\hbar\omega_{LJ} = 8.5$ (Lennard–Jones pair interaction, simple harmonic oscillator singlet potential, one dimension). The solid curves are the many-body expansion (pair), the dashed curves are the local state expansion (triplet), and the dotted curves are the classical Maxwell–Boltzmann exponent, $-\beta\mathcal{H}(\mathbf{q}, \mathbf{p})$. From top to bottom in each series, the classical kinetic energy for the middle particle is $\beta p_2^2/2m = 0$, 2 and 8. An arbitrary normalization constant has been added to each surface so that they coincide at $\{q_2, p_2\} = \{0, 0\}$.

with zero momentum. Three approaches are compared: the present many-body expansion (pair level), the local state expansion given in chapter 10 (singlet, first order; one variable and two fixed particles), and the classical Maxwell–Boltzmann exponent. (The imaginary part of the exponent, which also contributes to the real part of the phase space weight, is not shown.) Overall, the data in figure 11.1 reveal that the two quantum approaches qualitatively agree in that there is substantial cancelation of the classical core repulsion between the Lennard–Jones particles. This means that the quantum particles will approach each other and interpenetrate much more deeply than their classical counterparts. The pair many-body expansion shows more cancelation than the local state expansion. The two approaches also agree in predicting that higher values of momentum are more accessible in the quantum case than is classically predicted. However, for this, the real part of the exponent, the many-body expansion shows greater variation with momentum than does the local state expansion. Since the Monte Carlo simulation results for the average energy given by the local state expansion appear to be in good agreement with benchmark results for this system, section 10.5.2, the quantitative disagreement with that approach evident in figure 11.1 might give one pause before further pursuing the many-body expansion.

It should be stressed that the data in figure 11.1 are for a relatively low temperature, $\beta\hbar\omega_{LJ} = 8.5$, and a relatively high density, $\rho r_e = 1$. The many-body expansion can be expected to converge most rapidly at high temperatures and low densities.

Figure 11.2 shows the real part of the phase space weight, $\mathrm{Re}\,\Omega^{(3)}(\mathbf{q}, \mathbf{p}) = \mathrm{Re}\exp \tilde{W}^{(3)}(\mathbf{q}, \mathbf{p})$, when the neighboring particles have non-zero momentum. The results seem unphysical. In particular, the increase in magnitude of

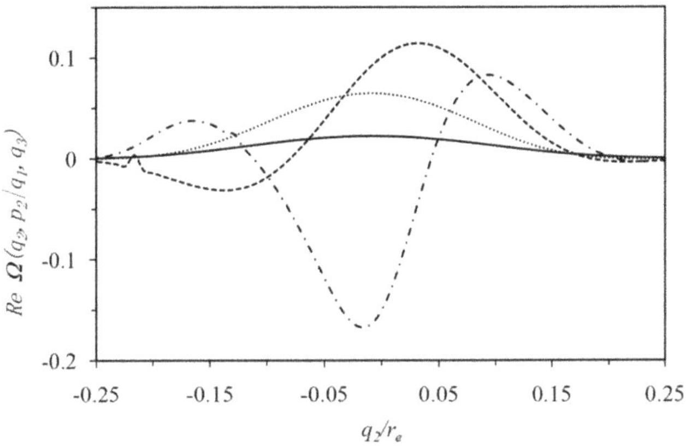

Figure 11.2. The real part of the phase space weight given by the many-body expansion (pair level) for three particles, with $q_1 = -r_e$, $p_1 r_e/\hbar = 8.38$ ($\beta p_1^2/2m = 5$), and $q_3 = r_e$, $p_3 r_e/\hbar = -11.85$ ($\beta p_3^2/2m = 10$) at $\beta\hbar\omega_{LJ} = 8.5$ (Lennard–Jones pair interaction, simple harmonic oscillator singlet potential, one dimension). The solid curve is $p_2 r_e/\hbar = 0$, the dotted curve is $p_2 r_e/\hbar = 5.24$, the dashed curve is $p_2 r_e/\hbar = 10.5$, and the dash-dotted curve is $p_2 r_e/\hbar = 18.7$.

the density with increase in momentum does not seem to be correct. It appears to be the source of an unrealistically high average kinetic energy when the pair many-body expansion commutation function is used in the Monte Carlo simulations (not shown). The dramatic oscillations with position and with momentum, and large negative values, also seem unphysical, particularly considering their relatively large magnitude. Although the local state expansion also gives region of phase space with negative weight (Attard 2020), the magnitude of the density in such regions is quite small relative to the maximum magnitude, and so for that approach they either cancel or contribute negligibly to any phase space average.

One can obtain a feeling for the limitations of the pair-terminated many-body expansion by writing out the $N = 3$, one-dimensional case explicitly. In this case equation (11.8), in which the singlet terms are absent, yields

$$\begin{aligned}
\nabla \tilde{W}^{(3)} \cdot \nabla \tilde{W}^{(3)} &= \nabla_1 \tilde{w}_{12}^{(2)} \cdot \nabla_1 \tilde{w}_{12}^{(2)} + 2\nabla_1 \tilde{w}_{12}^{(2)} \cdot \nabla_1 \tilde{w}_{13}^{(2)} + \nabla_1 \tilde{w}_{13}^{(2)} \cdot \nabla_1 \tilde{w}_{13}^{(2)} \\
&\quad + \nabla_2 \tilde{w}_{21}^{(2)} \cdot \nabla_2 \tilde{w}_{21}^{(2)} + 2\nabla_2 \tilde{w}_{21}^{(2)} \cdot \nabla_2 \tilde{w}_{23}^{(2)} + \nabla_2 \tilde{w}_{23}^{(2)} \cdot \nabla_2 \tilde{w}_{23}^{(2)} \\
&\quad + \nabla_3 \tilde{w}_{31}^{(2)} \cdot \nabla_3 \tilde{w}_{31}^{(2)} + 2\nabla_3 \tilde{w}_{31}^{(2)} \cdot \nabla_3 \tilde{w}_{32}^{(2)} + \nabla_3 \tilde{w}_{32}^{(2)} \cdot \nabla_3 \tilde{w}_{32}^{(2)} \\
&\approx \nabla_1 \tilde{w}_{12}^{(2)} \cdot \nabla_1 \tilde{w}_{12}^{(2)} + \nabla_2 \tilde{w}_{21}^{(2)} \cdot \nabla_2 \tilde{w}_{21}^{(2)} + \nabla_2 \tilde{w}_{23}^{(2)} \cdot \nabla_2 \tilde{w}_{23}^{(2)} \\
&\quad + \nabla_3 \tilde{w}_{32}^{(2)} \cdot \nabla_3 \tilde{w}_{32}^{(2)} + 2\nabla_2 \tilde{w}_{21}^{(2)} \cdot \nabla_2 \tilde{w}_{23}^{(2)}.
\end{aligned} \tag{11.55}$$

The second approximation retains only the nearest neighbor contributions. The final term here, $2\nabla_2 \tilde{w}_{21}^{(2)} \cdot \nabla_2 \tilde{w}_{23}^{(2)}$, which is a three-body term, is neglected in equation (11.9). There is no reason to suppose that this term is smaller in magnitude than those that are retained. (The commutation function is complex; the retained terms are not in general positive.)

One can conclude from this that for more than two particles, the many-body expansion of the total commutation function terminated at the pair level is inconsistent with its temperature derivative based on the above approximation to the non-linear term. It would mean, for example, that the commutation function at one temperature, $W^{(N)}(\mathbf{q}, \mathbf{p}; \beta)$, formed using the series of singlet and pair commutation functions obtained by summing over energy states at that temperature, would not be equal to that obtained by integrating the total commutation function from a higher temperature, $W^{(N)}(\mathbf{q}, \mathbf{p}; \beta_0)$, using the approximation for the temperature derivative. Alternatively, the correct temperature derivative implies the existence of three-body terms, which contradicts the termination of the many-body expansion at the pair level. This inconsistency does not appear to be negligible, and is no doubt responsible for the unphysical results exhibited in the figures above.

This result, together with those in the figures and the unpublished simulations, suggest that the many-body expansion advocated in the text is not viable at the pair level. Perhaps the local state expansion, chapter 10, represents the most viable path forward for quantum statistical mechanics in classical phase space.

11.5 Ursell clusters, Lee–Yang theory, classical phase space

In this section the Ursell cluster functions of classical statistical mechanics are reviewed, subsection 11.5.1. An identity for the Maxwell–Boltzmann operator is given in subsection 11.5.2. This is used to obtain an exponential series expansion, subsection 11.5.3, which is subsequently resummed in terms of the binary collision kernel. These sections more or less follow Pathria (1972, sections 9.6 and 9.7). The exponential series is transformed to classical phase space in subsection 11.5.4.

11.5.1 Ursell cluster theory

Consider a general classical many-particle system. It is often the case that one has or desires the particle probability densities. For example, for an N particle system, the singlet probability density, $\wp(\mathbf{q}_1)$, is the probability of finding a particle at $\mathbf{q}_1$, (more precisely, within $d\mathbf{q}$ of $\mathbf{q}_1$). The pair probability density, $\wp(\mathbf{q}_1, \mathbf{q}_2)$, is the probability of finding a particle at $\mathbf{q}_1$ and a particle at $\mathbf{q}_2$. And the n-probability density, $\wp(\mathbf{q}^n)$, is the probability of finding n particles at the respective positions. In general the various particle probability densities are distinguished by the number of their arguments.

For the case of the pair probability density, the particles become uncorrelated at large separations, so that the pair probability is just the product of the singlet probabilities,

$$\wp(\mathbf{q}_1, \mathbf{q}_2) \to \wp(\mathbf{q}_1)\wp(\mathbf{q}_2), \quad |\mathbf{q}_1 - \mathbf{q}_2| \to \infty. \tag{11.56}$$

In practical terms large separations means beyond the range of the interaction potential, or distances at which the interaction potential is negligible compared to the temperature. Under such circumstances it is useful to extract the short-ranged, non-trivial part of the pair density, by defining

$$\wp_{\mathrm{U}}(\mathbf{q}_1, \mathbf{q}_2) = \wp(\mathbf{q}_1, \mathbf{q}_2) - \wp(\mathbf{q}_1)\wp(\mathbf{q}_2). \tag{11.57}$$

This obviously has about the same range as the pair potential and goes to zero at large separations. (Actually at large separations it is linearly proportional to the pair potential if in three dimensions the latter decays at least as fast as the separation cubed (Attard *et al* 1991, 1993).)

This basic idea underlies the cluster expansion developed by Ursell (1927). In this generalized approach the uncorrelated part of the multi-particle probability is subtracted, leaving the 'connected' part that manifests the direct correlations between particles. For the singlet case the two are the same,

$$\wp_{\mathrm{U}}(\mathbf{q}_1) = \wp(\mathbf{q}_1). \tag{11.58}$$

For the pair case one has

$$\wp_{\mathrm{U}}(\mathbf{q}_1, \mathbf{q}_2) = \wp(\mathbf{q}_1, \mathbf{q}_2) - \wp(\mathbf{q}_1)\wp(\mathbf{q}_2), \tag{11.59}$$

and for the triplet case one has

$$\begin{aligned} \wp_{\mathrm{U}}(\mathbf{q}_1, \mathbf{q}_2, \mathbf{q}_3) = \wp(\mathbf{q}_1, \mathbf{q}_2, \mathbf{q}_3) &- \wp(\mathbf{q}_1, \mathbf{q}_2)\wp(\mathbf{q}_3) \\ &- \wp(\mathbf{q}_2, \mathbf{q}_3)\wp(\mathbf{q}_1) - \wp(\mathbf{q}_3, \mathbf{q}_1)\wp(\mathbf{q}_2) \\ &+ 2\wp(\mathbf{q}_1)\wp(\mathbf{q}_2)\wp(\mathbf{q}_3). \end{aligned} \tag{11.60}$$

Notice how the coefficients of the terms on the right-hand side add to zero.

Consider a system where the N particles experience just singlet and pair potentials $U_N(\mathbf{q}) = \sum_{j=1}^{N} u(\mathbf{q}_j) + \sum_{j<k}^{N} u(\mathbf{q}_j, \mathbf{q}_k)$. The singlet and pair potentials are distinguished by the number of their arguments. The phase space weight is just the Maxwell Boltzmann factor of the classical Hamiltonian,

$$\Omega^{(N),\,\mathrm{cl}}(\mathbf{q}, \mathbf{p}) = e^{-\beta\mathcal{H}(\mathbf{q},\,\mathbf{p})} = \prod_{j=1}^{N} e^{-\beta\mathcal{K}(\mathbf{p}_j)+u(\mathbf{q}_j)} \prod_{k<\ell}^{N} e^{-\beta u(\mathbf{q}_k,\,\mathbf{q}_\ell)}, \tag{11.61}$$

normalizing factors being ignored. The superscript N has been added here so that systems with different numbers of particles can be treated next. Hopefully use of the same symbol Ω for the phase space weight, as here, and for the grand potential, as in chapters 3 and 7, will not cause confusion.

In the pair case in a system with $N = 2$ one has

$$\begin{aligned} \Omega^{(2),\,\mathrm{cl}}(\mathbf{q}_{1,2}, \mathbf{p}_{1,2}) &= e^{-\beta\mathcal{K}(\mathbf{p}_1)+u(\mathbf{q}_1)}e^{-\beta\mathcal{K}(\mathbf{p}_2)+u(\mathbf{q}_2)}e^{-\beta u(\mathbf{q}_1,\,\mathbf{q}_2)} \\ &\to e^{-\beta\mathcal{K}(\mathbf{p}_1)+u(\mathbf{q}_1)}e^{-\beta\mathcal{K}(\mathbf{p}_2)+u(\mathbf{q}_2)}, \quad |\mathbf{q}_1 - \mathbf{q}_2| \to \infty \\ &= \Omega^{(1),\,\mathrm{cl}}(\mathbf{q}_1, \mathbf{p}_1)\,\Omega^{(1),\,\mathrm{cl}}(\mathbf{q}_2, \mathbf{p}_2), \end{aligned} \tag{11.62}$$

since $u(\mathbf{q}_1, \mathbf{q}_2) \to 0$ in the large-separation limit. From this one sees that the connected part is explicitly

$$\Omega_{\mathrm{U}}^{(2),\mathrm{cl}}(\mathbf{q}_{1,2}, \mathbf{p}_{1,2}) = \Omega^{(1),\,\mathrm{cl}}(\mathbf{q}_1, \mathbf{p}_1)\,\Omega^{(1),\,\mathrm{cl}}(\mathbf{q}_2, \mathbf{p}_2)\{e^{-\beta u(\mathbf{q}_1,\,\mathbf{q}_2)} - 1\}. \tag{11.63}$$

The quantity in braces is the pair Mayer f-function. This explicit pair result for a system of $N = 2$ particles is obviously analogous to the pair result given above for general N.

For a quantum system of N particles experiencing singlet and pair potentials, the position matrix of the probability operator is

$$\begin{aligned}\Omega^{(N)}(\mathbf{q}', \mathbf{q}) &= \langle \mathbf{q}' | e^{-\beta \hat{\mathcal{H}}^{(N)}(\mathbf{r})} | \mathbf{q} \rangle \\ &= \sum_{\mathbf{n}} e^{-\beta E_{\mathbf{n}}^{(N)}} \phi_{\mathbf{n}}^{(N)}(\mathbf{q}') \phi_{\mathbf{n}}^{(N)}(\mathbf{q})^*. \end{aligned} \tag{11.64}$$

This invokes unsymmetrized wave functions, so that the effects of particle statistics have to be added later. The diagonal elements of this, which may be denoted $\Omega^{(N)}(\mathbf{q})$, give the probability of a position configuration of the N particles.

In the pair case, $N = 2$, one has

$$\begin{aligned}\hat{\mathcal{H}}^{(2)}(\mathbf{r}_{1,2}) &= \hat{\mathcal{K}}(\mathbf{r}_1) + \hat{\mathcal{K}}(\mathbf{r}_2) + u(\mathbf{r}_1) + u(\mathbf{r}_2) + u(\mathbf{r}_1, \mathbf{r}_2) \\ &\to \hat{\mathcal{H}}^{(1)}(\mathbf{r}_1) + \hat{\mathcal{H}}^{(1)}(\mathbf{r}_2), \quad |\mathbf{r}_1 - \mathbf{r}_2| \to \infty. \end{aligned} \tag{11.65}$$

This means that in the large-separation limit the Maxwell–Boltzmann operator factorizes

$$e^{-\beta \hat{\mathcal{H}}^{(2)}(\mathbf{r}_{1,2})} \to e^{-\beta \hat{\mathcal{H}}^{(1)}(\mathbf{r}_1)} e^{-\beta \hat{\mathcal{H}}^{(1)}(\mathbf{r}_2)}, \quad |\mathbf{r}_1 - \mathbf{r}_2| \to \infty. \tag{11.66}$$

One can argue that the same holds for the probability matrix,

$$\Omega^{(2)}(\mathbf{q}_{1,2}) \to \Omega^{(1)}(\mathbf{q}_1) \Omega^{(1)}(\mathbf{q}_2), \quad |\mathbf{q}_1 - \mathbf{q}_2| \to \infty. \tag{11.67}$$

The validity of this is clearest in the form $\Omega^{(2)}(\mathbf{q}_{1,2}) = \langle \mathbf{q}_{1,2} | e^{-\beta \hat{\mathcal{H}}^{(2)}(\mathbf{r}_{1,2})} | \mathbf{q}_{1,2} \rangle$, because here the position eigenfunctions concentrate the scalar product integral, $\delta(\mathbf{r}_1 - \mathbf{q}_1)\delta(\mathbf{r}_2 - \mathbf{q}_2)$, and the above factorization of the Maxwell–Boltzmann operator holds. In the form of the sum over energy states, the asymptotic limit is less clear. Nevertheless, one can argue that $\phi_{\mathbf{n}}^{(2)}(\mathbf{q}_{1,2}) \to \phi_{\mathbf{n}_1}^{(1)}(\mathbf{q}_1)\phi_{\mathbf{n}_2}^{(1)}(\mathbf{q}_2)$, $|\mathbf{q}_1 - \mathbf{q}_2| \to \infty$, and also that the quantized energy levels must be restricted to $E_{\mathbf{n}}^{(2)} = E_{\mathbf{n}_1}^{(1)} + E_{\mathbf{n}_2}^{(1)}$.

From this asymptotic limit the connected part of the position probability for two quantum particles is

$$\begin{aligned}\Omega_{\mathrm{U}}^{(2)}(\mathbf{q}_{1,2}) &= \Omega^{(2)}(\mathbf{q}_{1,2}) - \Omega^{(1)}(\mathbf{q}_1)\, \Omega^{(1)}(\mathbf{q}_2) \\ &= \langle \mathbf{q}_{1,2} | e^{-\beta \hat{\mathcal{H}}^{(2)}} | \mathbf{q}_{1,2} \rangle - \langle \mathbf{q}_1 | e^{-\beta \hat{\mathcal{H}}^{(1)}} | \mathbf{q}_1 \rangle \langle \mathbf{q}_2 | e^{-\beta \hat{\mathcal{H}}^{(1)}} | \mathbf{q}_2 \rangle. \end{aligned} \tag{11.68}$$

Finally, the classical phase space formulation of quantum statistical mechanics that is the subject of this book gives the phase space weight for a quantum system as

$$\Omega^{(N)}(\mathbf{q}, \mathbf{p}) = \frac{\langle \mathbf{q} | e^{-\beta \hat{\mathcal{H}}^{(N)}} | \mathbf{p} \rangle}{\langle \mathbf{q} | \mathbf{p} \rangle}. \tag{11.69}$$

We have also denoted this $e^{-\beta \mathcal{H}^{(N)}(\mathbf{q}, \mathbf{p})} \omega_{\mathrm{p}}(\mathbf{q}, \mathbf{p})$, as $e^{-\beta \mathcal{H}^{(N)}(\mathbf{q}, \mathbf{p})} e^{W_{\mathrm{p}}(\mathbf{q}, \mathbf{p})}$, and as $e^{\tilde{W}_{\mathrm{p}}(\mathbf{q}, \mathbf{p})}$. Again these are for unsymmetrized wave functions, leaving the symmetrization function to be included later.

The subject of this chapter is a many-body expansion of the commutation function. Terminated at the pair level it reads

$$W_{\rm p}(\mathbf{q}, \mathbf{p}) = \sum_{j=1}^{N} w(\mathbf{q}_j, \mathbf{p}_j) + \sum_{k<\ell}^{N} w(\mathbf{q}_k, \mathbf{p}_k; \mathbf{q}_\ell, \mathbf{p}_\ell). \tag{11.70}$$

Again the singlet and pair commutation functions may be distinguished by the number of their arguments. It is implicitly assumed that these are independent of N. As discussed in connection with equation (11.2), at large separations the gradient of the potential goes to zero faster than the potential itself, which means that $w(\mathbf{q}_{k,\ell}, \mathbf{p}_{k,\ell})/u(\mathbf{q}_{k,\ell}) \to 0$, $|\mathbf{q}_k - \mathbf{q}_\ell| \to \infty$. Hence for $N = 2$ particles one has

$$\begin{aligned} \Omega^{(2)}(\mathbf{q}_{1,2}, \mathbf{p}_{1,2}) &= e^{-\beta \mathcal{H}^{(2)}(\mathbf{q}, \mathbf{p})} e^{w(\mathbf{q}_1, \mathbf{p}_1) + w(\mathbf{q}_2, \mathbf{p}_2) + w(\mathbf{q}_{1,2}, \mathbf{p}_{1,2})} \\ &\to e^{-\beta[\mathcal{H}^{(1)}(\mathbf{q}_1, \mathbf{p}_1) + \mathcal{H}^{(1)}(\mathbf{q}_2, \mathbf{p}_2)]} e^{w(\mathbf{q}_1, \mathbf{p}_1) + w(\mathbf{q}_2, \mathbf{p}_2)}, \quad q_{12} \to \infty \\ &= \Omega^{(1)}(\mathbf{q}_1, \mathbf{p}_1) \Omega^{(1)}(\mathbf{q}_2, \mathbf{p}_2). \end{aligned} \tag{11.71}$$

This can also be seen directly from the fundamental definition

$$\begin{aligned} \Omega^{(2)}(\mathbf{q}_{1,2}, \mathbf{p}_{1,2}) &= \frac{\langle \mathbf{q}_{1,2} | e^{-\beta \hat{\mathcal{H}}^{(2)}(\mathbf{r}_{1,2})} | \mathbf{p}_{1,2} \rangle}{\langle \mathbf{q}_{1,2} | \mathbf{p}_{1,2} \rangle} \\ &\to \frac{\langle \mathbf{q}_{1,2} | e^{-\beta \hat{\mathcal{H}}^{(1)}(\mathbf{r}_1)} e^{-\beta \hat{\mathcal{H}}^{(1)}(\mathbf{r}_2)} | \mathbf{p}_{1,2} \rangle}{\langle \mathbf{q}_{1,2} | \mathbf{p}_{1,2} \rangle}, \quad q_{12} \to \infty \\ &= \Omega^{(1)}(\mathbf{q}_1, \mathbf{p}_1) \Omega^{(1)}(\mathbf{q}_2, \mathbf{p}_2). \end{aligned} \tag{11.72}$$

Again then the connected part of the pair weight is

$$\begin{aligned} &\Omega^{(2)}_{\rm U}(\mathbf{q}_{1,2}, \mathbf{p}_{1,2}) \\ &= \Omega^{(2)}(\mathbf{q}_{1,2}, \mathbf{p}_{1,2}) - \Omega^{(1)}(\mathbf{q}_1, \mathbf{p}_1)\, \Omega^{(1)}(\mathbf{q}_2, \mathbf{p}_2) \\ &= \frac{\langle \mathbf{q}_{1,2} | e^{-\beta \hat{\mathcal{H}}^{(2)}} | \mathbf{p}_{1,2} \rangle}{\langle \mathbf{q}_{1,2} | \mathbf{p}_{1,2} \rangle} - \frac{\langle \mathbf{q}_1 | e^{-\beta \hat{\mathcal{H}}^{(1)}} | \mathbf{p}_1 \rangle}{\langle \mathbf{q}_1 | \mathbf{p}_1 \rangle} \frac{\langle \mathbf{q}_2 | e^{-\beta \hat{\mathcal{H}}^{(1)}} | \mathbf{p}_2 \rangle}{\langle \mathbf{q}_2 | \mathbf{p}_2 \rangle}. \end{aligned} \tag{11.73}$$

As before, one expects that these pair results that have been derived for an $N = 2$ system will also hold for the pair probability for general N.

11.5.2 Identities for the Maxwell–Boltzmann operator

With the energy operator being the sum of the kinetic energy operator $\hat{\mathcal{K}}$ and the potential energy U, $\hat{\mathcal{H}}(\mathbf{r}) = \hat{\mathcal{K}}(\mathbf{r}) + U(\mathbf{r})$, the Maxwell–Boltzmann operator may be expanded in what is called an exponential series (Pathria 1972, section 9.7). One begins with the identity,

$$e^{-\beta \hat{\mathcal{H}}} = e^{-\beta \hat{\mathcal{K}}} - \int_0^\beta {\rm d}\beta' \, e^{-(\beta - \beta')\hat{\mathcal{K}}} U e^{-\beta' \hat{\mathcal{H}}}. \tag{11.74}$$

This can be proven by induction: assume that it is true at β. Then at $\beta + \Delta_\beta$ the left-hand side is

$$e^{-[\beta+\Delta_\beta]\hat{\mathcal{H}}} = e^{-\beta\hat{\mathcal{H}}} - \Delta_\beta \hat{\mathcal{H}} e^{-\beta\hat{\mathcal{H}}}, \tag{11.75}$$

and the right-hand side is

$$\begin{aligned} \text{RHS} &+ \Delta_\beta \frac{\partial \text{RHS}}{\partial \beta} \\ &= e^{-\beta\hat{\mathcal{H}}} - \Delta_\beta \hat{\mathcal{K}} e^{-\beta\hat{\mathcal{K}}} - \Delta_\beta U e^{-\beta\hat{\mathcal{H}}} + \Delta_\beta \hat{\mathcal{K}} \int_0^\beta \mathrm{d}\beta' \, e^{-(\beta-\beta')\hat{\mathcal{K}}} U e^{-\beta'\hat{\mathcal{H}}} \\ &= e^{-\beta\hat{\mathcal{H}}} - \Delta_\beta U e^{-\beta\hat{\mathcal{H}}} - \Delta_\beta \hat{\mathcal{K}} \, e^{-\beta\hat{\mathcal{H}}} \\ &= e^{-\beta\hat{\mathcal{H}}} - \Delta_\beta \hat{\mathcal{H}} e^{-\beta\hat{\mathcal{H}}}. \end{aligned} \tag{11.76}$$

The putative identity is obviously true as $\beta \to 0$, where both sides go like $1 - \beta\hat{\mathcal{H}} + \mathcal{O}(\beta^2)$. Hence the identity must be true for all β.

Incidently, since the kinetic and potential energies play equivalent roles in the Hamiltonian operator, an alternative identity is

$$e^{-\beta\hat{\mathcal{H}}} = e^{-\beta U} - \int_0^\beta \mathrm{d}\beta' \, e^{-(\beta-\beta')U} \hat{\mathcal{K}} e^{-\beta'\hat{\mathcal{H}}}. \tag{11.77}$$

We shall mainly focus on systems where the particles experience just singlet and pair potentials,

$$U^{(N)}(\mathbf{q}) \equiv U^{(1;N)}(\mathbf{q}) + U^{(2;N)}(\mathbf{q}) = \sum_{j=1}^{N} u(\mathbf{q}_j) + \sum_{j<k}^{N} u(\mathbf{q}_j, \mathbf{q}_k). \tag{11.78}$$

In this case it appears convenient to associate the two singlet operators (the kinetic energy and the singlet potential) and to write their exponential as

$$\hat{v}(\beta) \equiv e^{-\beta[\hat{\mathcal{K}} + U^{(1;N)}]} = \prod_{j=1}^{N} \hat{v}_j(\beta). \tag{11.79}$$

With this the first identity may be refashioned as

$$e^{-\beta\hat{\mathcal{H}}} = \hat{v}(\beta) - \int_0^\beta \mathrm{d}\beta' \, \hat{v}(\beta - \beta') \, U^{(2;N)} \, e^{-\beta'\hat{\mathcal{H}}}. \tag{11.80}$$

We have not shown the N-dependence of $\hat{\mathcal{H}}$ or of $\hat{v}(\beta)$ because it is relatively trivial and it is clear from the context.

11.5.3 Exponential series expansion

Modifying slightly Pathria (1972, section 9.7) for the case that the singlet potential is included with the kinetic energy operator, the exponential series for the

Maxwell–Boltzmann operator can be extracted from equation (11.80) by successive substitution.

$$e^{-\beta\hat{\mathcal{H}}} = \hat{v}(\beta) - \int_0^\beta \mathrm{d}\beta'\ \hat{v}(\beta - \beta')\ U^{(2;N)}\ \hat{v}(\beta') \\ + \int_0^\beta \mathrm{d}\beta'\ \hat{v}(\beta - \beta')\ U^{(2;N)} \int_0^{\beta'} \mathrm{d}\beta''\ \hat{v}(\beta' - \beta'')\ U^{(2;N)}\ \hat{v}(\beta'') - \ldots \tag{11.81}$$

Each additional term in this series carries an additional factor of $U^{(2;N)}$. Hence a few or several terms of the expansion suffice in the weak potential regime. Unfortunately, in the core region the interaction potential becomes large or infinite, which poses difficulties for the application of the series expansion.

It is convenient to represent these and the following results by diagrams. As mentioned, Kahn and Uhlenbeck (1938) introduced the classical cluster functions of Ursell (1927) into quantum mechanics. Lee and Yang (1959a, 1959b, 1960a, 1960b, 1960c) developed the diagrammatic form, including the binary kernel resummation. The following treatment modifies slightly that of Pathria (1972, section 9.7).

For just one particle, $N = 1$, the pair potential is absent, $U^{(2;1)} = 0$ and one has

$$e^{-\beta\hat{\mathcal{H}}_1} = \hat{v}_1(\beta) = e^{-\beta[(\hbar^2/2m)\nabla_1^2 + u_1]}, \tag{11.82}$$

where the subscript is the label of the particle, $\hat{v}_1(\beta) \equiv \hat{v}(\beta; \mathbf{r}_1)$. This may be denoted by a single vertical line,

$\hat{v}_1(\beta) =$  (11.83)

(This is a little loose as we intend to signify the temperature β' by a vertical position on the line.) The attached label 1 makes the identity of the particle clear.

For two particles, $N = 2$, the total pair potential is $U^{(2;2)} = u_{12}$, and the exponential series yields

$$\begin{aligned}\hat{\Omega}_{12}^{(2)}(\beta) &\equiv e^{-\beta\hat{\mathcal{H}}^{(2)}(\mathbf{r}_1,\mathbf{r}_2)} \\ &= \hat{v}_1(\beta)\hat{v}_2(\beta) - \int_0^\beta \mathrm{d}\beta'\ \hat{v}_1(\beta-\beta')\hat{v}_2(\beta-\beta')u_{12}\hat{v}_1(\beta')\hat{v}_2(\beta') \\ &\quad + \int_0^\beta \mathrm{d}\beta'\ \hat{v}_1(\beta-\beta')\hat{v}_2(\beta-\beta')u_{12} \\ &\quad \times \int_0^{\beta'} \mathrm{d}\beta''\ \hat{v}_1(\beta'-\beta'')\hat{v}_2(\beta'-\beta'')u_{12}\hat{v}_1(\beta'')\hat{v}_2(\beta'') + \ldots\end{aligned}$$

$=$ 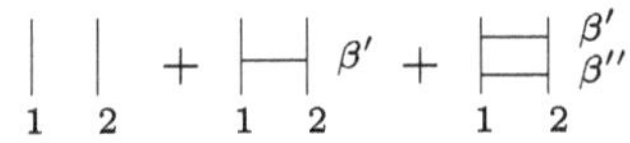 $+ \ldots$ (11.84)

One can identify the horizontal bar with the negative of the pair potential between the labeled particles. The height at which it is placed loosely represents the temperature dependence of the operator that acts upon it.

For $N = 3$, the total pair potential is $U^{(2;3)} = u_{12} + u_{23} + u_{31}$, and the exponential series is

$$\begin{aligned}\hat{\Omega}^{(3)}_{123}(\beta) &\equiv e^{-\beta\hat{\mathcal{H}}^{(3)}(\mathbf{r}_1,\mathbf{r}_2,\mathbf{r}_3)} \\ &= \hat{v}_1(\beta)\hat{v}_2(\beta)\hat{v}_3(\beta) - \int_0^\beta \mathrm{d}\beta'\, \hat{v}_1(\beta-\beta')\hat{v}_2(\beta-\beta')\hat{v}_3(\beta-\beta') \\ &\quad \times [u_{12}+u_{23}+u_{31}]\hat{v}_1(\beta')\hat{v}_2(\beta')\hat{v}_3(\beta') \\ &\quad + \int_0^\beta \mathrm{d}\beta' \int_0^{\beta'} \mathrm{d}\beta''\, \hat{v}_1(\beta-\beta')\hat{v}_2(\beta-\beta')\hat{v}_3(\beta-\beta') \\ &\quad \times [u_{12}+u_{23}+u_{31}]\hat{v}_1(\beta'-\beta'')\hat{v}_2(\beta'-\beta'')\hat{v}_3(\beta'-\beta'') \\ &\quad \times [u_{12}+u_{23}+u_{31}]\hat{v}_1(\beta'')\hat{v}_2(\beta'')\hat{v}_3(\beta'') - \ldots\end{aligned} \tag{11.85}$$

$=$ 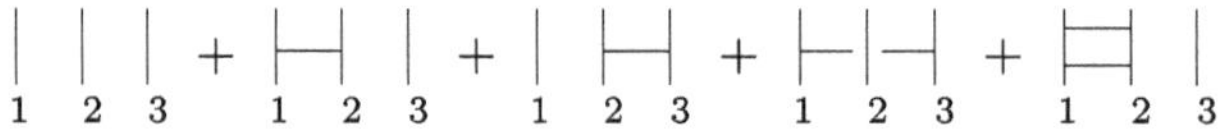 $+\ldots$

The disconnected diagrams commute. The pair potential plays the same role in this regard as it does in the above discussion of Ursell clusters. Therefore one can say that the Maxwell–Boltzmann operator $e^{-\beta\hat{\mathcal{H}}^{(N)}}$ is the sum of all diagrams with N particles and temperature β. And that the connected part of the Maxwell–Boltzmann operator is the sum of all connected diagrams with N particles and temperature β.

In section 11.5.1, the connected part of the Maxwell–Boltzmann operator for two particles was identified using the fact that the pair potential goes to zero at large separations. The full pair operator may be denoted $\hat{\Omega}^{(2)}_{12}(\beta) = e^{-\beta\hat{\mathcal{H}}^{(2)}(\mathbf{r}_1,\mathbf{r}_2)}$. The connected part is

$$\hat{\Omega}^{(2)}_{\mathrm{c},12}(\beta) = \hat{\Omega}^{(2)}_{12}(\beta) - \hat{v}_1(\beta)\hat{v}_2(\beta). \tag{11.86}$$

The binary kernel operator is defined as

$$\hat{B}_{12}(\beta) = -u_{12}\hat{\Omega}^{(2)}_{12}(\beta). \tag{11.87}$$

(Results analogous to those now derived follow from the commutated version of this, with the order of the two operators on the right-hand side swapped.) The two important points about the binary kernel are that it is connected (i.e., it has the same range as the pair potential), and that it is well-behaved in the core region (i.e., the pair Maxwell–Boltzmann operator can be expected to go to zero exponentially fast in the core, which swamps the power law divergence of the pair potential). This second point, amongst other things, means that the Fourier transform of the binary kernel phase function should exist.

Since the binary kernel operator goes to zero in the limit of large separations, it must be related to the connected part of the pair operator. In fact one has

$$\hat{B}_{12}(\beta) = \frac{\partial\hat{\Omega}^{(2)}_{\mathrm{c},12}(\beta)}{\partial\beta} + \left[\hat{\mathcal{H}}^{(1)}_1 + \hat{\mathcal{H}}^{(1)}_2\right]\hat{\Omega}^{(2)}_{\mathrm{c},12}(\beta). \tag{11.88}$$

In diagrammatic form the binary kernel operator is

$$\hat{B}_{12}(\beta) = -u_{12}\hat{\Omega}^{(2)}_{12}(\beta) =$$ 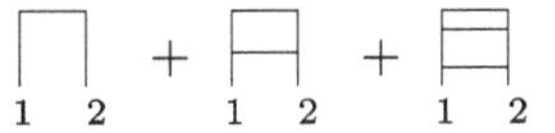 $$+\ldots$$

$$=$$ 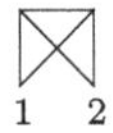 $$. \qquad (11.89)$$

The top horizontal bond represents the common factor u_{12}, and the X-bond represents the totality of diagrams with horizontal bonds at any height between 0 and β. Similarly we can write

$$\hat{B}_{12}(\beta' - \beta'') =$$ 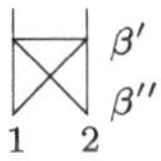 $$. \qquad (11.90)$$

The two-particle connected Maxwell–Boltzmann operator can be expressed as a temperature integral of the binary kernel,

$$\hat{\Omega}^{(2)}_{\mathrm{U},12}(\beta) = \int_0^\beta \mathrm{d}\beta'\, \hat{v}_1(\beta-\beta')\hat{v}_2(\beta-\beta')\hat{B}_{12}(\beta')$$

$$=$$ 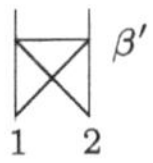 $$. \qquad (11.91)$$

This can be confirmed by differentiating with respect β and comparing the result with the second expression above for the binary kernel.

The three-particle connected Maxwell–Boltzmann operator with two bonds is given by the six connected terms of the nine terms in the total operator,

$$\begin{aligned}\hat{\Omega}^{(3)}_{\mathrm{U},123}(\beta) = &\int_0^\beta \mathrm{d}\beta' \int_0^{\beta'} \mathrm{d}\beta''\, \hat{v}_1(\beta-\beta')\hat{v}_2(\beta-\beta')\hat{v}_3(\beta-\beta') \\ &\times [u_{12}+u_{23}+u_{31}]\hat{v}_1(\beta'-\beta'')\hat{v}_2(\beta'-\beta'')\hat{v}_3(\beta'-\beta'') \\ &\times [u_{12}+u_{23}+u_{31}]\hat{v}_1(\beta'')\hat{v}_2(\beta'')\hat{v}_3(\beta'') \\ &- \hat{v}_3(\beta)\int_0^\beta \mathrm{d}\beta' \int_0^{\beta'} \mathrm{d}\beta''\, \hat{v}_1(\beta-\beta')\hat{v}_2(\beta-\beta') \\ &\times u_{12}\hat{v}_1(\beta'-\beta'')\hat{v}_2(\beta'-\beta'')u_{12}\hat{v}_1(\beta'')\hat{v}_2(\beta'') \\ &- 2 \text{ analogous terms} + \mathcal{O}(u_{jk}^3)\end{aligned}$$

$$=$$ 

$$+$$  $$, \qquad (11.92)$$

the particle labels being understood. The horizontal $u^{(2)}$-bonds may be resummed and replaced by $\hat{B}$-bonds: to leading order, $\hat{B}_{12}(\beta) = -u_{12}\hat{v}_1(\beta)\hat{v}_2(\beta)$. The first term given explicitly here, for example, transforms to

$$
\begin{aligned}
&= \int_0^{\beta} \mathrm{d}\beta' \int_0^{\beta'} \mathrm{d}\beta'' \, \hat{v}_1(\beta-\beta')\hat{v}_2(\beta-\beta')\hat{v}_3(\beta-\beta') \\
&\quad \times u_{12}\hat{v}_1(\beta'-\beta'')\hat{v}_2(\beta'-\beta'')\hat{v}_3(\beta'-\beta'') \\
&\quad \times u_{23}\hat{v}_1(\beta'')\hat{v}_2(\beta'')\hat{v}_3(\beta'') \\
&\Rightarrow \int_0^{\beta} \mathrm{d}\beta' \int_0^{\beta'} \mathrm{d}\beta'' \, \hat{v}_1(\beta-\beta')\hat{v}_2(\beta-\beta')\hat{v}_3(\beta-\beta'') \\
&\quad \times \hat{B}_{12}(\beta'-\beta'')\hat{B}_{23}(\beta'')\hat{v}_1(\beta'')
\end{aligned}
$$

$$=$$ 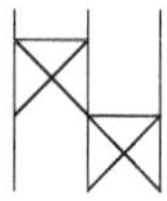 . (11.93)

Notice that the explicit temperature arguments for each particle add up to β. The six connected two-bond terms given above can be obtained from this by the 3! permutations of the labels. It is important that no uninterrupted (isolated) horizontal bonds between the same pair of particles remain, because these have already been counted in $\hat{B}_{jk}(\beta'')$: a diagram involving $\hat{B}_{jk}(\beta'-\beta'')\hat{B}_{\ell n}(\beta''-\beta''')\hat{B}_{jk}(\beta''')$ is allowed only if ℓ or $n = j$ or k and $\{\ell, n\} \neq \{j, k\}$.

11.5.4 Exponential expansion for the phase space weight

For a quantum system, the unsymmetrized part of the phase space weight is defined as

$$\Omega_{\mathrm{p}}(\mathbf{q}, \mathbf{p}) \equiv e^{-\beta\mathcal{H}(\mathbf{q},\mathbf{p})}\omega_{\mathrm{p}}(\mathbf{q}, \mathbf{p}) = \frac{\langle \mathbf{q}|e^{-\beta\hat{\mathcal{H}}}|\mathbf{p}\rangle}{\langle \mathbf{q}|\mathbf{p}\rangle} = e^{\mathbf{q}\cdot\mathbf{p}/i\hbar}e^{-\beta\hat{\mathcal{H}}}e^{-\mathbf{q}\cdot\mathbf{p}/i\hbar}. \tag{11.94}$$

The momentum eigenfunction appears here, $\hat{\mathbf{p}}e^{-\mathbf{q}\cdot\mathbf{p}/i\hbar} = \mathbf{p}e^{-\mathbf{q}\cdot\mathbf{p}/i\hbar}$.

Using the identity (11.80) this is

$$
\begin{aligned}
&\Omega_{\mathrm{p}}^{(N)}(\mathbf{q}, \mathbf{p}; \beta) \\
&= e^{\mathbf{q}\cdot\mathbf{p}/i\hbar}\hat{v}(\beta)e^{-\mathbf{q}\cdot\mathbf{p}/i\hbar} - e^{\mathbf{q}\cdot\mathbf{p}/i\hbar}\int_0^{\beta} \mathrm{d}\beta' \, \hat{v}(\beta-\beta') \, U^{(2;N)} \, e^{-\beta'\hat{\mathcal{H}}}e^{-\mathbf{q}\cdot\mathbf{p}/i\hbar} \\
&= e^{\mathbf{q}\cdot\mathbf{p}/i\hbar}\hat{v}(\beta)e^{-\mathbf{q}\cdot\mathbf{p}/i\hbar} - e^{\mathbf{q}\cdot\mathbf{p}/i\hbar}\int_0^{\beta} \mathrm{d}\beta' \, \hat{v}(\beta-\beta') \, U^{(2;N)} \, e^{-\mathbf{q}\cdot\mathbf{p}/i\hbar}\Omega_{\mathrm{p}}(\beta') \\
&= e^{\mathbf{q}\cdot\mathbf{p}/i\hbar}\hat{v}(\beta)e^{-\mathbf{q}\cdot\mathbf{p}/i\hbar} - e^{\mathbf{q}\cdot\mathbf{p}/i\hbar}\int_0^{\beta} \mathrm{d}\beta' \, \hat{v}(\beta-\beta') \, U^{(2;N)} \, \hat{v}(\beta')e^{-\mathbf{q}\cdot\mathbf{p}/i\hbar} \\
&\quad + e^{\mathbf{q}\cdot\mathbf{p}/i\hbar}\int_0^{\beta} \mathrm{d}\beta' \, \hat{v}(\beta-\beta')\int_0^{\beta'} \mathrm{d}\beta'' \, U^{(2;N)}\hat{v}(\beta'-\beta'')U^{(2;N)}\hat{v}(\beta'')e^{-\mathbf{q}\cdot\mathbf{p}/i\hbar} \\
&\quad - \ldots
\end{aligned}
\tag{11.95}
$$

Recall that $U^{(2;N)}(\mathbf{q}) = \sum_{j<k}^{N} u_{jk}$, where the pair potential is $u_{jk} = u(\mathbf{q}_j, \mathbf{q}_k)$.

If one defines

$$\tilde{U}^{(2;N)} \equiv e^{\mathbf{q}\cdot\mathbf{p}/i\hbar}U^{(2;N)}e^{-\mathbf{q}\cdot\mathbf{p}/i\hbar}, \quad \text{and} \quad \hat{\tilde{v}}(\beta) \equiv e^{\mathbf{q}\cdot\mathbf{p}/i\hbar}\hat{v}(\beta)e^{-\mathbf{q}\cdot\mathbf{p}/i\hbar}, \tag{11.96}$$

then this is more simply written as

$$\begin{aligned}
&\Omega_{\rm p}^{(N)}(\mathbf{q}, \mathbf{p}; \beta)\\
&\quad = \tilde{v}(\beta) - \int_0^\beta \mathrm{d}\beta'\, \hat{\tilde{v}}(\beta - \beta')\, \tilde{U}^{(2;N)}\, \Omega_{\rm p}(\beta')\\
&\quad = \tilde{v}(\beta) - \int_0^\beta \mathrm{d}\beta'\, \hat{\tilde{v}}(\beta - \beta')\, \tilde{U}^{(2;N)}\, \tilde{v}(\beta')\\
&\qquad + \int_0^\beta \mathrm{d}\beta'\, \hat{\tilde{v}}(\beta - \beta') \int_0^{\beta'} \mathrm{d}\beta''\, \tilde{U}^{(2;N)}\, \hat{\tilde{v}}(\beta' - \beta'')\, \tilde{U}^{(2;N)}\, \tilde{v}(\beta'')\\
&\qquad - \ldots
\end{aligned} \tag{11.97}$$

Note that $\tilde{v}(\beta) \equiv (e^{\mathbf{q}\cdot\mathbf{p}/i\hbar}\hat{v}(\beta)e^{-\mathbf{q}\cdot\mathbf{p}/i\hbar})$ denotes the phase space function rather than the operator.

From the binary kernel operator, equation (11.87), $\hat{B}_{12}(\beta) = -u_{12}\hat{\Omega}_{12}^{(2)}(\beta)$, the binary kernel operator in phase space may be defined as

$$\hat{\tilde{B}}_{12}(\beta) \equiv e^{\mathbf{q}_{1,2}\cdot\mathbf{p}_{1,2}/i\hbar}\hat{B}_{12}(\beta)e^{-\mathbf{q}_{1,2}\cdot\mathbf{p}_{1,2}/i\hbar}. \tag{11.98}$$

In general this is an operator. However in the event that it trails in an equation for a phase space function, it is to be understood to be a function. This is signified by the removal of the circumflex. To leading order this is

$$\begin{aligned}
\hat{\tilde{B}}_{12}(\beta) &\sim -\, e^{\mathbf{q}_{1,2}\cdot\mathbf{p}_{1,2}/i\hbar}u_{12}\hat{v}_{1,2}(\beta)e^{-\mathbf{q}_{1,2}\cdot\mathbf{p}_{1,2}/i\hbar} + \cdots\\
&= -\, u_{12}\hat{\tilde{v}}_{1,2}(\beta).
\end{aligned} \tag{11.99}$$

Hence we can carry out the same resummation as above to rewrite the unsymmetrized phase space weight as

$$\begin{aligned}
&\Omega_{\rm p}^{(N)}(\mathbf{q}, \mathbf{p}; \beta)\\
&\quad = \tilde{v}(\beta) + \sum_{j<k}^{N} \int_0^\beta \mathrm{d}\beta'\, \hat{\tilde{v}}(\beta - \beta')\, \tilde{B}_{jk}^{(2)}(\beta')\\
&\qquad + \sum_{j<k}^{N} \sum_{\ell<n}^{N}{}' \int_0^\beta \mathrm{d}\beta'\, \hat{\tilde{v}}(\beta - \beta') \int_0^{\beta'} \mathrm{d}\beta''\, \hat{\tilde{B}}_{jk}^{(2)}(\beta' - \beta'')\, \tilde{B}_{\ell n}^{(2)}(\beta'')\\
&\qquad + \cdots
\end{aligned} \tag{11.100}$$

The prime signifies that the term $\{\ell, n\} = \{j, k\}$ is excluded because this has already been counted in the binary kernel resummation.

The utility of this formulation in terms of the binary kernel, the exponential expansion, and also possibly Ursell clusters, is unclear.

References

Attard P, Bérard D R, Ursenbach C P and Patey G N 1991 The interaction free energy between planar walls in dense fluids. An Ornstein–Zernike approach with results for hard-sphere Lennard–Jones, and dipolar systems *Phys. Rev.* A **44** 8224

Attard P 1992 Pair hypernetted chain closure for fluids with three-body potentials. Results for argon with the Axilrod–Teller triple dipole potential *Phys. Rev.* A **45** 3659

Attard P 1993 Asymptotic analysis of primitive model electrolytes and the electrical double layer *Phys. Rev.* E **48** 3604

Attard P 2002 *Thermodynamics and Statistical Mechanics: Equilibrium by Entropy Maximisation* (London: Academic)

Attard P 2018a *Entropy Beyond the Second Law. Thermodynamics and Statistical Mechanics for Equilibrium, Non-Equilibrium, Classical, and Quantum Systems* (Bristol: IOP Publishing)

Attard P 2018b Algorithm for quantum many-particle systems by transformation to classical phase space with test results for quantum harmonic oscillators arXiv:1811.02032

Attard P 2020 Quantum statistical mechanics in classical phase space. V. Quantum local, average global arXiv:2005.06165

Hansen J P and McDonald I R 1986 *Theory of Simple Liquids* (London: Academic)

Harrison S F and Mayer J E 1938 The statistical mechanics of condensing systems. IV *J. Chem. Phys.* **6** 101

Kahn B and Uhlenbeck G E 1938 On the theory of condensation *Physica* **5** 399

Kirkwood J G 1933 Quantum statistics of almost classical particles *Phys. Rev.* **44** 31

Lee T D and Yang C N 1959a Many-body problem in quantum statistical mechanics. I. General formulation *Phys. Rev.* **113** 1165

Lee T D and Yang C N 1959b Many-body problem in quantum statistical mechanics. II *Phys. Rev.* **116** 25

Lee T D and Yang C N 1960a Many-body problem in quantum statistical mechanics. III. Zero-temperature limit for dilute hard spheres *Phys. Rev.* **117** 12

Lee T D and Yang C N 1960b Many-body problem in quantum statistical mechanics. IV. Formulation in terms of average occupation number in momentum space *Phys. Rev.* **117** 22

Lee T D and Yang C N 1960c Many-body problem in quantum statistical mechanics. V. Degenerate phase in Bose–Einstein condensation *Phys. Rev.* **117** 897

Mayer J E 1937a The statistical mechanics of condensing systems. I *J. Chem. Phys.* **5** 67

Mayer J E and Ackermann P G 1937b The statistical mechanics of condensing systems *J. Chem. Phys.* **5** 74

Mayer J E and Harrison S F 1938 The statistical mechanics of condensing systems. III *J. Chem. Phys.* **6** 87

Merzbacher E 1970 *Quantum Mechanics* 2nd edn (New York: Wiley)

Messiah A 1961 *Quantum Mechanics* vols 1 and 2 (North-Holland: Amsterdam)

Morita T and Hiroike K 1961 A new approach to the theory of classical fluids. III—General treatment of classical systems *Progr. Theor. Phys.* **25** 537

Pathria R K 1972 *Statistical Mechanics* (Oxford: Pergamon)

Stell G 1964 Cluster exansions for classical systems in equilibrium *The Equilibrium Theory of Classical Fluids II* ed H L Frisch and J L Lebowitz (New York: W A Benjamin) p 171

Ursell H D 1927 The evaluation of Gibbs phase-integral for imperfect gases *Proc. Camb. Phil. Soc.* **23** 685

Wigner E 1932 On the quantum correction for thermodynamic equilibrium *Phys. Rev.* **40** 749

IOP Publishing

Quantum Statistical Mechanics in Classical Phase Space

Phil Attard

Chapter 12

Density matrix and partition function

A first principles derivation of the density matrix is given. It is proven that the space of wave functions of an isolated system has uniform weight density, and that the states of a complete operator of an isolated system have equal weight. It is shown that conservation laws entangle otherwise independent wave functions, which leads to the collapse of the subsystem principle quantum states. Summing over the uniformly random phases of the subsystem degenerate states causes them to also collapse. The collapsed wave function represents the subsystem as a mixture of pure states and gives the von Neumann trace form for the partition function and for statistical averages. The Maxwell–Boltzmann form for the density operator results from the reservoir entropy. Accounting for wave function symmetrization constrains the trace to allowed unique subsystem states.

12.1 Introduction

This chapter derives the partition function of quantum statistical mechanics from first principles. Whereas previous chapters of the book have focused on the transformation of the partition function to classical phase space, the development of algorithms for computing the consequent phase functions, and the application to specific systems, the present chapter deals with the foundations of quantum statistical mechanics itself.

The von Neumann (1932) trace form for the partition function (and statistical averages) is arguably the central equation of quantum statistical mechanics. It gives the canonical equilibrium partition function as the trace of the density operator,

$$Z = \mathrm{TR}' \hat{\rho} = \sum_n{}' e^{-\beta \mathcal{H}_n}. \tag{12.1}$$

Two conceptual issues are raised by this expression: first, the trace treats the quantum system as a mixture of pure states, suppressing the superposition and interference effects that normally occur in an expectation value. And second, the density operator is identified with the Maxwell–Boltzmann operator, at least for a

doi:10.1088/978-0-7503-4055-7ch12

canonical equilibrium system. Starting from the postulates of quantum mechanics, this chapter derives both of these results. The presentation closely follows that of Attard (2018, chapter 7).

12.2 Quantum statistical average

In this section the basic notation is established, and the fundamental results of quantum statistical mechanics are set out, preparatory to their subsequent first principles derivation.

12.2.1 Expectation value

The state of a quantum system may be specified by a particular wave function ψ. We do not address symmetrization explicitly in this chapter, except for a brief discussion in section 12.4.6. A wave function will usually be denoted in the position representation, $\psi(\mathbf{r})$, or else in the bra–ket notation, $|\psi\rangle$ and $\langle\psi|$. The squared norm (or squared modulus) of the wave function is

$$N(\psi) \equiv \langle\psi|\psi\rangle = \int \mathrm{d}\mathbf{r}\ \psi(\mathbf{r})^*\ \psi(\mathbf{r}), \tag{12.2}$$

where the asterisk denotes the complex conjugate. It is usually the case that the wave function is normalized, $N(\psi) = 1$.

The expectation value of an operator $\hat{O}$ when the system is in the wave state ψ is (Messiah 1961, Merzbacher 1970)

$$O(\psi) = \frac{1}{N(\psi)}\langle\psi|\hat{O}|\psi\rangle = \frac{1}{N(\psi)}\int \mathrm{d}\mathbf{r}\ \psi(\mathbf{r})^*\hat{O}(\mathbf{r})\psi(\mathbf{r}). \tag{12.3}$$

The system can also be described as being in a particular quantum state $n = 0, 1, 2, \ldots$, which are the eigenstates of some operator. The corresponding eigenfunctions ζ_n are orthonormal, $\langle\zeta_n|\zeta_m\rangle = \delta_{n,m}$, and they form a complete basis for the Hilbert space. Hence the wave function can be written

$$\psi(\mathbf{r}) = \sum_n \psi_n\zeta_n(\mathbf{r}), \quad \psi_n = \langle\zeta_n|\psi\rangle. \tag{12.4}$$

An arbitrary operator can also be represented in this basis, $O_{mn} = \langle\zeta_m|\hat{O}|\zeta_n\rangle$, and its expectation value then takes the form (Messiah 1961, Merzbacher 1970)

$$O(\psi) = \frac{1}{N(\psi)}\sum_{m,n} \psi_m^* O_{mn}\psi_n. \tag{12.5}$$

Suppose $\zeta_n^{\mathrm{O}}(\mathbf{r})$ is a normalized eigenfunction of the operator with eigenvalue O_n, $\hat{O}\zeta_n^{\mathrm{O}} = O_n\zeta_n^{\mathrm{O}}$. The eigenstates may possibly be degenerate, but nevertheless the eigenfunctions can be organized to form a complete set, $\psi = \sum_n \psi_n^{\mathrm{O}}\zeta_n^{\mathrm{O}}$. In the eigenfunction basis the operator is diagonal, $O_{mn}^{\mathrm{O}} = O_n\delta_{mn}$, and the expectation value in the wave state ψ is

$$O(\psi) = \frac{1}{N(\psi)} \sum_n \psi_n^{\mathrm{O}*} \psi_n^{\mathrm{O}} O_n. \tag{12.6}$$

The quantity $\psi_n^{\mathrm{O}*}\psi_n^{\mathrm{O}}/N(\psi)$ has the interpretation as the probability of the system being in the quantum state n (given that it is in the wave state ψ).

In this form, the eigenvalues O_n are the possible outcomes of a measurement with the operator $\hat{O}$, and the square of the amplitude $\psi_n^{\mathrm{O}*}\psi_n^{\mathrm{O}}$ is proportional to the probability that that particular outcome will occur. In the more general form for the expectation value as the double sum in an arbitrary basis, equation (12.5), the complex numbers $\psi_m^*\psi_n$ cannot be interpreted in terms of classical probability theory because the latter is predicated upon real, non-negative weights (Attard 2002). The general form exhibits the phenomenon of quantum superposition and interference: the quantum system can be in more than one state at a time, and the different states effect each other.

The wave function collapses upon measurement, $\psi \Rightarrow \zeta_n^{\mathrm{O}}$, which is to say that the system no longer exists in a superposition of states but rather becomes a pure state. The wave function could be said to be decoherent in that there is no interference between states. In general a decoherent system is a mixture of pure states, and its expectation value is formally identical to a classical average, equation (12.6). The specifically quantum aspects of the system (interference, superposition) are no longer present.

One may use the bra–ket notation for the wave function to construct the (single-wave function) density operator,

$$\hat{\rho} \equiv \frac{1}{N(\psi)} |\psi\rangle \langle\psi|. \tag{12.7}$$

With this the expectation value may be written in the form of the trace of the operator product,

$$O(\psi) = \mathrm{TR}\, \hat{\rho}\, \hat{O} = \frac{1}{N(\psi)} \sum_{m,n} \psi_n \psi_m^* O_{mn}. \tag{12.8}$$

The single-wave function density operator is not like other quantum operators; due to its dyadic form, it cannot be diagonalized. The proof is by contradiction. Consider some representation in which it is diagonal. If $\rho_{n_1n_1} = |\psi_{n_1}|^2$ and $\rho_{n_2n_2} = |\psi_{n_2}|^2$ are two distinct non-zero diagonal elements, then the corresponding off-diagonal elements, $\rho_{n_1n_2} = \rho_{n_2n_1}^* = \psi_{n_1}^*\psi_{n_2}$, would also be non-zero. This contradiction proves that the single-wave function density operator cannot be diagonal in any representation. The exception to this is if it has precisely one non-zero element, as only occurs upon collapse due to measurement.

12.2.2 Statistical average

In the context of the statistical average in quantum statistical mechanics, the non-diagonal nature of the single-wave function density operator is directly relevant.

This is because conventionally the statistical average is expressed as a trace over the density operator (von Neumann 1932). We shall return to this matter shortly, but for the present let us assume the existence of a probability operator, $\hat{\wp}$. Let us further assume that the statistical average can be expressed as the trace of it and the observable operator

$$\begin{aligned}\langle \hat{O} \rangle_{\text{stat}} &= \text{TR}\, \hat{\wp}\, \hat{O} \\ &= \sum_{m,n} \wp_{mn}\, O_{nm} \\ &= \sum_{n} \wp^{\text{O}}_{nn}\, O^{\text{O}}_{nn} \\ &= \sum_{n} \wp^{\text{S}}_{nn}\, O^{\text{S}}_{nn}.\end{aligned} \tag{12.9}$$

In the second last equality the eigenstates of the observable operator have been used. This form only requires the diagonal elements of the probability operator, whether or not the latter is diagonalizable. This is identical to the classical probability form for a statistical average, namely the sum over states of the operator eigenvalue, $O^{\text{O}}_{nn} = O_n$, times the probability of the state, $\wp^{\text{O}}_{nn}$. From the point of view of classical probability this is rather appealing, but it does raise the question of how it arises from the quantum expectation value form that reflects a superposition of states and interference between states.

The final equality for the statistical average is also problematic. Here the eigenstates of the probability operator are invoked. Although not essential for the present discussion, and anticipating somewhat results to come, these can be assumed to be entropy eigenstates and denoted by the superscript S, $\hat{\wp}|\zeta^{\text{S}}_n\rangle = \wp_n|\zeta^{\text{S}}_n\rangle$, and $\wp^{\text{S}}_{mn} = \wp_n \delta_{m,n}$. Again this is apparently dissonant with the expectation value for a wave state, where the density operator cannot be diagonalized. This demands a more careful examination of the origin of the trace form for the statistical average and of the properties of the probability operator. One can already see that any such diagonalizable probability operator cannot be a single-wave function density operator.

The challenge is to get from the expectation value form to the statistical average form. In the former, the density operator always includes off-diagonal contributions (except in the observable operator representation itself). In the latter the off-diagonal contributions vanish in both the entropy and the observable operator representations. The qualitative difference between these two expressions can be seen in figure 12.1.

12.2.2.1 Ensemble image

The conventional view of quantum statistical mechanics is in terms of the density operator rather than the probability operator (von Neumann 1932, Messiah 1961, Merzbacher 1970). Because the density operator of a single-wave function cannot be diagonalized, most workers instead imagine an ensemble of systems, labeled

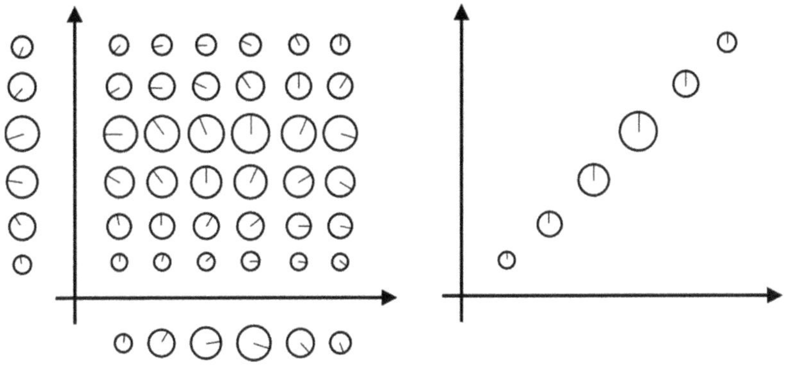

Figure 12.1. Density operator for a superposition of states (left) and for a mixture of pure states (right). The amplitude and phase of the wave function is represented on the axes. Reproduced from Attard (2018).

$a = 1, 2, \ldots, M$, each with its own wave function ψ_a. The ensemble average of the single-wave function density operators defines the many-wave function density operator,

$$\hat{\rho} \equiv \frac{1}{M}\sum_{a=1}^{M} \frac{|\psi_a\rangle\,\langle\psi_a|}{\langle\psi_a|\psi_a\rangle}. \tag{12.10}$$

(Usually the wave functions are normalized, $\langle\psi_a|\psi_a\rangle = 1$, and the denominator is redundant.) This can also be called the averaged density operator. This does not factorize into a dyadic product, which is the crucial difference from the single-wave function density operator.

There is no reason to suppose that this cannot be diagonalized. Hence the many-wave function density operator can be equated to the probability operator introduced above. In the representation in which it is diagonal (presumably the entropy representation), the entries equal the state probabilities,

$$\bar{\rho}_{mn}^{\mathrm{S}} = \frac{1}{M}\sum_{a=1}^{M} \frac{\psi_{a,\,m}^{\mathrm{S}*}\psi_{a,\,n}^{\mathrm{S}}}{\langle\psi_a|\psi_a\rangle} = \wp_n\delta_{m,n}. \tag{12.11}$$

One can see that the off-diagonal terms average to zero if the phase angles of the coefficients are uncorrelated (decoherent). The diagonal terms all have zero phase angle exactly and so they cannot cancel with each other. This shows that the crucial difference between the one- and many-wave function density operators is this averaging process.

In view of this, in so far as one or other exists, the averaged density operator is equal to the probability operator, $\hat{\rho} = \hat{\wp}$. This operator relationship holds in any representation, not just the entropy representation in which both were said to be diagonal. If the probability operator is an ordinary operator, then it can only be diagonal in the basis formed from the eigenfunctions of the entropy operator (or of any operator that commutes with the entropy operator). This would mean that the off-diagonal phase cancelation just discussed for the averaged density operator only occurred in the entropy basis and not in any other basis. It remains to prove this, either for the averaged density operator in the ensemble picture, or else for the probability operator directly.

With this definition of the many-wave function density operator, and its equality with the probability operator, the statistical average can be written as the trace of it and the observable operator,

$$\langle \hat{O} \rangle_{\text{stat}} = \text{TR}\, \hat{\wp}\, \hat{O} = \frac{1}{M} \sum_{a=1}^{M} \frac{\langle \psi_a | \hat{O} | \psi_a \rangle}{\langle \psi_a | \psi_a \rangle}. \tag{12.12}$$

The conventional presentation of the ensemble picture is that the wave functions ψ_a are each a pure quantum state, in which case the ensemble comprises a mixture of pure quantum states. This means that each system in the ensemble has a fully collapsed wave function. The above discussion departs slightly from this in that it is sufficient for equation (12.11) to hold, which is to say that the phases of the entropy states are decoherent. In other words we need not insist that each member of the ensemble be in a pure quantum state.

12.2.2.2 Imaginary ensemble

The ensemble picture has been criticized in classical statistical mechanics and more generally (Attard 2018, section 1.4). The problem is that an ensemble of systems is nothing but a mental image that has no physical realization. As such ensembles cannot provide the mathematical or physical basis for the development of statistical mechanics, whether classical or quantum. In the real world the application of statistical mechanics is to a single system in each case, and any justification for the equations of statistical mechanics must therefore be based on the properties of such single systems.

Since the many-wave particle density operator is a direct manifestation of the ensemble, it is similarly unsatisfactory. The problem is that such a collection of independent replica systems has no physical reality. The fact of the matter is that an actual measurement is made on a single system. This is not to say that the trace form of the quantum statistical average (with a probability operator equivalent to an averaged density operator) is incorrect; in fact this trace form has long been successfully used. Rather it says that the explanation and justification for it does not lie in the ensemble picture. And of course, without the ensemble picture there is no need or basis for the many-wave function density operator.

Instead of an ensemble average one might consider that the decoherent system arises from a time average for the density operator. Let us imagine that a measurement extends over a brief time interval, τ. In this period the wave function evolves, $\psi(t)$, and one can define the averaged density operator as

$$\hat{\wp} \equiv \frac{1}{\tau N_0} \int_0^{\tau} \mathrm{d}t\, |\psi(t)\rangle\, \langle \psi(t)|, \tag{12.13}$$

assuming unitary evolution of the wave function, $N(\psi(t)) \equiv \langle \psi(t)|\psi(t)\rangle = N_0$. Again assuming that this is the same as the probability operator, $\hat{\bar{\rho}} = \hat{\wp}$, it must be diagonal in the entropy representation,

$$\bar{\rho}_{ln}^{\mathrm{S}} = \frac{1}{\tau N_0} \int_0^\tau \mathrm{d}t\, \psi_l^{\mathrm{S}*}(t)\psi_n^{\mathrm{S}}(t) = \wp_n \delta_{l,n}. \tag{12.14}$$

Presumably, the phase factors of the entropy modes evolve independently, which makes them effectively random. This means that the off-diagonal elements average to zero over the measurement time.

It is not clear that this putative temporal averaging over the measurement time is any more realistic than the ensemble averaging. Both explanations have the appearance of *post facto* rationalization. Under the circumstances it would perhaps be more honest not to pretend to justify equation (12.9), but rather simply to assert it as the definition of the statistical average, taking the existence of the probability operator as a given. In my opinion it is better not to introduce or to invoke the averaged density operator as it has no physical reality.

In the following two sections, a derivation from first principles is given of the von Neumann trace form for the statistical average equation (12.9). The derivation gives the explicit form for the probability operator and shows that it is diagonal in the entropy representation. It also demonstrates the passage from the wave function and expectation value formulation of quantum mechanics to the statistical average formulation as a weighted sum over quantum states.

12.3 Uniform weight density of wave space

In classical statistical mechanics the weight density of points in phase space of an isolated system is uniform. That the weight density is uniform is essential to the derivation of, for example, the Maxwell–Boltzmann probability distribution for the canonical equilibrium system. The proof of uniformity is based on the incompressibility of a classical Hamiltonian trajectory, and the equivalence of phase space averages and time averages (Attard 2002, section 5.1).

For the present quantum problem one needs to prove a similar fundamental uniformity for an isolated system, which may be achieved in an analogous fashion to the classical case. In section 12.3.1, it is simply asserted that the quantum states of an isolated system are equally probable. In section 12.3.2, it is shown that the probability density is constant on a trajectory. And in section 12.3.3, it is proven that the probability density is uniform in wave space based on time uniformity and energy and norm hypersurface density.

12.3.1 Equal state probability hypothesis

The simplest presentation of the probability operator focuses on the quantum states, and, as in the classical case, it begins with the microstates of an isolated system. From their weight the required quantum state probabilities follow, first for an isolated system, and thence for a subsystem interacting with a reservoir. By analogy with the ergodic hypothesis of classical statistical mechanics, one can take it as an axiom that

$$\text{all quantum states of an isolated system have equal weight.} \tag{12.15}$$

As will be seen, the quantum states are the eigenstates of any complete operator. These are the diagonal elements in any representation.

This hypothesis is justified by mathematical derivation from more formal considerations in the two following subsections; it appears as equation (12.40) below. That analysis is more fundamental than the above assertion regarding the equality of the weights of quantum states. The analysis that follows in sections 12.3.2 and 12.3.3 is based on the Hilbert space of the wave functions of an isolated system. The aim is to show that the wave space probability density for an isolated system is uniform.

12.3.2 Trajectory uniformity

Consider an isolated system with wave function ψ. Schrödinger's equation for the time rate of change of the latter is

$$i\hbar\dot{\psi}^0 = \hat{\mathcal{H}}\psi. \tag{12.16}$$

Here $\hat{\mathcal{H}}$ is the Hamiltonian or energy operator. This is here taken to be independent of time. The superscript 0 is used for adiabatic quantities, which are those of the isolated system. In this section only the total isolated system appears.

One can construct a trajectory by integrating this

$$\psi(t) \equiv \psi^0(t|\psi_0, t_0) = \hat{U}^0(t - t_0)\psi_0. \tag{12.17}$$

Here the adiabatic time propagator is $\hat{U}^0(t) = e^{t\hat{\mathcal{H}}/i\hbar}$. For a time-dependent Hamiltonian, this is $\hat{U}^0(t - t_0) = e^{\int_{t_0}^{t} \mathrm{d}t' \, \hat{\mathcal{H}}(t')/i\hbar}$, with time-ordering assumed (Merzbacher 1970 section 18.7).

Each point ψ in wave space has a real, non-negative weight density $\omega(\psi, t)$. As usual, the normalized form of this is the probability density $\wp(\psi, t)$. In section 12.4, these will be related to the probability operator and to the statistical average.

As for any density, $\mathrm{d}\psi \; \wp(\psi, t)$ is the probability of the system being within $|\mathrm{d}\psi|$ of ψ. This is a real non-negative number: $\mathrm{d}\psi \; \wp(\psi, t) = |\,\mathrm{d}\psi \; \wp(\psi, t)| = |\,\mathrm{d}\psi\,|\,|\wp(\psi, t)|$. Since it is always the product that occurs, without loss of generality one may take each to be individually real.

The probability density is normalized

$$\int \mathrm{d}\psi \; \wp(\psi, t) = \int \mathrm{d}\underline{\psi} \; \wp(\underline{\psi}, t) = 1. \tag{12.18}$$

Here has been invoked an arbitrary representation, $\underline{\psi} = \{\psi_n\}$, $n = 1, 2, \ldots$ The coefficients are complex, $\psi_n = \psi_n^{\mathrm{r}} + \mathrm{i}\psi_n^{\mathrm{i}}$, and the infinitesimal volume element can be written as

$$d\underline{\psi} = d\underline{\psi}^{r}\, d\underline{\psi}^{i} \equiv d\psi_1^{r}\, d\psi_1^{i}\, d\psi_2^{r}\, d\psi_2^{i} \ldots \tag{12.19}$$

With this all the integrations are over the real line, $\psi_n^{r} \in [-\infty, \infty]$, and $\psi_n^{i} \in [-\infty, \infty]$, $n = 1, 2, \ldots$.

As is the case for Hamilton's classical equations of motion, Schrödinger's equation gives an incompressible trajectory,

$$\begin{aligned} \frac{d\dot{\underline{\psi}}^{0}}{d\underline{\psi}} &= \underline{\partial}_{\psi^{r}} \cdot \dot{\underline{\psi}}^{0,\,r} + \underline{\partial}_{\psi^{i}} \cdot \dot{\underline{\psi}}^{0,\,i} \\ &= \underline{\partial}_{\psi} \cdot \dot{\underline{\psi}}^{0} + \underline{\partial}_{\psi^{*}} \cdot \dot{\underline{\psi}}^{0*} \\ &= \frac{1}{i\hbar}\mathrm{TR}\, \underline{\underline{\mathcal{H}}} - \frac{1}{i\hbar}\mathrm{TR}\, \underline{\underline{\mathcal{H}}} \\ &= 0. \end{aligned} \tag{12.20}$$

The first two equalities are in fact the general expression for the trajectory compressibility. They hold as well for the total time derivative (for the subsystem wave function, with the subscript 0 dropped) when the subsystem is open to a reservoir, heat bath, or environment. The final two equalities only hold when the evolution is given by Schrödinger's equation.

As in classical phase space, the compressibility gives the logarithmic rate of change of the volume element. Since the compressibility vanishes, the volume element is a constant of the motion of the isolated system,

$$d\psi^{0}(t) = d\psi_0. \tag{12.21}$$

The total time derivative of the probability density on the adiabatic trajectory is

$$\begin{aligned} \frac{d^{0}\wp(\psi, t)}{dt} &= \frac{\partial\wp(\psi, t)}{\partial t} + \dot{\underline{\psi}}^{0,\,r} \cdot \underline{\partial}_{\psi^{r}}\wp(\psi, t) + \dot{\underline{\psi}}^{0,\,i} \cdot \underline{\partial}_{\psi^{i}}\wp(\psi, t) \\ &= \frac{\partial\wp(\psi, t)}{\partial t} + \dot{\underline{\psi}}^{0} \cdot \underline{\partial}_{\psi}\wp(\psi, t) + \dot{\underline{\psi}}^{0*} \cdot \underline{\partial}_{\psi^{*}}\wp(\psi, t) \\ &= \frac{\partial\wp(\psi, t)}{\partial t} + \underline{\partial}_{\psi} \cdot \left[\dot{\underline{\psi}}^{0}\wp(\psi, t)\right] + \underline{\partial}_{\psi^{*}} \cdot \left[\dot{\underline{\psi}}^{0*}\wp(\psi, t)\right]. \end{aligned} \tag{12.22}$$

The first and second equalities are just the general definition of the total derivative. The final equality is valid on an adiabatic trajectory, upon which the compressibility vanishes. The probability flux can be seen to be $J_{\wp}(\psi, t) \equiv \dot{\psi}\wp(\psi, t)$.

For the isolated system, the probability density evolves adiabatically,

$$\wp(\psi_1, t_1) = \int d\psi_0\, \wp(\psi_0, t_0)\, \delta(\psi_1 - \psi^{0}(t_1|\psi_0, t_0)). \tag{12.23}$$

This invokes Bayes' theorem for the transition probability, $\wp(\psi_1, t_1; \psi_0, t_0) = \wp(\psi_1, t_1|\psi_0, t_0)\, \wp(\psi_0, t_0)$, together with the type 1 reduction condition (Attard 2018, equations (3.14) and (3.17)). The conditional transition probability in the present adiabatic case is $\wp(\psi_1, t_1|\psi_0, t_0) = \delta(\psi_1 - \psi^0(t_1|\psi_0, t_0))$, which reflects the deterministic nature of Schrödinger's equation.

From the incompressibility of wave space under Schrödinger's equation, one has that $d\psi_0 = d\psi_1$. With $t_1 = t_0 + \Delta_t$, expanding to linear order in the time step yields

$$\begin{aligned} \wp(\psi_1, t_1) &= \int d\psi_0\, \wp(\psi_0, t_0)\, \delta(\psi_1 - \psi^0(t_1|\psi_0, t_0)) \\ &= \wp(\psi_1 - \Delta_t \dot{\psi}^0, t_0) \\ &= \wp\big(\psi_1, t_0\big) - \Delta_t \underline{\dot{\psi}}^0 \cdot \underline{\partial}_\psi \wp\big(\psi, t_0\big) - \Delta_t \underline{\dot{\psi}}^{0*} \cdot \underline{\partial}_{\psi^*} \wp\big(\psi, t_0\big). \end{aligned} \tag{12.24}$$

Rearranging, this gives the partial time derivative as

$$\frac{\partial \wp(\psi, t)}{\partial t} = -\underline{\dot{\psi}}^0 \cdot \underline{\partial}_\psi \wp(\psi, t) - \underline{\dot{\psi}}^{0*} \cdot \underline{\partial}_{\psi^*} \wp(\psi, t). \tag{12.25}$$

This holds for an isolated system evolving under Schrödinger's equation.

Combining this with the second equality of equation (12.22), the total adiabatic time derivative of the probability density vanishes, $d^0\wp(\psi, t)/dt = 0$. Hence for an isolated system, the probability density is a constant of the motion,

$$\wp(\psi(t), t) = \wp(\psi_0, t_0), \tag{12.26}$$

where $\psi(t) \equiv \psi^0(t|\psi_0, t_0)$.

This is the quantum analog of the classical result that the total time derivative of the probability density in the classical phase space of an isolated system vanishes under Hamilton's equations of motion (Attard 2018, equation (5.20)). This is not surprising as the physical picture underlying the original classical result is the same as that which holds for the present quantum result. Like Hamilton's equations, Schrödinger's equation is deterministic, which means that trajectories do not cross, and they are neither created nor destroyed. In both cases trajectories within a given volume remain inside throughout its evolution. In so far as the probability of the volume is proportional to the number of trajectories it contains, this is conserved by both Hamilton's and Schrödinger's equations. What the two equations additionally have in common is that they are incompressible, which means that the volume is also a constant of the motion. These two facts mean that the probability density must be conserved moving along the trajectory.

For the present equilibrium system, the Hamiltonian operator does not depend upon time. In this case the probability density must be independent of time, $\wp(\psi)$. One therefore has the stronger result

$$\wp(\psi^0(t|\psi_0, t_0)) = \wp(\psi_0). \tag{12.27}$$

This says that the probability density is the same everywhere on a trajectory.

Finally, by analogy with the classical ergodic hypothesis, one may suppose that a single trajectory passes sufficiently close to all relevant points of the state space. The hypothesis says that one can take any two points in state space, ψ_1 and ψ_2 with the same norm and energy expectation, to lie on a single trajectory. (It is well-known that Schrödinger's equation conserves the norm and energy of the wave function during its evolution—see below.) By the above, they therefore have the same probability density

$$\wp(\psi_2) = \wp(\psi_1), \text{ if } E(\psi_2) = E(\psi_1) \text{ and } N(\psi_2) = N(\psi_1). \tag{12.28}$$

Hence the wave space probability density must be of the form

$$\wp(\psi|E, N) = \frac{\delta(E(\psi) - E)\,\delta(N(\psi) - N)}{W(E, N)}, \tag{12.29}$$

with the normalizing factor being

$$W(E, N) = \int \mathrm{d}\psi \; \delta\Big(E(\psi) - E\Big)\,\delta\Big(N(\psi) - N\Big). \tag{12.30}$$

The problem now is to transform from the conditional wave space probability, $\wp(\psi|E, N)$, to the unconditional wave space probability, $\wp(\psi)$. To do this one has to make some hypothesis about the dependence of the wave space weight density on the hypersurface, $W(E, N)$. If one assumes that this is constant, or at worse has negligible variation with energy and norm, then one can conclude that an isolated system has uniform weight in wave space,

$$\wp(\psi) = \frac{1}{W}, \quad W = \int \mathrm{d}\psi. \tag{12.31}$$

Obviously it is a little unsatisfactory to simply assume this. It is also unsatisfactory to assume that a single trajectory completely fills each hypersurface. Now an alternative derivation of the result is given, which avoids these assumptions. The approach is the direct analog of the classical derivations (Attard 2002, section 5.1.3, Attard 2018, section 5.2.3).

12.3.3 Time average and hypersurface density

Schrödinger's equations conserves energy and norm. These are well-known results, which are straightforward to prove directly.

The norm squared of the wave function is $N(\psi) \equiv \langle\psi|\psi\rangle$. Its adiabatic rate of change is

$$\begin{aligned} \dot{N}^0(\psi) &= \langle\dot{\psi}^0|\psi\rangle + \langle\psi|\dot{\psi}^0\rangle \\ &= \left\langle \frac{1}{\mathrm{i}\hbar}\hat{\mathcal{H}}\psi \middle| \psi \right\rangle + \left\langle \psi \middle| \frac{1}{\mathrm{i}\hbar}\hat{\mathcal{H}}\psi \right\rangle \\ &= \frac{-1}{\mathrm{i}\hbar}\langle\psi|\hat{\mathcal{H}}^\dagger|\psi\rangle + \frac{1}{\mathrm{i}\hbar}\langle\psi|\hat{\mathcal{H}}|\psi\rangle \\ &= 0. \end{aligned} \tag{12.32}$$

The final equality follows because the Hamiltonian operator, like all physical operators, is Hermitian conjugate, $\hat{\mathcal{H}}^{\dagger} = \hat{\mathcal{H}}$. In fact the Hamiltonian operator obeys a stronger symmetry: it is both real, $\hat{\mathcal{H}}^{*} = \hat{\mathcal{H}}$, and symmetric, $\hat{\mathcal{H}}^{\mathrm{T}} = \hat{\mathcal{H}}$.

The energy of the wave state ψ is $E(\psi) \equiv \langle\psi|\hat{\mathcal{H}}|\psi\rangle/N(\psi)$. Hence by a similar argument it is also a constant of the adiabatic motion,

$$\begin{aligned}\dot{E}^{0}(\psi) &= \frac{1}{N(\psi)}\langle\dot{\psi}^{0}|\hat{\mathcal{H}}|\psi\rangle + \frac{1}{N(\psi)}\langle\psi|\hat{\mathcal{H}}|\dot{\psi}^{0}\rangle \\ &= \frac{1}{N(\psi)}\langle\frac{1}{\mathrm{i}\hbar}\hat{\mathcal{H}}\psi|\hat{\mathcal{H}}|\psi\rangle + \frac{1}{N(\psi)}\langle\psi|\hat{\mathcal{H}}|\frac{1}{\mathrm{i}\hbar}\hat{\mathcal{H}}\psi\rangle \\ &= \frac{-1}{\mathrm{i}\hbar N(\psi)}\langle\psi|\hat{\mathcal{H}}^{\dagger}\hat{\mathcal{H}}|\psi\rangle + \frac{1}{\mathrm{i}\hbar N(\psi)}\langle\psi|\hat{\mathcal{H}}\hat{\mathcal{H}}|\psi\rangle \\ &= 0.\end{aligned} \tag{12.33}$$

These results show that the trajectory of an isolated system, $\psi(t)$, is confined to a hypersurface of constant norm and energy. In view of this let $\tilde{\psi}$ denote a normalized wave function. Let χ denote a wave function on a hypersurface of constant energy E and norm squared N. This hypersurface can be labeled $\{N, E\}$, and it is obviously a subspace of the Hilbert space of all wave functions. In this notation the normalized wave function is $\tilde{\psi} = \tilde{\psi}(\chi, E)$, and the full wave function is $\psi = \psi(\tilde{\psi}, N) = \psi(\chi, E, N)$.

Ultimately what is required is the weight density of the full wave space of the isolated system, $\omega(\psi)$. Initially the weight density on the hypersurface, $\omega(\chi|N, E)$, is derived. This is then combined with the density of the $\{N, E\}$-hypersurface in the wave space to obtain the full weight density.

It is taken as axiomatic that time is homogeneous, which means that a statistical average is a simple time average over a trajectory. This implies that the weight density on the $\{N, E\}$-hypersurface must be inversely proportional to the speed of the trajectory,

$$\begin{aligned}\omega(\chi|N, E) &\propto |\dot{\psi}^{0}|^{-1} \\ &= \langle\psi|\hat{\mathcal{H}}\hat{\mathcal{H}}|\psi\rangle^{-1/2},\end{aligned} \tag{12.34}$$

since the Hamiltonian operator is Hermitian. Notice that this depends upon ψ. This is just the time that the system spends in a volume element $|\mathrm{d}\chi|$, which is the same as the time spent in the volume element of the full wave space $|\mathrm{d}\psi|$. The proportionality factor is an immaterial constant on the hypersurface. This constant cannot vary with energy E or with norm squared N because this would violate the axiom that a statistical average is a simple time average over the trajectory irrespective of energy or norm squared.

Now we transform this hypersurface weight density to the full wave space. The Jacobean for the transformation $\chi \Rightarrow \tilde{\psi}$ is

$$| \tilde{\nabla} E | = \left[\frac{\partial E(\psi)}{\partial |\tilde{\psi}\rangle} \frac{\partial E(\psi)}{\partial \langle\tilde{\psi}|} \right]^{1/2} = \langle\tilde{\psi}|\hat{\mathcal{H}}\hat{\mathcal{H}}|\tilde{\psi}\rangle^{1/2}, \tag{12.35}$$

and that for the transformation $\tilde{\psi} \Rightarrow \psi$,

$$| \nabla N | = \left[\frac{\partial N(\psi)}{\partial |\psi\rangle} \frac{\partial N(\psi)}{\partial \langle\psi|} \right]^{1/2} = \langle\psi|\psi\rangle^{1/2}. \tag{12.36}$$

With these the full weight density is

$$\begin{aligned} \omega(\psi|N, E) &= \omega(\chi|N, E) \frac{|\tilde{\nabla} E|}{\Delta_E} \frac{|\nabla N|}{\Delta_N}, \quad |N(\psi) - N| < \Delta_N, \quad |E(\psi) - E| < \Delta_E \\ &\propto \frac{\langle\tilde{\psi}|\hat{\mathcal{H}}\hat{\mathcal{H}}|\tilde{\psi}\rangle^{1/2} \langle\psi|\psi\rangle^{1/2}}{\langle\psi|\hat{\mathcal{H}}\hat{\mathcal{H}}|\psi\rangle^{1/2}} \delta(N(\psi) - N)\, \delta(E(\psi) - E) \\ &= \delta(N(\psi) - N)\, \delta(E(\psi) - E). \end{aligned} \tag{12.37}$$

Integrating this over all energies and norm squared yields the unconditional weight density for the wave space of the isolated system,

$$\omega(\psi) = 1. \tag{12.38}$$

The interpretation of the origin of this result is essentially the same as that for classical phase space (Attard 2002, section 5.1.3, Attard 2018, section 5.2.3). The weight density is inversely proportional to the speed of the trajectory (i.e., low speed equates to more time in a volume element), and proportional to the hypersurface density, (i.e., for fixed spacing between discrete hypersurfaces, Δ_E and Δ_N, large gradients correspond to more hypersurfaces per unit wave space volume). These ideas are depicted in in figure 12.2.

Just as in the classical case for Hamilton's equations of motion, it is a remarkable consequence of Schrödinger's equation that the speed is identical to the magnitude of the gradient of the hypersurface, so that these two cancel to give a

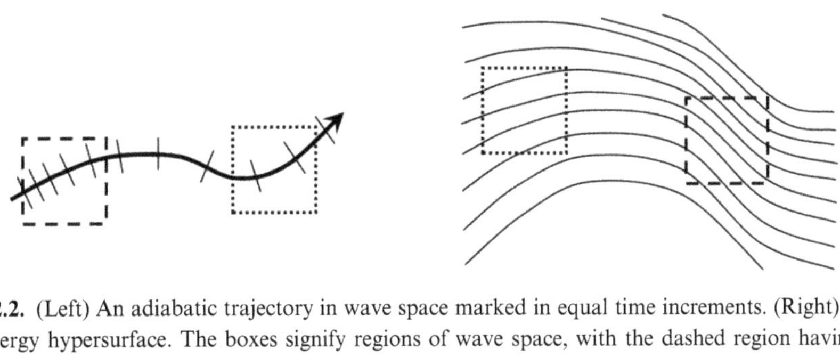

Figure 12.2. (Left) An adiabatic trajectory in wave space marked in equal time increments. (Right) Contours of the energy hypersurface. The boxes signify regions of wave space, with the dashed region having greater weight than the dotted region in each case, all other things being equal. Reproduced from Attard (2018).

weight density that is uniform in wave space. This result is the same as that given above as equation (12.31). It has a more satisfactory basis in that the derivation proves rather than assumes that the hypersurface weight $w(E, N)$ is independent of energy or norm.

This weight density is a physical observable, and as such there must be a corresponding weight operator whose expectation value it is. Hence one must have

$$\omega(\psi) \equiv \frac{\langle\psi|\hat{\omega}|\psi\rangle}{\langle\psi|\psi\rangle} = 1. \tag{12.39}$$

Since this must hold for all wave states, the weight operator for the isolated system must be the identity operator, and one must have

$$\hat{\omega} = \hat{\mathrm{I}}, \quad \text{or} \quad \omega_{mn} = \delta_{m,n}. \tag{12.40}$$

The result holds in any representation.

One conclusion that follows from this is that the quantum microstates of the system are not unique. Rather any complete operator gives a collective of microstates. Also a particular set of microstates is the set of diagonal elements of a particular representation. Equivalently, they are the eigenstates of any complete operator. This result implies that these microstates all have equal weight, which is the axiom given as equation (12.15) at the start of this section.

A second conclusion is that the off-diagonal elements in any representation have zero weight. It should be clear from this that the weight operator is of no use for the expectation value, since the off-diagonal elements of the single-wave function density matrix, the dyadic $\psi_m^*\psi_n$, $m \neq n$, certainly contribute to the expectation value. Nevertheless, it will be shown that for two interacting systems, the expectation value of an operator on one system has to be weighted by the states of the other system, and for this a weight operator is required (albeit one derived from the weight operator of the present isolated system).

12.4 Canonical equilibrium system

Now the canonical equilibrium system is analyzed in detail. As usual, the total isolated system consists of a subsystem and a thermal reservoir that can exchange energy with each other. In the literature the reservoir is often called the bath, or, particularly in the quantum field, the environment. Also the phrase 'the system' is often used to mean what is here called 'the subsystem'. Here, and less pedantically throughout the book, the terms used are total system, subsystem, and reservoir. These are denoted by superscripts or subscripts tot, s, and r where any ambiguity exists.

12.4.1 Entropy of energy states

As described by Attard (2002, section 1.3, 2018, section 1.1.2), the classical theory of probability is formulated in terms of microstates and macrostates, and these now need to be defined for a quantum system. A microstate, which is labeled by a single lower case Roman letter, usually corresponds to an eigenstate of a complete set of commuting operators. In this case the eigenvalue is unique. However, it can also correspond to an eigenstate of an incomplete operator. In such a case the microstates are degenerate and microstates with different labels can share the same eigenvalue of the incomplete operator. A macrostate, which is labeled by a lower case Greek letter, corresponds to the principal quantum number of an operator. Each macrostate has a unique label and corresponds to a unique eigenvalue of the operator. A Greek and Roman letter paired together is used to signify a microstate, with the Greek letter labeling the principal quantum number, and the Roman letter labeling the degenerate quantum states.

For example, writing the eigenfunctions with a single microstate label, one has $\hat{O}|\zeta_n^{\rm O}\rangle = O_n|\zeta_n^{\rm O}\rangle$. In this case different values of n may yield the same eigenvalue (assuming that $\hat{O}$ is incomplete operator). Alternatively, the eigenfunctions may be written as the combination of principal and degenerate state labels, $\hat{O}|\zeta_{\alpha g}^{\rm O}\rangle = O_\alpha|\zeta_{\alpha g}^{\rm O}\rangle$. In this case different values of α necessarily yield different eigenvalues.

For the present canonical equilibrium system, it is energy that is exchanged between the subsystem and the reservoir. Hence the focus is here on energy states. Since the microstates of an isolated system must have equal weight, equation (12.40), the weight of the energy microstates is $w_{\alpha g}^{\rm E} = 1$, since without loss of generality this can be set to unity. This means that the energy microstates of the total system have no internal entropy, $S_{\alpha g}^{\rm E} = k_{\rm B} \ln w_{\alpha g}^{\rm E} = 0$. (This does not take into account wave function symmetrization, which gives an internal entropy for the microstates, e.g., equations (2.25), (3.6), and (5.57).)

In the general formulation of classical probability (Attard 2002, section 1.3, Attard 2018, section 1.1.2), the weight of a macrostate is the sum of the weights of the microstates contained within it. In the present case the energy macrostate weight is

$$w_\alpha^{\rm E} = \sum_g^{({\rm E},\,\alpha)} w_{\alpha g}^{\rm E} \equiv N_\alpha^{\rm E}. \tag{12.41}$$

This is just the number of degenerate states with energy E_α. Hence the entropy of an energy macrostate is the logarithm of the degeneracy,

$$S_\alpha^{\rm E} = k_{\rm B} \ln N_\alpha^{\rm E}. \tag{12.42}$$

Analogous to the analysis for classical thermodynamics (Attard 2002, section 2.2.1, 2018, section 2.2.1), the entropy of the reservoir when the subsystem is in a particular energy microstate will be required. Since energy is conserved one must have

$$E^{\rm tot} = E^{\rm s} + E^{\rm r}. \tag{12.43}$$

As mentioned, tot means the total system, s means the subsystem, and r means the reservoir. The interaction energy between the subsystem and the reservoir is included in the reservoir energy. In the thermodynamic limit (i.e., macroscopic systems) the interaction energy is relatively negligible.

Now the reservoir entropy can be written in two different ways, which will be equated to each other below. The first expression directly invokes the energy degeneracy discussed above,

$$S^{\mathrm{r}}(E_{\alpha}^{\mathrm{r}}) = k_{\mathrm{B}} \ln N_{\alpha}^{\mathrm{Er}}, \text{ with } N_{\alpha}^{\mathrm{Er}} \equiv \sum_{g\in\alpha}{}^{(\mathrm{Er})}. \tag{12.44}$$

This is the entropy of the isolated reservoir in this energy macrostate.

The second expression for the reservoir entropy invokes the definition of the thermodynamic temperature, $T^{-1} \equiv \partial S(E)/\partial E$. By definition the reservoir is much larger than the subsystem, and so the subsystem energy may be used as an expansion variable. The Taylor expansion yields

$$S^{\mathrm{r}}(E^{\mathrm{r}}) = S^{\mathrm{r}}(E^{\mathrm{tot}} - E^{\mathrm{s}}) = \text{const.} - \frac{E^{\mathrm{s}}}{T^{\mathrm{r}}}. \tag{12.45}$$

The higher order terms are negligible in the thermodynamic limit. This gives the reservoir entropy as a function of the subsystem energy. Henceforth the superscript on the reservoir temperature will be dropped as it is the only temperature that will appear.

12.4.2 Wave function entanglement

For the total system, the wave function lies in the Hilbert space that comprises that of the subsystem and the reservoir. It is assumed that the subsystem and the reservoir interact so weakly that they may be treated as quasi-independent. This means that for any state of one all permitted states of the other are available. The total number of states is essentially the product of the two individual totals, subject to the conservation laws. Also the basis functions for the total system are the product of the basis functions of the subsystem and of the reservoir, each considered as isolated.

With $\{|\zeta_n^{\mathrm{s}}\rangle\}$ an orthonormal basis for the subsystem and $\{|\zeta_m^{\mathrm{r}}\rangle\}$ an orthonormal basis for the reservoir, the most general wave state of the total system can be expanded in terms of these basis functions,

$$|\psi_{\mathrm{tot}}\rangle = \sum_{n,m} c_{nm}|\zeta_n^{\mathrm{s}}, \zeta_m^{\mathrm{r}}\rangle. \tag{12.46}$$

If the coefficient matrix is dyadic, $c_{nm} = c_n^{\mathrm{s}} c_m^{\mathrm{r}}$, then the wave state is separable, $|\psi_{\mathrm{tot}}\rangle = |\psi_{\mathrm{s}}, \psi_{\mathrm{r}}\rangle$ with $|\psi_{\mathrm{s}}\rangle = \sum_n^{(\mathrm{s})} c_n^{\mathrm{s}}|\zeta_n^{\mathrm{s}}\rangle$, and $|\psi_{\mathrm{r}}\rangle = \sum_m^{(\mathrm{r})} c_m^{\mathrm{r}}|\zeta_m^{\mathrm{r}}\rangle$. Alternatively, if the coefficients are not dyadic, then the wave state is inseparable, which is called an entangled state (Messiah 1961, Merzbacher 1970).

In general entangled states arise from the conservation laws. This can be seen explicitly for the present case of energy exchange, in which the total energy is fixed, $E^{\mathrm{tot}} = E^{\mathrm{s}} + E^{\mathrm{r}}$. (Recall that the interaction energy is included in the reservoir energy.) Using the respective energy eigenfunctions as a basis for the subsystem and the reservoir, the most general expansion of the total wave function is

$$|\psi_{\mathrm{tot}}\rangle = \sum_{\alpha g, \beta h} c_{\alpha g, \beta h} |\zeta^{\mathrm{Es}}_{\alpha g}, \zeta^{\mathrm{Er}}_{\beta h}\rangle. \tag{12.47}$$

From energy conservation one must have that

$$c_{\alpha g, \beta h} = 0 \text{ if } E^{\mathrm{s}}_{\alpha} + E^{\mathrm{r}}_{\beta} \neq E_{\mathrm{tot}}. \tag{12.48}$$

This condition cannot be satisfied for a system able to exchange energy if the coefficient is in dyadic from. To see this, suppose the contrary, that $c_{\alpha g, \beta h} = c^{\mathrm{s}}_{\alpha g} c^{\mathrm{r}}_{\beta h}$. Denote one of the occupied subsystem energy macrostates by, α_1, so that $c^{\mathrm{s}}_{\alpha_1 g} \neq 0$. Because the macrostate labels are distinct, there can be only one non-zero reservoir macrostate, say β_1, that conserves energy, $E^{\mathrm{s}}_{\alpha_1} + E^{\mathrm{r}}_{\beta_1} = E_{\mathrm{tot}}$. Hence one must have that $c^{\mathrm{r}}_{\beta h} = 0$, $\beta \neq \beta_1$. But a unique occupied reservoir energy macrostate implies that α_1 must also be unique, which is to say that there can be no other occupied subsystem energy macrostate.

This shows that if the total wave function is in product form, then the subsystem and reservoir each can only have a single fixed energy. This in turn means that neither wave function, ψ_{s} or ψ_{r}, can contain a superposition of principal energy states (because these are non-degenerate). It follows from this that the subsystem is unable to change energy by exchange with the reservoir, because during such an exchange it would be in a superposition of the initial and final energy states. Therefore a dyadic form for the expansion coefficient corresponds to the subsystem and the reservoir being isolated from each other. But this contradicts the basis of a canonical equilibrium system, namely that the energy of the subsystem fluctuates over time as it exchanges energy with the reservoir. One concludes from this contradiction that if the subsystem and the reservoir can exchange energy, then the total wave function must be entangled.

Accordingly, for the present canonical equilibrium system, the total wave function cannot be factorized. Instead it must have a representation of the form

$$|\psi_{\mathrm{tot}}\rangle = \sum_{\alpha, g, h} c_{\alpha g, h} |\zeta^{\mathrm{Es}}_{\alpha g}, \zeta^{\mathrm{Er}}_{\beta_\alpha h}\rangle. \tag{12.49}$$

Here the reservoir principal energy label β_α depends upon the subsystem principal energy α and is defined implicitly by $E^{\mathrm{s}}_{\alpha} + E^{\mathrm{r}}_{\beta_\alpha} = E_{\mathrm{tot}}$. For brevity, the coefficients are written, $c_{\alpha g, h} \equiv c^{\mathrm{s}}_{\alpha g} c^{\mathrm{r}}_{\beta_\alpha h}$. Because β_α depends upon α, this is entangled (i.e., the sum of the products is not equal to the product of the sums).

The most well-known case of entanglement is that of particles with spin that result from radioactive decay (figure 12.3). It is the conservation of spin that is responsible for the entanglement. Measurement of the spin of either daughter particle collapses the superposed wave function into one or other of the two wave functions that form it. If it were not for spin conservation, the daughter wave function would be the superposition of four possible wave functions, which are the independent (i.e., unentangled) product of two spinors for each particle (figure 12.4). It is spin conservation that entangles the two daughter particles and that reduces the superposition basis set from four to two possibilities.

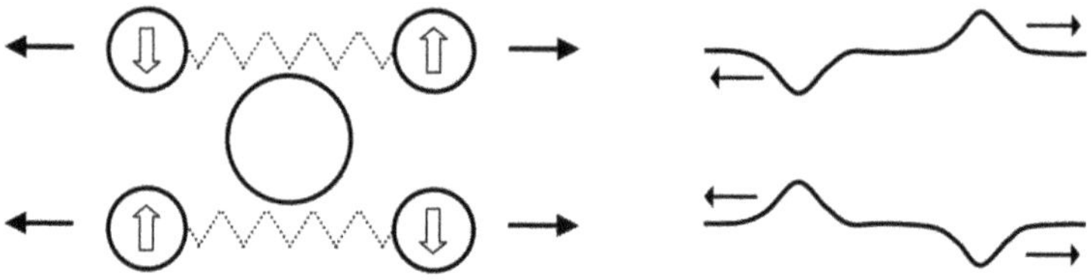

Figure 12.3. The two possible decays (left, upper and lower) of a spin 0 particle (left, center) into two spin half particles. The two wave functions (right) that are superposed for the final state each represents the entanglement of two single particle wave functions. Reproduced from Attard (2018).

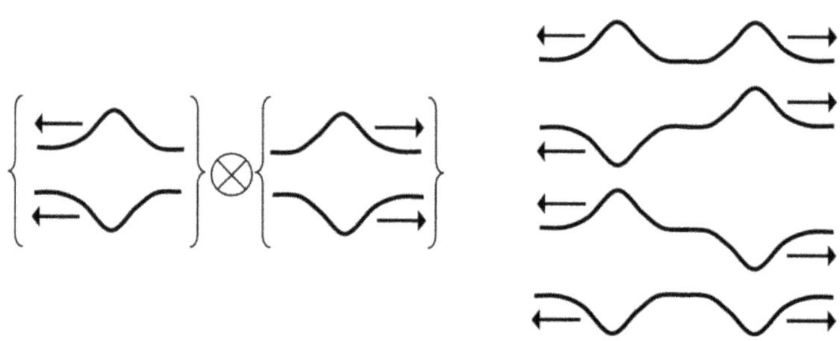

Figure 12.4. The dyadic product of two spin half particle wave functions (left), and the four wave functions this gives for superposition, each representing two unentangled single particle wave functions (right). Reproduced from Attard (2018).

It should be clear in this example that the entangled and unentangled wave functions are very different. For the entangled case, a measurement of the product of the spins of the two daughter particles yields $\langle\psi|\hat{\sigma}_l\hat{\sigma}_r|\psi\rangle = -1$. The same measurement on the unentangled wave function yields $\langle\psi|\hat{\sigma}_l\hat{\sigma}_r|\psi\rangle = 0$. This example is but one of many that could be offered, since quite generally entanglement has far-reaching consequences.

12.4.3 Expectation values and wave function collapse

For the present canonical equilibrium system, conservation of energy causes the entanglement of the subsystem and reservoir wave functions. It will now be shown that this causes the principal energy states of the subsystem to collapse, which is to say that a subsystem expectation value is diagonal in these. This is quite significant for quantum statistical mechanics in view of the discussion in section 12.2 that the single-wave function density operator cannot be diagonalized. In contrast, if a probability operator existed in the usual sense of an observable operator, then it can be diagonalized. Therefore the collapse of the principal energy states due to entanglement that will now be established already goes a long way to proving the existence of the probability operator.

In view of the expansion for the wave function of the total system, equation (12.49), and assuming a normalized wave function, $\langle\psi_{\rm tot}|\psi_{\rm tot}\rangle = 1$, the expectation value of an operator on the subsystem is

$$\begin{aligned}
\langle \psi_{\rm tot} | \hat{O}^{\rm s} | \psi_{\rm tot} \rangle &= \sum_{\alpha' g'; h'} \sum_{\alpha g; h} c^*_{\alpha' g', h'} \, c_{\alpha g, h} \left\langle \zeta^{\rm Es}_{\alpha' g'}, \zeta^{\rm Er}_{\beta_{\alpha'} h'} \right| \hat{O}^{\rm s} \left| \zeta^{\rm Es}_{\alpha g}, \zeta^{\rm Er}_{\beta_\alpha h} \right\rangle \\
&= \sum_{\alpha' g'; h'} \sum_{\alpha g; h} c^*_{\alpha' g', h'} \, c_{\alpha g, h} \left\langle \zeta^{\rm Es}_{\alpha' g'} \middle| \hat{O}^{\rm s} \middle| \zeta^{\rm Es}_{\alpha g} \right\rangle \left\langle \zeta^{\rm Er}_{\beta_{\alpha'} h'} \middle| \zeta^{\rm Er}_{\beta_\alpha h} \right\rangle \\
&= \sum_{\alpha, g, g'} \sum_{h \in \beta_\alpha}^{(\rm Er)} c^*_{\alpha g', h} \, c_{\alpha g, h} \left\langle \zeta^{\rm Es}_{\alpha g'} \middle| \hat{O}^{\rm s} \middle| \zeta^{\rm Es}_{\alpha g} \right\rangle \\
&= \sum_{\alpha, g, g'} \sum_{h \in \beta_\alpha}^{(\rm Er)} c^*_{\alpha g', h} \, c_{\alpha g, h} \, O^{\rm s, E}_{\alpha g', \alpha g}.
\end{aligned} \tag{12.50}$$

The third equality follows because the relationship between α and β_α is bijective (the macrostates are non-degenerate), $\langle \zeta^{\rm Er}_{\beta_{\alpha'} h'} | \zeta^{\rm Er}_{\beta_\alpha h} \rangle = \delta_{\alpha', \alpha} \, \delta_{h', h}$. Only the principal diagonal elements of the subsystem matrix operator in the energy representation contribute to this. Equivalently, the principal diagonal elements of the density matrix. The projection onto the subsystem in this case has the effect of converting a superposition of the subsystem basis states into a mixture of pure subsystem basis states, as arises from a collapsed wave function.

In this result energy conservation and entanglement have reduced and collapsed the original four sums over independent principal energy states into a single sum over principal energy states, from α', α, β', β then to two sums, α', α, and then to a single sum, α. The degenerate energy states of the reservoir have also collapsed, since the orthogonality of the reservoir basis functions have reduced the sums over h and h' to a single sum over h. Specifically, whereas for an isolated system the non-diagonal entries of the operator matrix in the energy representation would contribute to the expectation value, when energy exchange with a reservoir occurs, orthogonality of the reservoir energy basis functions together with energy conservation eliminate the off-diagonal subsystem energy terms. The principal energy states have at this stage collapsed, whilst the degenerate energy states of the subsystem remain in superposition form.

Here collapse refers to the collapse of the wave function and to the collapse into principal energy states. Collapse means that the number of possible states has been reduced, and also that the interference between superposition states has been reduced by the elimination of off-diagonal energy terms.

12.4.3.1 Random phase approach

The microstates of the total isolated system have equal weight. This can be taken as an hypothesis, equation (12.15), or as a theorem that follows from the uniform density of wave space, equation (12.40). In view of this, the coefficients in the representation of the total wave function all have the same magnitude, which can be set to unity, $|c_{\alpha g, h}| = 1$. (With this choice, the expectation value, which is actually the statistical average, has to be explicitly normalized.) The coefficients therefore are of the form

$$c_{\alpha g, h} = e^{{\rm i} \theta_{\alpha g, h}}, \tag{12.51}$$

with the phase θ being real and ${\rm i} \equiv \sqrt{-1}$.

This form is reminiscent of the so-called EPR (Einstein–Podolsky–Rosen) state in which entangled qubits are often described in similar terms (cf figure 12.3).

A given total wave functions correspond to a specific sets of phases, $\{\theta_{\alpha g,h}\}$. Therefore averaging over the phases makes the product of coefficients that appears in equation (12.50), zero unless $g' = g$,

$$\left\langle c^*_{\alpha g',\,h}\, c_{\alpha g,h} \right\rangle_{\text{stat}} = \left\langle e^{i\left[\theta_{\alpha g,h}-\theta_{\alpha g',h}\right]} \right\rangle_{\text{stat}} = \delta_{g,g'}. \tag{12.52}$$

This will be explicitly shown in the next section by averaging over the total wave space. With this, the reservoir sum in the expectation value becomes

$$\begin{aligned} \left\langle \sum_{h\in\beta_\alpha}^{(\text{Er})} c^*_{\alpha g',\,h}\, c_{\alpha g,h} \right\rangle_{\text{stat}} &= \delta_{g,g'} \sum_{h\in\beta_\alpha}^{(\text{Er})} \\ &= \delta_{g,g'}\, N^{\text{Er}}_{\beta_\alpha} \\ &= \delta_{g,g'}\, e^{S^{\text{r}}_{\beta_\alpha}/k_{\text{B}}} \\ &= \delta_{g,g'}\, e^{-E^{\text{s}}_\alpha/k_{\text{B}}T}. \end{aligned} \tag{12.53}$$

The statistical average on the left-hand side means the average over total wave functions, which means the average over randomly and uniformly distributed phases. The penultimate equality gives the entropy of the reservoir macrostate β_α as the logarithm of the number of microstates it contains, equation (12.44), and the final equality uses the thermodynamic relationship, equation (12.45), with the constant part effectively incorporated into the normalization factor.

The effect of averaging the expectation value over the wave function is to collapse the degenerate subsystem energy states, $g = g'$. This both reduces the number of elements contributing to the expectation value and removes the interference between the degenerate energy states (i.e., they are no longer in superposition). Hence both the principal and degenerate states in the energy representation of an operator on the subsystem are in diagonal form. It will be recalled that this result had to be established in order to derive the mixture form for the statistical average, equation (12.9). Equivalently, the probability operator is fully diagonal in the energy representation. Since in the canonical equilibrium system, the entropy operator is proportional to the energy operator, $\hat{S} = -\hat{\mathcal{H}}/T$, this means that the probability operator is fully diagonal in the entropy representation. (In general the entropy operator is derived from the variables that are exchangeable between the subsystem and the reservoir, and so in all equilibrium and non-equilibrium systems the probability operator will be diagonal in the entropy representation.)

The diagonal nature of the probability operator signifies the elimination of the interference between energy states, both principal and degenerate. This transforms the subsystem from a superposition of states to a mixture of states.

The statistical average also counts the degenerate reservoir energy states. Summing over these weights each subsystem energy state by the number of degenerate reservoir microstates. But by equation (12.44), this is just the exponential of the corresponding reservoir entropy. It is this second effect, combined with equation (12.45), that gives the Maxwell–Boltzmann form for the probability operator.

The average over the phases of the expectation value, equation (12.50), is the statistical average of the operator. With the above result it is

$$\langle \hat{O}^{\mathrm{s}} \rangle_{\mathrm{stat}} = \frac{1}{Z} \sum_{\alpha,g} e^{-E_\alpha/k_{\mathrm{B}}T} O^{\mathrm{s,E}}_{\alpha g,\, \alpha g}. \tag{12.54}$$

Here Z is just the normalization constant (actually partition function), and the sum is over the primary and degenerate energy states of the subsystem. This has the same functional form as the final equality in equation (12.9), which is the von Neumann trace form for the statistical average. This gives explicitly the probability matrix as the Maxwell–Boltzmann form, $\wp^{\mathrm{S}}_{\alpha g,\, \alpha g} = Z^{-1} e^{-E_\alpha/k_{\mathrm{B}}T}$.

12.4.3.2 Wave space approach

The statistical average of an operator on the subsystem can also be expressed as an integral over the total wave space of the expectation value with respect to the entangled wave function. The weight density of the total wave space is uniform, section 12.3. The integral over wave space is carried out by integrating the coefficients in the energy basis state expansion. The allowed coefficients respect energy conservation, equation (12.49), which means that $\mathrm{d}\psi_{\mathrm{tot}} \equiv \mathrm{d}\underline{\underline{c}} \equiv \prod_{\alpha g,h} \mathrm{d}c^{\mathrm{r}}_{\alpha g,\, h}\, \mathrm{d}c^{\mathrm{i}}_{\alpha g,\, h}$. The real and imaginary parts of the coefficient each belong to the real line, $\in(-\infty, \infty)$. Entanglement has reduced the number of indices from four to three. With the expectation value given by equation (12.50), $O^{\mathrm{s}}(\psi_{\mathrm{tot}}) \equiv \langle \psi_{\mathrm{tot}} | \hat{O}^{\mathrm{s}} | \psi_{\mathrm{tot}} \rangle / \langle \psi_{\mathrm{tot}} | \psi_{\mathrm{tot}} \rangle$, the statistical average is

$$\begin{aligned}
\langle O^{\mathrm{s}} \rangle_{\mathrm{stat}} &= \frac{1}{Z'} \int \mathrm{d}\psi_{\mathrm{tot}}\, O^{\mathrm{s}}(\psi_{\mathrm{tot}}) \\
&= \frac{1}{Z'} \sum_{\alpha,g,g'} \sum_{h\in\beta_\alpha}^{(\mathrm{Er})} O^{\mathrm{s,E}}_{\alpha g',\, \alpha g} \int \mathrm{d}\underline{\underline{c}}\, \frac{c^{*}_{\alpha g',\, h}\, c_{\alpha g,h}}{N(\psi_{\mathrm{tot}})} \\
&= \frac{1}{Z'} \sum_{\alpha,g} \sum_{h\in\beta_\alpha}^{(\mathrm{Er})} O^{\mathrm{s,E}}_{\alpha g,\, \alpha g} \int \mathrm{d}\underline{\underline{c}}\, \frac{|c_{\alpha g,h}|^2}{N(\psi_{\mathrm{tot}})} \\
&= \frac{1}{Z'} \sum_{\alpha,g} e^{S^{\mathrm{r}}_{\beta_\alpha}/k_{\mathrm{B}}} O^{\mathrm{s,E}}_{\alpha g,\, \alpha g} \times \text{const.} \\
&= \frac{1}{Z} \sum_{\alpha,g} e^{-E^{\mathrm{s}}_\alpha/k_{\mathrm{B}}T} O^{\mathrm{s,E}}_{\alpha g,\, \alpha g}.
\end{aligned} \tag{12.55}$$

Here the norm squared is $N(\psi_{\mathrm{tot}}) = \sum_{\alpha g,h} |c_{\alpha g,h}|^2$. The third equality follows since the terms in which the integrand is odd with respect to any coefficient vanish, and so the only non-vanishing terms have $g = g'$. The fourth equality follows because all the integrations are identical and give a constant that is independent of α, g, and h, which can be incorporated into the normalizing partition function. The fourth and fifth equalities use the two forms for the reservoir entropy, equations (12.44) and (12.45). The partition function ensures that $\langle \hat{\mathrm{I}}^{\mathrm{s}} \rangle_{\mathrm{stat}} = 1$.

This agrees with the random phase approach, equation (12.54), which should not be surprising since the uniformity of wave space of an isolated system implies the equal weight of isolated system microstates. Both methods show that the reduction and collapse of the energy macrostates is due to energy conservation and the consequent entanglement of the total wave function. They both also show that the collapse of the subsystem degenerate energy states is due to the fact that the off-diagonal contributions average to zero.

This equivalence demonstrates that quantum statistical mechanics can be as well formulated in terms of the wave function as in terms of quantum states.

12.4.4 Maxwell–Boltzmann probability operator

The statistical average given by both approaches agree with the von Neumann trace form, specifically the final equality in equation (12.9). This is in the form of a classical average: it is the sum over states of the probability of a state times the value of the observable in the state. It is essential to the result that the off-diagonal elements in the entropy representation of the probability (density) matrix vanish, which is to say that the wave function in this basis is decoherent.

One can easily convert the present result for the average to the generic von Neumann trace form, the first equality in equation (12.9). Of course, the probability operator replaces the conventional many-wave function density operator. In the energy basis, the representation of any function of the Hamiltonian operator is diagonal, and the diagonal entries are just the function of the eigenvalues,

$$e^{-E_{\alpha}^{\mathrm{s}}/k_{\mathrm{B}}T}\delta_{\alpha',\alpha}\delta_{g',g} = \left\{e^{-\hat{\mathcal{H}}/k_{\mathrm{B}}T}\right\}^{\mathrm{E}}_{\alpha'g',\, \alpha g}, \tag{12.56}$$

with $\hat{\mathcal{H}}$ being the subsystem Hamiltonian operator. For the canonical equilibrium system, the reservoir entropy operator is $\hat{S} = -\hat{\mathcal{H}}/T$, a consequence of equation (12.45). This means that in the present canonical equilibrium system entropy eigenfunctions are identical to the energy eigenfunctions. More generally, the entropy representation is always the representation in which an open quantum system is diagonal. With this the quantum statistical average of a subsystem operator can be written

$$\begin{aligned}\langle \hat{O} \rangle_{\mathrm{stat}} &= \frac{1}{Z}\sum_{\alpha,g} e^{-E_{\alpha}/k_{\mathrm{B}}T} O^{\mathrm{s,E}}_{\alpha g,\, \alpha g} \\ &= \frac{1}{Z}\sum_{\alpha g, \alpha' g'} \left\{e^{-\hat{\mathcal{H}}/k_{\mathrm{B}}T}\right\}^{\mathrm{E}}_{\alpha'g',\, \alpha g} O^{\mathrm{s,E}}_{\alpha g,\, \alpha' g'} \\ &= \mathrm{TR}\, \wp\, \hat{O}.\end{aligned} \tag{12.57}$$

The final form is independent of any specific representation or basis. One can conclude that the probability operator for the canonical equilibrium system is just the Maxwell–Boltzmann operator,

$$\hat{\wp} \equiv \frac{1}{Z} e^{\hat{S}/k_{\mathrm{B}}} \equiv \frac{1}{Z} e^{-\hat{\mathcal{H}}/k_{\mathrm{B}}T}. \tag{12.58}$$

The normalizing partition function is $Z = \mathrm{TR}\, e^{-\hat{\mathcal{H}}/k_{\mathrm{B}}T}$. The first equality holds in general, whereas the second equality is specific for the canonical equilibrium system. This result for the canonical equilibrium probability operator (or many-wave function density operator) is directly analogous to the classical result that is so well known that it is usually invoked directly without derivation (e.g., Pathria 1972).

The Maxwell–Boltzmann probability operator is the direct quantitative consequence of the sum over the degenerate energy microstates of the reservoir. This gives the exponential of the reservoir entropy for each particular subsystem energy macrostate. This is just the matrix representation of the Maxwell–Boltzmann probability operator in the energy basis. Although the energy representation was used to derive this result, the final expression as a trace of the product of the two operators is invariant with respect to the representation.

One sees in this derivation that it is the statistical average that causes the superposition of the degenerate energy microstates of the subsystem to collapse into a mixture. The reduction and collapse of the energy macrostates occurred at the level of the expectation value due to the energy conservation law and the entanglement of the reservoir and the subsystem. As has been mentioned above, both collapses are necessary for the mixture form for the statistical average, equation (12.9). The superposition entropy states, both principle and degenerate, had to collapse so that the probability operator was diagonal in the entropy representation. Although we do not work with a density operator, if we were to, the result (i.e., equating it to the probability operator) would imply that the density operator is an average over multiple wave functions.

For the present formulation, the statistical average can be written as an integral of the expectation value over the subsystem wave space

$$\langle \hat{O} \rangle_{\mathrm{stat}} = \int \mathrm{d}\psi \, \frac{\langle \psi | \hat{\wp} \, \hat{O} | \psi \rangle}{\langle \psi | \psi \rangle}. \tag{12.59}$$

It is no longer necessary to show that these belong to the subsystem, the superscript s, because the reservoir has been integrated out. This is a continuum version of the trace. The equivalence of this with the above discrete expression for the trace can be shown by expanding the subsystem wave function in any basis, $|\psi\rangle = \sum_n \psi_n |\zeta_n\rangle$, and integrating,

$$\int \mathrm{d}\psi \, \frac{\langle\psi|\hat{\wp}\,\hat{O}|\psi\rangle}{\langle\psi|\psi\rangle} = \int \mathrm{d}\underline{\psi} \, \frac{\sum\limits_{m,n,l} \psi_m^* \psi_l \wp_{mn} O_{nl}}{\langle\psi|\psi\rangle}$$
$$= \frac{1}{Z''} \sum_{m,n} \{e^{\hat{S}/k_B}\}_{mn} O_{nm} \int \mathrm{d}\underline{\psi} \, \frac{\psi_m^* \psi_m}{\langle\psi|\psi\rangle} \tag{12.60}$$
$$= \frac{\text{const.}}{Z''} \sum_{m,n} \{e^{\hat{S}/k_B}\}_{mn} O_{nm}$$
$$= \mathrm{TR}\, \hat{\wp}\, \hat{O}.$$

In the second equality $l = m$ because odd powers of ψ_l vanish. Since the integration is the same for all m, it is a constant that can be taken outside of the sums and incorporated into the partition function, as in the third equality.

In section 12.3, the probability density in wave space for an isolated total system was shown to be uniform in wave space. In contrast, the probability density in the wave space of a subsystem of a total system is non-uniform. For the present canonical equilibrium subsystem that can exchange energy with a reservoir, it is given by the expectation value of the Maxwell–Boltzmann probability operator. The probability density is the expectation value of the probability operator,

$$\wp(\psi) = \frac{\langle\psi|\hat{\wp}|\psi\rangle}{\langle\psi|\psi\rangle}, \tag{12.61}$$

where ψ is the wave function of the subsystem. It should be clear that the statistical average is *not* the integral over the subsystem wave space of the probability density and the expectation value of the operator,

$$\langle\hat{O}\rangle_{\text{stat}} \neq \int \mathrm{d}\psi \, \wp(\psi) O(\psi). \tag{12.62}$$

The physical reason that this does not hold is that in general statistical averages are the value of a variable in a state times the probability of the state, summed over the states. The problem with this expression, is that both $\wp(\psi)$ and $O(\psi)$ are expectation values, which include the superposition of states.

12.4.5 Environmental selection

In the quantum literature, the present subsystem that can exchange energy with a thermal reservoir is generally called an open quantum system. These have been treated in detail from a number of different perspectives (Davies 1976, Breuer and Petruccione 2002, Weiss 2008). The present technique of exchange of a conserved quantity with a reservoir and wave function collapse due to entanglement has some similarity to the theory of quantum measurement that is known as environmental selection or einselection (Zeh 2001, Zurek 2003, Schlosshauer 2005). Einselection explicitly includes the influence of the environment on the system of interest, which is an obvious point in common with the above analysis of the subsystem and reservoir.

The specific goal of einselection is to explain the apparent collapse of the wave function upon measurement, which is collapse is another point of similarity with the present derivation of equilibrium quantum statistical mechanics.

In view of these two points of similarity, it seems worthwhile to summarize the einselection approach in a way that links it to the approach used above based on the entropy representation, entanglement, and conservation laws. Although this departs from the presentations of the advocates of einselection in the theory of quantum measurement, it does convey the spirit of the approach particularly in invoking the quasi-orthogonality of the reservoir, as is briefly explained below.

As above the total system labeled tot consists of a subsystem s and a reservoir r that interact with each other. More conventionally, the reservoir is usually called the environment, and the subsystem is usually called the system. Let $\{\zeta_n^{\rm Ss}\}$ and $\{\zeta_n^{\rm Sr}\}$ be the entropy eigenfunctions that form a basis for the subsystem and for the reservoir, respectively. For the most common case, the canonical equilibrium system, energy is the exchanged conserved variable, and entropy eigenfunctions are just energy eigenfunctions. In the general case, the entropy eigenfunctions account for all conserved quantities that are exchanged between the subsystem and reservoir.

The subsystem and the reservoir may be regarded as quasi-independent since the volume of the boundary region over which they interact is much smaller than the volume of either. Quasi-independent means that for each state of the subsystem with a given energy, all the states of the reservoir permitted by the remaining energy can occur. This implies that the total number of states of the total system is the product of the allowed number of states of each considered as isolated. In view of this, the Hilbert space of the total system is the direct product of that of the subsystem and that of the reservoir. Therefore the wave function of the total system may be written as

$$|\psi_{\rm tot}\rangle = \sum_{n,m} c_{nm}|\zeta_n^{\rm Ss}\rangle \otimes |\zeta_m^{\rm Sr}\rangle = \sum_{n,m} c_{nm}|\zeta_n^{\rm Ss}, \zeta_m^{\rm Sr}\rangle. \tag{12.63}$$

In section 12.4.2, it was shown that the conservation law created an entangled total wave function that cannot be the product of a subsystem wave function and a reservoir wave function. However, since the entropy basis directly accounts for the conservation laws, which is its main virtue, the total wave function can be written

$$|\psi_{\rm tot}\rangle = \sum_{\alpha,g,h} c^{\rm s}_{\alpha g}c^{\rm r}_{\beta_\alpha h}|\zeta_{\alpha g}^{\rm Ss}\rangle \otimes |\zeta_{\beta_\alpha h}^{\rm Sr}\rangle \equiv \sum_{\alpha,g,h} c_{\alpha g,h}|\zeta_{\alpha g}^{\rm Ss}, \zeta_{\beta_\alpha h}^{\rm Sr}\rangle. \tag{12.64}$$

The sum over the common label precludes factorization into subsystem and reservoir wave functions, and so this is the proper form for an entangled wave function.

We now analyze the time evolution of two wave functions of the total system that have the same starting reservoir wave function but differ in the starting wave function of the subsystem. At time $t = 0$ let the normalized wave function of the total system be $|\psi_{\rm tot}\rangle = |\zeta_{\alpha g}^{\rm Ss}, \psi_0^{\rm r}\rangle$. This is not entangled; the subsystem has wave function

$\zeta_{ag}^{\rm Ss}$ (i.e., it is in the entropy state αg), and the reservoir has wave function $\psi_0^{\rm r}$. These are arbitrary except for the fact that the respective expectation values of the energy (or other relevant conserved transferable quantities) must add up to the total conserved value. After time τ this evolves to

$$|\psi_{\rm tot}\rangle \overset{\tau}{\rightarrow} \sum_{\beta,h} c_{\beta h}^{{\rm s}(\alpha g)} |\zeta_{\beta h}^{\rm Ss}, \psi_{\beta h}^{{\rm r}(\alpha g)}\rangle. \tag{12.65}$$

The dependence on the initial αg and the final βh subsystem states of the subsystem expansion coefficients and of the evolved reservoir wave function, is shown. The dependence on the initial reservoir wave function $\psi_0^{\rm r}$ and the time interval τ is not shown. These dependencies arise from the interactions between the reservoir and the subsystem during the evolution of the total wave function.

Now for the second initial normalized total wave function. This has a different subsystem basis function but the same reservoir initial wave function, $|\zeta_{\gamma l}^{\rm Ss}, \psi_0^{\rm r}\rangle$. The evolution of the inner product of this and the first wave function is given by

$$\begin{aligned}\left\langle \psi_0^{\rm r}, \zeta_{\gamma l}^{\rm s} \middle| \zeta_{\alpha g}^{\rm s}, \psi_0^{\rm r} \right\rangle &\overset{\tau}{\rightarrow} \sum_{\beta,h} \sum_{\eta,p} c_{\eta p}^{{\rm s}(\gamma l)*} c_{\beta h}^{{\rm s}(\alpha g)} \langle \psi_{\eta p}^{{\rm r}(\gamma l)}, \zeta_{\eta p}^{\rm Ss} | \zeta_{\beta h}^{\rm Ss}, \psi_{\beta h}^{{\rm r}(\alpha g)}\rangle \\ &= \sum_{\beta,h} c_{\beta h}^{{\rm s}(\gamma l)*} c_{\beta h}^{{\rm s}(\alpha g)} \langle \psi_{\beta h}^{{\rm r}(\alpha g)} | \psi_{\beta h}^{{\rm r}(\alpha g)}\rangle \delta_{\alpha,\gamma} \delta_{l,g} \\ &\equiv C_{\alpha g} \delta_{\alpha,\gamma} \delta_{l,g}.\end{aligned} \tag{12.66}$$

The second equality arises from the orthonormality of the subsystem basis functions, and the quasi-orthogonality of the reservoir wave functions. Quasi-orthogonality arises from the fact that in a d-dimensional space, the amount of overlap between two random unit vectors decreases with increasing dimensionality. The idea is illustrated in figure 12.5. Since a typical reservoir or environment comprises on the order of Avogadro's number of particles, the overlap between any two random wave functions is effectively a delta function. In the present case the different subsystem starting wave functions mean that the two reservoir wave functions rapidly become uncorrelated.

The isolated total system has unitary time propagator, namely the adiabatic time propagator, which preserves the scalar product of two total wave functions during their evolution. Since the left-hand side is unity for the case that $\gamma = \alpha$ and $l = g$, one sees that $C_{\alpha g} = 1$.

Now we turn to the evolution of the expectation value of a subsystem operator $\hat{A}$. Suppose that the initial total wave function is in a superposition of entropy states, $|\psi_{\rm tot}\rangle = \sum_{\alpha,g} a_{\alpha g} |\zeta_{\alpha g}^{\rm Ss}, \psi_0^{\rm r}\rangle$. After time τ, which may be interpreted as the time over which the measurement is performed or, equivalently, the operator is applied, the expectation value evolves to

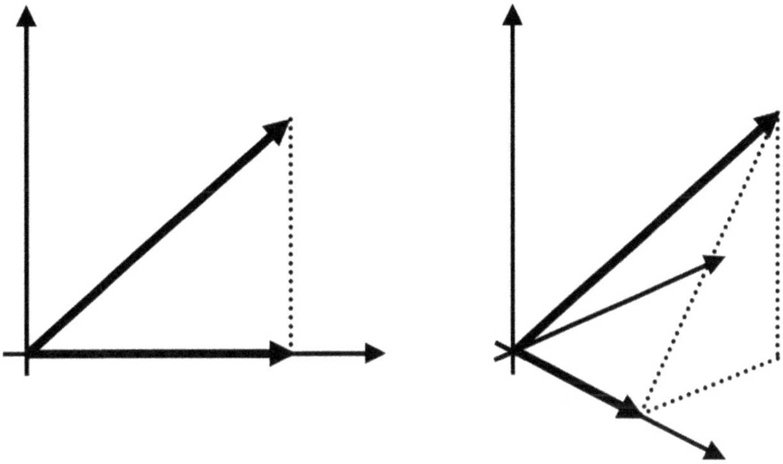

Figure 12.5. The magnitude of overlap between a unit vector symmetrically oriented and a random axis. In two dimensions (left) the projection has length $1/\sqrt{2}$, and in three dimensions (right) the projection has length $1/\sqrt{3}$. The average overlap magnitudes over all orientations are $2/\pi$ and 1/2, respectively. Reproduced from Attard (2015).

$$\begin{aligned}\langle\psi_{\rm tot}|\hat{O}^{\rm s}|\psi_{\rm tot}\rangle &= \sum_{\alpha,g}\sum_{\beta,h} a_{\alpha g}^{*}a_{\beta h}\langle\zeta_{\alpha g}^{\rm Ss}|\zeta_{\beta h}^{\rm Ss}\rangle \\ &\overset{\tau}{\to} \sum_{\alpha,g}\sum_{\beta,h} a_{\alpha g}^{*}a_{\beta h}\sum_{\eta,p}\sum_{\gamma,l} c_{\eta p}^{{\rm s}(\alpha g)*}c_{\gamma l}^{{\rm s}(\beta h)} \\ &\quad\times\langle\psi_{\eta p}^{{\rm r}(\alpha g)},\,\zeta_{\eta p}^{\rm Ss}|\hat{O}^{\rm s}|\zeta_{\gamma l}^{\rm Ss},\,\psi_{\gamma l}^{{\rm r}(\beta h)}\rangle \\ &= \sum_{\alpha,g}\sum_{\gamma,l} a_{\alpha g}^{*}a_{\alpha g}\langle\zeta_{\gamma l}^{\rm Ss}|\hat{O}^{\rm s}|\zeta_{\gamma l}^{\rm Ss}\rangle c_{\gamma l}^{{\rm s}(\alpha g)*}c_{\gamma l}^{{\rm s}(\alpha g)}\langle\psi_{\gamma l}^{{\rm r}(\alpha g)}|\psi_{\gamma l}^{{\rm r}(\alpha g)}\rangle \\ &\equiv \sum_{\gamma,l} A_{\gamma l}^{\rm S*}A_{\gamma l}^{\rm S}O_{\gamma l,\,\gamma l}^{\rm Ss},\end{aligned} \tag{12.67}$$

with $A_{\gamma l}^{\rm S} \equiv \sum_{\alpha,g} a_{\alpha g}c_{\gamma l}^{{\rm s}(\alpha g)}\sqrt{\langle\psi_{\gamma l}^{{\rm r}(\alpha g)}|\psi_{\gamma l}^{{\rm r}(\alpha g)}\rangle}$. The third equality follows from the quasi-orthogonality of the reservoir wave functions. By setting $\hat{O} = \hat{\rm I}$, which shows that $\sum_{\gamma,l}A_{\gamma l}^{\rm S*}A_{\gamma l}^{\rm S} = 1$, one can see that the wave functions are correctly normalized.

This result is more akin to a statistical average than to a quantum expectation value. The present result involves only the diagonal elements of the operator matrix in the entropy representation. The expectation value derived above taking into account wave function entanglement due to the conservation of transferable quantities, equation (12.50), has off-diagonal, degenerate state contributions. The statistical average of the expectation value, equation (12.55), consists solely of diagonal elements, as here. The decoherent averaging used in the earlier analysis is the analog of the time evolution in the present case. Taking the results of the two approaches to be the same, the statistical mechanical analysis may be used to quantify the einselection result, namely $A_{\gamma l}^{\rm S*}A_{\gamma l}^{\rm S} = Z^{-1}e^{-\beta E_{\gamma}}$.

The physical interpretation of this einselection result for the expectation value of a subsystem operator is the same as that given above for a statistical average. The evolved expectation value appears as that of a mixture of entropy states of the subsystem, with the $A^{\rm S}_{\gamma l}$ being the coefficients of the wave functions. The entropy states are the eigenstates of the operators corresponding to the exchangeable variables. The mixture of decoherent wave functions is signified by the fact that only diagonal terms appear in the evolved expectation value. In contrast, a coherent wave function that corresponds to a superposition of states contributes off-diagonal terms to the expectation value in all representations other than that of the operator itself. The coherent case is the usual form of an expectation value in quantum mechanics; for a subsystem wave function in the entropy representation, it would be $O(\psi) \equiv \langle\psi|\hat{O}|\psi\rangle = \sum_{m,n}\psi^{\rm S*}_m\psi^{\rm S}_n O^{\rm S}_{mn}$, assuming normalization. Here the off-diagonal terms clearly contribute, whereas they don't in the case of environmental selection where time evolution, decorrelation, and quasi-orthogonality create a mixture of pure entropy eigenstates.

As has been mentioned, environmental selection and the reservoir analysis above are closely related. Both show that in the expectation value of a subsystem operator averaged over the total wave function the off-diagonal terms are suppressed (in the entropy representation). This is variously called wave function collapse, or decoherence, or a mixture of pure quantum states, and is thought to be relevant to the so-called measurement problem. Both show that this occurs due to the interactions with the reservoir or environment. Arguably, the reservoir formalism has the advantage of being exact, since it does not invoke quasi-orthogonality. And also, the reservoir formalism yields an explicit formula for the probability operator and for the statistical average.

12.4.6 Symmetrization

The above analysis neglects the symmetrization of the wave function. In fact little changes when this is taken into account. The proof of the uniform density of wave space remains unchanged with the replacement $\psi \Rightarrow \psi^{\pm}$. The number of degenerate energy states becomes $N^{\rm E,\pm}_\alpha$, which is much smaller than the unsymmetrized case, but in any case is just a number. The thermodynamic expression for the reservoir entropy is formally unchanged.

For the canonical equilibrium system particles are not permitted to exchange between the subsystem and the reservoir. Therefore the energy eigenfunction of the total system is the product of the individually symmetrized energy eigenfunctions, $|\zeta^{\rm Es,\pm}_{\alpha g}, \zeta^{\rm Er,\pm}_{\beta h}\rangle$. The collapse of the wave function due to entanglement from energy conservation follows as in the text. The subsystem energy states that are summed over for the partition function and for the statistical average, are weighted by the same Maxwell–Boltzmann factor that derives from the thermodynamic expression for the reservoir entropy, and are those allowed by the symmetrization rule for the subsystem alone. This is signified by attaching a prime to the trace, as was used throughout the preceding chapters.

As mentioned, summing over the allowed symmetrization states for the subsystem means that the energy microstates have an internal entropy, e.g., equations (2.25), (3.6), and (5.57).

References

Attard P 2002 *Thermodynamics and Statistical Mechanics: Equilibrium by Entropy Maximisation* (London: Academic)

Attard P 2015 *Quantum Statistical Mechanics: Equilibrium and Non-Equilibrium Theory from First Principles* (Bristol: IOP Publishing)

Attard P 2018 *Entropy Beyond the Second Law. Thermodynamics and Statistical Mechanics for Equilibrium, Non-Equilibrium, Classical, and Quantum Systems* (Bristol: IOP Publishing)

Breuer H P and Petruccione F 2002 *The Theory of Open Quantum Systems* (Oxford: Oxford University Press)

Davies E B 1976 *Quantum Theory of Open Systems* (London: Academic)

Merzbacher E 1970 *Quantum Mechanics* 2nd edn (New York: Wiley)

Messiah A 1961 *Quantum Mechanics* vols 1 and 2 (Amsterdam: North-Holland)

von Neumann J 1932 *Mathematische Grundlagen der Quantenmechanik* (Berlin: Springer)

Pathria R K 1972 *Statistical Mechanics* (Oxford: Pergamon)

Schlosshauer M 2005 Decoherence, the measurement problem, and interpretations of quantum mechanics arXiv:quant-ph/0312059v4

Weiss U 2008 *Quantum Dissipative Systems* 3rd edn (Singapore: World Scientific)

Zeh H D 2001 *The Physical Basis of the Direction of Time* 4th edn (Berlin: Springer)

Zurek W H 2003 Decoherence, einselection, and the quantum origins of the classical arXiv:quant-ph/0105127v3

www.ingramcontent.com/pod-product-compliance
Ingram Content Group UK Ltd.
Pitfield, Milton Keynes, MK11 3LW, UK
UKHW051904100926
8556IPUK00005B/760

9 780750 340533